ELECTRIC DISTRIBUTION SYSTEMS

IEEE Press
445 Hoes Lane
Piscataway, NJ 08854

ELECTRIC DISTRIBUTION SYSTEMS

ABDELHAY A. SALLAM
OM P. MALIK

Mohamed E. El-Hawary, *Series Editor*

A JOHN WILEY & SONS, INC., PUBLICATION

Published by John Wiley & Sons, Inc., Hoboken, New Jersey.
Published simultaneously in Canada

Library of Congress Cataloging-in-Publication Data:

Sallam, A. A. (Abdelhay. A.)
Electric distribution systems / A.A. Sallam.
p. cm.—(Ieee press series on power engineering ; 45)
ISBN 978-0-470-27682-2 (hardback)
1. Electric power distribution. I. Title.
TK3001.S325 2010
621.319—dc22

2010033573

Printed in Singapore

oBook ISBN: 978-0-470-94389-4
ePDF ISBN: 978-0-470-94384-7
ePub ISBN: 978-1-118-00221-6

10 9 8 7 6 5 4 3 2 1

CONTENTS

PREFACE

The main consideration of distribution systems, as intermediate media between the subtransmission systems and the customer's premises, is to maximize the utilization of electric energy to supply the end users with energy in a secure and efficient manner. Several circuits feed customers at different locations, in comparison to the transmission and subtransmission systems, which have only a few circuits. Distribution systems have to cater to a large variety of customers with significantly different demand patterns.

In addition, developments in sustainable and renewable generation (commonly referred to as distributed generation) application of a large class of power electronics-based devices demand response programs feasible for use with smart grid technologies, and so on have added new complexities in the planning, design, and operation of distribution systems. This has made the analysis of distribution systems rather complex.

Due to the large variety of customers and demands, electric distribution systems cover a very broad spectrum of topics. The topics covered in this book are relevant from both the academic and practical aspect. They are of interest for electric utilities and industry as well as individuals working with distribution systems.

The operator or utility engineer who is interested in studying or working on distribution systems needs to know the topics addressed in this book and their practical implementation. Different aspects of system planning should be studied to define the system structure that feeds present and future demands. The protection system and switchgear based on short-circuit calculations and earthing systems must be designed. Power quality, system management, and automation as well as distributed generation are essential for the reader's awareness since they play a prominent role in system operation.

Various major topics are grouped together in this book in five parts.

PART I: FUNDAMENTAL CONCEPTS

The fundamental concepts of distribution systems are the subject of Chapter 1. The duties of distribution engineers including the factors affecting the planning process are introduced here. It is aimed at identifying the key steps in planning. The layout of the distribution system for both small and big cities and examples of structures used in distribution systems at medium and low voltages are presented.

The primary function of the distribution system is to feed electric loads. Therefore, it is necessary to determine during the planning process not only the present load and its makeup but also the expected load growth in the near future. Definitions of load forecast terms and different methods of estimating the demand forecast are explained in Chapter 2 with application examples.

PART II: PROTECTION AND DISTRIBUTION SWITCHGEAR

This part includes earthing, protection systems, and distribution switchgear. Earthing in distribution systems is an important subject that deserves to be studied, especially as the protection system is based on it. Various methods of earthing and a general description of the types of protection used in distribution systems are presented in Chapters 3 and 5, respectively. The design of protection necessitates some explanation of short-circuit calculation methods, and these are presented in Chapter 4.

Automation and measuring equipment for distribution systems is installed in the switchgear (indoor or outdoor). Therefore, details about switchgear devices and the major factors affecting the design of switchboards are included in Chapter 6.

PART III: POWER QUALITY

It is not sufficient to just plan the distribution system to meet the load demand with minimum interruptions (number and duration). It is of crucial importance to emphasize the quality of supply, in particular, when feeding sensitive loads. Therefore, the key elements of power quality (voltage quality, power factor, and harmonics) and means of their improvement are explained in Chapters 7–11.

PART IV: MANAGEMENT AND AUTOMATION

It is desirable to achieve a plan of a distribution system that takes into account the economics, that is, reducing the expenses and investments. How to verify these requirements is explained in Chapter 12 by applying demand-side management and energy-efficiency policies.

In addition, more attention should be given to the enhancement of distribution system performance. Methodologies applied to improve the performance of the distribution systems, such as distribution system automation and monitoring where automation helps to decrease the system interruptions, increase the reliability, and enhance the performance, are also discussed.

Monitoring helps in timely decision making. The difference between the system automation and monitoring, using supervisory control and data acquisition (SCADA) systems, is illustrated with the aid of examples. SCADA definitions and components, architectures of SCADA systems, and the conditions of using various architectures are given in Chapter 13. In addition, the smart grid vision is illustrated as a recent trend for the development of system automation and SCADA applications.

PART V: DISTRIBUTED GENERATION

Electricity produced using local generation including small renewable sources with the goal of feeding local loads or as backup sources to feed critical loads in case of emergency and utility outage is often referred to as "distributed generation" in North American terms and "embedded generation" in European terms. Therefore, distributed generation produces electricity at or near the place where it is used to meet all or a part of the customers' power needs. It ranges in size from less than 1 kW to tens or, in some cases, hundreds of kilowatts. On the other hand, demand for electric energy continues to grow and a large investment is required to develop both the distribution and transmission systems accordingly. Thus, great attention is being paid to utilizing private and distributed energy sources to be able to meet the load demand. Different types of distributed energy sources and the benefits gained from interconnecting these sources with the distribution system are described in Chapter 14.

Electric power distribution systems cover a broad spectrum of topics that need to be included in such a book. To keep the overall length of the book within a reasonable limit, many of these topics could not be covered in depth. Therefore, all material is supported by an extensive list of references where the interested reader can get more details for an in-depth study.

Abdelhay A. Sallam
Om P. Malik

ACKNOWLEDGMENTS

No work of any significance can be accomplished without the help received from many sources. In that respect, this book is no exception. The authors are grateful for the invaluable help received from many sources. We wish to express our gratitude to the following, in particular, without whose help it would not have been possible to put this book together:

- Mr. Hany Shaltoot of Schneider Electric, Egypt, for providing access to a number of relevant articles and company practices relating to the distribution systems. He also helped with obtaining permission from Square D to include in the book information on AccuSine® product.
- Square D North America for permission to include the AccuSine® product photos.
- Technical and sales staff members of ABB, Egypt, for making available manuals describing the company practices and a number of illustrations included in the book with permission.
- Dr. Azza Eldesoky for the information on load forecasting that is included in the book, and Dr. Ahmed Daoud for editing some of the illustrations.
- Dr. Tamer Melik, Optimal Technologies (Canada) Inc., for making available the report on which a part of the material in Chapter 14 is based.
- European Commission, Community Research, Smart Grids technology platform for making available the report on which a part of the material in Chapter 13 is based.

In addition, help has been received from a number of other sources to which we are indebted and wish to express our sincere thanks.

All this work requires the moral support of the families and we wish to recognize with our warm appreciation. We dedicate this book:

To our wives,
Hanzada Sallam and Margareta Malik.

A. A. S
O. P. M

PART I

FUNDAMENTAL CONCEPTS

CHAPTER 1

MAIN CONCEPTS OF ELECTRIC DISTRIBUTION SYSTEMS

1.1 INTRODUCTION AND BACKGROUND

To achieve a good understanding of electric distribution systems, it is necessary to first get acquainted with the appropriate background. A description of the main concepts of electric distribution systems is given in this chapter followed by a more detailed discussion of the various aspects in the following chapters.

1.1.1 Power System Arrangements

A power system contains all electric equipment necessary for supplying the consumers with electric energy. This equipment includes generators, transformers (step-up and step-down), transmission lines, subtransmission lines, cables and switchgear [1]. As shown in Figure 1.1, the power system is divided mainly into three parts. *The first part* is the *generation* system in which the electricity is produced in power plants owned by an electric utility or an independent supplier. The generated power is at the generation voltage level. The voltage is increased by using step-up power transformers to transmit the power over long distances under the most economical conditions. *The second part* is the *transmission* system that is responsible for the delivery of power to load centers through cables or overhead transmission lines. The transmitted power is at extra high voltage (EHV) (transmission network) or high voltage (HV) (subtransmission network). *The third part* is the *distribution* system where the voltage is stepped down at the substations to the medium voltage (MV) level.

Electric Distribution Systems, First Edition. Abdelhay A. Sallam, Om P. Malik.

Published 2011 by John Wiley & Sons, Inc.

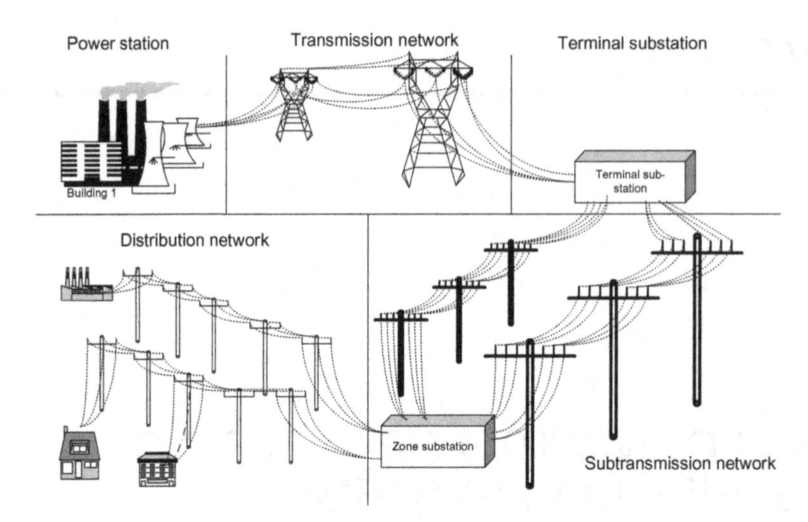

Figure 1.1 Electricity supply system [2].

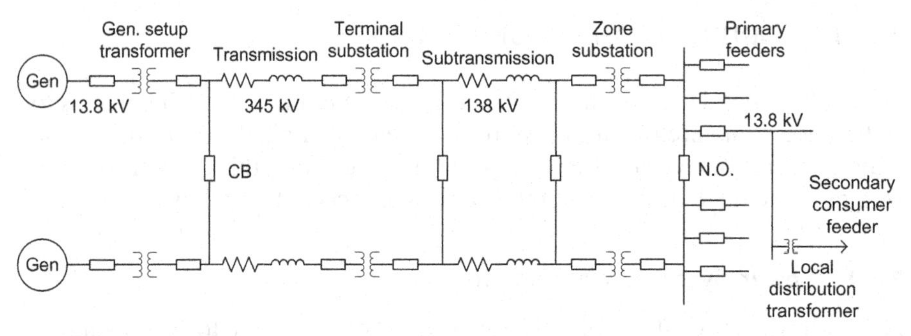

Figure 1.2 A typical electric supply system single-line diagram. CB = circuit breaker; N.O. = normal open.

The power is transmitted through the distribution lines (or cables) to the local substations (distribution transformers) at which the voltage is reduced to the consumer level and the power lines of the local utility or distribution company carry electricity to homes or commercial establishments.

The physical representation given in Figure 1.1 needs to be expressed by a schematic diagram adequate for analyzing the system. This is done by drawing a single-line diagram (SLD) as shown in Figure 1.2. This figure illustrates two power systems connected together by using tie-links as they exist in real practice to increase system reliability and decrease the probability of load loss. The voltage values shown in this figure are in accordance with the standards of North American power systems.

Each system contains *generators* delivering power at generation voltage level, say 13.8 kV. By using step-up transformers, the voltage is stepped up to

345 kV and the power is transmitted through the *transmission* system. The transmission lines are followed by 138 kV subtransmission lines through terminal substations. The subtransmission lines end at the zone substations where the voltage is stepped down to 13.8 kV to supply the MV distribution network at different distribution points (DPs) as primary feeders. Then the electricity is delivered to the consumers by secondary feeders through local distribution transformers at low voltage (LV) [3, 4].

To get a better understanding of the physical arrangement of the power system, consider how electricity is supplied to a big city. In the first part of the arrangement, the power stations are often located far away from the city zones and sometimes near the city border. According to how big the city is, the second part of the arrangement (transmission and subtransmission systems) is determined. Overhead transmission lines and cables can be used for both systems. They are spanned along the boundary of the city where the terminal and zone substations are located as well. This allows the planner to avoid the risk of going through the city by lines that operate at HV or EHV. For the third part, the distribution system, the total area of the city is divided into a number of subareas depending on the geographic situation and the load (amount and nature) within each subarea. The distribution is fed from the zone substation and designed for each subarea to provide the consumers with electricity at LV by using local transformers.

As an illustrative example, consider the total area of a big city is divided into three residential areas and two industrial areas as shown in Figure 1.3. Power station #1, terminal substations #2 (345/138/69 kV), and the zone substations #3 (138/69/13.8 kV) are located at the boundary of the city. The transmission system operates at 138 and 69 kV. Both of these systems are around the city and do not go through the city subareas. Of course, the most economical voltage for the transmission and subtransmission systems is determined in terms of the transmitted power and the distance of power travel. Also, the supply network to the industrial zones is operating at 69 kV because of the high power demand and to avoid the voltage drop violation at the MV level [5].

Substation #4 (69/13.8 kV) is located at a certain distance inside the city boundary where the distribution system starts to feed the loads through DPs. The outgoing feeders from DPs are connected to local distribution transformers to step down the MV to LV values.

For small cities, the main sources on the boundary are either power stations or substations 138/13.8 kV or 69/13.8 kV to supply the distribution system including various DPs in different zones of the city. The outline of this arrangement is shown in Figure 1.4.

1.2 DUTIES OF DISTRIBUTION SYSTEM PLANNERS

The planners must study, plan, and design the distribution system 3–5 years and sometimes 10 or more years ahead. The plan is based on how the system

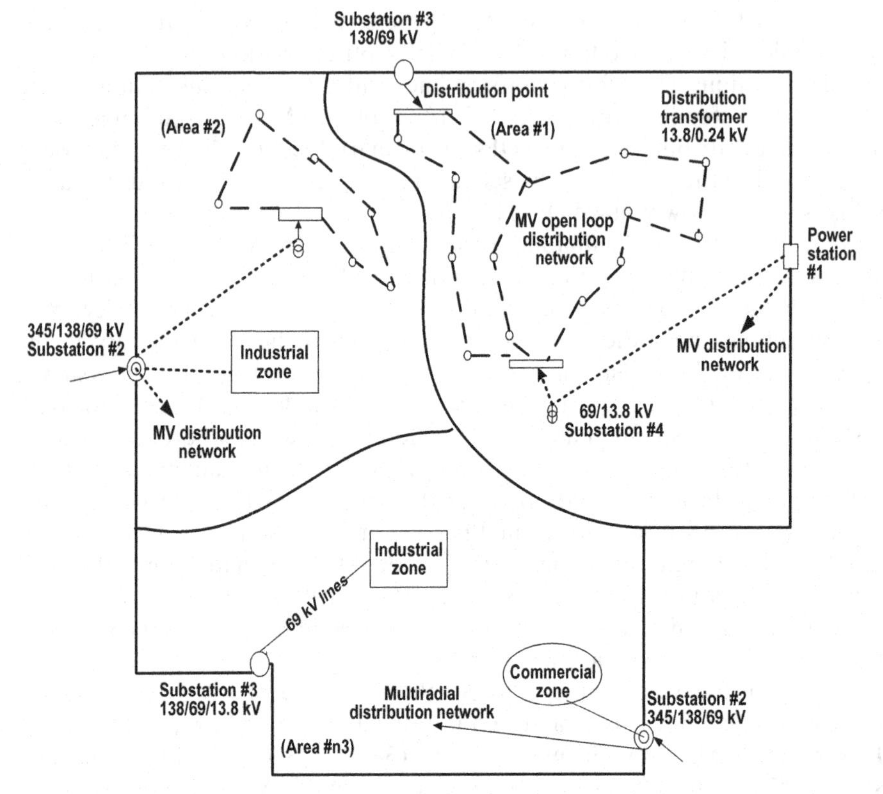

Figure 1.3 Electric supply system to a big city.

can meet the predicted demand for electricity supplied through its subtransmission lines and zone substations, and on improving the reliability of supply to the customers.

This necessitates gathering the following information:

- The history, demand forecasts, and capacity of each zone substation.
- Evaluation of probable loss of load (LOL) for each subtransmission line and zone substation. This requires an accurate reliability analysis including the expected economic and technical impact of the load loss.
- Determination of standards applied to the distributor's planning.
- Studying the available solutions to meet forecast demand including demand management and the interaction between power system components and embedded generation, if any.
- The choice and description of the best solution to meet forecast demand including estimated costs and evaluation of reliability improvement programs undertaken in the preceding year. The benefits of improving the

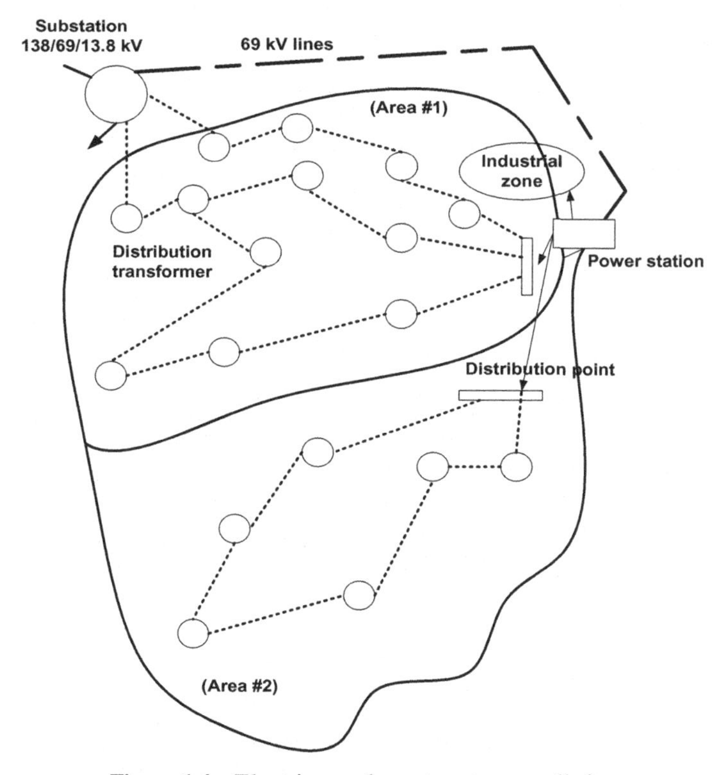

Figure 1.4 Electric supply system to a small city.

system reliability and the cost of applying the best solution to enhance the system performance must be compiled; that is, a cost wise study must be done.

The main steps of electric distribution system planning can be depicted by the flowchart shown in Figure 1.5. The flowchart starts with identifying the system capacity to enable the planner to model the network loading and performance, and to identify system inadequacies and constraints. This is done as a second step with the aid of information about demand forecasts, standards, asset management system, and condition monitoring (CM). As a third step, all feasible network solutions are identified, and the cost of each in addition to the lead time of implementation is estimated. Consequently, it leads to the preparation of a capital plan and investment in major works for specific years ahead as a fourth step. The next procedure is concerned with detailed economic and technical evaluation of feasible solutions. It is obvious to expect the next step to be the selection of the preferred solution, and then to review the compliance with standard requirements and obtaining the approval of authorized boards to start the implementation of the plan as the

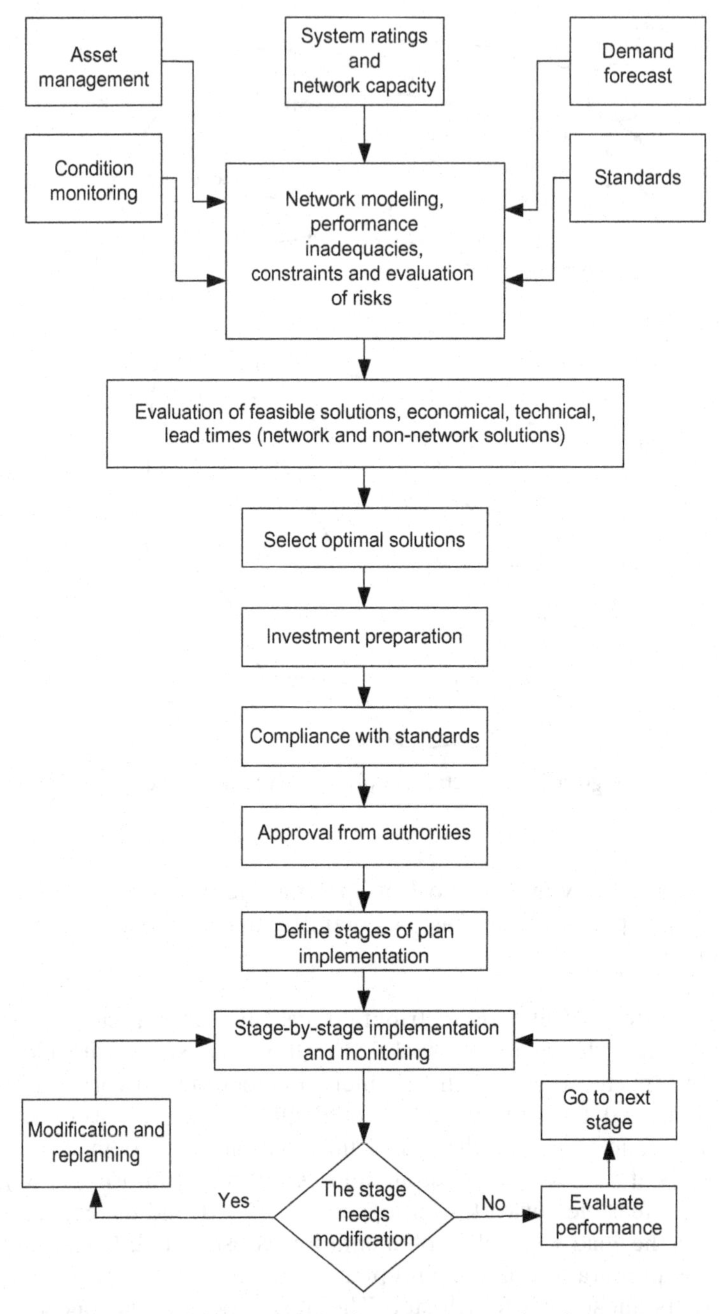

Figure 1.5 Flowchart of distribution system planning process.

last step of the flowchart. The planner is required to monitor the plan during implementation. Usually, the implementation includes multistages. At each stage, it is probable to receive feedback that may necessitate modification and replanning.

1.3 FACTORS AFFECTING THE PLANNING PROCESS

As described in Section 1.2, the planning process depends mainly on the factors mentioned below.

1.3.1 Demand Forecasts

For distribution systems, the study of demand forecasts concerns mainly with the estimation of expected peak load in the short term. The peak load is affected by several factors such as social behavior, customer activity, and customer installations connected to the network and weather conditions.

In general, no doubt that the study of load forecasting is very important as it provides the distribution planners with a wide knowledge domain. This domain encompasses not only the expected peak load but also the nature and type of loads, for example, commercial, industrial, and residential. This knowledge domain helps the planners to identify to what extent the distribution system is adequate. It also helps when proposing the plan of meeting the load growth and choosing the optimal solution that may be network augmentation or no network augmentation. The network augmentation solutions mean that additional equipment will be added to the system to increase its capacity, while no network augmentation solutions mean maximizing the performance of the existing system components.

1.3.2 Planning Policy

The suggested distribution system plan must be evaluated as investment process. Its fixed and running costs are estimated as accurately as possible. The plan may include the replacement of some parts of the network and/or adding new assets in addition to increasing the lifetime of present system components in accordance with an asset management model. Thus, asset management has a prominent role in the planning process. It aims to manage all distribution plant assets through their life cycle to meet customer reliability, safety, and service needs. The asset management model consists of an asset manager who is functionally separated within the company from the service providers. The asset manager decides what should be done and when, based on the assessment of asset needs, and then retains service providers to perform those tasks. Consequently, the asset manager develops distribution plant capital investment programs, develops all distribution plant maintenance programs, and ensures execution of programs by service providers (Fig. 1.6).

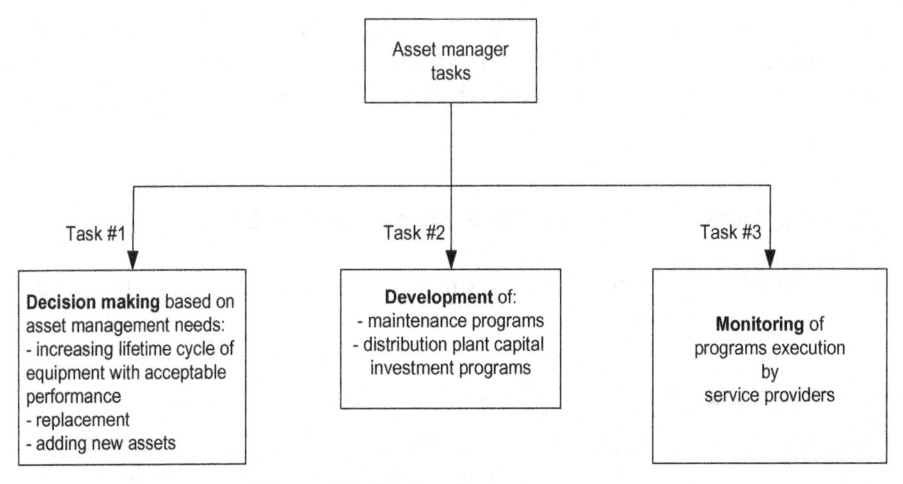

Figure 1.6 Tasks of an asset manager.

There is no doubt that utilities have to find ways to reduce maintenance cost, avoid sudden breakdown, minimize downtime, and extend the lifetime of assets. This can be achieved by CM with the capability to provide useful information for utilizing distribution component in an optimal fashion (more explanation is given in the next section). It can be concluded as both the investment and management of planning process must be integrated to achieve maximum revenue and efficiency for the customers and utilities as well.

1.3.3 CM

CM is a system of regular scheduled measurements of distribution plant health. It uses various tools to quantify plant health, so that a change in condition can be measured and compared. CM can also be an effective part of both a plant maintenance program, including condition-based maintenance (CBM) and performance optimization programs.

Time-based maintenance (TBM) has been the most commonly used maintenance strategy for a long time. TBM, to examine and repair the assets offline according to either a time schedule or running hours, may prevent many failures. However, it may also cause many unnecessary shutdowns, and unexpected accidents will still occur in between maintenance intervals. Manpower, time, and money were wasted because the activity of maintenance was blind with little information of the current condition of the assets. In contrast, CBM lets operators know more about the state of the assets and indicates clearly when and what maintenance is needed so that it can reduce manpower consumption as well as guarantee that the operation will not halt accidentally. CBM can be an optimal maintenance service with the help of a CM system to provide correct and useful information on the asset condition [6, 7].

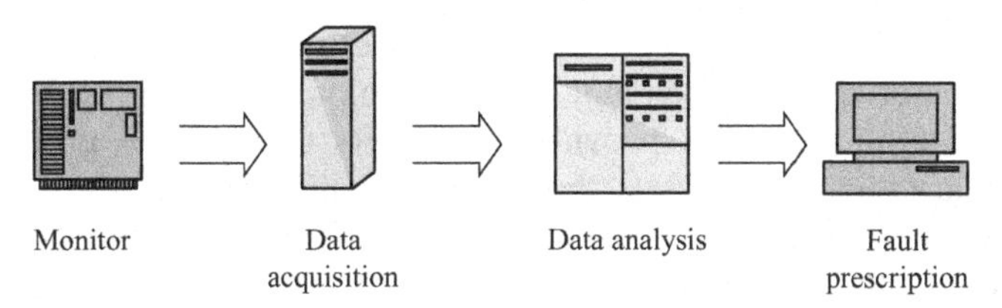

Figure 1.7 Main parts of a CM system.

CM should be capable of performance monitoring, comparing the actual measured performance to some design or expected level. When conditions slowly degrade over time, simple trend analysis can be used to raise alerts to operators that attention is required. For instance, in distribution systems, temperatures, pressures, and flows can be monitored and the thermal performance computed from these measurements. This can be compared with design conditions, and if negative trends develop over time, they can be indicative of abnormal or other performance-related problems [8].

The more difficult challenge is to identify when imminent equipment or component failure will cause an unplanned outage or will otherwise produce a change in plant performance. In some cases, simply trending the right parameter may be effective in avoiding this scenario, but usually degradation is due to a combination of several factors that cannot be predicted a priori or detected from a casual review of the trend data.

CM contains four parts as shown in Figure 1.7. The first is to monitor and measure the asset physical parameters (usually by using sensors) if their detectable changes can reveal incipient faults long before catastrophic failures occur. It converts the physical quantities into electric signals. The second is a data acquisition unit, which is built for amplification and preprocessing of the output signals from monitors, for example, conversion from analogue to digital. The third part is to analyze the collected data for fault detection by comparing the results of measurements with design conditions. Based on detected abnormal signals and existing expert systems, the fourth part presents to the operator full prescriptions, for example, fault location, fault type, status of asset, and advice for maintenance.

1.3.4 Reliability Planning Standards

The various assets of a power system (generation, transmission, and distribution) must follow the standards that ensure the continuity of supply in the event of system component outages. Components outage may be either a maintenance outage or a contingency outage such as external disturbances, internal faults, component failures, and lightning strokes.

Reliability standards provide a criterion for decision making toward the continuity and availability of power supply at any time and at different operating conditions. The decision may include an increase of operation automation and monitoring and/or adding some of the following equipment:

- Automatic circuit reclosers (ACRs) that result in a significant reduction of customer interruptions and customer minutes off supply. In addition, the fast fault clearance provided by ACRs reduces the probability of secondary damage to assets, thereby increasing the chances of successful recloser attempts for transient faults. As a result, customers experience fewer sustained outages.
- Fault locators to provide system control and network operators with the approximate location of faults enabling operators to locate the faults faster, thereby reducing supply restoration times.
- Ultrasound leakage detectors to detect leakage current on assets enabling corrective action to be carried out before pole fires develop.
- Thermovisions to detect hot spots on assets to enable corrective actions to be carried out before they develop into faults due to thermal breakdown of components.

The planner must establish an acceptable compromise between the economical and technical points of view with the goal of supplying electricity to customers at prices as low as possible and at accepted category of reliability level. Categorization of customer reliability levels in regard to distribution systems is explained in the next section.

1.3.5 Categories of Customer Reliability Level

The distribution system is reliable when the interruption periods are as small as possible, that is, less LOL. Therefore, the distribution system structure must be designed in such a way that the continuity of supply at a desired level of quality is satisfied. Different structures are explained in Section 1.6.

As a standard, it is common to classify the customers into three levels of reliability:

- *Level 1*: For high priority loads such as hospitals, industries, water pump plants, emergency lighting, and essential commercial loads, the system reliability must be as high as possible. This can be achieved by feeding the load through two independent sources (one in service and another as a standby). The interruption time is very short. It is just the time for transferring from one source to another and isolating the faulty part in the network automatically.
- *Level 2*: For moderate priority loads such as domestic loads, the interruption time is sufficient for manually changing the source feeding the loads.
- *Level 3*: For other loads having low priority, the interruption time is longer than the former two levels. This time is sufficient for repairing or replacing the faulty equipment in the distribution system.

1.4 PLANNING OBJECTIVES

The distribution system planning objectives are summarized below:

- meeting the load growth at desired quality,
- providing efficient and reliable supply,
- maximizing the performance of system components,
- satisfying the most cost-effective means of distribution system development, and
- minimizing the price of electricity to customers by:
  - choosing the most cost-effective solution and
  - minimizing total life cycle costs.

Therefore, distribution system planning is based on the following main key aspects:

- load forecasting,
- power quality,
- compliance with standards,
- investment with highest revenue,
- power loss, and
- amount of LOL.

1.4.1 Load Forecasting

Load forecast study is one of the most important aspects in planning because the loads represent the final target of the power system. Generation and transmission systems planning depends on long-term load forecast, while the distribution system planning depends on the short-term load forecast. The function of the power system is to feed the loads. So, load forecasting is the main base for estimating the investment.

The difficulty in load forecasting results from its dependence on uncertain parameters. For instance, the load growth varies from time to time and from one location to another. Various techniques of demand forecast estimation are given in the next chapter.

1.4.2 Power Quality

Meeting the demand forecasts by distribution system planning is necessary but not a sufficient condition to achieve a good plan. The power quality is a complementary part. It must be at a desired level to be able to supply customers with electricity. The power quality is determined by the electric parameters:

voltage, power factor, harmonic content in the network, and supply frequency. More details are explained in the forthcoming chapters (Part III).

1.4.3 Compliance with Standards

The distribution system planner takes into account the rules and standards that must be applied to system design. The system infrastructure, such as lines, cables, circuit breakers, and transformers; system performance; and system reliability, must all be in compliance with the international codes. Supervisory control and data acquisition (SCADA) systems have been employed for distribution automation (DA) and distribution management systems (DMS) in order to achieve high operational reliability, to reduce maintenance costs, and to improve quality of service in distribution systems. Moreover, once reliable and secure data communication for the SCADA system is available, the next step is to add intelligent application operation at remote sites as well as at the DA/DMS control centers. Use of intelligent application software increases the operating intelligence, supports smart grid initiatives, and achieves a greater return on investment. More details are given in Chapter 13 (Part IV).

1.4.4 Investments

Investments required to establish the system infrastructure must be estimated before implementing the plan. It is associated with financial analysis. As is mentioned in Reference 1, financial analysis, including life cycle costs, should be performed for the solutions that satisfy the required technical and performance criteria. Individual components within the network may have a life span, in some cases, in excess of 60 years, and life cycle costs can be a significant issue.

Investments of distribution systems should be guided by the principles of efficient reliability, power quality, and least cost [9]. They can be divided into new investments and replacement investments.

In *new investments*, where the existing network is expanded, a new network is constructed or the present network that may need to add some components is reconstructed.

In *replacement investments*, an existing component is replaced by a new identical component. This is usually done for maintenance purposes due to aging or malfunction of the old component.

The major target of the investment strategy is the minimization of the total cost within technical boundaries during the whole lifetime of the distribution network. The total cost for a network lifetime is considered to be comprised of three components: capital cost, operational cost including losses, and interruption cost [10]:

$$C_{\text{total}} = \int_0^T (C_{\text{cap}} + C_{\text{oper}} + C_{\text{intp}})dt, \tag{1.1}$$

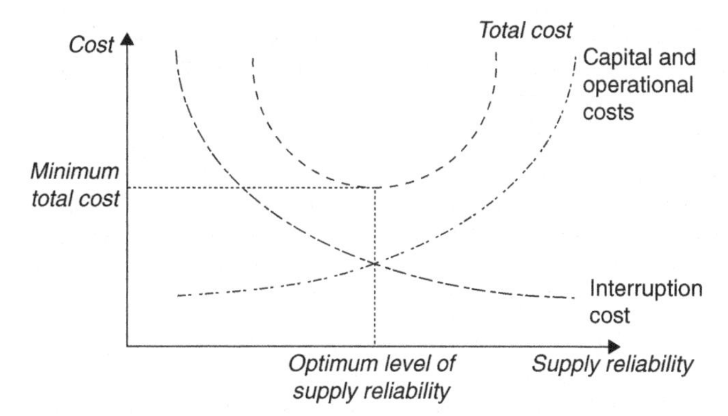

Figure 1.8 Balancing of the direct cost of service and the indirect cost of interruption [2].

where

C_{total} = total cost,
C_{oper} = operational cost,
T = network lifetime,
C_{cap} = capital cost, and
C_{intp} = interruption cost.

An ideal example of the determination of supply reliability (one of the technical requirements) at which the total cost is minimum is depicted in Figure 1.8. It is seen that as the supply reliability increases, the interruption cost decreases while the sum of the other two components, $C_{cap} + C_{oper}$, increases. Thus, the total cost has a minimum value providing the optimum level of supply quality.

In general, the problem is not as simple as illustrated in the above example. A detailed and efficient planning is necessary to supply the demand growth, to accomplish reliability and power quality requirements, and, at the same time, to optimize the use of the financial resources. In addition, the planning depends on many aspects such as consumers and regulatory agency requirements, environmental issues, and technological evolution as well as budget constraints. Therefore, it is a rather complex optimization problem because of its dependence on an enormous number of variables and constraints.

Different available expansion and improvement projects as solutions of such problem can be applied. They must be analyzed and prioritized to optimize the plan considering their costs and benefits. Therefore, the project prioritization problem is aimed at a search for the formation of a network strategic plan, for a specific period ahead, that best accomplishes the technical requirements and best improves the system performance subjected to budget constraints.

New tools and smart search methods, such as genetic algorithms and pareto optimality, have been used in the distribution system planning problem to generate and to test some alternatives for the network expansion. They can solve complex and discrete objective function problems, with large search space, that cannot be solved by other optimization techniques [11–13].

Introduction of deregulation and power markets has brought new challenges for the optimal investment strategies where the significance of the total cost components varies depending on the regulation model. For example, the importance of the power quality and thereby incentives for investments that improve power quality are strongly dependent on the regulation model and how power quality is included in the regulation. For instance, the interruptions as a measure of power quality can be identified by their number (number of interruptions) and/or duration (interruption time). The investments are focused on developing the distribution network to decrease the number of interruptions if it has more weight than the interruption time. On the contrary, where the interruption time has greater importance to be decreased, the investments are directed to increase the DA. Therefore, it can be said that the prioritization of the investments depends on the parameters of regulation.

Investments can also gain economic benefits not only by reduced total cost but also by increased allowed return. Allowed return in many cases is dependent on the current value of the distribution network assets, which can be increased by investments and based on the regulation method used (e.g., rate-of-return, price cap and revenue cap, and yardstick regulations) [10].

In addition, the distribution network investment decisions aim to minimize the cost to customers. Such alternatives include, but are not necessarily limited to, demand-side management and embedded generation.

1.4.5 Distribution Losses

Distribution losses are inevitable consequences of distributing energy between the zone substations and consumers. Losses do not provide revenues for the utilities and are often one of the controlling factors when evaluating alternative planning and operating strategies. The distribution utilities concern themselves with reducing the losses in the distribution systems according to the standard level. The level of losses will be influenced by a number of factors, technical and operational, such as network configuration, load characteristics, substations in service, and power quality required. It is important to manage these factors by appropriate incentives and thus optimize the level of losses.

Losses in distribution networks can be broken down into technical losses and nontechnical losses.

Technical losses comprise of variable losses and fixed losses.

Variable losses (load losses) are proportional to the square of the current, that is, depending on the power distributed across the network. They are often referred to as copper losses that occur mainly in lines, cables, and copper parts of transformers. Variable losses can be reduced by

- increasing the cross-sectional area of lines and cables for a given load;
- reconfiguring the network, for example, by providing more direct and/or shorter lines to where demand is situated;
- managing the demand to reduce the peaks on the distribution network;
- balancing the loads on three-phase networks;
- encouraging the customers to improve their power factors; and
- locating the embedded generating units as close as possible to demand.

Fixed losses (no-load losses) occur mainly in the transformer cores and take the form of heat and noise as long as the transformer is energized. These losses do not vary with the power transmitted through the transformer and can be reduced by using high-quality raw material in the core (e.g., special steel or amorphous iron cores incur lower losses). Another way to reduce fixed losses is to switch off transformers operating at low demand. Of course, this depends on the network configuration that enables the operator to switch some loads to other sources in the distribution network.

Nontechnical losses (commercial losses) comprise of units that are delivered and consumed, but for some reason are not recorded as sales. They are attributed to metering errors, incorrect meter installation, billing errors, illegal abstraction of electricity, and unread meters. Use of electronic meters will help reduce those losses since the accuracy is high. Also, incentives and obligation on participants should be as correct as possible to reduce the illegal abstraction of electricity. The chart shown in Figure 1.9 depicts a summary of types of distribution losses and the factors affecting them.

Reducing losses may have an added value to the cost of capital expenditure. It, on the other hand, will help reduce the amount of electricity production required to meet demand, and this will have wider benefits. Therefore, it yields the necessity of direct trade-off between the cost of capital expenditure and the benefits gained from loss reduction. To do that, the losses should be estimated as accurately as possible. More explanation of loss estimation in distribution systems is given in Chapter 8.

1.4.6 Amount of LOL

The distribution system components are exposed to unexpected failure and thus being out of service. If the failed component is a major component in the system, a shortage of capacity will occur and the system will not be able to provide some customers with electricity. The power demand of those customers is expressed as the amount of LOL.

For example, a zone substation includes N transformers at normal conditions. Assuming one transformer is out of service under contingency conditions, the total rating of the substation is decreased as shown in Figure 1.10. According to the line representing the demand forecast, the shaded area is the expected energy loss for a specific period.

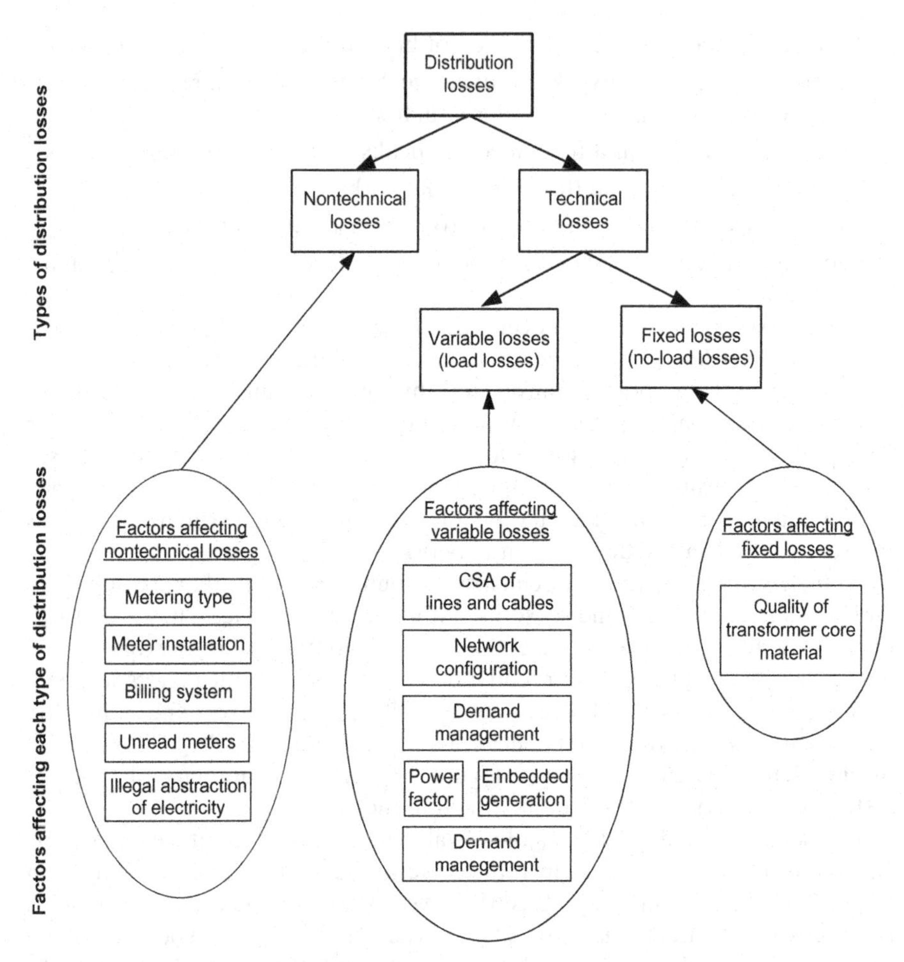

Figure 1.9 Types and factors affecting distribution losses.

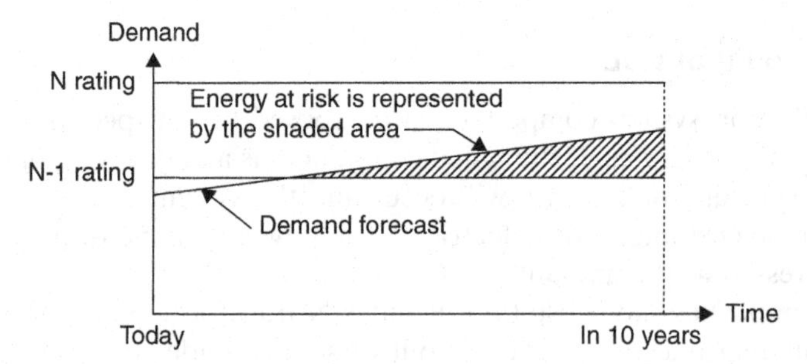

Figure 1.10 Relationship between N rating and N-1 rating and energy at risk [2].

1.5 SOLUTIONS FOR MEETING DEMAND FORECASTS

The planners should think about the available solutions that can be applied to modify the distribution network installations to meet the load growth in the next 3–5 years.

The solutions can be classified into two types: network solutions and non-network solutions. Network solutions are the solutions that need adding and/or replacing some network assets, while non-network solutions are concerned with optimizing the performance of existing network assets.

1.5.1 Network Solutions

According to the power system arrangement shown in Figure 1.1, the network solutions start from the load points to determine the adequacy of the size of the load. If it is not the standard size for the expected loads, the feeders must be resized. This is as a first solution, but if it is not sufficient, the planner must look at the DP switchboard design because it may be necessary to add a panel with circuit breaker for a new feeder or rearrange the present load loops.

Then, the testing of adequacy goes toward the infrastructure at higher voltage level to decide the best solution, which may be adopted or a combination of the following solutions:

- adding new subtransmission lines,
- adding new transformer to zone substation, and
- adding new zone substation.

In addition to the former solutions related to the system structure, the planner may need to add capacitors to improve the power factor or regulators to enhance the voltage profile as alternative solutions.

1.5.2 Non-Network Solutions

These types of solutions have a priority of application, particularly, if they are feasible, because these solutions are mostly more economic. The following are different alternative solutions possible:

a) *Embedded Generation*: The embedded generation would be connected to the distribution networks. Possible embedded generators could include the following types:
   - gas turbine power stations,
   - cogeneration from industrial processes, and
   - generation using renewable energy.

More details are described in Chapter 14 (Part V).

b) *Demand Management and Demand Response*: *Demand management* schemes have the potential to substantially reduce energy use for a given energy service, thereby reducing long-term energy and capacity needs, and so defer augmentation projects. This can be achieved by shifting customers' usage to off-peak and/or by using high-efficiency, low-energy appliances and reducing energy wastage. Demand management schemes could include

   - peak clipping,
   - valley filling,
   - load shifting,
   - strategic conservation,
   - load building, and
   - flexible load shape.

If such schemes were established, their effectiveness would depend on the extent of customer uptake.

Demand can also be reduced through encouragement for the use of high-efficiency appliances and energy-efficient designed homes and buildings (insulation, natural lighting, etc.). The subject of demand-side management is dealt with in more detail in Chapter 12 (Part IV).

Demand response, which means "actions voluntarily taken by a consumer to adjust the amount or timing of his energy consumption," results in short-term reductions in peak energy demand. Actions are generally in response to an economic signal and comprise of three possible types [14, 15]:

- price response, which refers to situations where consumers voluntarily reduce energy demand due to high prices during times of peak demand;
- demand bidding, where large consumers could "sell" their reductions in demand to the utility in times of peak demand; and
- voluntary load shedding, which refers to situations where consumers voluntarily reduce energy demand during times of high demand and/or constrained supply.

Therefore, demand management is focused on achieving sustained energy use reductions and is often driven by incentives, whereas demand response is market driven and results in temporary reductions or temporal shifts in energy use.

Non-network solutions of distribution system planning may enable the deferment or avoidance of major distribution system augmentation projects. Part of the avoided costs of the augmentation projects may be passed on to the owner of the embedded generator (or other means of demand reduction) as annual network support payment.

1.6 STRUCTURE OF DISTRIBUTION NETWORKS

1.6.1 Distribution Voltage Levels

North American and European are the two systems of distribution voltage levels mostly used around the world. The primary and secondary voltages (MV and LV, respectively) are given in Table 1.1 for both systems. The voltage choice depends on the type of load (residential, commercial, industrial), load size, and the distance at which the load is located. This is illustrated by a typical block diagram shown in Figure 1.11.

It is seen that the distribution voltage in the European system is higher than that in the North American system. It has advantages and disadvantages as below:

- *Advantages:* The system can carry more power for a given ampacity and has less voltage drop and less line losses for a given power flow. Consequently, the system can cover a much wider area. Because of longer reach, the system needs fewer substations.
- *Disadvantages:* More customer interruptions because the circuits are longer, that is, less level of reliability. Therefore, a major concern is to keep reliability at the desired level depending on the load category (more details are given in Section 1.6.2.2). From the cost point of view, the system equipment (transformers, cables, insulators, etc.) is more expensive.

1.6.2 Distribution System Configurations

Configuration of the distribution networks follows one or a combination of the following standard systems:

- radial system where the load is supplied through one radial feeder;
- open-ring system where the load is supplied through one of two available feeders, that is, one side of the ring;

TABLE 1.1 Distribution Voltages for North American and European Systems

Type of Voltage	North American System	European System
Primary distribution voltage (line to line)	From 4 to 35 kV	From 6.6 to 33 kV
Three-phase secondary voltage (line to line)	208, 480, or 600 V	380, 400, or 416 V
Single-phase secondary voltage (line to neutral)	120/240, 277, or 347 V	220, 230, or 240 V

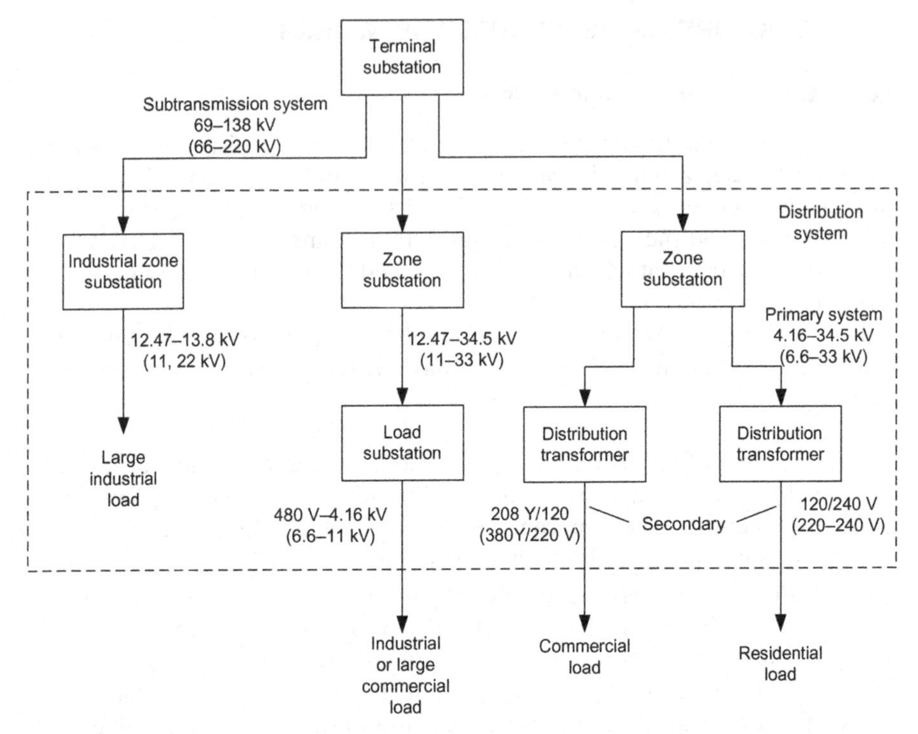

Figure 1.11 Distribution voltages for different loads (values written between parenthesis belong to the European system; otherwise, the values belong to the North American system).

- closed-ring system where the load is supplied through the two sides of the ring simultaneously;
- dual-ring system; in this case, the load is connected with two rings at the same time, that is, it has four incoming feeders; and
- multiradial system, which means supplying the load by more than one radial feeder.

These systems can be applied to establish the distribution network at either the MV or the LV.

1.6.2.1 MV Distribution Networks The MV distribution networks are the intermediate networks between the sources and the DPs in different zones of the load area.

The DP can be connected to the source through two in-service parallel feeders. At the same time, one feeder can operate at full load if a fault occurs in the second feeder as shown in Figure 1.12a. For more reliability, two DPs can be connected to each other as shown in Figure 1.12b. In this case, DP2 is

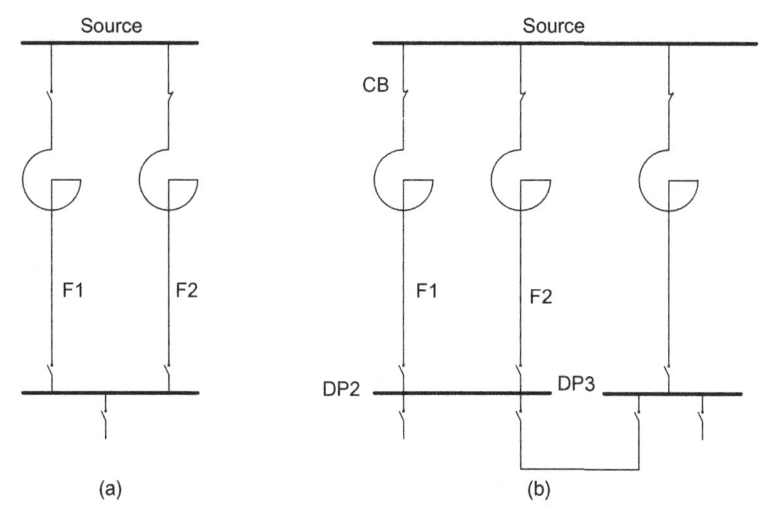

Figure 1.12 MV supply network to DP with parallel feeders. CB = circuit breaker with overcurrent protection; F = feeder; DP = distribution point.

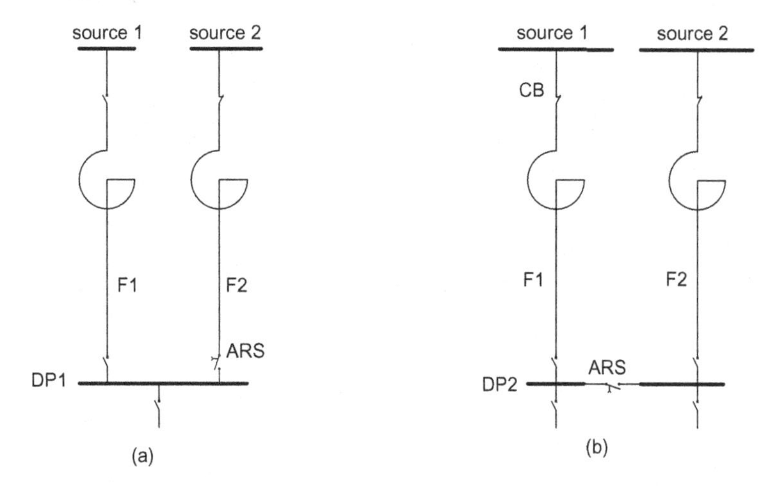

Figure 1.13 MV supply network to DP with independent feeders. ARS = automatic reserve switching; F = feeder; CB = circuit breaker; DP = distribution point.

connected to the source through two parallel feeders, while DP3 is connected to the same source through one feeder.

Of course, it is better and more reliable to use different sources feeding the DPs. In Figure 1.13a, the DP1 has two incoming feeders; the first is connected to source #1, while the second is connected to source #2. In this case, one feeder only is in-service. The other feeder is a standby and feeds DP1 through automatic reserve switching (ARS). Another way is to divide the bus bar of DP2

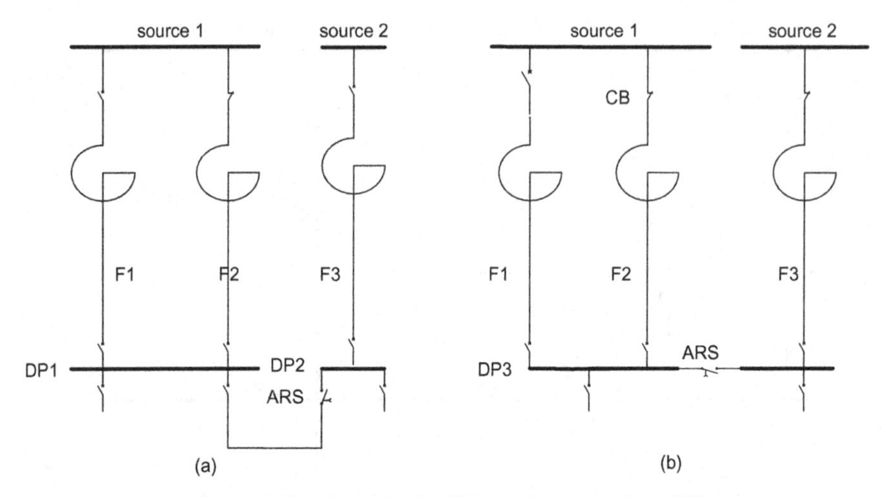

Figure 1.14 Combined MV supply network to DP.

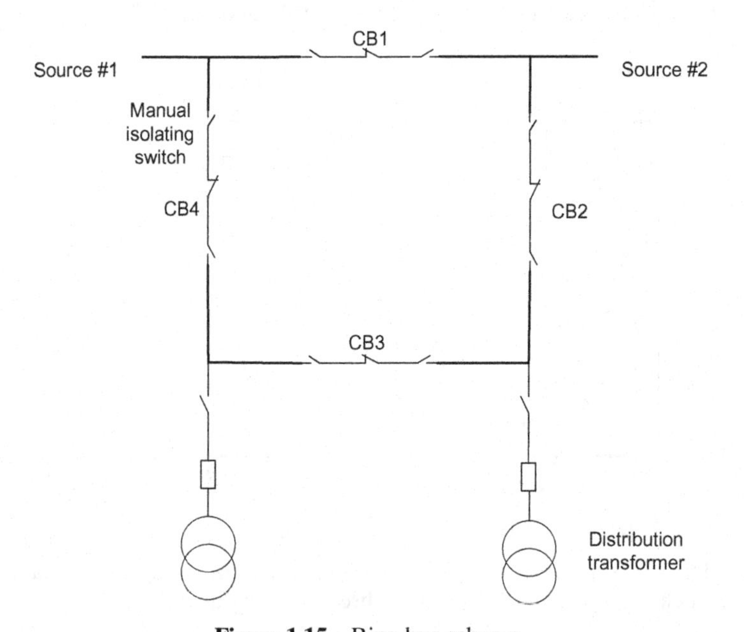

Figure 1.15 Ring bus scheme.

(Fig. 1.13b) into two sections connected to each other by a bus-coupler circuit breaker that is normally open. Each section has an incoming feeder connected to an independent source as shown in Figure 1.13b. Different configurations can be applied as given in Figure 1.14a,b, respectively. A ring bus scheme shown in Figure 1.15 can also be applied (popular in the United States). In this scheme, a fault anywhere in the ring results in two circuit breakers opening

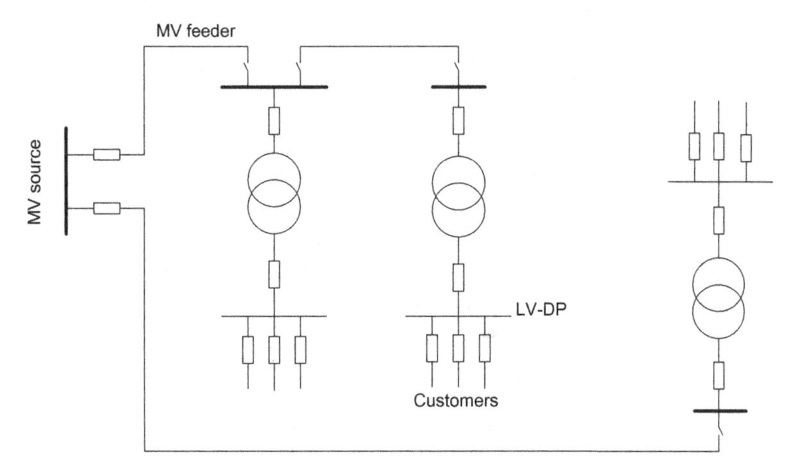

Figure 1.16 Radial nonreserved distribution network.

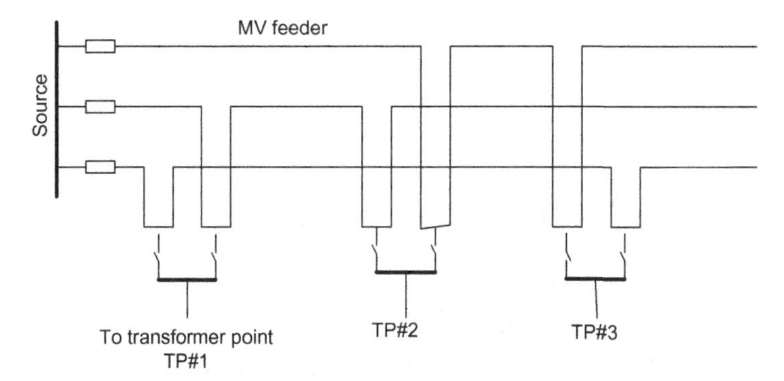

Figure 1.17 Multiradial distribution network with automatic reserve switching on MV side.

to isolate the fault, for example, when a fault occurs in source #1, circuit breakers CB1 and CB4 would operate to isolate the fault while source #2 would feed the loads. Circuit breakers are installed with two manual isolating switches on both sides to perform maintenance safely and without service interruption. Alternatively, radial and multiradial configurations can be used as shown in Figures 1.16 and 1.17, respectively. Two additional configurations, an integrated open-ring and radial distribution networks, are shown in Figures 1.18 and 1.19, respectively.

1.6.2.2 LV Distribution Networks The radial structure is the most commonly used in North America. The secondary feeders transmit power to the loads through distribution transformers (MV/LV). Single-phase power at voltage level 120/240V is usually supplied to residences, farms, small offices,

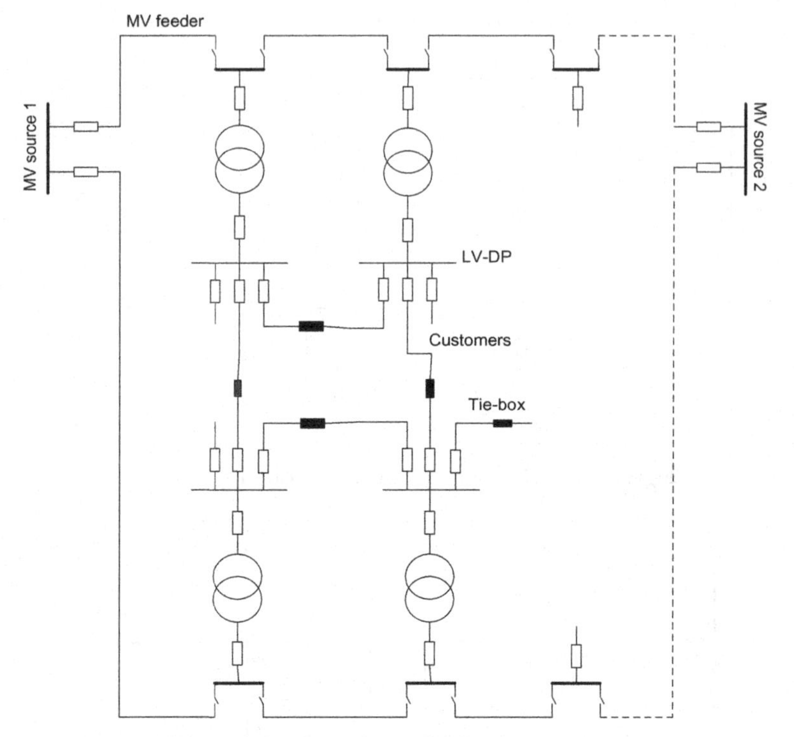

Figure 1.18 Open-loop distribution network.

and small commercial buildings (Fig. 1.20a). Three-phase power is usually supplied to large farms, as well as commercial and industrial customers. Typical voltage levels for three-phase power are 208Y/120 V, 480Y/277 V, or 600Y/347 V (Fig. 1.20b).

To increase the reliability, in particular, for LV applications with a very high load density, two or more circuits operate in parallel to transmit the power into a secondary (LV) bus at which the load is connected. Each circuit includes a separate primary (MV) feeder, distribution transformer, and secondary feeder that is connected to secondary bus through a network protector. This scheme is known as "secondary spot network" (Fig. 1.21). If a fault occurs on a primary feeder or distribution transformer of one of the circuits, its network protector receives a reverse power from the other circuits. This reverse power causes the network protector to open and disconnect the faulty circuit from the secondary bus. The load is only interrupted in case of simultaneous failure of all primary feeders or at fault occurrence on the secondary bus. Of course, this scheme is more expensive because of the extra cost of network protectors and duplication of transformer capacity. In addition, it requires a special construction of the secondary bus to reduce the potential of arcing fault escalation as well as a probable increase of secondary equipment rating since the short circuit current capacity increases in case of transformer parallel operation.

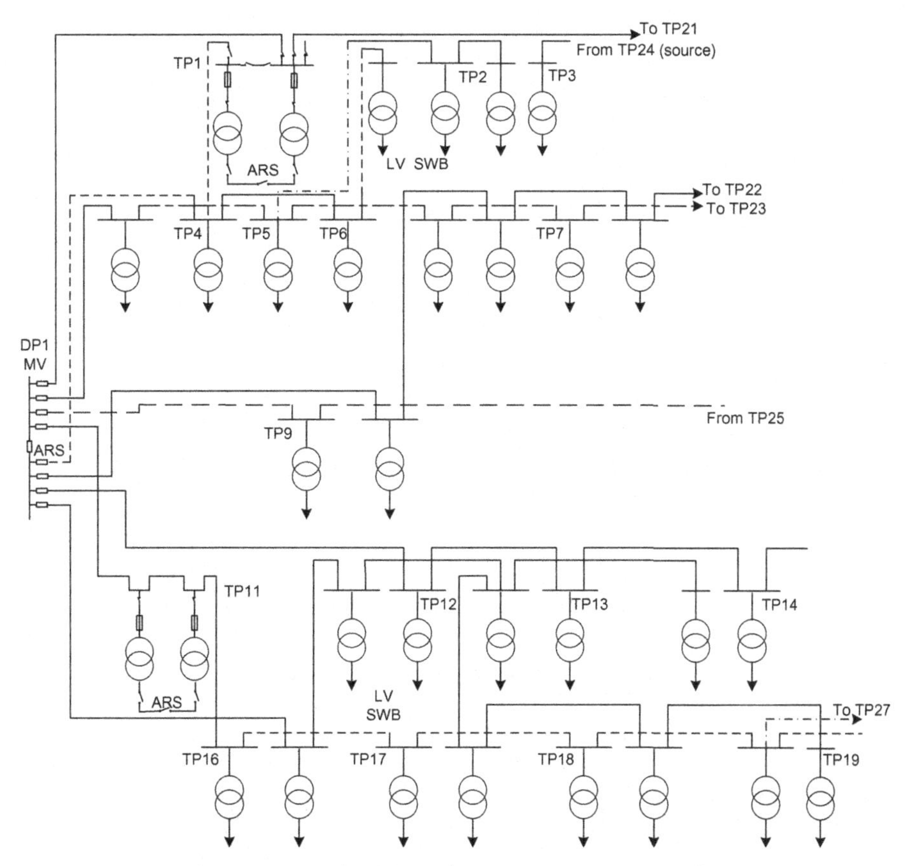

Figure 1.19 Radial MV network with DP (ARS on LV bus bar of TP). TP = transformer point; DP = distribution point; ARS = automatic reserve switching; SWB = switchboard.

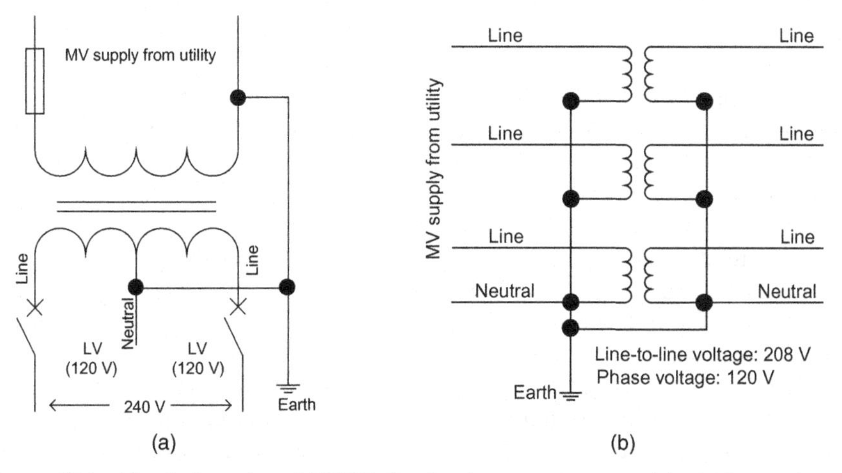

Figure 1.20 North American MV/LV distribution transformers. (a) 120/240 V single-phase service. (b) Typical 208 V, three-phase Y connected service.

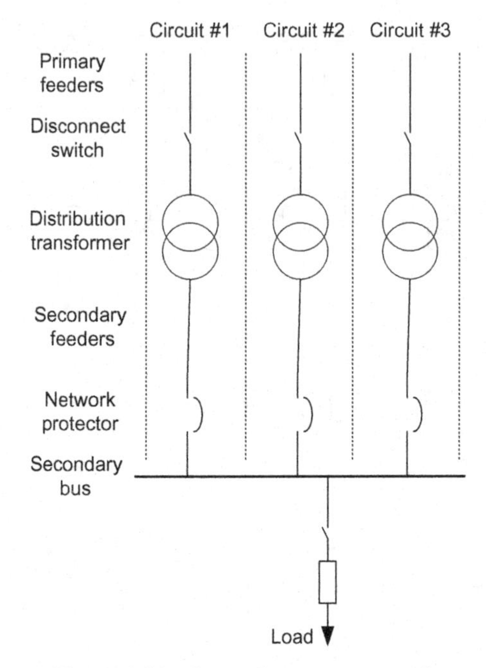

Figure 1.21 Secondary spot network.

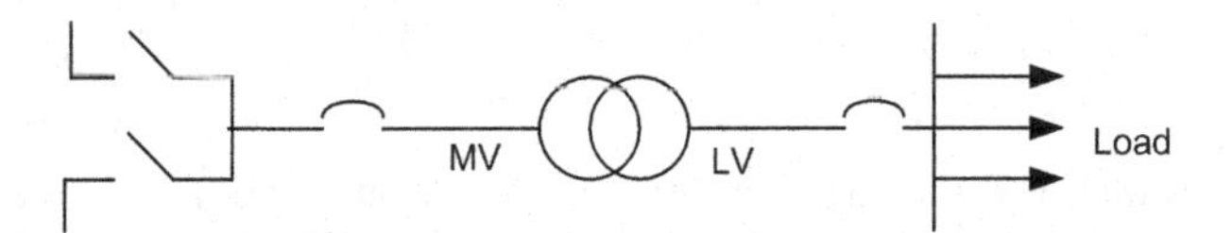

Figure 1.22 Supply system to a third category customer.

For European distribution systems, the choice of any of the standard structures above (Section 1.5.2) to establish the LV distribution networks depends on the type of the majority of loads and the reliability level required supplying these loads.

For reliability level 3, the distribution transformers are supplied by two incoming feeders (one is a standby to the other) on the MV side. The outgoing feeders at LV side are radial feeders (cables or overhead lines) as shown in Figure 1.22. If a fault occurs on one of these outgoing feeders, the concerned load is disconnected until the fault is repaired.

For reliability level 2, the MV side of the distribution transformer is connected to the source through two feeders (one in-service and the other is a standby). The outgoing feeders on the LV side are structured by one of the following:

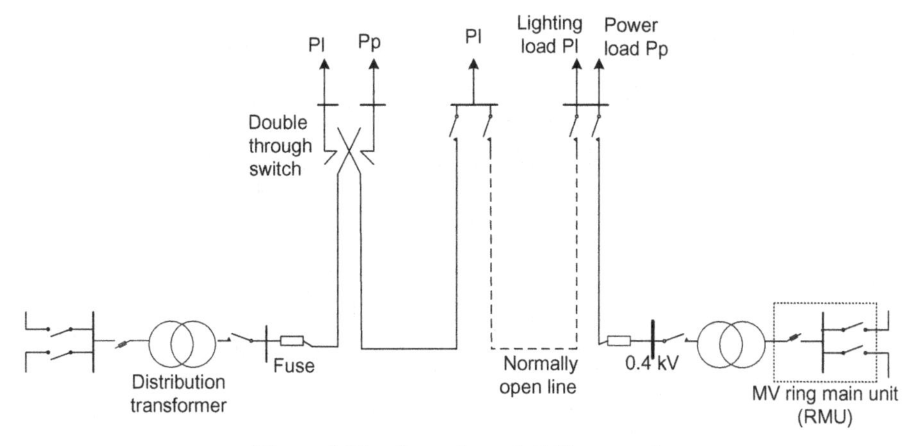

Figure 1.23 Open-loop 0.4 kV network.

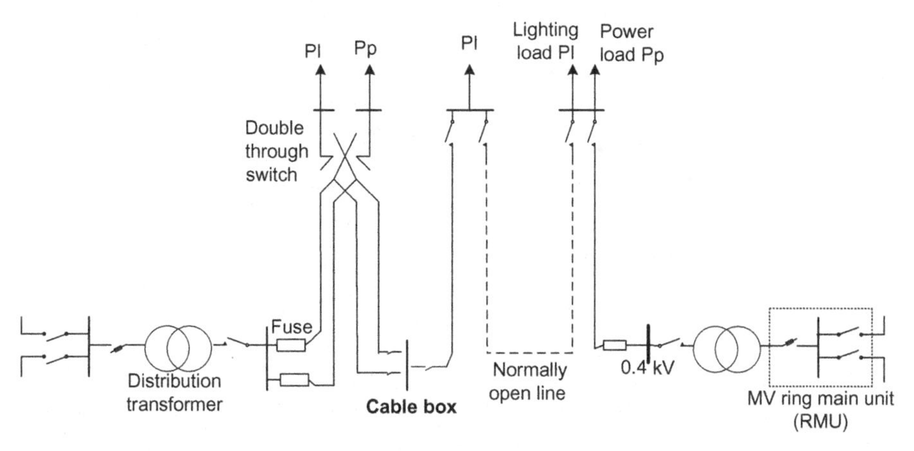

Figure 1.24 Combined open/double radial LV network.

- open-ring network (Fig. 1.23),
- combined open/double radial network (Fig. 1.24),
- double radial/double transformer cabinet network as in Figure 1.25, and
- double radial network where each radial feeder is connected to a single transformer cabinet (Fig. 1.26).

These structures can reduce the interruption periods significantly.

For reliability level 1, the same structures as for level 2 are used but equipped with fast automation techniques. In addition, a local generator with rating sufficient to the very critical parts of the load is located at the load location.

Also, the closed-ring or semiclosed ring systems can be used. They need more complex protection and control systems.

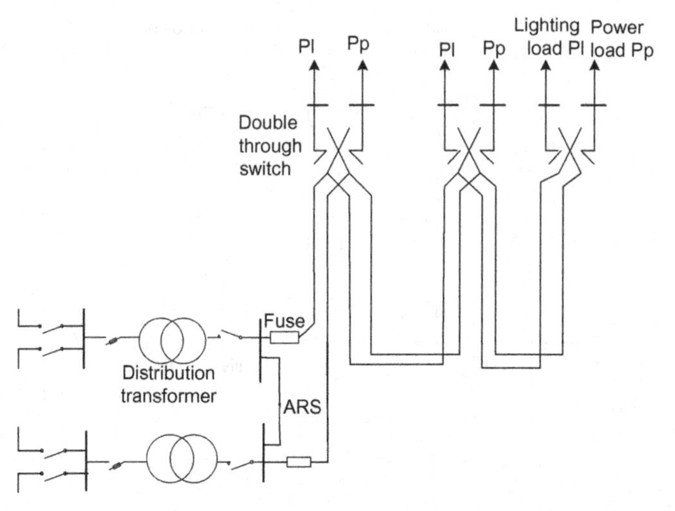

Figure 1.25 Double radial LV network. ARS = automatic reserve switching.

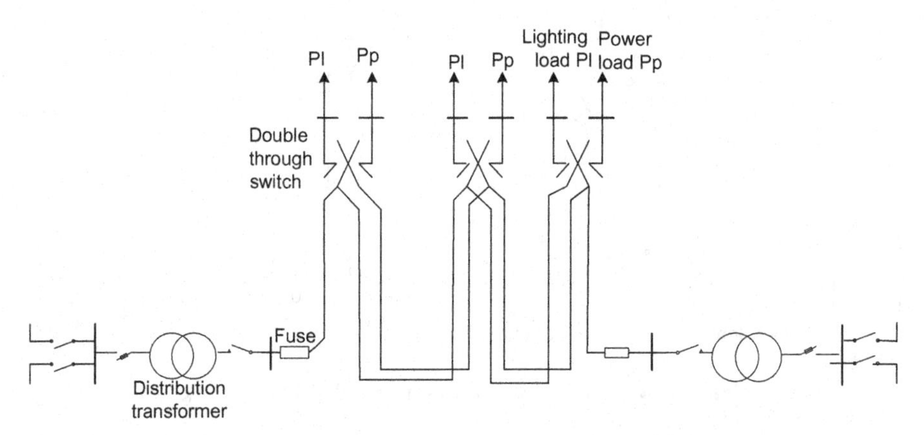

Figure 1.26 Double radial LV network. Each radial is connected to a single distribution transformer point.

Two typical examples of both North American and European design aimed to highlight the difference of configurations and structures of distribution networks, MV and LV, are shown in Figures 1.27 and 1.28, respectively.

It is seen that the North American configuration is based on Reference 16:

- maximizing network primaries by reducing the length of secondaries in order to reduce losses;
- regular earthing of MV neutral distribution (three-phase, four-wire multi-earthed primary);

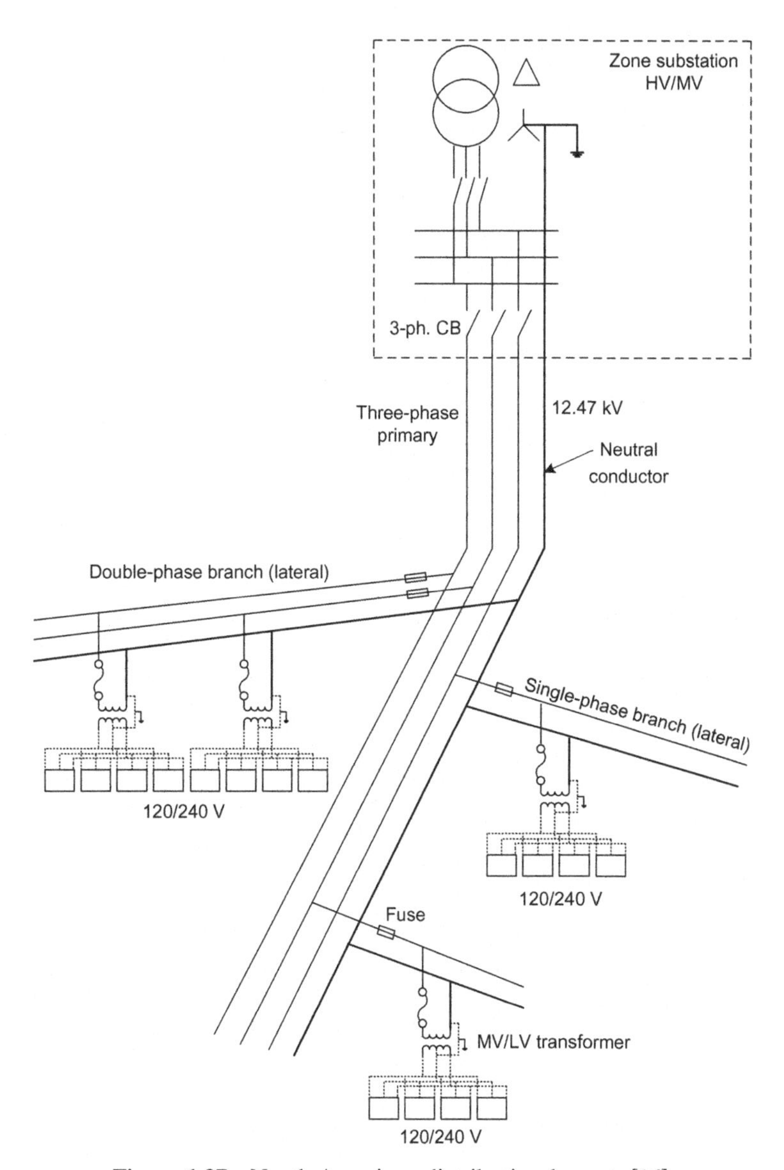

Figure 1.27 North American distribution layouts [16].

- using three-phase main primary with three-phase, two-phase, or single-phase shunting for branches (MV/LV connections); and
- radial structure.

On the other hand, the European configuration applies the following principles:

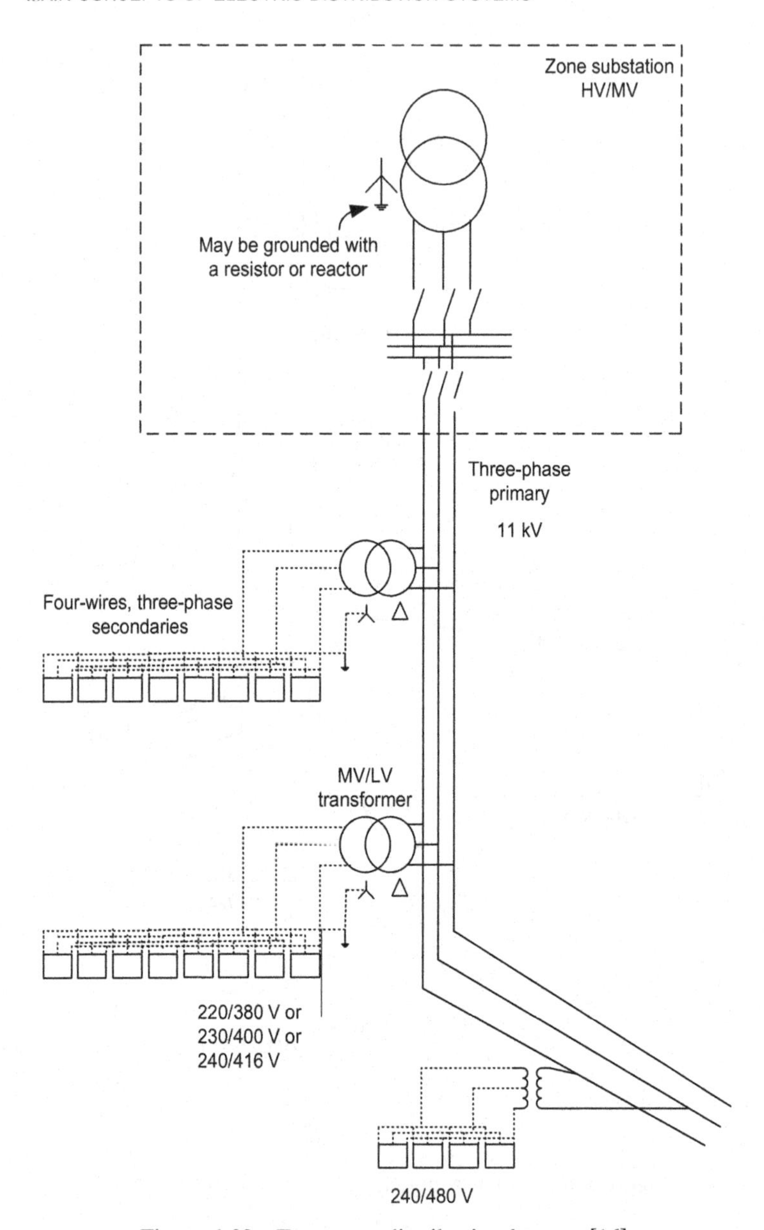

Figure 1.28 European distribution layouts [16].

- At the zone substation (HV/MV substation), neutral earthing is either solid earthing or earthing via an impedance limiting the phase-to-earth short circuit current.
- Three-phase primaries without distributed neutral are used.
- Radial structure.

CHAPTER 2

LOAD DEMAND FORECASTING

2.1 INTRODUCTION

Customer load demand in electric distribution systems is subject to change because human activities follow daily, weekly, and monthly cycles. The load demand is generally higher during the daytime and early evening when industrial loads are high, lights are on, and so forth, and lower from late evening to early morning when most of the population is asleep. Estimating the distribution system load expected at some time in the future is an important task in order to meet exactly any network load at whatever time it occurs [17].

On the other hand, the distribution system planning is a multistep process as described by the flowchart (Fig. 1.5) in Chapter 1. The most important key element, on which all steps are based, is load forecast. This defines the distribution system capabilities that need to be achieved by the future system. If it is done inappropriately, all subsequent steps will be directed at planning for future loads different from the load that will develop, and the entire planning process is at risk.

Therefore, load forecast plays a crucial role in all aspects of planning, operation, and control of an electric power system. It is an essential function for operating a power network reliably and economically [18]. So, the need and relevance of forecasting demand for an electric utility has become a much discussed issue in the recent past. It is not only important for distribution or power system planning but also for evaluating the cost-effectiveness of investing in the new technology and the strategy for its propagation.

Electric Distribution Systems, First Edition. Abdelhay A. Sallam, Om P. Malik.

Published 2011 by John Wiley & Sons, Inc.

According to the time horizon, load forecast can be classified as short term, midterm, and long term [19]. Short-term load forecasting (STLF) over an interval ranging from an hour to a week is important for different functions as unit commitment, economic dispatch, energy transfer scheduling, and real-time control. The midterm load forecast (MTLF), ranging from 1 month to 5 years and sometimes 10 or more years, is used by the utilities to purchase enough fuel and for the calculation of various electricity tariffs. Long-term load forecast (LTLF) covering from 5 to 20 years or more is used by planning engineers and economists to plan for the future expansion of the system, for example, type and size of generating plants and transmission lines, that minimize both fixed and variable costs.

STLF involves the prediction of the electric load demand on an hourly, daily, and weekly basis. It is an area of significant economic value to the utilities. It is also a problem that has to be tackled on a daily basis. Reliable forecasting tools would enable the power companies to plan ahead of time for peak demands and better allocate their resources to avoid any disruptions to their customers. It also helps the system operator to efficiently schedule spinning reserve allocation and provides information that enables the possible energy interchange with other utilities. In addition to these economic reasons, it plays an important role in the real-time control and the security function of an energy management system. The amount of savings in the operating costs and the reliability of the power supply depend on the degree of accuracy of the load prediction.

There is an essential need for accuracy in the demand forecast [20, 21]. This is because the underestimated demand forecast could lead to undercapacity resulting in poor quality of service where blackouts may occur. On the other hand, overestimation could lead to overcapacity, that is, excess capacity not needed for several years ahead. Consequently, the utility has to cover the cost of such overcapacity without revenues, and this is not favorable. In addition to ensuring the accuracy of forecast, the rationalization of pricing structures and design of demand-side management programs must be emphasized as well. Demand-side management is implemented using different methods explained in Chapter 12. These methods are based on an hour-by-hour load forecast and the end-use components with a goal of changing the system load shape.

Most forecasting methods use statistical techniques or artificial intelligence algorithms such as regression, neural networks, fuzzy logic (FL), and expert systems. Two methods, the so-called end-use and econometric approaches, are broadly used for medium- and long-term forecasting. A variety of other methods, which include the so-called similar-day approach, various regression models, time series, neural networks, FL, and expert systems, have been developed for short-term forecasting [22].

There is no single forecasting method that could be considered effective for all situations. The selection of a method of load forecast depends on the nature of the data available, and the desired nature and level of detail of the forecasts. Sometimes, it may be even appropriate to apply more than one method and then compare the forecasts to decide the most plausible one. Hence, every utility must find the most suitable technique for its application.

2.2 IMPORTANT FACTORS FOR FORECASTS

Several factors should be considered for STLF, such as time factors, weather data, and possible customers' classes. The medium- and long-term forecasts take into account the historical load and weather data, the number of customers in different categories, the appliances in the area and their characteristics including age, the economic and demographic data and their forecasts, the appliance sales data, and other factors.

Three principal time factors affect the load pattern. These are seasonal effects, weekly/daily cycle, and legal/religious holidays, which play an important role in load patterns. The seasonal fluctuations rely on the climatic influences (temperature, length of the day, etc.) and varying human activities (holidays, seasonal work, etc.). There are seasonal events that produce an important structural modification in the electricity consumption patterns such as the shifts to and from daylight saving time, start of the school day, and significant reductions in activities during holidays. Weekly fluctuations, type of day, are due to the presence of working days and weekends. The existence of holidays and weekends has the general effect of significantly lowering the load values to levels well below normal. Daily fluctuations, the day shape during the day, depend on the human activities such as work, school, and entertainment [23].

Most utilities have large components of weather sensitive loads such as those due to air conditioning. Thus, weather factors, such as temperature, wind speed, and humidity, have a significant effect on the variation in the load patterns. In many systems, temperature and humidity are the most important weather variables for their effect on the load.

Most electric utilities serve customers of different types such as residential, commercial, and industrial. The electric usage pattern is different for customers that belong to different classes but is somewhat alike for customers within each class. Therefore, most utilities distinguish load behavior on a class-by-class basis.

2.3 FORECASTING METHODOLOGY

It is evident that forecasting must be a systematic process dependent on the time period for which it is going to be used. Forecasting methodologies can be classified on the method used. In a more definitive way, they can be categorized as deterministic or probabilistic. A third approach is a combination of both deterministic and probabilistic methods. Accordingly, the categories are mathematically based on extrapolation, correlation, or a combination of both.

Over the last few decades a number of forecasting methods have been developed and applied to the problem of load forecast. Two of the methods, so-called end-use and econometric approaches, are broadly used for medium- and long-term forecasting. A variety of methods used for short-term forecasting fall in the realm of statistical techniques such as time series and regression method, or artificial intelligence algorithms such as expert systems, FL, and neural networks.

2.3.1 Extrapolation Technique

This is simply a "fitting a trend" approach that depends on an extrapolation technique. The mode of load variation can be obtained from the pattern of historical data and hence the curve fitting function is chosen. Selected well-known functions for the curve fitting are listed below:

- Straight line: $Y = a + bx$
- Parabola: $Y = a + bx + cx^2$
- S-curve: $Y = a + bx + cx^2 + dx^3$
- Exponential: $Y = ce^{dx}$
- Double exponential curve: $Y = \ln^{-1}(a + ce^{dx})$.

The exponential form has a special application and that is when the abscissa is obtained from a logarithmic scale as $\ln Y$. To explain, it is assumed that a simple straight line fits. Then,

$$\ln Y = a + dx.$$

From which $Y = e^{a+dx}$

$$= e^a e^{dx}.$$

Thus, $Y = ce^{dx}$ is the exponential form, where $c = e^a$.

Clearly, this does not exclude the use of exponential form with linear abscissa if the data fit the assumption.

The most common curve fitting technique to evaluate the coefficients a and d is the well-known "least squares method."

2.3.2 Correlation Technique

This technique is based on the calculation of correlation coefficient, which necessitates the calculation of what is known as variance and covariance as below.

Assume two random independent variables x and y, that is, the events $x = x_i$ and $y = y_i$, are independent events. The product of two random independent variables, xy, is a random variable $x_i y_i$, that is, the expectation of a product is the product of the expectations (only for independent type). That is,

$$E(xy) = E(x)E(y),$$

where E represents the "expectation of."

Assume $E(x) = \mu_x$ and $E(y) = \mu_y$, where μ is defined as the number of events/number of possible outcomes.

Therefore,

$$E(xy) = \mu_x \mu_y.$$

To measure the deviation from its expected value μ, the quantity σ, defined as standard deviation, is introduced:

$$\sigma = \sqrt{E(x-\mu)^2}.$$

The variance is defined as

$$\sigma^2 = E(x-\mu)^2. \tag{2.1}$$

The covariance of two independent variables is given by

$$\sigma_{xy}^2 = E(x-\mu_x)(y-\mu_y). \tag{2.2}$$

It should be noted that when $y = x$, the covariance proves to be the generalized form of variance.

Expanding the above relationship,

$$\begin{aligned}\sigma_{xy}^2 &= E(xy - y\mu_x - x\mu_y + \mu_x\mu_y)\\ &= E(xy) - E(y)\mu_x - E(x)\mu_y + \mu_x\mu_y\\ &= E(xy) - E(x)E(y),\end{aligned}$$

where

$$\begin{aligned}E(x)E(y) &= 0 \text{ if } x \text{ and } y \text{ are independent variables}\\ &\neq 0 \text{ if } x \text{ and } y \text{ are dependent variables.}\end{aligned}$$

The quantitative measure of the strength of dependability is called the correlation coefficient Γ.

Thus,

$$\Gamma = \frac{\sigma_{xy}^2}{\sigma_x \sigma_y}. \tag{2.3}$$

Sample variance:

If $x_1, x_2, x_3, \ldots, x_n$ are n independent observations of a variable x, the sample variance is defined by

$$S_{sv}^2 = (i/n)\sum(x_i - \overline{x})^2,$$

where $\overline{x}$ is the arithmetic mean of the variables, that is,

$$\begin{aligned}\overline{x} &= (x_1 + x_2 + x_3 + \ldots + x_n)/n \\ &= (1/n)\sum x_i \qquad \text{for } i = 1, 2, 3, \ldots, n. \qquad (2.4)\end{aligned}$$

It should be noted that unlike the theoretical σ^2, the sample variance is computed from the observed samples and hence it is actually available. Methods based on correlation technique relate the load to its dependent factors such as weather conditions (humidity, temperature, etc.) and economic and general demographic factors. This could clearly include population data, building permits, and so on.

2.3.3 Method of Least Squares

In Section 2.3.1, a number of functions, to be used as curve fitting for load curves, are introduced. To generalize, a curve fit in a polynomial form is represented as below:

$$Y = a_o + a_1x + a_2x^2 + \ldots + A_mx^m.$$

Start with "simple regression" to fit the straight line $Y = a_1 + a_2x$ to rectangular points shown in Figure 2.1.

The deviation of observation points from the assumed straight line should be minimized. Assuming a set of data (x_i, y_i), $i = 1, 2, 3, \ldots, n$ is represented by some relationship $Y = f(x)$, containing r unknowns $a_1, a_2, a_3, \ldots, a_r$, the deviations, sometimes called residuals R_i, are formed by

$$R_i = f(x_i) - y_i.$$

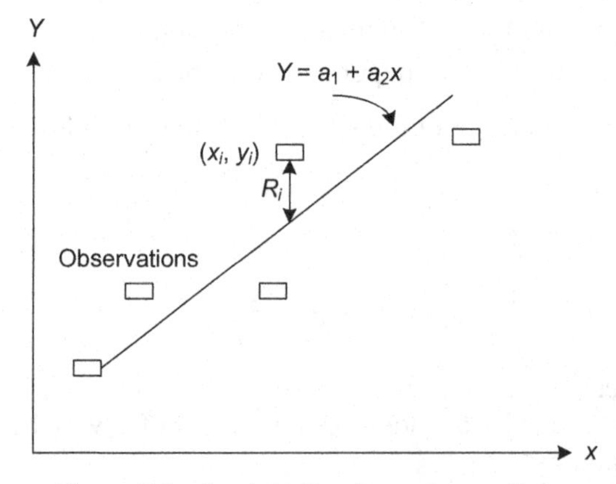

Figure 2.1 Straight line fit to observations.

The sum of squares of residuals is

$$S = \sum R_i^2 = \sum [f(x_i) - y_i]^2.$$

The unknown polynomial coefficients, a's, can be determined so that S is minimum by applying the relations

$$\frac{\partial S_i}{\partial a_i} = 0 \qquad i = 1, 2, \ldots, r. \tag{2.5}$$

This particular form of Equation 2.5 is known as the "normal equations" and the technique is called the "principle of least squares."

The residuals in this case (straight line curve fit) are

$$R_i = (a_1 + a_2 x_i) - y_i,$$

so that $S = \Sigma R_i^2 = \text{minimum}, i = 1, 2, 3, \ldots, n.$

Thus,

$$S = [(a_1 + a_2 x_1) - y_1]^2 + [(a_1 + a_2 x_2) - y_2]^2 + \ldots + [(a_1 + a_2 x_n) - y_n]^2.$$

Differentiating S partially with respect to a_1 and a_2, the two relations below are deduced:

$$\frac{\partial S}{\partial a_1} = 0 = 2[(a_1 + a_2 x_1) - y_1] + 2[(a_1 + a_2 x_2) - y_2] + \ldots + 2\,[(a_1 + a_2 x_\mathrm{n}) - y_\mathrm{n}],$$

$$\frac{\partial S}{\partial a_2} = 0 = 2x_1[(a_1 + a_2 x_1) - y_1] + 2x_2[(a_1 + a_2 x_2) - y_2] + \ldots$$
$$+\, 2x_n[(a_1 + a_2 x_n) - y_n].$$

Rearrange the two relations to be written in the form

$$n a_1 + \left(\sum x_i\right) a_2 = \sum y_i,$$

$$\left(\sum x_i\right) a_1 + \left(\sum x_i^2\right) a_2 = \sum x_i y_i \qquad i = 1, 2, 3, \ldots, n,$$

and in a matrix form

$$\begin{bmatrix} K_{11} & K_{12} \\ K_{21} & K_{22} \end{bmatrix} \begin{bmatrix} a_1 \\ a_2 \end{bmatrix} = \begin{bmatrix} C_1 \\ C_2 \end{bmatrix},$$

where $K_{11} = n$, $K_{12} = \Sigma x_i$, $K_{12} = K_{21}$, $K_{22} = \Sigma x_i^2$, $C_1 = \Sigma y_i$, $C_2 = \Sigma x_i y_i$, and $i = 1, 2, 3, \ldots, n$.

By matrix inversion,

$$\begin{bmatrix} a_1 \\ a_2 \end{bmatrix} = \frac{1}{D} \begin{bmatrix} K_{22} & -K_{12} \\ -K_{21} & K_{11} \end{bmatrix} \begin{bmatrix} C_1 \\ C_2 \end{bmatrix},$$

where the determinant of matrix $D = K_{11}K_{22} - K_{12}K_{21}$.

The same procedure can be applied when the curve fit is a polynomial of the mth order, where $(m + 1)$ linear equations are obtained to be solved for $(m + 1)$ unknowns.

Example 2.1:

Hourly data for the load on a cable feeding an industrial system and the corresponding atmospheric temperature for a 24-h period are given in Table 2.1.

TABLE 2.1 Observed Hourly Data

Hourly Observation	Load (Y) (MW)	Temperature (x) (°C)
1	10.98	1.83
2	11.13	−1.28
3	12.51	−0.67
4	8.40	14.89
5	9.27	16.33
6	8.73	21.83
7	6.36	23.55
8	8.50	24.83
9	7.82	21.5
10	9.14	14.17
11	8.24	8.00
12	12.19	−1.72
13	11.88	−2.17
14	9.57	3.94
15	10.94	8.22
16	9.58	9.17
17	10.09	15.17
18	8.11	21.11
19	6.83	21.11
20	8.88	23.61
21	7.68	22.28
22	8.47	14.50
23	8.86	7.00
24	10.36	0.78
25	11.08	−1.89

It is required to develop an equation to fit these data to be used for forecasting.

Apply the above technique for the two correlated events, load and temperature, to fit a straight line with the equation

$$Y_i = a_1 + a_2 x_i + \varepsilon_i \qquad i = 1, 2, 3, \ldots, 25,$$

where ε_i is the deviation from the true line.

The sum of squares of deviations from the true line is

$$S = \sum_{i=1}^{25} \varepsilon_i^2 = \sum_{i=1}^{25} (Y_i - a_1 - a_2 x_i)^2.$$

Thus, the normal equations are

$$\frac{\partial S}{\partial a_1} = -2 \sum_{i=1}^{25} (Y_i - a_1 - a_2 x_i),$$

$$\frac{\partial S}{\partial a_2} = -2 \sum_{i=1}^{25} x_i (Y_i - a_1 - a_2 x_i).$$

Equating the two equations (normal equations) to zero, a_1 and a_2 can be obtained:

$$25a_1 + a_2 \sum_{i=1}^{25} x_i = \sum_{i=1}^{25} Y_i,$$

$$a_1 \sum_{i=1}^{25} x_i + a_2 \sum_{i=1}^{25} x_i^2 = \sum_{i=1}^{25} x_i\, Y_i.$$

From which

$$a_1 = 11.0649, a_2 = -0.143782,$$

and the straight line equation for the estimated Y is

$$Y_{\text{est}} = 11.0649 - 0.143782x.$$

2.3.4 STLF Techniques

Stochastic time series approach is the most popular prediction technique. It is still used today by many power companies because of the ease of understanding and implementation and the accuracy of its results. The idea was originally proposed in Reference 24 for handling a special class of nonstationary processes. These algorithms are primarily based on applying autoregressive moving average (ARMA) or autoregressive integrated moving average (ARIMA) models to the historical data. Statistical analyses are then employed to estimate model parameters. Extensive analysis has been performed to select the method of estimating the model parameters [25]. However, a drawback of this technique is that it requires a large computational time for the parameter identification.

In the *multiple regression* method [26], the load is found in terms of explanatory variables such as weather and nonweather variables, which influence the electric load. The explanatory variables are identified on the basis of correlation analysis between each of these (independent) variables with the load (dependent) variable. Experience about the load to be modeled helps an initial identification of the suspected influential variables. The estimation of the regression coefficients is usually found using the least square estimation technique. Statistical tests such as F-statistic test [27] are performed to determine the significance of the model. The problem encountered in this technique is that the relationship between the load and weather variables is nonlinear and hence leads to a large prediction error. Also, finding functional relationship between weather variables and current load demand is difficult.

Expert systems [28] incorporates rule and procedures used by human experts in the field of interest into software that is then able to automatically make forecasts without human assistance. Expert systems work best only when a human expert is available to work with software developers for a considerable amount of time in imparting the expert's knowledge to the expert system software developers. Also, an expert's knowledge must be appropriate for codification into software rules.

FL [29] enables uncertain or ambiguous data, often encountered in real life, to be modeled. It is able to simultaneously handle numerical data and linguistic knowledge. However, FL requires a thorough understanding of the fuzzy variables of the input/output relationships as well as good judgment to select the fuzzy rules and membership functions that influence most of the solution of the application.

The advantage of *artificial neural networks* (ANNs) [30, 31] for load forecasting can be attributed to the fact that they combine both time series and regression approach. As for the time series, the ANN traces previous load patterns and predicts a load pattern using recent load data. Moreover, it uses weather information for modeling. It is able to perform nonlinear modeling and adaptation, and does not require assumption of any functional relationship between load and weather variables. In applying a neural network to electric load forecasting, one must select one of a number of architectures (e.g., Hopfield, back propagation, etc.), the number and connectivity of layers and elements, use of bidirectional or unidirectional links, and the number format (e.g., binary or continuous) to be used by inputs and outputs, and internally.

The most popular ANN architecture for electric load forecasting is back propagation. Back propagation neural networks use continuously valued functions and supervised learning. That is, under supervised learning, the actual numerical weights assigned to element inputs are determined by matching historical data (such as time and weather) to desired outputs (such as historical electric loads) in a preoperational "training session." ANNs with unsupervised learning do not require preoperational training.

However, the inability of an ANN to provide an insight into the nature of the problem being solved and to establish rules for the selection of optimum network topology remains a practical disadvantage of this technique.

The integration of ANN and FL is referred to as a fuzzy neural network (FNN) [32]. With this integrated approach, some of the uncertainties in the input/output pattern relationships are removed by the FL thereby increasing the effectiveness of the ANN. Thus, the drawbacks of the two techniques when used on their own are overcome. Various methods of short-term load forecasting with their advantages and disadvantages are given in Table 2.2.

2.3.4.1 Stochastic Time Series This forecasting procedure was developed almost three decades ago and is still very popular among forecasters. The methodology proposed in Reference 24 assumes three general classes of a model that can describe any type or pattern of stationary time series data. A time series is said to be stationary if there is no systematic change in the mean value and its variance. Examples of such models are the autoregressive (AR) model, the moving average (MA) model, and the ARMA model, which is obtained by combining together components of the first two types mentioned.

In the autoregression process, the current value of the time series $y(t)$ is expressed linearly in terms of its previous values $(y(t-1), y(t-2),\ldots)$ and a random noise $a(t)$. The order of this process depends on the oldest previous value at which $y(t)$ is regressed on. For an AR process of order p (i.e., AR (p)), time series can be written as

$$y(t) = a(t) + \phi_1 y(t-1) + \phi_2 y(t-2) + \ldots + \phi_p y(t-p), \tag{2.6}$$

where $\phi_1,\ldots, \phi_p$ are the autoregression parameters. By introducing the backshift operator B that defines $y(t-1) = By(t)$ and, consequently, $y(t-m) = B^m y(t)$, Equation 2.6 can now be written in the alternative form

$$\phi(B)y(t) = a(t), \tag{2.7}$$

where $\phi(B) = 1 - \phi_1 B - \phi_2 B^2 - \ldots - \phi_P B^P$.

MA model assumes that the current value of the time series $y(t)$ can be expressed linearly in terms of the current and previous values of a white noise series $a(t), a(t-1), \ldots$. The noise series is constructed from the forecast errors or residuals when signal observations become available. The order of the

TABLE 2.2 Short-Term Load Forecasting (STLF) Techniques

STLF Technique	Advantage	Disadvantage
Stochastic time series	Ease of understanding and implementation and accuracy of its results	Longer computational time for the parameter identification
Multiple regression	Model the relationship of load consumption and other factors such as weather, day type, and customer class	Finding functional relationship between weather variables and current load demand is difficult.
Expert system	Incorporates rules and procedures used by human experts into software that is then able to automatically make forecasts without human assistance	Works best only when a human expert is available. Also, an expert's knowledge must be appropriate for codification into software rules.
FL	Model uncertain data often encountered in real life. It is able to simultaneously handle numerical data and linguistic knowledge.	Requires a thorough understanding of the fuzzy variables as well as good judgment to select the fuzzy rules and membership functions
ANNs	It combines both time series and regression approach. It is able to perform nonlinear modeling and adaptation and does not require assumption of any functional relationship between load and weather variables.	The inability of an ANN to provide an insight into the nature of the problem being solved and to establish rules for the selection of optimum network topology
Fuzzy neural networks	Some of the uncertainties in the input/output pattern relationships are removed by the FL thereby increasing the effectiveness of the ANN	—

process depends on the oldest noise value at which $y(t)$ is regressed on. For an MA of order q, MA(q), the model can be cxpressed as

$$y(t) = a(t) - \theta_1 a(t-1) - \theta_2 a(t-2) - \ldots - \theta_q a(t-q), \tag{2.8}$$

where $\theta_1, \ldots, \theta_q$ are the MA parameters. Similar application of the back shift operator on the white noise would allow Equation 2.8 to be written as

$$y(t) = \theta(B) a(t), \tag{2.9}$$

where $\theta(B) = 1 - \theta_1 B - \theta_2 B^2 - \ldots - \theta_q B^q$.

The third model considered by the Box–Jenkins approach is a mixed model. It includes both the autoregression and MA terms in the model. This leads to a model described as

$$y(t) = \phi_1 y(t-1) + \ldots + \phi_p y(t-p) + a(t) - \theta_1 a(t-1) - \ldots - \theta_q a(t-q). \quad (2.10)$$

The process defined by Equation 2.10 is called autoregression MA of order (p, q), ARMA (p, q). The model can also be represented as

$$\phi(B)y(t) = \theta(B)a(t). \quad (2.11)$$

The basic theory behind AR, MA, and ARMA models applies to stationary data only. As mentioned before, this implies that the mean, the variance, and the autocovariance of the process are invariant under time translations. If the observed time series is nonstationary and does not vary about the fixed mean, then the series has to be differenced until a stationary situation is achieved. This model is called an "integrated" model because the stationary model, which is fitted to the differenced data, has to be summed, or integrated, to provide a model for the nonstationary data. The Box–Jenkins models for nonstationary time series are autoregression integrated (ARI), moving average integrated (MAI), and ARIMA process. The general form of the ARIMA process of order (p, d, q) is

$$\phi(B)\nabla^d y(t) = \theta(B)a(t), \quad (2.12)$$

where ∇ is the difference operator and d is the degree of differencing involved to reach a stationary stage.

As a result of daily, weekly, yearly, or other periodicities, many time series exhibit periodic behavior in response to one or more of these periodicities. Therefore, different classes of models, which have this property, are designated as seasonal processes. Seasonal time series could be modeled as an AR, MA, ARMA or an ARI, MAI, and ARIMA seasonal process similar to the nonseasonal time series discussed before. It is shown in Reference 33 that the general multiplicative model $(p, d, q) \times (P, D, Q)_s$ for a time series model can be written as

$$\phi(B)\Phi(B^s)\nabla^d\nabla_s^D y(t) = \theta(B)\Theta(B^s)a(t), \quad (2.13)$$

where ∇^d, $\phi(\mathrm{B})$, and $\theta(\mathrm{B})$ have been defined previously. Similar definitions of ∇_s^D, $\Phi(B^s)$, and $\Theta(B^s)$ are given in the following relationships:

$$\nabla_s^D = (y(t) - y(t-s))^D = (1 - B_s)^D y(t), \quad (2.14)$$

$$\Phi(B^s) = 1 - \Phi_1 B^s - \Phi_2 B^{2s} - \ldots - \Phi_P B^{P_s}, \quad (2.15)$$

$$\Theta(B^s) = 1 - \Theta_1 B^s - \Theta_2 B^{2s} - \ldots - \Theta_Q B^{Q_s}. \quad (2.16)$$

An example demonstrating the seasonal time series modeling is the model for an hourly data with a daily cycle. Such a model can be expressed using the model of Equation 2.12 with $s = 24$.

The procedure for Box–Jenkins techniques [25] depends first, on the identification of the model, which is determined by analyzing the raw load data. This analysis includes the plot of autocorrelation and partial autocorrelation functions. The use of these tools leads to the initial estimation of the required data transformation and the degree of differencing to obtain a stationary process. Second, an estimation of the parameter of the identified load-forecasting model is usually achieved through the use of efficient estimation methods such as maximum likelihood and unconditional least squares. Finally, the model is checked by testing whether the residual series is a white noise. If the residual is not white noise, the inadequacy of the model has to be corrected in view of the autocorrelation and partial autocorrelation functions of the residual.

Example 2.2:

The shape of typical electric load data for 1 week is shown in Figure 2.2. It is required to forecast the electric load with 1-h lead time using ARIMA model.

By analyzing the characteristic of the electric load, it is found that this electric load exhibits daily and weekly variations. The load behavior for weekdays (Sunday to Thursday) has the same pattern but with small random variations. The weekday load pattern is different from Thursday, Friday, and Saturday. Comparing weekday loads with Thursday loads, the level of Thursday loads is relatively low during the p.m. hours because of its proximity to the

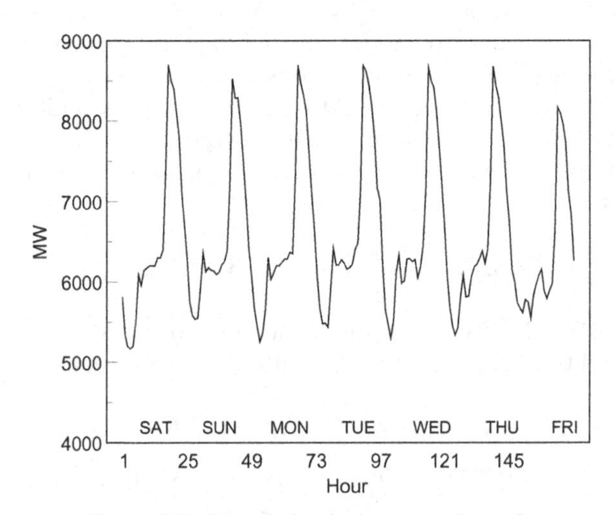

Figure 2.2 Hourly load curve over 1 week.

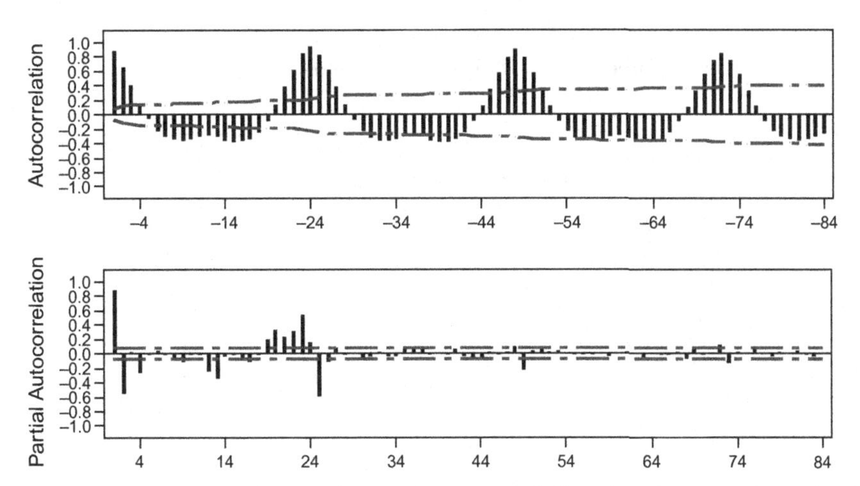

Figure 2.3 Autocorrelation and partial autocorrelation functions of 5 weeks in summer [33].

weekend. The level of Saturday load during a.m. hours is influenced by Friday, a holiday, and thus is low due to pickup loads when all businesses and industries just start work. In addition to the daily and weekly variations, the electric load exhibits seasonal variations over the year.

As an illustration, five consecutive weeks of hourly load data before the day for which load is to be predicted are used. The autocorrelation and the partial autocorrelation functions of 5 weeks are shown in Figure 2.3. This figure demonstrates that these data are not stationary and some differencing is required. In addition, these functions (in particular autocorrelation) show peaks at 24-h intervals for the previous 5 weeks of data.

Several models can be investigated by analyzing the residuals and diagnostic checking of all suggested models. The most suitable ARIMA model to model the hourly load demand of such load profile is found to be ARIMA $(2, 1, 0) \times (1, 1, 2)_{24}$ that can be represented by the following equation:

$$(1-\Phi_1 B-\Phi_2 B^2)(1-\phi_1 B^{24})(1-B)(1-B^{24})y(t)=(1-\Theta_1 B^{24}-\Theta_2 B^{48})a(t). \tag{2.17}$$

Having selected the model, the next step is to forecast the hourly load. The forecast results are analyzed and compared based on the mean absolute percentage error (MAPE) and the root mean square error (RMSE) as defined by the following two equations:

$$\text{MAPE}=\sum_{i=1}^{N}\frac{e_i}{N}, \tag{2.18}$$

TABLE 2.3 Average MAPE and RMSE of Hourly Load Forecast over 1 Week in the Summer

	ARIMA	
Day	MAPE (%)	RMSE (%)
Sunday	1.39	1.17
Monday	0.86	0.76
Tuesday	1.17	1.02
Wednesday	1.34	1.30
Thursday	1.16	1.01
Friday	1.42	1.60
Saturday	1.60	1.49
Average	1.28	1.19

where $e_i = \frac{|\text{actual load} - \text{forecasted load}|}{\text{actual load}} \times 100,$

$$\text{RMSE} = \sqrt{1/N} * \sum (\text{actual load} - \text{forecasted load})^2. \qquad (2.19)$$

The weekly averages of MAPE and RMSE of the weekdays and weekends over the forecast period are shown in Table 2.3, while the hourly load forecast over 1 week using the ARIMA model is illustrated in Figure 2.4.

2.3.5 Medium- and Long-Term Load Forecasting Methods

The end-use modeling, econometric modeling, and their combinations are the most often used methods for medium- and long-term load forecasting. Descriptions of appliances used by customers, the sizes of the houses, the age of equipment, technology changes, customer behavior, and population dynamics are usually included in the statistical and simulation models based on the so-called end-use approach. In addition, economic factors such as per capita incomes, employment levels, and electricity prices are included in econometric models. These models are often used in combination with the end-use approach. Long-term forecasts include the forecasts on the population changes, economic development, industrial construction, and technology development [34].

In *end-use method*, the impact of energy usage patterns of different devices and systems is analyzed. The end-use models for electricity demand focus on the different characteristics of energy use in the residential, commercial, and industrial sectors of the economy. For instance, in the residential sector, electricity is used for water heating, air-conditioning, refrigeration, and lighting, whereas in the industrial sector the major use is in electric motors.

The end-use method is based on the premise that energy is required for the service that it delivers and not as a final goal. More details about this method are explained later in this chapter.

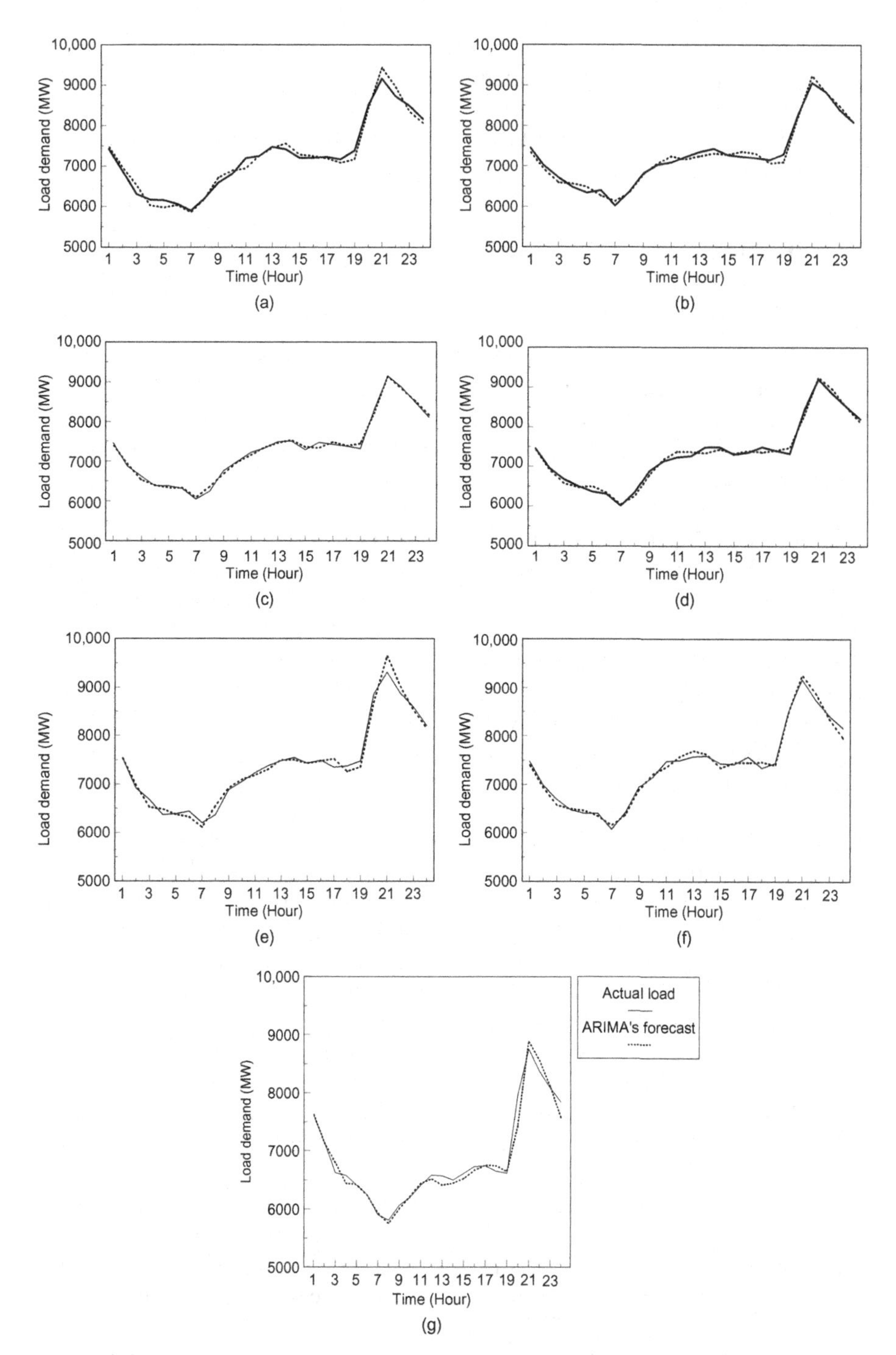

Figure 2.4 Forecast and actual load profile for 1 week in summer: (a) Saturday, (b) Sunday, (c) Monday, (d) Tuesday, (e) Wednesday, (f) Thursday, and (g) Friday [33].

The *econometric models* approach combines economic theory and statistical techniques for forecasting electricity demand. The approach estimates the relationships between energy consumption (dependent variables) and factors influencing consumption. The relationships are estimated by the least squares method or time series methods.

2.4 SPATIAL LOAD FORECASTING (SLF)

2.4.1 Main Aspects of SLF

SLF is based on three aspects as main requirements: (1) how well the forecast supports the planning process, (2) where will the future load develop, and (3) when will the expected load growth occur? The three aspects are explained to indicate their effect on estimating the load forecast and consequently on the planning process [34].

2.4.1.1 First Aspect What is the estimated value of the forecast load? The answer to this question is essential. So, different techniques can be applied such as trending and time series methods. Accurate estimation of the amount of load growth helps the planners to serve the expected load pattern as ideally as possible by correctly identifying the future equipment additions. The planners need to study the adequacy of equipment capacity such as transformers, circuit breakers, transmission lines, and feeders to meet the load growth in both transmission and distribution systems. The plan may include replacement of some equipment, addition of equipment, or management of the existing system.

2.4.1.2 Second Aspect Where is the geographic location of the future load growth? Providing information on the location of future load growth is one of the spatial load forecast requirements because it may change the plan. For example, consider a city in which an institute will be located. The location of institute is suggested to be either in the north or in the south of the city. In the north site, the available land area is just sufficient for the institute, while in the south site additional land is available. So, the second suggestion (south site) is attractive as it will create other activities such as building a shopping center, park, and industries. Then, the predicted load is the load of the institute plus other activities in addition to the normal annual growth rate. For the first suggestion, the future load is just the load of the institute plus the annual growth rate.

Therefore, two scenarios are presented with uncertainty of deciding which one will be implemented. Of course, the planner meets some difficulty in this situation. If the planner considers the forecast based on the first scenario, the plan may have a lack of capacity and if the forecast based on the second scenario is considered, the plan may have extra wasted capacity. In both cases, the forecasting is poor, leading the planner to a wrong decision, and the cost will be high. This is because the first scenario necessitates adding additional generation to substitute the lack of capacity and the second scenario has unused additional capacity, that is, waste of money. Also, the worst mistake that can be made is to try to circumvent the need for multiscenario planning by using "average" or "probabilistic" forecasts. This too leads to a poor forecast.

The solution of this uncertainty problem is to try to design a plan with a fixed direction for the short-term forecast and a changeable direction for the long-term forecast keeping in mind that the distribution planning is based on the short-term forecast, while the transmission planning is based on the medium- and long-term forecast.

2.4.1.3 Third Aspect When will the expected load growth take place? The time at which the load growth takes place is quite important. This is because, if the plan needs adding some equipment, for example, transformer or building a new substation, it needs time. So, it necessitates an early start with enough time to carry out the plan so that the infrastructure is available when needed.

In summary, the foregoing explanation highlights the main aspects of SLF. These aspects provide the information needed to plan transmission and distribution systems in a way that fits the planning process. This means describing as accurately as possible the *amount*, *location*, and *timing* of future load growth in a way suited to the short- and long-term planning needs.

2.4.2 Analysis Requirements

2.4.2.1 Spatial Resolution It represents the number of service areas where the size of each service area provides information on the future load growth in that area (location). So, the spatial resolution in a forecast must match the system planning needs. For distribution system planning, service areas of small size are needed, that is, the entire area is divided into a large number of service areas (subareas) getting a high spatial resolution. Thus, the utilities may pay more to collect the data satisfying the desired solution. As the system voltage level gets higher, the size of the service area becomes large, i.e. the spatial resolution is getting less. It can be seen from Table 2.4 that the

TABLE 2.4 Typical Service Area Sizes for Transmission, Distribution, and Generation System Planning [34]

Level	Service Area (km^2)	No. of Service Subareas
Generation	Entire area (300)	1
Transmission	256	3
Subtransmission	128	6
Distribution substation	76.8	10
Primary feeders	12.8	60
Single-phase laterals	0.64	1200
Secondary feeders	0.0256	30,000

planning at the highest system level (generation) for load to be served needs the entire area as a service area, while the planning of transmission system needs the load in a service area of 256 km^2 in size. As the system voltage level gets lower, the size of the service area also gets smaller; that is, both the number of service subareas and spatial resolution are getting high.

Therefore, the summation of small area forecasts in distribution system planning compared with the entire area forecasts is a useful way to check the accuracy and reasonableness. Also, the forecast of the entire area is a step toward assuring that nothing is missed in the utility's transmission and distribution planning.

2.4.2.2 Time and Peak Load Forecasts The transmission and distribution planning is based on annual peak. This needs year-by-year expansion studies during short-term period and plans for selected years in long-term period. It is common to set planning periods of 1–7 years ahead for short-term forecast and longer period until 25 years ahead for long-term forecast.

The load forecast seeks the annual peak load and in some cases the seasonal peak load, particularly when taking into account the weather conditions. This is because some equipment may be overloaded in the summer due to the high ambient temperature in spite of getting its peak load less than that in the winter.

For accurate peak load forecast, the coincidence of service area peaks must be studied. So, forecast of hourly loads of the peak day is usually performed not only for the peak load as a value but also for the peak duration. It is important to study the peak duration because it may allow the system equipment to operate on overload for a short period without failure, that is, no need to replace or add new equipment.

2.4.2.3 Type of Load It is worth classifying the loads into classes where each class has loads of the same nature. For instance, the most used classes for the customers are residential, commercial, and industrial. In some cases, each class can be classified into subclasses, but it must be assured that this subclassification gives more information to what the plan needs. General shape of the load curves for the traditional customer classes is shown in Figure 2.5. For

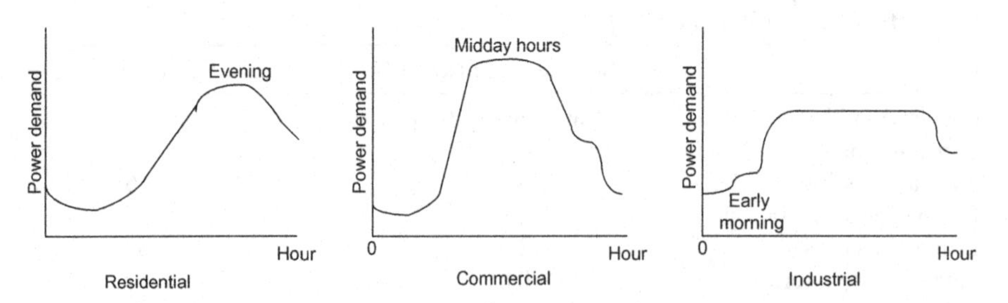

Figure 2.5 Load curves for traditional customer classes.

residential class, the peak load is almost in the evening while for commercial class it is at midday hours. The industrial class has low load at early morning, then approximately constant load until shutting down in the afternoon.

Distribution system planners can deduce three major points related to each customer in each class: (1) the peak load per customer, (2) the time and duration of this peak, and (3) the total energy consumption represented by the area under the curve.

The load curves are recorded and measured over some specific period (1 day, 1 week, 1 month). This period is divided into time intervals where the average power of an interval is calculated by dividing the area under the curve at this interval by the length of interval. This average power is called "demand" and the time interval is called "demand interval."

2.4.2.4 Sensitivity Analysis In general, sensitivity analysis means that the analyzed system must be studied under conditions different than the assumed conditions to see the impact of varied assumptions on the system performance.

Sensitivity analysis for long-term spatial forecast can be done by multiscenario planning. Distribution system planning is based on short-term spatial forecast. In this case, the "where" aspect poses some difficulty for the sensitivity analysis in determining the various locations. It may cause an error in forecast. The planner should try to decrease this error by multiscenario approach with the possibility of repeating the planning process according to the feedback information.

2.4.3 Load, Coincidence, and Diversity Factors (DFs)

Load factor (LF) is defined as the ratio of average to peak load. The value of the average load is the total energy consumed during an entire period divided by the number of period hours. Thus,

$$\begin{aligned} \text{LF} &= \frac{\text{kW (average)}}{\text{kW (maximum demand)}} \\ &= \frac{\text{kWh}}{\text{kW (maximum demand)} \times \text{hours}}. \end{aligned} \tag{2.20}$$

The pattern of load curves of Figure 2.5 describes the average load behavior of each class. Each class includes a number of customers. Therefore, the average load behavior of each customer can be represented by a curve that looks like the curve of its class with values equal to the values of the class divided by the number of customers in this class. In particular, the load curve of a customer (e.g., residential customer) is not smooth; it has needles and erratic shifts as shown in Figure 2.6. This is due to the nature of the residential customer loads that mostly include appliances such as water heater, refrigerators, and air

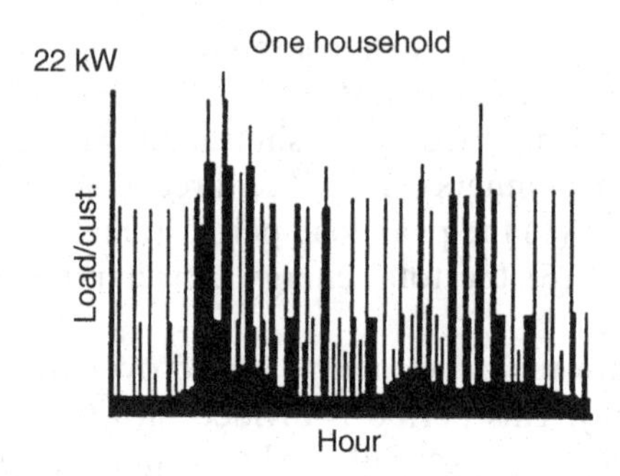

Figure 2.6 Load curve of one residential customer [34].

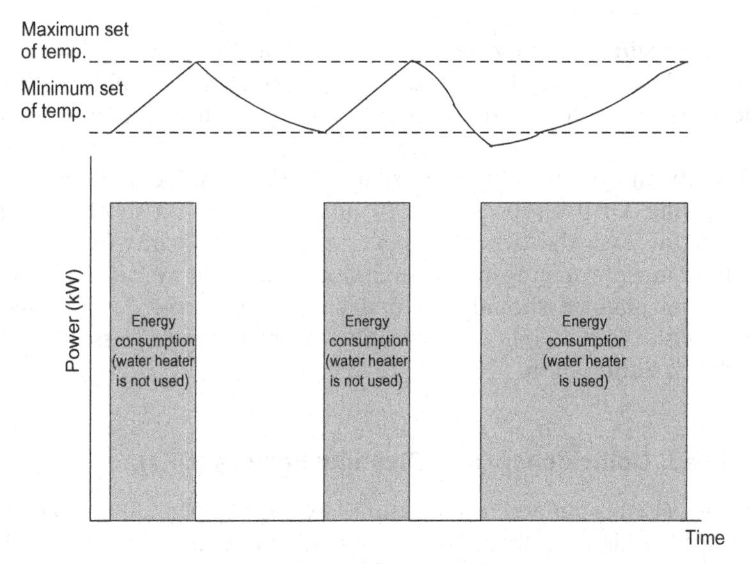

Figure 2.7 Water heater load behavior.

conditioners. These appliances are working with on/off duty cycle controlled by a thermostat. The demand power pattern of a water heater over an hour, where the thermostat is set at desired temperature, is shown in Figure 2.7. It is noted that the water heater is switched on when the temperature is less than the set value. The amount of energy consumption is proportional to the difference between water temperature and temperature setting. This difference is small if the water heater is not used and the energy consumed is just to substitute the leakage effect. On the other hand, the electric energy consumed is getting larger if the water heater is used because of the large difference of temperature. The air conditioners have the same nature of operation. It is controlled by the thermostat, and the electric energy consumption is proportional to the difference between the room temperature and the temperature setting.

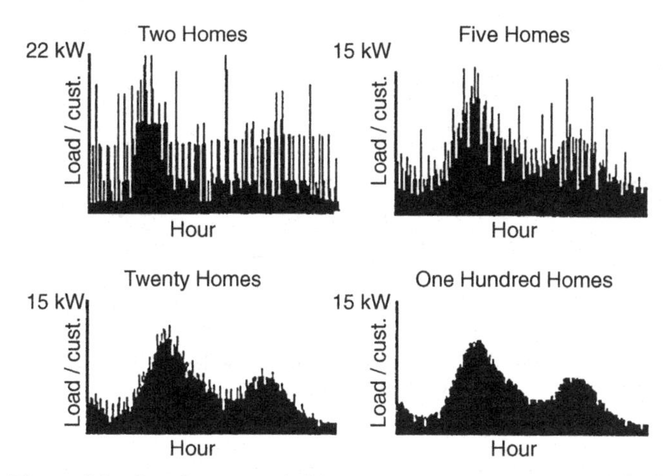

Figure 2.8 Load curves of different numbers of customers [34].

To measure and record the daily load curve of two customers, it will not be the algebraic sum of the two customers' individual load curves. This is because the appliances of the two customers do not work in synchronism. It is found that as the number of customers is increased, the load curve gets smoother and the peak value is decreased because of the noncoincidence of individual peaks (Fig. 2.8).

The ratio of measured peak load of a number of customers to the summation of individual peaks is defined as the coincidence factor (CF), that is,

$$\mathrm{CF} = \frac{\text{(measured peak of load)}}{\sum \text{(individual peaks)}}.$$

For example, for a group of N customers,

$$\mathrm{CF}\ (\text{group}) = \frac{\text{(measured peak load of a group of } N \text{ customers)}}{\sum_{i=1}^{i=N} P_i}, \qquad (2.21)$$

where P_i is the peak of the ith load.

The DF is defined as the reciprocal of the CF, that is,

$$\mathrm{DF} = \frac{1}{\mathrm{CF}}. \qquad (2.22)$$

It is noted from Figures 2.6 and 2.8 that in the case of one customer (one home) load curve, the erratic up and down may reach very high peak several times and last a short duration for each peak (few minutes). As the number

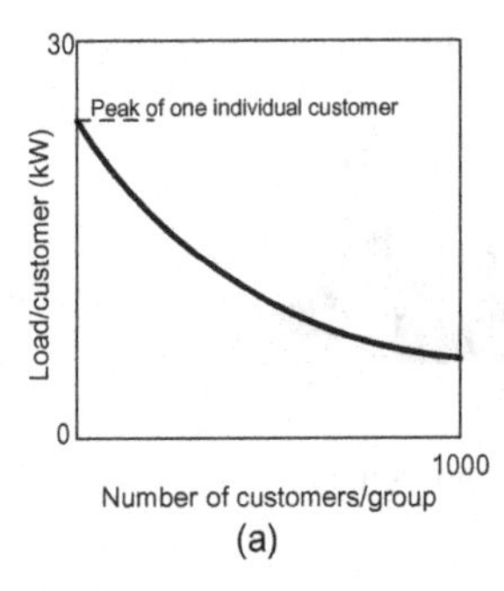

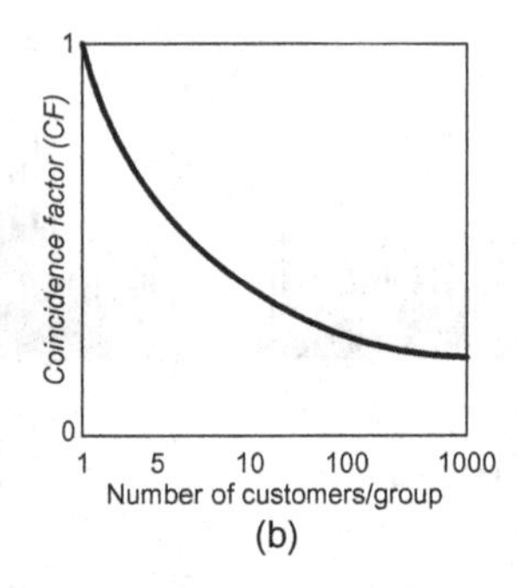

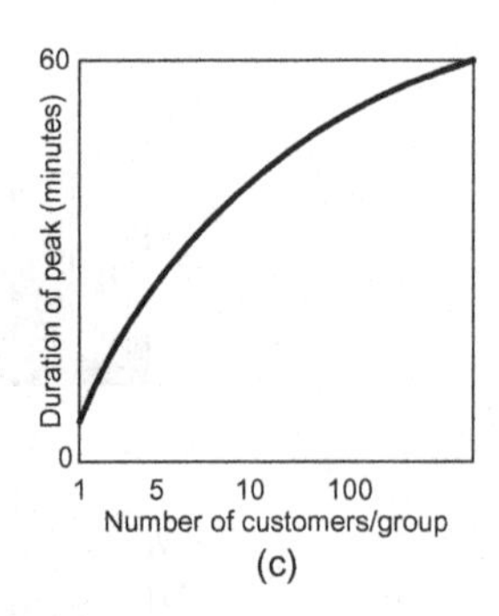

Figure 2.9 Effect of number of customers per group on (a) load per customer, (b) coincidence factor, and (c) duration of peak.

of customers in a group increases, the peak load of that group decreases compared with the sum of peak loads of individual customers and lasts a longer duration (over an hour) with its occurrence once a day.

Therefore, the relation between the number of customers in a group and the load per customer is as depicted in Figure 2.9a. The load per customer in this relation means the participation of each customer in the peak load of the group. Consequently, it is seen from Equation 2.21 that the CF decreases as the number of customers in a group increases (Fig. 2.9b). Similarly, the relation between the peak value duration in minutes versus the number of customers per group is drawn in Figure 2.9c.

2.4.4 Measuring and Recording Load Behavior

The load behavior represented by daily, monthly, or annual load curves is an important tool for forecasting. So, the data of these curves must be collected accurately as the load behavior is based on them. The load curve shape can change depending on the method of measuring and cyclic recording of these data.

The measuring method determines the quantity to be measured: Is it the instantaneous load or the energy consumed during a period? This is referred by what is called "sampling method." The data cyclic recording means the number of time intervals (load points) per period at which the measured data are recorded, for example, 96 load points (length of time interval 15 min) and 48 load points (length of 30 min) a day. This is referred by what is called "sampling rate."

The sampling method and sampling rate must be chosen in an appropriate way for the planning purposes.

2.4.4.1 Sampling Methods Two sampling methods are used:

1. *Sampling by Integration*: In this method, the measuring instrument measures and records the energy consumed during the sampling interval, that

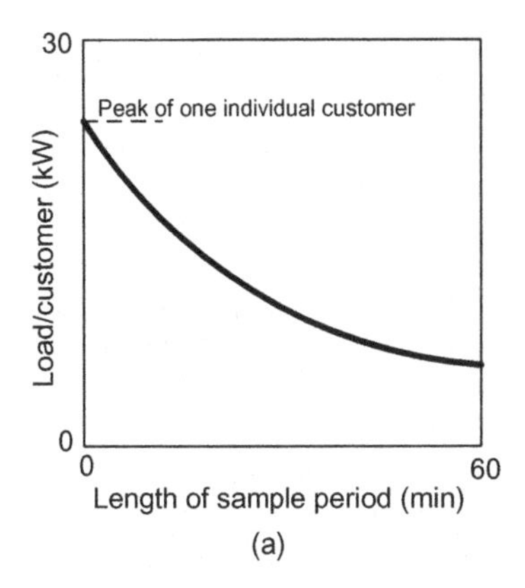

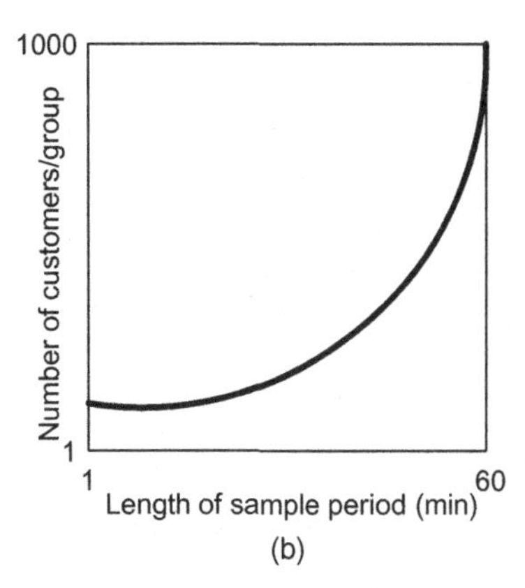

Figure 2.10 Variation of length of sample period versus (a) load/customer and (b) number of customers per group.

is, integrates the area under the load curve of each interval to be stored and recorded.

2. *Discrete Instantaneous Sampling Method*: It measures and records the load at the beginning of each sample interval. So, the changes of load during the interval are missed, that is, not measured.

Thus, for the erratic load curve, the sampling by integration is preferred where the error will be large in case of using discrete instantaneous method.

2.4.4.2 Sampling Rate The sampling rate, which is the frequency of measuring and recording the load data, depends on the number of customers in a group and the method of sampling.

If the number of customers in a group is increased, the peak load of the group is decreased (i.e., the load per customer is going down) and the peak duration is getting longer. So, the length of sampling period can be longer (i.e., lower sampling rate can be applied). In this case, the discrete instantaneous sampling may be used without a risk of big error (Fig. 2.10a). Of course, by decreasing the number of customers in a group, the peak load per customer and its duration decrease. The load curve becomes more erratic. So, the sampling rate should be high, that is, small sampling period (Fig. 2.10b). The preferred sampling method in this case is sampling by the integration method.

2.5 END-USE MODELING

End-use means the different products of electric usage; for example, lighting is one of these products. It necessitates a device that uses electricity to provide this particular product. This device may be an incandescent lamp, fluorescent lamp, sodium vapor lamp, and so on, and each type has its own load behavior.

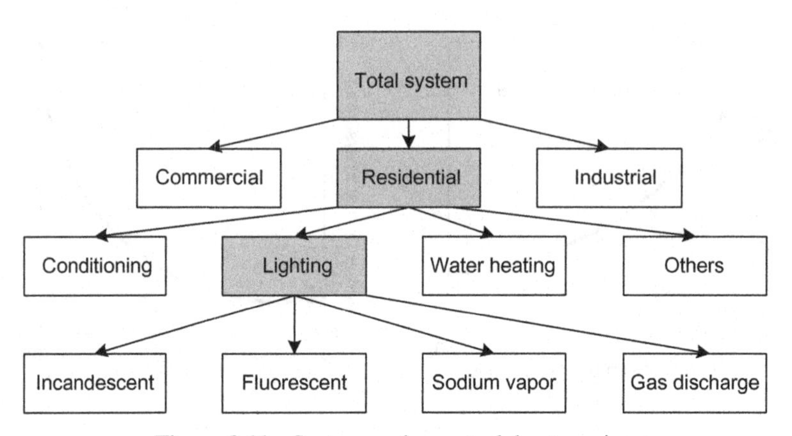

Figure 2.11 System end-use model categories.

So, the end-use models are based on a "bottom-up" approach to load modeling. This modeling of electric load demand as a function of time depends on customer classes. Three categories for total system load are specified, first category as customer classes, second category as end-use classes within each customer class, and third category as appliances within each end-use class.

Applying the "bottom-up" approach, the end-use analysis commences with the lowest category level that is "appliances" category to study the type of devices and determining how much their product is needed. For the example of lighting, and as shown in Figure 2.11, the total system load is classified into three classes (most common use): residential, commercial, and industrial classes as a first category that is broken down into end-use classes, for example, residential class is broken down into lighting, water heating, air-conditioning, and others as a second category. Each end-use class is broken down into incandescent, fluorescent, sodium vapor, and gas-discharge lamps.

Therefore, forecasting by applying the end-use model results in

- determining the future load curve shape,
- specifying the peak load, and
- identifying the energy usage developments that are predicted from expected future changes in appliance mixture, and appliance technology and efficiency.

2.6 SPATIAL LOAD FORECAST METHODS

Spatial electric load forecasting methods can be grouped into three categories: nonanalytic, trending, and simulation. *Nonanalytic* methods include all forecasting techniques, computerized or not, which perform no analysis of historical or base year data during the production of their forecast. Such a forecasting method depends entirely on the judgment of the user, even if it employs a computer program to accept "input data" consisting of the user's forecast trends as well as to output maps and tables derived from that bias. Despite the skill and experience of the planners who may be using them, these methods fall far short of the forecast accuracy produced by analytic methods.

Trending methods forecast future load growth by extrapolating past and present trends into the future. Most utilize some form of univariate or multivariate interpolation and extrapolation, most notably multiple linear regression, but a wide variety of other methods, some nonalgebraic, have been applied with varying degrees of success. Trending methods are most often categorized according to the mathematical technique used to perform the trend extrapolation.

Simulation methods analyze and attempt to replicate the process by which electric load changes in both location and time. All use some form of consumer-class or land-use framework to assess location of demand, and most represent usage over time on a per-consumer basis using load curve distinctions defined on that same consumer-class basis. Most are computerized, using software developed specifically or primarily for electric power planning purposes.

However, most simulation methods apply something between a limited and a very comprehensive analysis of the local geography, location economy, land use, population demographics, and electric load consumption. Categorization of simulation methods is difficult because of the great variety of urban models, land-use bases, pattern recognition algorithms, end-use load models that have been developed and applied, and differences in how they are put together into complete programs.

2.6.1 Trend Methods

Trending methods work with historical load data, extrapolating past load growth patterns into the future. In this type, the variable to be predicted is expressed purely as a function of time, rather than by relating it to other economic, demographic, policy, and technological variables. This function of time is obtained as the function that best explains the available data and is observed to be the most suitable for short-term projections. It also does not concern with methodology applied to determine the values of variables to be projected.

The trend methods have the advantage of simplicity and ease of use. The main disadvantage is that they ignore possible interaction of the variables under study with other economic factors such as incomes, prices, population growth, urbanization, and policy changes. The main concern of trend analysis is that time is the factor that determines the variable value under study or the pattern of the variable in the past will continue into the future. The most common trending method is polynomial curve fitting, using multiple regressions to fit a polynomial function to historical peak load data and then extrapolating that function into the future to provide the forecast.

2.6.1.1 Polynomial Curve Fit Polynomial form is the general formula to fit the curves (straight line, parabola, S-curve). It depends on the order of the polynomial. Polynomial coefficients can be calculated by using multiple regressions and deducing the normal equations as described in Section 2.3.3.

In forecasting a variable that the utilities are concerned with, such as the annual peak load on feeders, substation bank, and major customer loads, planners and forecasters should first choose or determine a suitable order of the polynomial. The coefficients of this polynomial are then calculated and from that the value of the peak load at any time in the future can be determined. This needs historical data that represent the peak load values at equal spaced time intervals (hourly, weekly, monthly, or annual). The data points must exceed the number of coefficients.

Assuming that a cubic polynomial curve fit is used to forecast the peak load of one of the feeders in a distribution system, its form is

$$\begin{aligned} L_n(t) &= a_3t^3 + a_2t^2 + a_1t + a_o \\ &= \text{annual peak load estimate for feeder } n \text{ for year } t, \end{aligned} \tag{2.23}$$

where t represents the year, beginning with $t = 1$ for the first time period of load history.

a_3, a_2, a_1, and a_o are the coefficients of the particular polynomial for feeder n and can be determined by using the multiple regressions method. They may vary from one feeder to another according to the historical data.

The same polynomial curve fit could be used for all feeders in the distribution system. On the other hand, a planner and forecaster could apply different forms of polynomial equations to different feeders. It depends on the feeder's historical data and the best polynomial to choose.

The first step to apply multiple regressions technique is to determine the parameter matrix P. The number of rows of this matrix is equal to the number

TABLE 2.5 Construction of Parameter Matrix P

	Polynomial Terms			
Past Years	Cubic Term (a_3t^3)	Quadratic Term (a_2t^2)	First-Order Term (a_1t)	Absolute Term (a_o)
1	1	1	1	1
2	8	4	2	1
3	27	9	3	1
4	64	16	4	1
5	125	25	5	1
6	216	36	6	1
7	343	49	7	1
8	512	64	8	1
9	729	81	9	1
10	1000	100	10	1

of years of data, while the number of columns is equal to the number of polynomial coefficients.

For instance, if the feeder's historical data is for the past 10 years, and a cubic polynomial curve fit (Eq. 2.23) is applied, then, the dimension of matrix P is 10×4. The cubed, squared, actual, and unity values of the past period (from 1 to 10) are fitted into the first, second, third, and fourth columns, respectively. They correspond to cubic, quadratic, first-order term, and absolute terms. The shaded part of Table 2.5 shows the elements of matrix P.

That is,

$$P = \begin{bmatrix} 1 & 1 & 1 & 1 \\ 8 & 4 & 2 & 1 \\ 27 & 9 & 3 & 1 \\ 64 & 16 & 4 & 1 \\ 125 & 25 & 5 & 1 \\ 216 & 36 & 6 & 1 \\ 343 & 49 & 7 & 1 \\ 512 & 64 & 8 & 1 \\ 729 & 81 & 9 & 1 \\ 1000 & 100 & 10 & 1 \end{bmatrix}.$$

The second step is to determine the load vector L_n whose elements are the peak load readings of the past 10 years:

$$L_n = \begin{bmatrix} l_n(1) \\ l_n(2) \\ l_n(3) \\ l_n(4) \\ l_n(5) \\ l_n(6) \\ l_n(7) \\ l_n(8) \\ l_n(9) \\ l_n(10) \end{bmatrix}.$$

The third step is to determine the coefficients a_3, a_2, a_1, and a_o by applying the equation

$$C_n = [P^T P]^{-1} \cdot P^T L_n, \tag{2.24}$$

where $C_n^T = [a_3\ a_2\ a_1\ a_o]$.

Substituting coefficients' values into Equation 2.23 (polynomial formula), the projected value of the peak load for any year following the last historical data point can be calculated. For example, for $t = 12$, the polynomial equation gives the peak load for 2 years beyond the last year of the historical data.

If data for any year of historical data are missing, it can be left out of the curve fit analysis, and the dimensions of the matrix P and the vector are changed accordingly. For example, if the data for the second and seventh years are missing, then P and L_n will be as below and the third step of analysis is the same as above by applying Equation 2.24:

$$P = \begin{bmatrix} 1 & 1 & 1 & 1 \\ 27 & 9 & 3 & 1 \\ 64 & 16 & 4 & 1 \\ 125 & 25 & 5 & 1 \\ 216 & 36 & 6 & 1 \\ 512 & 64 & 8 & 1 \\ 729 & 81 & 9 & 1 \\ 1000 & 100 & 10 & 1 \end{bmatrix} \text{ and } L_n = \begin{bmatrix} l_n(1) \\ l_n(3) \\ l_n(4) \\ l_n(5) \\ l_n(6) \\ l_n(8) \\ l_n(9) \\ l_n(10) \end{bmatrix}.$$

In another example, to illustrate the elements of matrix P when using data points of the past 10 years, consider a polynomial fit of the form

$$L_n = a_3t^3 + a_2t^2 + a_1t + a_o + a_{-1}t^{-1}.$$

Implementation of the same steps above gives Table 2.6, where the shaded area represents the elements of matrix P.

TABLE 2.6 *P* Matrix Elements for Four Coefficients of a Cubic Equation

	Polynomial Terms				
Past Years	Cubic Term (a_3t^3)	Quadratic Term (a_2t^2)	First-Order Term (a_1t)	Absolute Term (a_o)	Term of Order ($-1a_{-1}t^{-1}$)
1	1	1	1	1	1
2	8	4	2	1	0.5
3	27	9	3	1	0.333
4	64	16	4	1	0.25
5	125	25	5	1	0.20
6	216	36	6	1	0.166
7	343	49	7	1	0.143
8	512	64	8	1	0.125
9	729	81	9	1	0.111
10	1000	100	10	1	0.10

Thus,

$$P = \begin{bmatrix} 1 & 1 & 1 & 1 & 1 \\ 8 & 4 & 2 & 1 & 0.50 \\ 27 & 9 & 3 & 1 & 0.333 \\ 64 & 16 & 4 & 1 & 0.25 \\ 125 & 25 & 5 & 1 & 0.20 \\ 216 & 36 & 6 & 1 & 0.166 \\ 343 & 49 & 7 & 1 & 0.143 \\ 512 & 64 & 8 & 1 & 0.125 \\ 729 & 81 & 9 & 1 & 0.111 \\ 1000 & 100 & 10 & 1 & 0.10 \end{bmatrix}.$$

The fitting error is minimized by determining the set of coefficients that minimize the fitted RMSE in fitting to the historical data. This is the role of the multiple regressions technique regardless of polynomial order. Of course, increasing the number of coefficients, a higher-order polynomial, is likely to yield less fitting error, but it may not affect load forecast. Passing the fitted curve through all historical data points does not mean highly accurate forecast. The forecasting accuracy is determined by the extent of coincidence of actual behavior with the forecasted one.

2.6.1.2 Saturation Growth Curve (S-Curve) Some curves are characterized as having their slope from an increasing rate of growth to a decreasing one. In general, the nature of these curves exhibits the following features (Fig. 2.12):

1. an initial period of relatively slow but gradually increasing growth,
2. an intermediate period of rapid growth, and
3. this is followed by a final period where the rate of growth declines and the observed data appear to reach a saturation level.

Among the trend growth models that have saturation levels are the modified exponential model, the Gompertz model, and the logistic model. S-curve is represented mathematically by a polynomial of third order as given in Section 2.3.1:

$$Y = a + bx + cx^2 + dx^3.$$

Depending on the period of historical data and how much it lies on the S-curve, the shape of growth behavior of S-curve may be widely changed as shown in Figure 2.13, giving different results with different levels of error.

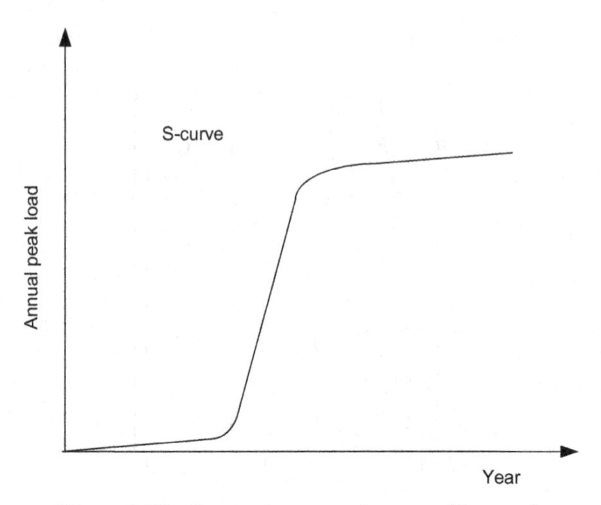

Figure 2.12 Saturation growth curve (S-curve).

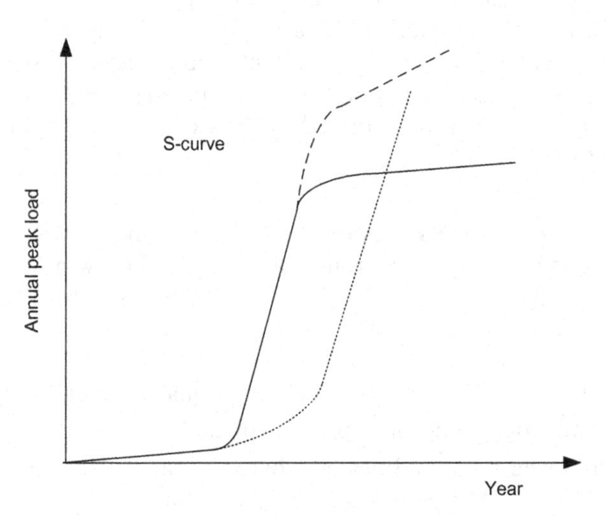

Figure 2.13 S-curves at different periods of historical data.

The solution of this problem is carried out by assuming a horizon year load, for example, 10 years after the last year of historical data. This horizon point is manipulated exactly as the historical points and added to matrix P as a time and to vector L as a load. The estimated load value for this added point is determined by the experience of the planners and the forecasters. Therefore, P and L for four coefficient polynomials are given as

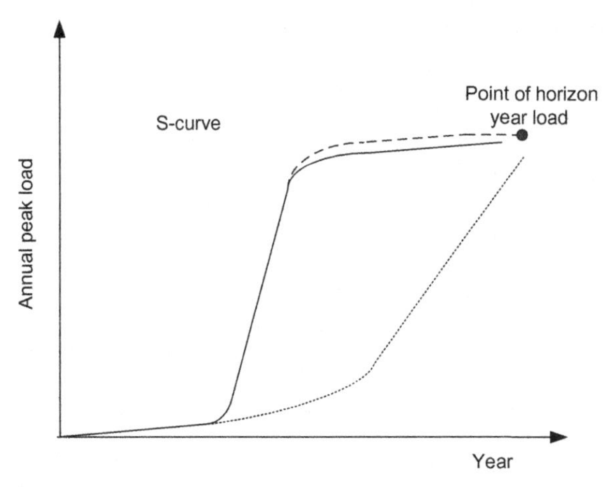

Figure 2.14 S-curves with horizon year load at different periods of historical data.

$$P = \begin{bmatrix} 1 & 1 & 1 & 1 \\ 8 & 4 & 2 & 1 \\ 27 & 9 & 3 & 1 \\ 64 & 16 & 4 & 1 \\ 125 & 25 & 5 & 1 \\ 216 & 36 & 6 & 1 \\ 343 & 49 & 7 & 1 \\ 512 & 64 & 8 & 1 \\ 729 & 81 & 9 & 1 \\ 1000 & 100 & 10 & 1 \\ 8000 & 400 & 20 & 1 \end{bmatrix} \text{ and } L_n = \begin{bmatrix} l_n(1) \\ l_n(2) \\ l_n(3) \\ l_n(4) \\ l_n(5) \\ l_n(6) \\ l_n(7) \\ l_n(8) \\ l_n(9) \\ l_n(10) \\ l_n(20) \end{bmatrix}.$$

The corresponding S-curve is shown in Figure 2.14. It is found that the error is highly decreased and does not affect short-term forecast.

PART II

PROTECTION AND DISTRIBUTION SWITCHGEAR

CHAPTER 3

EARTHING OF ELECTRIC DISTRIBUTION SYSTEMS

3.1 BASIC OBJECTIVES

Earthing plays an important role in the safe and reliable operation of an electric network. In a medium- (MV) or low-voltage (LV) installation, the neutral may be directly connected to earth (directly earthed) or connected through a resistor or a reactor (impedance-earthed). The neutral is unearthed when a connection has not been made between the neutral point and earth.

On the occurrence of an insulation failure or accidental earthing of a phase, the values taken by the fault currents, the touch voltages, and overvoltages are closely linked to the type of neutral earthing connection. A directly earthed neutral or low-impedance-earthed neutral strongly limits overvoltages, although it causes very high fault currents and tripping as soon as the first fault occurs to reduce the probability of equipment damage.

On the other hand, an unearthed neutral or high-impedance-earthed neutral limits the fault current to very low values but encourages the occurrence of high voltages (HVs). So, the system may be operated with earth fault present, improving system continuity, but the equipment must have an insulation level compatible with the level of overvoltages to reduce the risk of damage.

A safe earthing design, in principle, has two objectives:

1. To assure that a person in the vicinity of grounded facilities is not exposed to the danger of critical shock. This can be covered by earthing the electric equipment.

Electric Distribution Systems, First Edition. Abdelhay A. Sallam, Om P. Malik.

Published 2011 by John Wiley & Sons, Inc.

2. To provide means to carry electric currents into the earth under normal and fault conditions without exceeding any operating and equipment limits or adversely affecting continuity of service. This can be covered when studying the power system earthing from the viewpoint of controlling the fault or system currents and voltages.

The choice of earthing system, in both MV and LV networks, depends on the type of installation as well as the network configuration. It is also influenced by the type of loads and service continuity required. Different methods of earthing equipment, distribution substations, and MV and LV networks are explained in the forthcoming sections.

3.2 EARTHING ELECTRIC EQUIPMENT

3.2.1 General Means

The first objective is concerned with covering the earthing requirements to meet the safety regulations for personnel and equipment. These requirements must satisfy the following [35]:

1. Safety
   a) *Protection Against Electric Shocks*: An accidental contact between an energized electric conductor and the metal frame that encloses it can occur due to insulation breakdown. As a result of this contact, the frame tends to be energized to the voltage level of the conductor causing electric-shock injuries to the person who may touch the frame. Therefore, to minimize such personal injuries, the equipment should be earthed through a low-impedance conductor to form a low-impedance path. Thus, the line-to-earth fault current flows through this path into the earth without creating a hazardous impedance voltage drop.
   b) *Protection of Conductors Against Physical Damage*: A physical damage of conductors can be a result of excessive temperature rise or arcing, which may cause fire in the establishment as well. To avoid such accidents, the earthing conductor and all parts of the fault circuit including terminations, joints, and connectors must be capable of carrying the fault current (magnitude and duration) without distress. In addition, effective design, installation, and maintenance of equipment-earthing systems are vital elements in reducing these accidents.
   c) *Protection Against Overloads*: One of the main elements on which the design of overcurrent protection system is based, is the equipment–earth system. The overcurrent protection system requires a low-impedance earth return path in order to operate properly. Therefore, the impedance of the earthing conductor must be low enough that

sufficient earth-fault current will flow to operate the overcurrent protective device and clear the fault rapidly.

d) *Protection Against Hostile Environment*: Particular hazards may be met within certain occupancies due to the accumulation of static charges on equipment, on materials being handled or processed, and by operating personnel where flammable or explosive liquids, gases, dusts, or fibers are present.

The discharge of an accumulation of static charges (called "static electricity") from an object to earth or to another object at different voltage can be the cause of a fire or an explosion if it takes place in the presence of flammable materials or combustible vapor and air mixtures. Such fires and explosions that may result from a static spark have caused injury to personnel and loss of life as well as a big financial loss due to property damage and business interruption.

Therefore, the static electricity must be mitigated to protect human life and prevent the losses in equipment and buildings. One of the methods used for this goal is bonding of various parts of the equipment together and earthing the entire system. Bonding minimizes voltage differences between conductive objects, thus preventing sparking between two bodies, while earthing minimizes voltage differences between objects and earth. The other methods (e.g., humidity control, ionization, conductive floors) are explained in Reference 36.

2. Conductor voltage drop

If a current-carrying conductor, even though nominally at earth potential, is connected to earth at more than one location, part of the load current will flow through the earth because it is then in parallel with the earth conductor. Since there is impedance in both the conductor and the earth, a voltage drop will occur both along the earth and the conductor. Most of the voltage drop in the earth will occur in the vicinity of the point of connection to earth. Because of this nonlinear voltage drop in the earth, most of the earth will be at a different potential than the earth conductor due to the load current flowing from this conductor to earth.

An equipment earthing conductor connected to the same electrode as the earthed load conductor will also have a potential difference from most of the earth due to the potential drop caused by the load current. In most instances, the potential difference will be too low to present a shock hazard to persons or affect operation of conventional electric load equipment. This should not exceed 1–2% in normal operating conditions. However, in many instances, it has been of sufficient level to be detected by livestock, either by coming in contact with noncurrent-carrying enclosures to which an equipment earthing conductor is connected, or where sufficient difference in potential exists between the earth contacts of the different hoofs [36].

3. Economy
   Installation costs should be kept to a minimum while observing the pertinent standards.

Safety of personnel covers from home appliances to substation equipment. As an example, considering the voltages as in the North American distribution systems, Figure 3.1 shows the reasons and importance of the three-pin receptacles being used in the homes. Figure 3.1a,b shows the effect of the capacitive coupling between the primary and secondary windings of a 14.4kV/120V distribution transformer. Although there is no earthing point in the secondary winding itself, the human body will complete the path to earth and hence personnel life is at risk where a current of 15mA is enough to shock a person. This indicates that at least one side of the secondary 120V winding must be earthed to reduce the probability of electric shocks, that is, instead of possible shocks from either end of the secondary windings, it is now from one end only, which would also be connected to a switch.

Figure 3.1c is for a person touching a faulty appliance that is connected to the above improved system with an earthed end but no earth return from the appliance to the source. As the appliance body is not earthed, the human body will complete the path of the leakage current I_L. Clearly, the first-hand solution is to connect the appliance body (case) to the earthed neutral and for simplicity via the neutral as shown in Figure 3.1d. On the other hand, it is required that the live side of the source must be connected to the unearthed side, but this cannot always be guaranteed because if the incoming source connection is truly reversed with this arrangement, the appliance (the motor) will still operate normally until a fault takes place and then electric shocks will be felt by personnel. This is shown in Figure 3.1e. For that reason, alternating current (AC) connectors with three pins (one phase + neutral + earth) should be used where a protective earth wire is used between the connector and the mains as shown in Figure 3.1f. Different styles of AC connectors in compliance with standards (e.g., National Electrical Manufacturers Association [NEMA] standards in the United States and IEC60320 and 60309 in Europe) exist in order to address different wiring systems [37] and to ensure safety. It is noted that the rated voltage of appliances in Figure 3.1 is 120V as in United States. For the European system, the same figure can be applied with 220V instead.

Finally, most modern premises are equipped with earth leakage protection shown in Figure 3.1g. In this, a current transformer (CT) is used to sum the live-wire current and the neutral current ($I_w + I_n$) to feed an earth leakage overcurrent relay, which would operate with a typical setting of 5–15mA. In some cases, the same CT is used for fault currents as well, but this arrangement could be expensive. It is normal to use two CTs to improve reliability and sensitivity.

Another example of the importance of earthing is in avoiding the impact of lightning overvoltage on MV networks. The induced overvoltages are either electromagnetic or electrostatic in origin and concern, in particular, unearthed

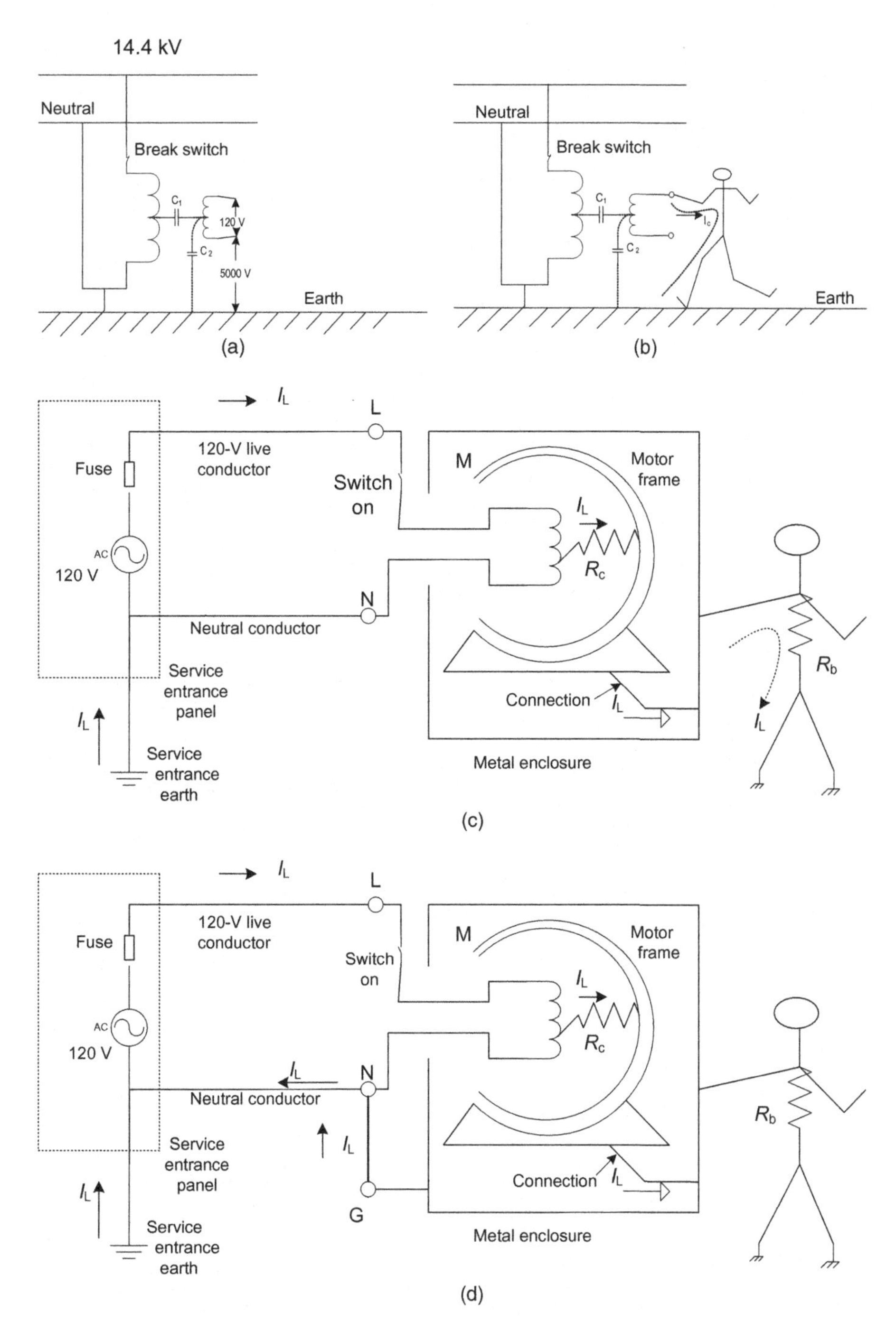

Figure 3.1 (a,b) Effect of capacitive coupling. (c–e) Importance of appliance earthing. (f,g) Appliance earthing by earth wire.

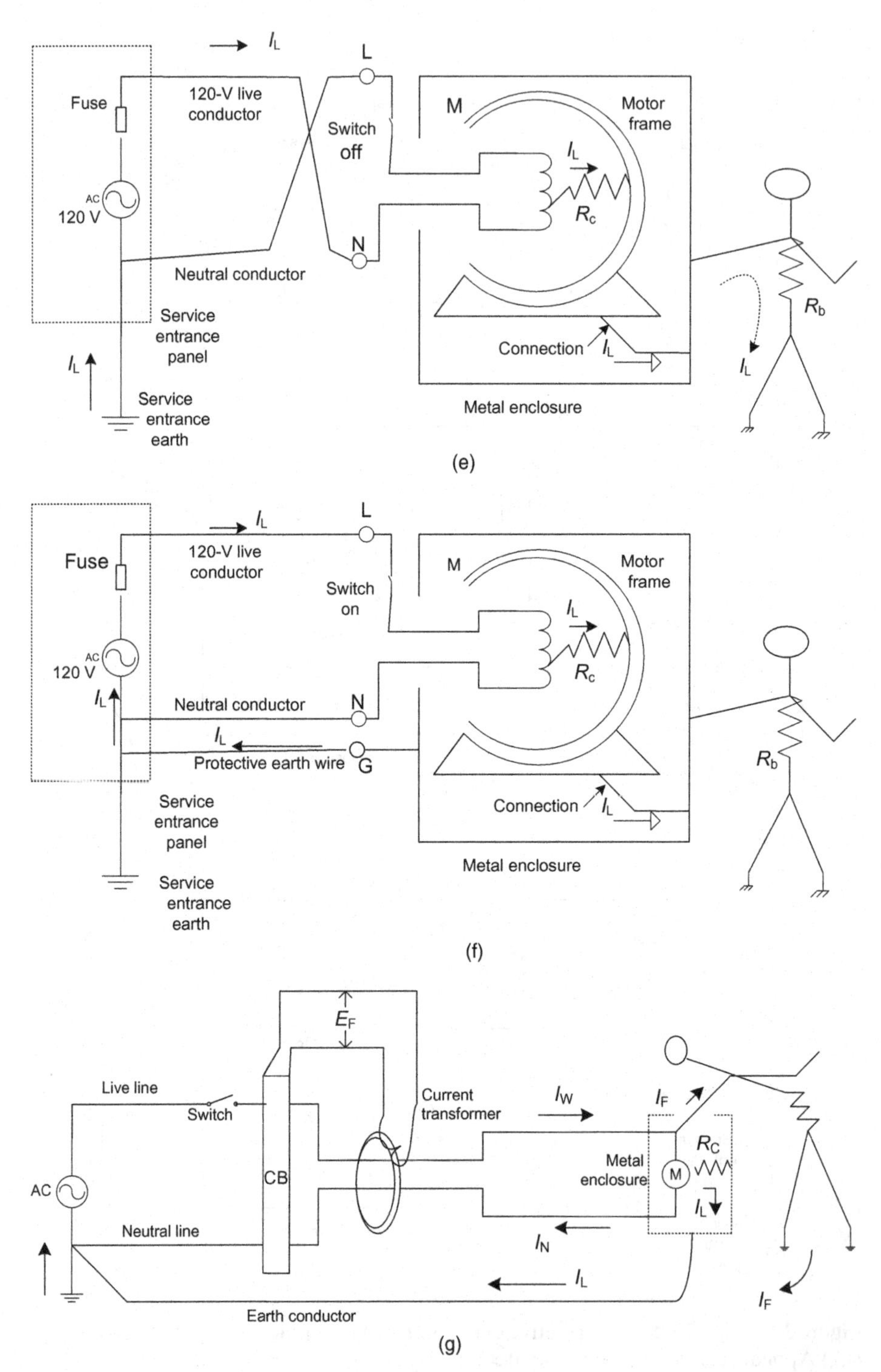

Figure 3.1 *Continued*

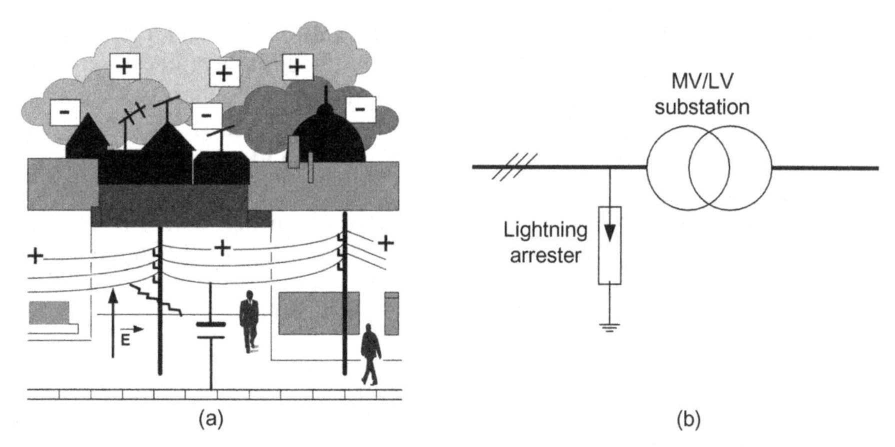

Figure 3.2 (a) Origin of an electrostatic overvoltage [38]. (b) Diverting the lightning stroke into the earth.

networks. Considering the electrostatic overvoltage in this example, if the MV network is an overhead one, in minutes preceding a lightning stroke, the line takes on a charge of opposite direction to that of the cloud, when a cloud charged at a certain potential is above this line (Fig. 3.2a). Before the lightning strikes, thus discharging the cloud, an electric field "E" exists between the line and earth, which can reach 30 kV/m [38]. Under the effect of this field, the line/earth capacitor is charged to a potential of around 150–500 kV according to how high the line is from the earth. Unenergetic breakdown may then occur in the least well-insulated components of the network. The best protection, in particular of human life, is provided by directing a maximum of the disturbance to earth at the closest possible point to the source of the disturbance.

Therefore, lightning arresters are commonly placed at the entrance of MV/LV substations. Their function is to create a controlled weak point in network insulation so that any arcing will systematically occur between the cloud and the earth through the arrester (Fig. 3.2b).

3.2.2 Substation Earthing

An earthing mat is used in a substation to secure the safety of personnel and equipment. A properly designed earthing mat must ensure the safety of maintenance personnel, operators, and generally any member of the public, who comes near an electric object tied to the earth mat under the influence of a major earth fault. Any protection equipment that is added or could be added to the system must not be accounted for during the design of the mat.

3.2.2.1 Step and Touch Voltage Regulations Because of the high fault currents, which may be produced during system earth faults, the current in the earth return, which is received by a substation, could also cause earth potential rise (EPR) high enough to be dangerous to the station personnel. Two cases of the current produced in a body, (i) due to a voltage induced between two feet of a technician while walking in the station, called the step voltage, and (ii) due to touching a panel, called the touch voltage, are shown in Figure 3.3a,b, respectively. The body current I_b can be calculated by using the impedance diagram for each case.

The step voltage is defined as "the difference in surface potential experienced by a person bridging a distance of 1 m with the feet without contacting

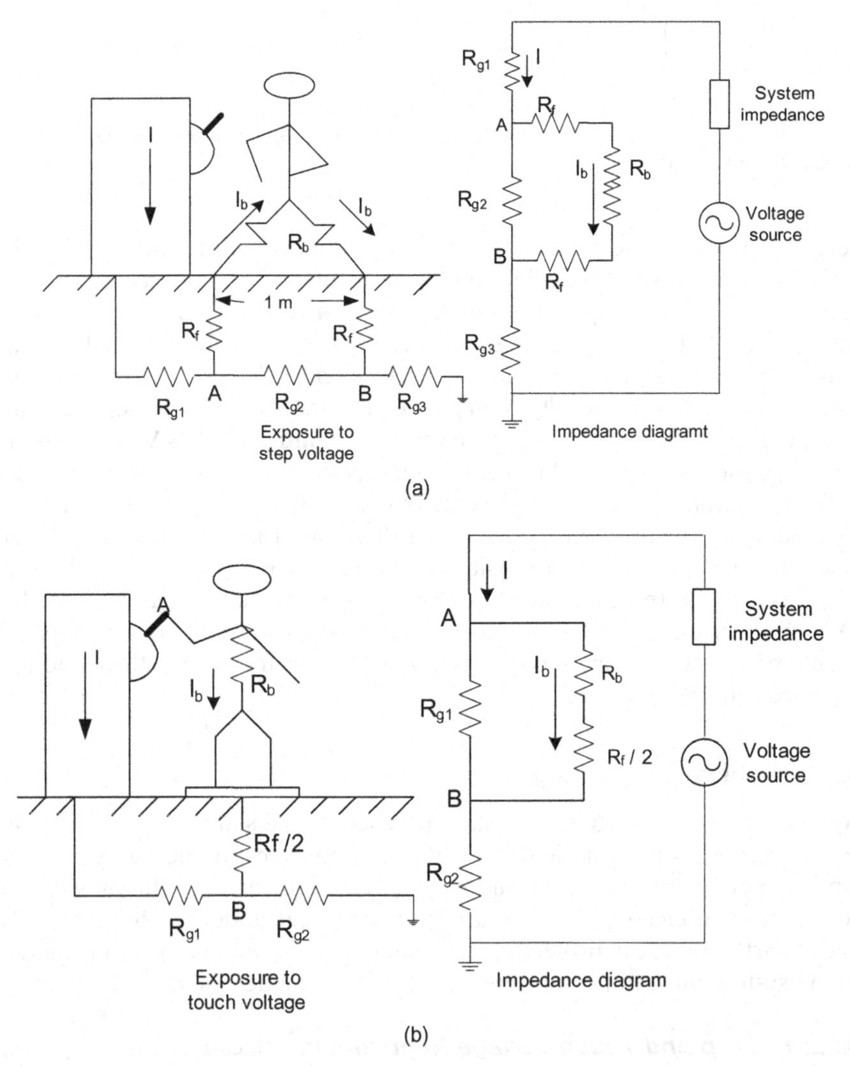

Figure 3.3 Step and touch potentials. (a) Step voltage; (b) touch voltage. R_f = contact foot resistance; R_b = body resistance; R_g = ground resistance.

any earthed object," and the touch voltage is defined as "the potential difference between the EPR and the surface potential at the point where a person is standing while at the same time having a hand in contact with an earthed structure" [39].

3.2.2.2 *The Human Factor* While 1 mA current can be felt by a normal healthy person [40], an average current of 9 mA does not impair the ability of a person holding an energized object to control his or her muscles and release it [41]. In a range of 9–25 mA, currents may be painful and can cause lack of muscular control. With higher currents, muscular contractions could make breathing difficult [42]. In most humans, currents of 60–100 mA will cause ventricular fibrillation [43]. Higher currents can stop the heart completely or cause electric burns. The phenomenon of fibrillation occurs when the heartbeats are not in synchronism with the body pulses, which maintain the blood pressure and in turn the flow of blood. Studies have shown that 99.5% of all healthy humans can tolerate a *current* through the heart region for a certain time, which can be determined by the following formula:

$$I_b = \sqrt{(S_B/t)A} \quad \text{for} \quad 0.03\,\text{s} \le t \le 3.0\,\text{s}, \tag{3.1}$$

where

I_b = maximum body current in amps
t = time duration in seconds
S_B = empirical constant related to the electric shock energy tolerated by a certain percent of a given population
= 0.0135 for body weight of 50 kg
= 0.0246 for body weight of 70 kg.

Thus, $I_b = 0.116/\sqrt{t}$, for a body weight of 50 kg, and $I_b = 0.157/\sqrt{t}$, for a body weight of 70 kg.

This formula (Eq. 3.1) is only valid for a time limited to a range between 0.03 and 3.0 s. In addition, one can clearly observe that for a time of 1 s, a human body can withstand an electric AC (50 Hz) of only 116 mA. This figure is found for a body weight of 50 kg. For 70 kg body weight, a current of 157 mA is to be used. Alternatively, one would assume 116 mA for women and 157 mA for men [39].

The International Electromechanical Commission (IEC) standard [44] has defined different time/current zones of electric shocks that a person may experience as shown in Figure 3.4. It should be noted that the square of this current is inversely proportional to the time. Exceeding this current would cause ventricular fibrillation. This is apparently true for the nominal 50 or 60 Hz power frequencies. For higher frequencies, for instance at a frequency of 3 kHz, the current magnitude reaches nearly 25 times that of the 50 Hz. It is not exactly the contrary for direct current. It is reported to be still higher but only five times.

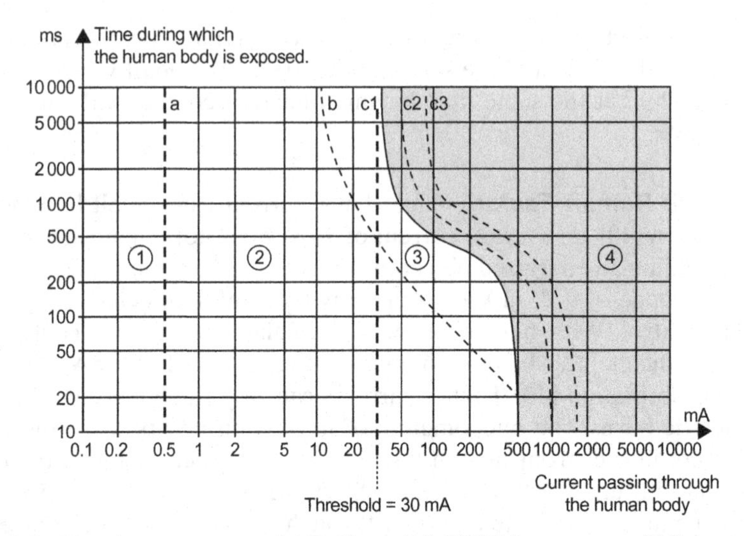

Figure 3.4 Time/current zones of AC effects (15–100 Hz) on persons [44]. 1 = perception zone; 2 = considerable discomfort; 3 = muscular contractions zone; 4 = risk of ventricular fibrillation zone (cardiac arrest); C1 = likelihood 5%; C2 = likelihood >50%.

On the other hand, following a severe shock, the heart requires about 5 min returning to its normal condition. Automatic reclosures, which implies two successive closely spaced shocks, would tend to produce cumulative effect. This effect is treated as one equivalent shock whose time is the sum of the two intervals.

In Equation 3.1, the fibrillation current is assumed to be a function of the individual body weight as the empirical constant S_B depends on it. Relationship between the critical current and body weight for several species of animals (claves, sheep, dogs, and pigs), and a 0.5% common threshold region for mammals is shown in Figure 3.5 [39].

The path of the electric current in a body for the two cases of a shock from just walking around in a high-voltage substation or a shock from touching equipment is shown in Figure 3.3a,b, respectively. These are known, as mentioned above, to be the step and touch potentials. The equivalent circuits are also shown in the diagrams. Use of 1000 Ω as an approximation for the body resistance is recommended in the IEEE Std. 80-2000 [39].

This yields the following total branch resistance:

1. For foot-to-foot resistance,

$$R = 1000 + 6\rho_s \Omega \tag{3.2}$$

2. For hand-to-foot resistance,

$$R = 1000 + 1.5\rho_s \Omega \tag{3.3}$$

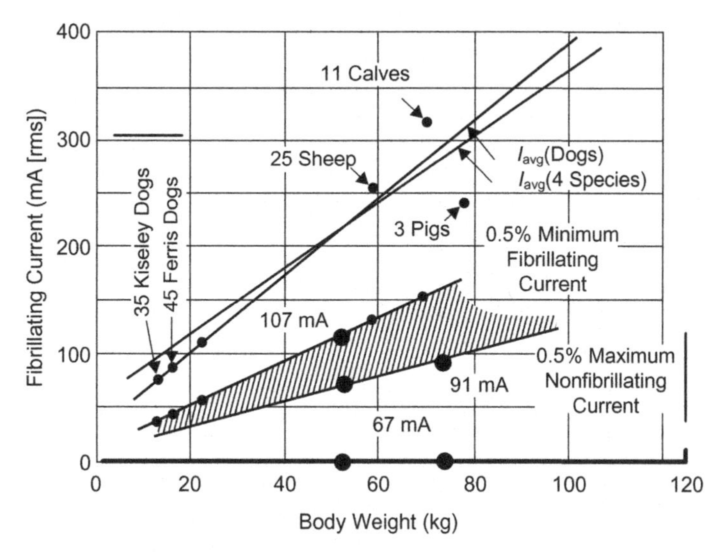

Figure 3.5 Fibrillation current versus body weight for various animals based on electric shock of 3-s duration [39].

where ρ_s is the soil resistivity in ohm meter. Actual measured resistances for a body are 2300 and 1100 Ω for foot to foot and hand to foot, respectively. It should be noted that the multipliers to ρ_s in Equations 3.2 and 3.3 are obtained based on the assumption of a circular electrode(s) in place of the foot or a hand.

Using the current and resistance equations given above, more convenient-related voltage formulas are obtainable. These are shown below:

$$E_{\text{step}} = 0.116(1000 + 6\rho_s)/\sqrt{t}\,\text{V}, \tag{3.4}$$

$$E_{\text{touch}} = 0.116(1000 + 1.5\rho_s)/\sqrt{t}\,\text{V}. \tag{3.5}$$

Equations 3.4 and 3.5 are now accepted worldwide and are known as *the permissible step and touch potential criteria.* When accounting for materials to increase earth resistivity, such as special surface layers of the common type crushed rocks, these equations become

$$E_{\text{step}} = [1000 + 6c_s(h_s, k)\rho_s] \times 0.116/\sqrt{t}\,\text{V}, \tag{3.6}$$

$$E_{\text{touch}} = [1000 + 1.5c_s(h_s, k)\rho_s] \times 0.116/\sqrt{t}\,\text{V}, \tag{3.7}$$

where $c_s = 1$ for no protective surface layers or to be taken from the diagram in Figure 3.6, which accounts for the depth (thickness) of the protective layer h_s. Reflection factor k, which accounts for the multilayer earth resulting from the added crushed rocks layer, is given as

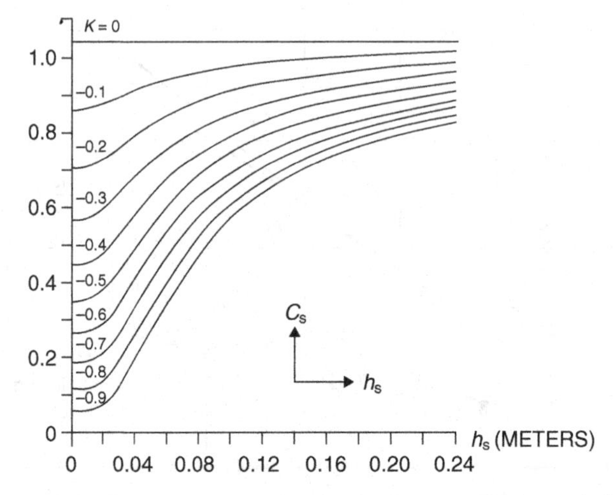

Figure 3.6 Factor c_s versus the depth of protective layer h_s [45].

$$k = (\rho - \rho_s)/(\rho + \rho_s), \tag{3.8}$$

where ρ is the resistivity of the original earth (i.e., the bottom permanent layer) Ω-m and ρ_s is the resistivity of the surface crushed rock layer Ω-m.

3.2.2.3 Measuring and Controlling Earth Resistance The allowable magnitude of earth resistance is normally related to the type and size of the substation or the industrial/commercial system to be fed. The following examples are for selected power systems with the corresponding recommended earth resistances.

Recommended Acceptable Values of Earthing Resistance [46] As it is mentioned in a 1958 American Institute of Electrical Engineers (AIEE) Committee Report, the most elaborate earthing system that can be designed may prove to be inadequate unless the connection of the system to the earth is adequate and has a low resistance. The earth connection is one of the most important parts of the earthing system. It is also the most difficult part to design and to obtain.

The perfect connection to earth would have zero resistance, but this is impossible to obtain. Earth resistances of less than 1 Ω can be obtained, although such a low resistance may not be necessary. The resistance required varies inversely with the fault current to earth. The larger the fault current, the lower the resistance must be.

For larger substations, the earth resistance should not exceed 1 Ω. For smaller substations and for industrial plants, in general, a resistance of less than 5 Ω should be obtained, if practical. The National Electrical Code (NEC), Article 250 approves the use of a single made electrode for the system earthing electrode, if its resistance does not exceed 25 Ω [47].

Consideration for Lightning Protection (When Required) When significant lightning threat is expected, lightning arresters would be installed and coupled with a maximum earth resistance of 5Ω.

It is clear from the above that measuring earth resistance plays a vital part in designing a safe power system. Most systems are earthed by cylindrical rods with a standard length of 3m (10ft) and a diameter of 1.59cm (5⁄8 in.). The number of rods together with the resistivity of earth determines the earth resistance. Three important formulas have now been considered:

- one earth rod with length b and radius r

$$R = \rho[\ln(4b/r) - 1]/2\pi b\Omega \tag{3.9}$$

- two earth circular rods with length b and radius r spaced by s where $b >> s$

$$R = \rho[\ln(4b/r) + \ln(4b/s) - 2 + (s/2b) - (s^2/16b^2) + (s^4/512b^4)]/4\pi b\Omega \tag{3.10}$$

- two earth circular rods with length b, radius r, and spaced by s where $b << s$

$$R = (\rho/4\pi b)(\ln(4b/r) - 1) + (\rho/4\pi s)\,[1 - (b^2/3s^8) + (2b^4/5s^4)]\,\Omega \tag{3.11}$$

where ρ in the above equations is the resistivity of the earth.

It is now evident that the resistivity of earth is of prime importance and is best to be measured. However, some figures of earth resistivity should help in the design.

	Resistivity, Ω-m		
Soil	Minute	Average	Maximum
Ashes, cinders, brine, waste	5.9	23.7	70
Clay, shale, loam	3.4	40.6	163
Same with varying proportions of sand and gravel	10.2	158.0	1350
Crushed rocks, gravel, sand, stones with little clay or loam	590	940.0	4580

It is possible that a significant variation in earth resistivity can take place due to the variable moisture and temperature. However, in most cases, the average values can be used.

A simple conventional method called "fall-of-potential method" is commonly used to measure the earth resistance of earth rod or small grid. It uses two auxiliary electrodes: current electrode and potential electrode in addition to the earthing electrode under test. By passing a current of known magnitude

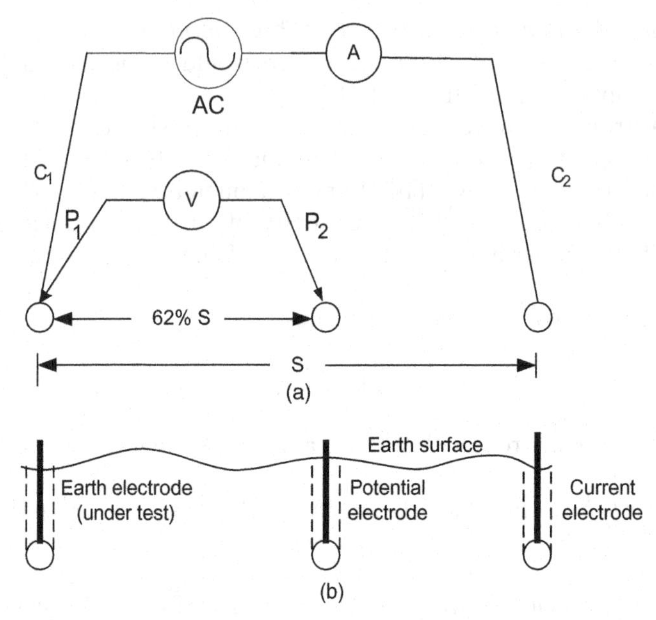

Figure 3.7 Fall-of-potential method for measuring earth resistance of earth rod or small grid.

through the earthing electrode under test and the auxiliary current electrode, then measuring the voltage between the electrode under test and the auxiliary potential electrode, the earth resistance can be calculated. The current electrode is recommended to be placed about 30–45 m (100–125 ft) from the electrode under test while the potential electrode is placed at 62% of that distance as shown in Figure 3.7.

Another simple method to measure the resistance and subsequently the resistivity of earth is shown in Figure 3.8. Instead of the use of the standard 3-m rods, the use of four equally spaced 1.5 m electrodes serves the requirement as shown in Figure 3.8. With the spacing "s" of 15.25 m (50 ft), the resistance equation given above for $b << s$ (Eq. 3.11) applies, and with this dimension it yields

$$\rho = 12.6\ \mathrm{s}\ \Omega\text{-m}.$$

Therefore, when applying a voltage source and connecting the circuit as shown in Figure 3.8,

$$R = V/I\ \Omega$$

and

$$\rho = 12.6\ \mathrm{s}\ V/I\ \Omega\text{-m}.$$

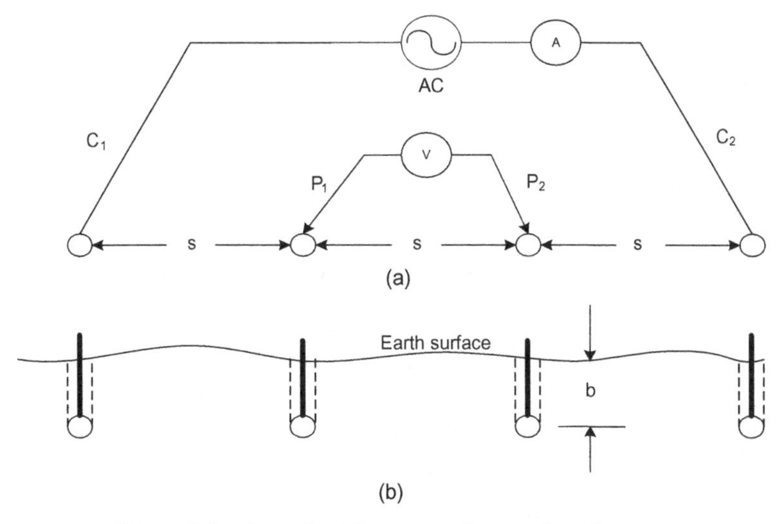

Figure 3.8 A method for measuring earth resistance.

Typical example when $I = 1\,\text{A}$, $V = 2\,\text{V}$, $s = 15.25\,\text{m}$, and the resistivity of the earth is

$$\rho = 384.3\ \Omega\text{-m}.$$

3.2.2.4 Substation Earthing Mats Consider a small distribution switching or transformer station where earthing by rods proves to be impractical due to the rocky substrate nature of the ground. However, if the rock substrate is covered with a thin overburden of soil, then earthing by a number of horizontal meshes is an acceptable substitute providing the earthing resistance would meet the standard recommended magnitudes previously mentioned in correspondence to the type of power system.

A nine-mesh earthing mat of square shape, whose area $A = w^2$ and $w >> b$, where w is the overall width of the mat in meters and b is the depth of the mat under the surface of the earth in meters, is shown in Figure 3.9a. The number of meshes inside that square has to be optimized to produce the overall earthing resistance; that is, it could be 1 or 4 or 9 or 16 and so on.

A practical relationship between $(2rb/A)$ and $(\pi wR/\rho)$ for a practical number of meshes 1, 4, and 9, where R is the required earth resistance in Ω and ρ is the soil resistivity in Ω-cm, is given in Figure 3.9b. The knowledge of the abscissa parameters is clearly met with the corresponding figures on the ordinate depending on the number of meshes, which is arbitrary.

For instance, assume $w = 6.1\,\text{m}$ (20 ft), $b = 0.3\,\text{m}$ (1 ft), and $r = 0.0064\,\text{m}$ (1/4 in.) then,

$$A = 37.2\ \text{m}^2,$$

$$(2rb/A) = 1.03 \times 10^{-4}.$$

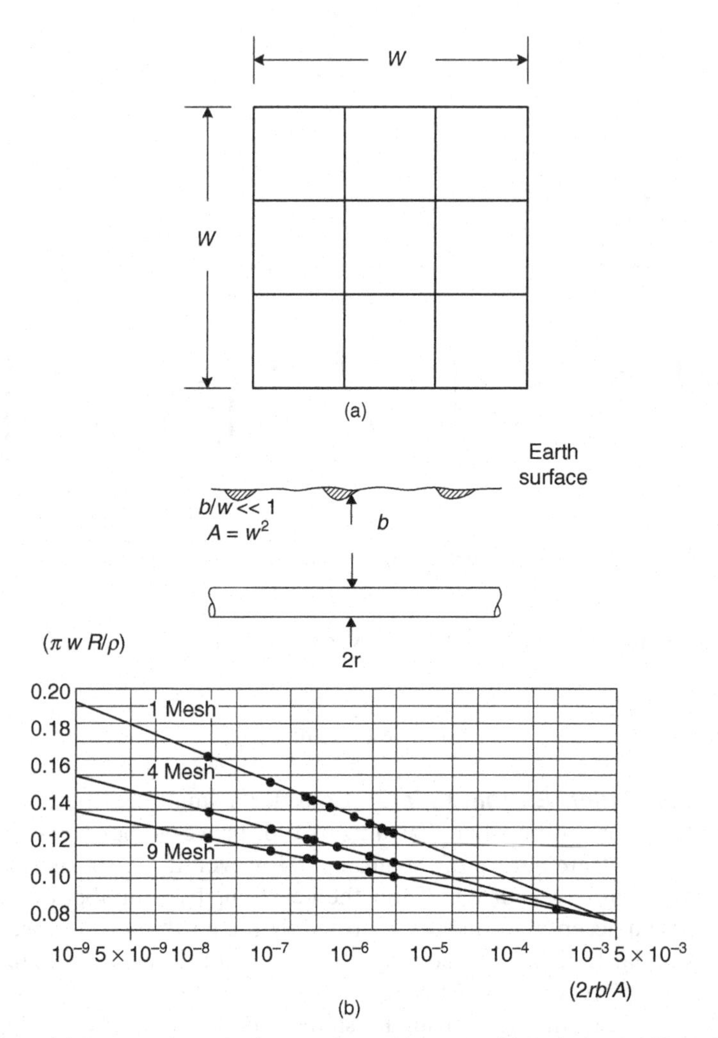

Figure 3.9 (a) Square-shaped, nine-mesh earthing mat. (b) Relation of $(2rb/A)$ versus $(\pi wR/\rho)$ [45].

From Figure 3.9b, a four mesh would produce

$$6.56\pi wR/\rho = 0.097,$$

and for a typical earth resistivity of $\rho = 10{,}000\,\Omega$-cm, the earth resistance becomes $R = 7.7\,\Omega$. It indicates that, when considering the assumed approximate average used value for ρ, the earth resistance R is close enough for the recommended $5\,\Omega$ for a light industrial system.

3.2.2.5 Design of Substation Earthing Mats to Meet the Step and Touch Voltage Regulations The load flow analysis can be used to design the earthing mat with a goal of limiting both step and touch voltages to be at safe values. When the above-mentioned step and touch potential criteria are to be met, the number of horizontal bars in both the x- and y-axes of Figure 3.9 would vary. Other considerations for having distributed earth rods together with special soil with suitable resistivity must be tried to achieve the required criteria.

This multidimensional problem is best optimized step by step using computers. However, to clarify the importance of such a problem, consider the following example.

Design an earthing mat for a primary 132/11 kV distribution substation of 2000 MVA short-circuit capacity on the 132 kV side. The mat must cover an area of $100 \times 100\,\text{m}^2$. Two hundred fifty milliseconds is estimated for earth fault maximum clearing time. When the earth resistance was measured at the site, average value was found to be 100 Ω-m. The contractor is prepared to deliver to the site crushed rocks to provide resistivity of 2000 Ω-m for a depth of 0.5 m.

Solution:

1. Assume a mat consisting of copper-clad steel wires (melting point 700°C) of diameter 0.03 m (1.242 in.). Now check for the current-carrying capacity:

$$\text{Fault current} = 2000/(\sqrt{3}\times 132) = 8.74\ \text{kA}.$$

   Current-carrying capacity = $8740/(\pi r^2)$ = 12.36 A/mm^2, which according to the universal standards would be acceptable for a 30-s fault duration. Therefore, when later current diversity within the mesh is accounted for, design and check for the other safety requirements is continued.

2. Check for the step and touch criteria with and without the crushed rocks soil surface:

$$E_{\text{step}} = 0.116[1000 + 6\times 100]/\sqrt{0.25} = 371\ \text{V},$$
$$E_{\text{touch}} = 0.116[1000 + 1.5\times 100]/\sqrt{0.25} = 267\ \text{V}.$$

3. Assume the earthing mat to have crossed wires spaced 5 m. Then, the total length of the wires is

$$L = 2\times 20\times 100 = 4000\ \text{m}.$$

   The total earthing area is $100 \times 100\,\text{m}^2$.

4. Now one can use a simplified formula for the total mat resistance given in IEEE Std. 80-1986 (American National Standards Institute, ANSI),

$$R = (\rho/4r) + \rho/L\Omega, \tag{3.12}$$

where

R = grid resistance in ohms,
ρ = soil resistivity in ohm meters,
r = radius of a circle with area equal to the grid in meters, and
L = total length of grid conductors in meters.

The first term gives the resistance of a circular plate with the same area as the grid. The second term compensates for the grid's departure from the idealized plate model. The more the length of the grid conductors increases, the smaller this term becomes. This equation is surprisingly accurate and is ideal for the initial stages of a study where only the basic data about the earth mat are available.

Therefore, using the above data,

$$\begin{aligned} R &= [100/(4\times\sqrt{10000}/\sqrt{\pi})] + 100/4000 \\ &= 0.4431 + 0.25 \\ &= 0.4681\ \Omega. \end{aligned}$$

5. Check for the EPR:

$$\begin{aligned} \text{EPR} &= R\times I_{sc} \\ &= 0.4681\times 8740 \\ &= 4091\ \text{V}. \end{aligned}$$

This value is higher than the step and touch voltage criteria. However, first, considering the effect of current diversity and a factor of 0.6 I_{sc} as practical for this mesh size:

$$\text{EPR} = 0.6\times 4091 = 2454\ \text{V}.$$

This is still higher than the criteria of both the step and touch voltages.

6. At this stage, the following alternative solutions are possible:
   a) Change spacing between grid mat wires from 5 to 2.5 m. This will decrease R and hence the EPR.
   b) Use crushed rocks for 0.5-m depth to cover the sand/stone soil. This will eventually increase both voltage criteria. However, it will also increase the resistance of the mat.
   c) Reinforce the mat with earthing rods at the sides of the grid mat.
7. Consider solution b first and then go to the other solutions for further refinements.

Therefore, with $\rho = 2000\ \Omega$-m,

$$E_{\text{step}} = 3016\ \text{V},$$
$$E_{\text{touch}} = 928\ \text{V}.$$

Also,

$$R = 9.362\ \Omega,$$
$$\text{EPR} = 9.362 \times 0.6 \times 8740 = 49{,}094\ \text{V},$$
$$E_{\text{mesh}} = 0.2 \times \text{EPR} = 9818\ \text{V}.$$

8. Reinforcing the mesh with thirty-six 7-m-long earthing rods would bring down the E_{mesh} to nearly 12 times less, approximately 818V.

Clearly, reinforcing the grid by earth rods may have helped without the use of crushed rock soil. Generally, a computer program would help a lot in determining the appropriate variables when required.

When power flow programs are used to design the substation earthing mat, refinements can be more flexible. For example, the spacing between the bars within the mat could be irregular to meet the step and touch voltage criteria.

3.2.2.6 Design of Substation Earthing Mats Using Computer Algorithms Computer algorithms can be used by utilities and industry to model the earthing systems and to provide an accurate design of earthing mats. The algorithms for modeling are mainly based on the following [39]:

- modeling of earthing system components such as earth rods, mat conductors, and soil;
- mathematical model formulation to describe system components interaction;
- calculation of earth-fault current flowing from each component into the earth; and
- potential computation at any point on the desired surface due to all the individual components.

The accuracy of computer algorithms is of great concern and the distribution engineers should pay more attention to it for various reasons, such as

- complexity of earthing system in case of including buried metallic structures or conductors not connected to that system;
- significant variations in soil resistivity, in particular, for multilayer soil model;

- earthing systems depend on different parameters that exceed the limitation of the equations; and
- flexibility in determining local danger points may be desired.

3.3 SYSTEM EARTHING

The second objective of earthing design, mentioned at the beginning of this chapter, can be covered by studying the system earthing. In general, the system is either unearthed or earthed [36, 45].

3.3.1 Unearthed Systems

Unearthed systems are the systems operated without intentional earth connection to the system conductors, except through potential indicating or measuring devices or other very high-impedance devices (Fig. 3.10). In an unearthed or high-impedance-earthed network, the equipment damage is reduced but it must have an insulation level compatible with the level of overvoltages able to develop in this type of network. In fact, these systems are earthed through the system capacitance to ground. Mostly, this is extremely high impedance, and the resulting system relationships to ground are weak and easily distorted.

The unearthed systems have two advantages, one is technical and the other is economical:

- *Technically*: The earth fault on a system causes a small earth current to flow, so the system may be operated even in the presence of earth fault, thus improving the system continuity.
- *Economically*: There are no expenditures required for earthing equipment or earthed system conductors.

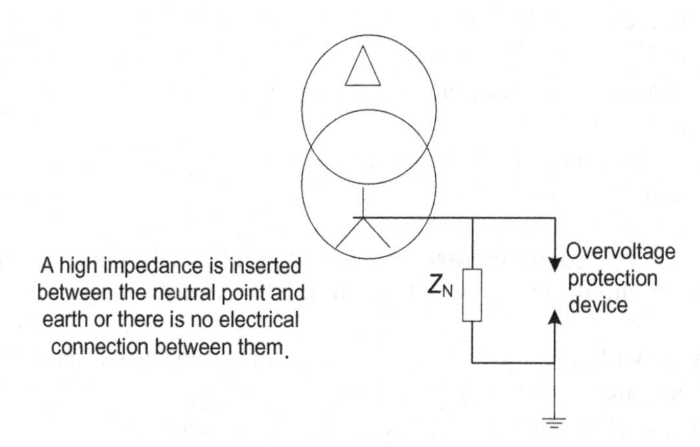

Figure 3.10 Unearthed system.

Therefore, to detect, locate, and clear the first earth fault, special devices with a specific scheme should be inserted into the system. A permanent insulation monitoring and switching upon second fault occurrence (double fault) are carried out as well.

On the other hand, a risk of high internal overvoltages makes it advisable to reinforce the equipment insulation. Also, difficulties of implementing selective protection upon occurrence of the first fault are involved. The phase-to-earth insulation level of equipment must be at least equal to the phase-to-phase level. This is because in the event of the occurrence of a permanent phase-to-earth fault, the voltage of both unaffected phases in relation to earth takes on the value of the phase-to-phase voltage if tripping is not triggered on the occurrence of the first fault. Consequently, the cables, rotating machines, transformers, and loads must be chosen taking this into consideration.

3.3.2 Earthed Systems

In the earthed systems, at least one conductor or point (usually the middle wire or neutral point of generators or transformers) is intentionally earthed, either solidly or through an impedance. The earthed systems have multiple advantages:

- greater safety;
- no excessive system overvoltages that can occur on unearthed systems during arcing, resonant, or near-resonant earth faults; and
- easier detection and location of faults when they occur.

3.3.3 Purpose of System Earthing

System earthing or intentional connection of a phase or neutral conductor to earth is for the purpose of controlling the voltage to earth within predictable limits. It also provides for a flow of current that will allow detection of an undesired connection between the system conductors and the earth, and which may initiate the operation of automatic devices to remove the source of voltage from conductors with such undesired connections to earth. The American NEC prescribes certain system earthing connections that must be made to be in compliance with the code. The control of voltage to earth limits the voltage stress on the insulation of conductors so that insulation performance can be predicted more readily. The control of voltage also allows for the reduction of shock hazard to any living body who might come in contact with the live conductors.

3.3.4 Definitions [36]

The reader should be aware of some definitions when studying system earthing.

Effectively Earthed: The system is said to be effectively earthed when satisfying the conditions

$$0 < (X_o/X_1) < 3 \quad \text{and} \quad 0 < (R_o/R_1) < 1,$$

where R_o and X_o are the zero sequence resistance and reactance, respectively. R_1 and X_1 are the positive sequence resistance and reactance, respectively.

Earthed: It is the connection to earth or to some extended conducting body that serves instead of the earth, whether the connection is intentional or accidental.

Per-Phase Charging Current (I_{Co}): It is the current (V_{LN}/X_{Co}) that passes through one phase of the system to charge the distributed capacitance per phase to earth of the system. V_{LN} is the line-to-neutral voltage and X_{Co} is the per-phase distributed capacitive reactance of the system.

Resonance: The enhancement of the response of a physical system (electric system or circuit) to a periodic excitation when the excitation frequency (f) is equal to the natural frequency of the system.

For an R-L-C series circuit, the series resonance occurs when $\omega L = 1/\omega C$ or $f = 1/(2\pi\sqrt{LC})$ Hz, where the total series reactance equals 0.

For an R-L-C parallel circuit, the parallel resonance occurs at the frequency at which the total admittance is the lowest, $X_l = X_c$, that is, $\omega L = 1/\omega C$ or $f = 1/(2\pi\sqrt{LC})$.

Static Charge: The electricity generated when two dissimilar substances come into contact. Conveyer belts are active producers of static electricity.

Switching Surge: A transient wave of voltage in an electric circuit caused by the operation of a switching device interrupting load current or fault current.

Transient Overvoltage: The temporary overvoltage of short duration associated with the operation of the switching device, a fault, a lightning stroke, or during arcing earth faults on the unearthed system.

System: An earthed system consists of all interconnected earthing connections in a specific power system and is defined by its isolation from adjacent earthing systems. The isolation is provided by transformers primary and secondary windings that are coupled only by magnetic means. Thus, the system boundary is defined by the lack of a physical connection that is either metallic or through a significantly high impedance. The limits and boundaries of earthing systems are illustrated in Figure 3.11.

System Charging Current: The total distributed capacitive charging current ($3V_{LN}/X_{Co}$) of a three-phase system.

Three-Phase, Four-Wire System: A system of AC supply comprising four conductors, three of which are connected as in a three-phase, three-wire system, and the fourth being connected to the neutral point of the supply or mid-point of one phase in case of Δ-connected transformer secondary, which may be earthed.

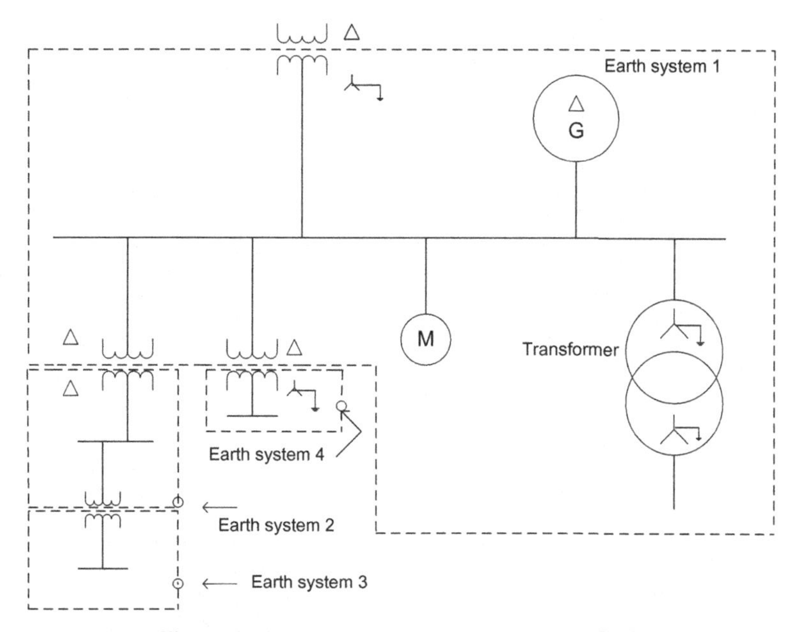

Figure 3.11 Earthing systems boundaries [36].

Three-Phase, Three-Wire System: A system of AC supply comprising three conductors, between successive pairs of which are maintained alternating differences of potential successively displaced in phase by one-third of a period.

3.3.5 Methods of System Neutral Earthing [36]

The earthing of the system can be done by either solid earthing or earthing through an impedance (reactive or resistive or earth-fault neutralizer) (Fig. 3.12).

Solid Earthed: The neutral point is connected directly through an adequate earth connection in which no impedance is intentionally inserted. The direct neutral earthing is either distributed or nondistributed.

Reactance Earthed: The neutral point is earthed through impedance, the principle element of which is an inductive reactance.

In a reactance-earthed system, the available earth-fault current should be at least 25% and preferably 60% of the three-phase fault current to prevent serious transient overvoltages ($X_o \leq 10X_1$). This is higher than the level of fault current desirable in a resistance-earthed system. Therefore, the resistance-earthed system is not an alternative to reactance-earthed system. The coil does not have to dissipate a high heat load because its resistance is low.

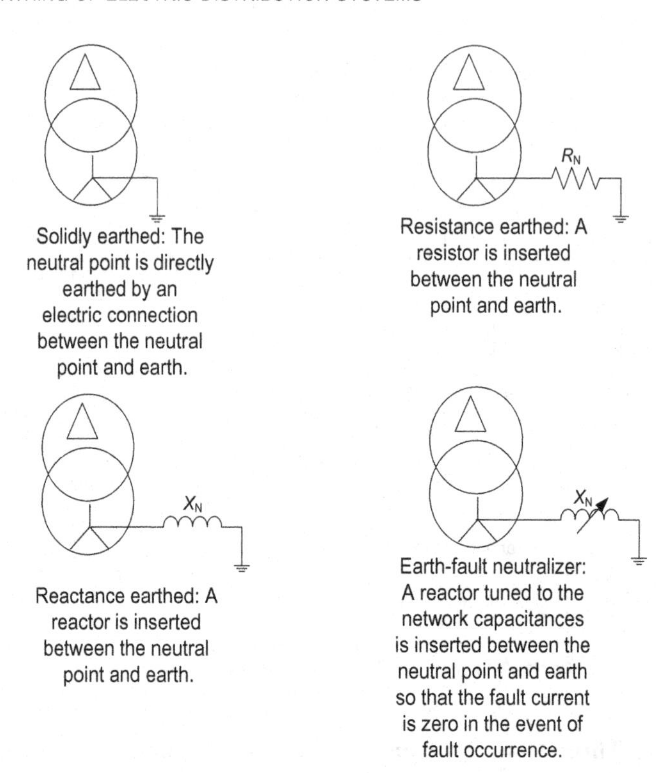

Figure 3.12 Different methods of earthing.

This type of earthing allows the implementation of simple selective protections if the current in the neutral earthing reactor I_L is much greater than the current in the phase-earth capacitances I_C, that is, $I_L >> I_C$.

Resistance Earthed: The neutral point is earthed through impedance, the principle element of which is a resistance. This resistance limits earth-fault current such that the current can flow for an extended period without causing damage. This level of current is commonly thought to be 10 A or less for high-resistance-earthed systems, which are designed to meet the criterion $R_o \leq X_{Co}$ to limit the transient overvoltages due to arcing earth faults.

For the low ohmic value of the resistance (low-resistance-earthed systems), it would be less than that meeting the high-resistance earthing criterion. The resistance in this case is selected to provide the desired relaying current.

Either low or high resistance (it depends on the magnitude of earth-fault current permitted), the earthing resistance limits the fault current that is needed for the following reasons:

- to reduce electric shocks and flash hazards to personnel,
- to reduce the momentary line voltage dip associated with the earth-fault occurrence and clearing,
- to reduce the mechanical stresses in circuits and apparatus carrying fault currents,
- to reduce burning and melting effects in faulted electric equipment, and
- to secure control of transient overvoltages.

There are no deterministic values for defining low and high resistances. Usually and in practice, the high resistance limits the earth-fault current to be within a level less than or equal to 10A, while it will be at least 100A in case of applying low-resistance earthed. Resistance earthing does not require the use of extra equipment and, in particular, cables having a special phase/earth insulation level.

Earth-Fault Neutralizer: It is a reactor with a selected high-resistance value. The reactance can be tuned to the system charging current so that the resulting phase-to-earth fault current is resistive and of a low magnitude. It will be 0 if $3L_0C_0\omega^2 = 1$. The current and V_{LN} are in phase; that is, they reach the zero value simultaneously. So, the tuned reactance helps to extinguish the flashover if the earth fault occurs in air. The installation continues to operate in spite of there being a permanent fault, with tripping necessarily occurring on the second fault. The first fault is indicated by the detection of the current flowing through the reactor. The reactor is dimensioned so that permanent operation is possible.

This system has little applications in industrial or commercial power systems. It has difficulties in establishing the condition $3L_0C_0\omega^2 = 1$ due to uncertain knowledge of the network's capacitance. The result is that throughout the duration of the fault, a residual current circulates in the fault. Care must be taken to make sure this current is not dangerous for personnel and equipment. In addition, it is impossible to provide selective protection upon first fault occurrence if the reactor has been tuned to $3L_0C_0\omega^2 = 1$. If it is systematically out of tune, $3L_0C_0\omega^2$ is not equal to 1 and selective protection upon occurrence of the first fault is complex and costly.

3.3.6 Creating Neutral Earthing

To provide the MV installation with neutral earthing, it should be determined first if the neutral is accessible or not accessible.

Resistance Earthed When the Neutral Is Accessible: A resistor is connected between the neutral outlet and the earth, either directly, (Fig. 3.13a), or through a single-phase transformer connected to the secondary via an

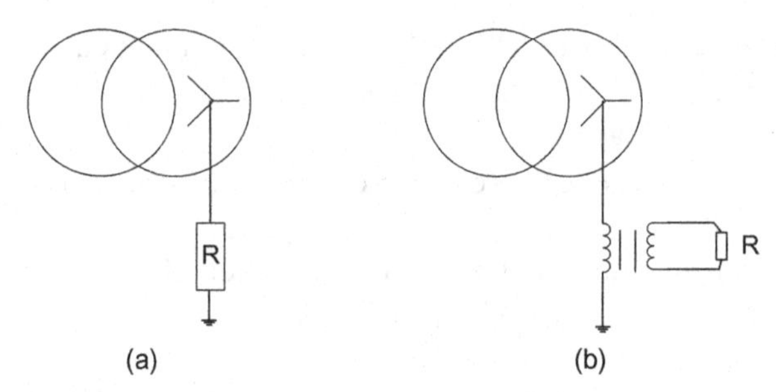

Figure 3.13 Earthing when the neutral is accessible. (a) Direct connection; (b) connection via single-phase transformer.

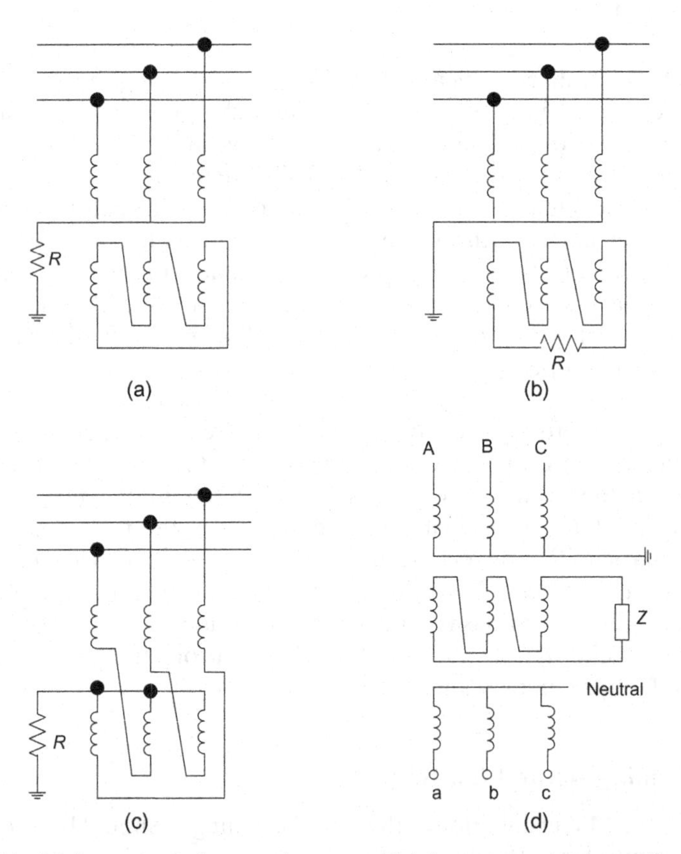

Figure 3.14 Main methods for creating an artificial neutral point in medium voltage networks. (a) Use of Y-Δ transformer with neutral earthing resistor; (b) use of Y-Δ transformer with resistor in the Δ; (c) use of a zigzag coil; (d) neutral point transformer with compensation delta.

equivalent resistor (Fig. 3.13b). This system is applied when the network is fed by a transformer with a star-connected secondary and an accessible neutral, or by a generator with an accessible neutral.

Earthing by Creating an Artificial Neutral: When the source neutral is not accessible (delta winding), the system is earthed by creating an artificial neutral (referred to as an earthing transformer). The artificial neutral can also be used when there are several parallel sources.

The earthing transformer can be made through the following ways:

- Using a Y-Δ transformer, the limiting resistor is connected between the primary neutral point and earth as the delta side is closed on itself (Fig. 3.14a).
- Using a Y-Δ transformer whose primary neutral point is directly earthed, a fault current limiting resistor is inserted in the secondary delta (Fig. 3.14b). This system is economically preferable to the former case since the resistor is in the LV side and not in the MV side.
- Using a zigzag coil, the limiting resistor is connected between the neutral point of the coil and earth (Fig. 3.14c).
- Using a neutral point transformer. This type of earthing transformer includes an extra winding, which creates a power outlet. For instance, a primary star winding with earthed neutral, a secondary delta winding closed by a fault current limiting resistor and another secondary star winding allowing the loads as well as the auxiliaries of an HV/MV substation to be fed (Fig. 3.14d).

  The systems adopted the most often are those shown in Figure 3.14b,c.

3.4 MV EARTHING SYSTEMS

The choice of neutral MV systems (or neutral earthing methods) defines among other things the voltage surge ratings and earth-fault currents that could be found on a network. It must be noted that these two parameters are contradictory, in view of the fact that obtaining a low fault current level leads to an HV surge and vice versa. These values thus pose electric constraints that the equipment must be capable of withstanding. However, the possible solutions for protection of the electric network are selected simultaneously with choosing the earthing method.

Referring to the system earthing described in the former section, the neutral earthing methods used in MV networks can be classified into five categories:

- direct distributed neutral earthing,
- direct nondistributed neutral earthing,
- neutral earthing via impedance,

TABLE 3.1 MV Neutral Earthing Methods and Their Applications throughout the World [16]

	Method				
	HV/MV	HV/MV	HV/MV	HV/MV	HV/MV
Country	Neutral Distributed with Multi-Earthing Points	Neutral Earthed Directly and Undistributed	Neutral Earthed via an Impedance	Neutral Earthed via a Designated Circuit	Neutral Not Connected to Earth
Australia	■				
Canada	■				
United States		■			
Spain		■	■	■	■
France			■		
Japan					■
Germany				■	

- neutral earthing via a designated circuit (e.g., a circuit to create an artificial neutral), and
- unearthed neutral (i.e., the neutral is not connected to earth).

None of the categories are dominant throughout the world: some solutions are specific to some countries, and several categories can be found within a single country. The choice of neutral earthing method is always the result of a compromise between installation and costs. A typical example for different countries using different categories is given in Table 3.1.

The main differences between the five categories lie in the behavior of the network in an earth-fault situation. These differences translate in real terms to the degree of

- ease of detection of these faults,
- security achieved for living beings, and
- impact on the requirements of electric equipment.

For instance, the distributed neutral earthing method that allows single-phase distribution can be considered in certain countries (e.g., United States and Canada) on the basis of its reduced installation costs, reduced losses, and

TABLE 3.2 Strengths and Weaknesses of MV Neutral Earthing Methods

Neutral Earthing Method	Strengths	Weaknesses
Direct distributed earthing	Authorizes one-phase and three-phase distributions	Requires numerous high-quality earthing points (safety) Requires a complex protection system Leads to high values of earth fault currents
Direct undistributed earthing	Eases detection of earthing faults	Leads to high values of earth-fault currents
Earthing via an impedance Compared with direct earthing	Limits earth-fault currents	Requires more complex protection system
Compared with unearthed neutral	Reduces surge overvoltage	Leads to higher earth-fault currents
Earthing via a designated circuit	Favors auto-extinction of earth-fault currents	Requires complex protection system
Unearthed	Limits earth-fault current	Leads to surge overvoltage

surges due to faults. However, it requires a high quality of neutral earthing, and to obtain a satisfactory degree of personal safety, it is necessary to include numerous MV devices (fuses, reclosers, sectionalizers).

A summary of the strengths and weaknesses of these categories (Table 3.2) shows why none of these categories are predominantly used throughout the world.

3.4.1 Influence of MV Earthing Systems

In all industrialized countries, the objectives of earthing systems are always the same, mainly for safety reasons to guarantee protection against electric current for persons. At MV/LV substation level, an MV phase-to-frame or a fault between MV and LV windings inside the transformer may present a risk for equipment and users of the LV network. Thus, limiting rises in potential due to MV faults has considerable repercussions on safety of personnel and property in LV.

In public and industrial MV systems, except in certain special cases, the neutral is not distributed, and there is no protective earth conductor between the MV load and the substation or between substations. A phase-to-earth fault thus results in a single-phase short-circuit current limited by earth connection resistance and the presence of limitation impedances.

For instance, on MV-frame disruptive breakdown inside the transformer and when the transformer frame and LV neutral are connected to the same earth connection (Fig. 3.15a), an MV fault current can raise the frame of the

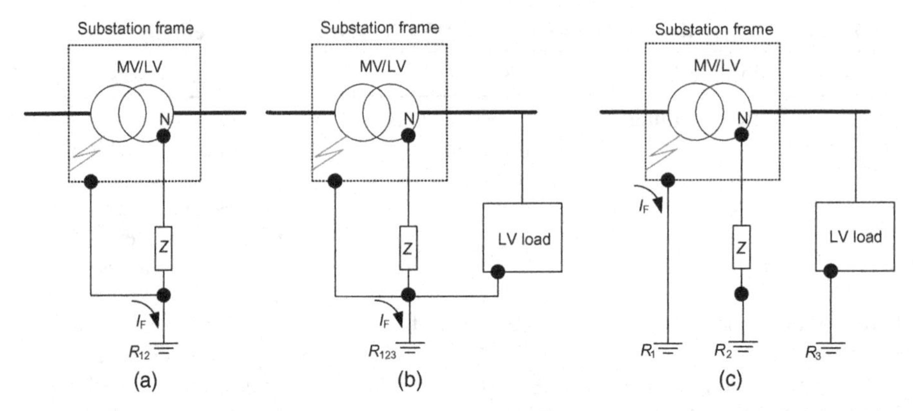

Figure 3.15 (a) Substation frame and LV neutral are connected to the same earth. (b) All earthing connections are grouped to a single earth. (c) All earthing connections are earthed separately.

TABLE 3.3 MV Earthing Systems for Public Distribution Examples [48]

Country	MV Earthing System	Frame Connection
Germany (10–20 kV)	Unearthed	Connected
Australia (11 and 12 kV)	Directly earthed	Separated except if $R < 1\,\Omega$
Belgium (6.3–11 kV)	Limitation impedance earthed	Separated
France (20 kV)	Direct or limitation impedance earthed	Separated except if $R < 3\,\Omega$ (overhead line [OHL])
United Kingdom (11 kV)	Unearthed	Separated except if $R < 1\,\Omega$
Italy (10–15 and 20 kV)	Unearthed on 10 kV and compensated on 38 kV	Separated
Ireland (10 and 38 kV)	Unearthed	Separated except if $R < 10\,\Omega$
Japan (6.6 kV)	Limitation impedance earthed	Connected $R < 65\,\Omega$
Portugal (10–30 kV)	Limitation impedance earthed	Separated except if $R < 1\,\Omega$
United States (4–25 kV)	Directly earthed or by low impedance	Connected

R is the equivalent resistance of earthing connection.

transformer and LV neutral to a dangerous potential ($I_F R_{12}$, where I_F is the fault current and R_{12} is the resistance of earth connection) with respect to the earth. If all the earth connections (substation frame, LV neutral, and load frames) have been grouped into a single one (Fig. 3.15b), a rise in potential of LV frames may be observed, which can be dangerous ($I_F R_{123}$, where R_{123} is the resistance of earth connection). On the other hand, if all earthing connections are earthed separately (Fig. 3.15c), the touch voltage in the substation can be raised to $I_F R_1$

(R_1 is the resistance of the earth connection of transformer frame). In all cases, the fault current depends on the MV earthing system and MV-frame disruptive breakdowns give rise to constraints, which can be severe. The application of these cases in different countries is explained in the next section.

3.4.2 MV Earthing Systems Worldwide

A few examples for public distribution systems worldwide are given in Table 3.3. It shows that, in many countries, the earth connections of the substation frame and neutral must be separate if their resulting value is not less than 1 Ω. It should be noted that in the United States, the earths of the source of the MV/LV substation and of the LV neutral are connected.

3.5 EARTHING SYSTEMS IN LV DISTRIBUTION NETWORKS

In compliance with IEC standard [49], three main earthing systems are used: IT, TT, and TN systems. Each system is named by two letters. The first letter defines the state of neutral (earthed or unearthed) and the second letter defines the state of the exposed conductive parts of electric installations [50]. The definitions used are as below:

- The letter "I" means that the neutral is unearthed or high impedance grounded.
- The letter "T" means that the neutral is directly earthed.
- The letter "N" means that the exposed conductive parts of electric installations are directly connected to the neutral conductor.

3.5.1 IT Earthing System

According to the above definitions, in this system, the neutral point is unearthed or connected to ground via a high impedance (an impedance of 1700 or 2000 Ω is often used) as denoted by the first letter "I." The exposed conductive parts of the loads are interconnected and earthed via a conductor of protective earthing (PE) as denoted by the second letter "T" [51].

This system is shown in Figure 3.16. It must be provided with an overvoltage limiter (P_s) between the MV/LV transformer neutral point and earth, in particular, when the neutral point is high impedance earthed (Z_N). The overvoltage limiter runs off external overvoltages, transmitted by the transformer, to the earth, and protects the LV network from a voltage increase due to flashover between the transformer's MV and LV windings.

The loads can be grouped and each group can be earthed individually if it is situated far away from the others. Also, if the short-circuit current is not high enough to activate protection against phase-to-phase faults, for example, if the loads are far away, protection should be provided with residual current devices (RCDs) to protect each group of loads.

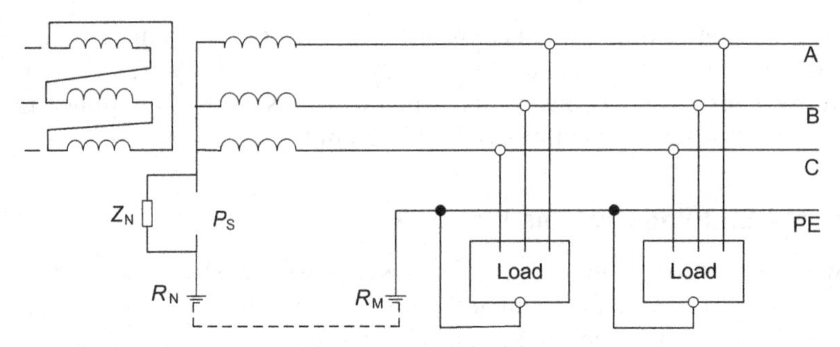

Figure 3.16 Unearthed or high-impedance-earthed neutral (IT system) in LV networks.

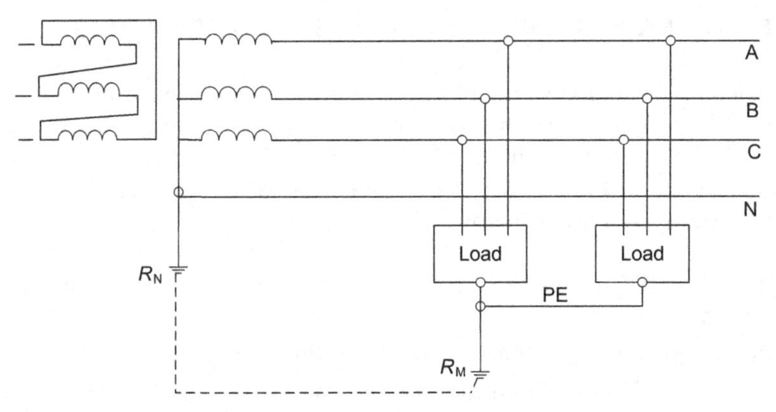

Figure 3.17 Directly earthed neutral (TT system) in LV networks.

3.5.2 TT Earthing System

In this system, the neutral is directly earthed as it is denoted by the first letter "T" and the exposed conductive parts of the loads are directly earthed via a conductor of PE individually or altogether as it is denoted by the second letter "T" (Fig. 3.17).

The earth of both neutral conductor and protective conductor may or may not be interconnected or combined. On the other hand, all exposed conductive parts protected by the same protective device should be connected to the same earth.

3.5.3 TN Earthing System

This system has directly earthed neutral, which is denoted by the first letter "T" while the exposed conductive parts of the loads are connected via a PE conductor to the neutral conductor.

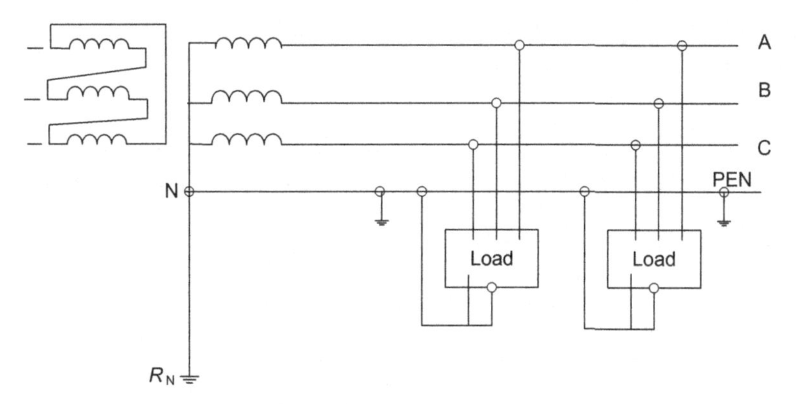

Figure 3.18 TNC earthing system.

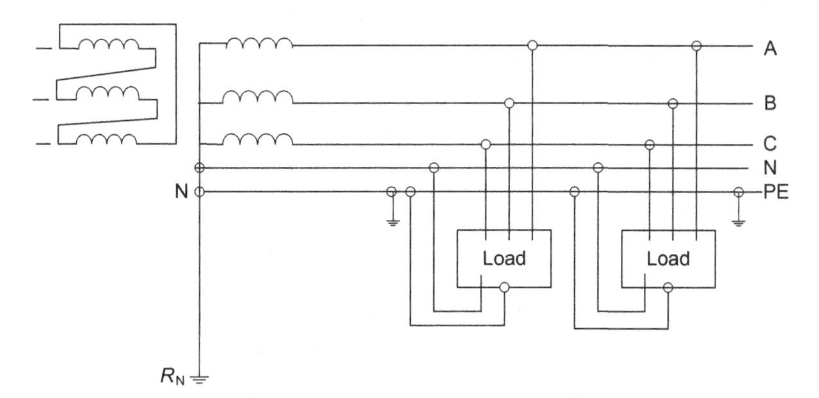

Figure 3.19 TNS earthing system.

The TN earthing system is further classified into two types named by a third letter. The first and second letters refer to the main system, TN, and the third letter typifies the nature of PE and neutral conductor interconnection.

Type 1: It is called TNC earthing system where the type is identified by the third letter, C. The neutral and protective earth conductors are combined in a single conductor called PEN, signifying protective earth and neutral. The earthing connections must be evenly placed along the length of the PEN conductor to avoid potential rise in the exposed conductive parts if a fault occurs (Fig. 3.18).

Type 2: It is called TNS earthing system. The type is identified by the third letter, S. The neutral and protective conductors are separate as shown in Figure 3.19.

In the case of the occurrence of a fault and to avoid potential rise in the exposed conductive parts, earthing connections must be placed along the length of the PE conductor, the same as in the TNC system.

TABLE 3.4 Techniques for Operating and Protecting Persons for Different Earthing Systems

Earthing System	IT	TT	TN
Operating technique	Insulation monitoring Indication of first insulation fault Switching if double fault occurs	Switching if first insulation fault occurs	Switching if first insulation fault occurs
Technique for protecting persons	Earthing of exposed conductive parts	Earthing of exposed conductive parts combined with RCDs	Interconnection and earthing of exposed conductive parts are imperative.
	First fault monitoring by insulation monitor	The exposed conductive parts protected by the same RCD must be connected to the same earth	Switching on occurrence of first fault by an overcurrent device
	Switching upon second fault occurrence by overcurrent protective devices		

The three LV earthing systems are different in the way they operate and protect living beings as given in Table 3.4. Each has its advantages and drawbacks as shown in Table 3.5.

3.5.4 LV Earthing Systems Worldwide

Most countries apply or derive inspiration from the three earthing systems described in Sections 3.5.1–3.5.3 for public and private distribution systems.

3.5.4.1 Public Distribution Systems The MV/LV transformers used are generally Dy11 (delta/star). However, the use of midpoint single-phase distribution for public distribution in the United States and Japan should be pointed out (Fig. 3.20).

The most common systems are TT and TN; a few countries, in particular Norway, use the IT system. Some examples of public distribution systems (LV consumers) are listed in Table 3.6. It also shows that Anglo-Saxon countries use the TNC, whereas the TT is used in the rest of the world.

TABLE 3.5 Advantages and Drawbacks of Different Earthing Systems

Earthing System	Advantages	Drawbacks
IT	The system provides best service continuity.	It requires a good level of network insulation.
	The short-circuit current is very low.	Tripping of two simultaneous faults should be carried out.
		Overvoltage limiters must be installed.
		Locating faults is difficult on large network.
		The exposed conductive parts must be equipotentially bonded.
		Distributing the neutral conductor is avoided.
TT	It is the simplest system to design, implement, monitor, and use.	Use of RCD on each outgoing feeder is necessary to obtain total selectivity.
	The presence of RCDs prevents the risk of fire if their sensitivity is $\leq$500 A (IEC 364-4 Std.).	Switching upon first insulation fault occurrence.
	The insulation fault current is low.	Special measures must be taken for the loads to avoid spurious tripping.
	Easy to locate the fault.	
TN	The installation of TNC-type system is less costly.	The fire risk is high and may cause damage to equipment.
	Use of overcurrent devices ensures protection against indirect contact.	The third harmonic and its multiples circulate in the PEN conductor.
		The PE conductor must remain at the same earth potential.
		Switching on first insulation fault occurrence.

3.5.4.2 Earthing Systems of Private LV Networks In some countries, the earthing systems of private LV networks, such as factories and mines, are illustrated below as examples [48].

In the United States, all the various earthing systems are used: The TNS is the most common (Fig. 3.21), but the IT and impedance-earthed IT are used in process factories.

In the Republic of South Africa (RSA), all three earthing systems are used with a preference for the TNS, which is characterized by the following:

TABLE 3.6 LV Earthing Systems for Public Distribution Examples Worldwide (LV Consumers) [48]

Country	LV Earthing System
Germany (230/400V)	TNC (mostly) and TT
Belgium (230/400V)	TT
Spain (230/400V)	TT
France (230/400V)	TT
United Kingdom (240/415V)	TNC town areas and TT rural areas
Italy (230/400V)	TT
Japan (100/200V)	TT
Norway (230/400V)	IT
Portugal (230/400V)	TT
United States (120/240V)	TNC

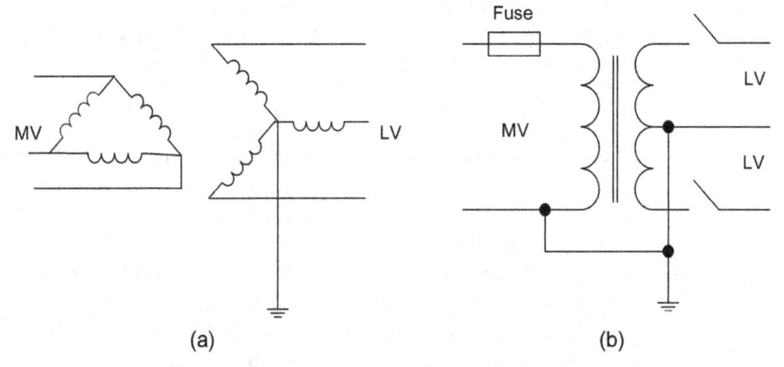

Figure 3.20 MV/LV public distribution transformers. (a) Three-phase, MV/LV, Dy11 distribution transformer in compliance with the European system. (b) Single-phase distribution transformer in compliance with the North American system.

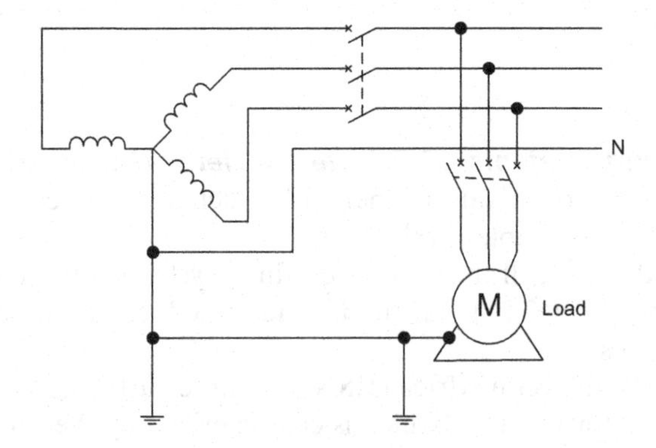

Figure 3.21 TNS earthing system in the United States.

- The protective conductor is distributed.
- The load frames are connected to the PE, which is earthed at MV/LV substation level.
- A resistor placed between the transformer neutral and the earth connection limits the insulation fault current.

In China, the three earthing systems are used to varying degrees. IT is used when continuity of service is vital and there is a real risk for persons (e.g., hospitals). TT is used in industry and TNS is increasingly chosen by design institutes for large projects.

CHAPTER 4

SHORT-CIRCUIT STUDIES

4.1 INTRODUCTION

Distribution systems are exposed to temporary or permanent faults. Temporary faults are the faults that disappear due to the action of protective devises and do not reappear when the equipment is restarted. Equipment with a permanent fault is isolated by protective devices until repaired and reenergized by operator intervention. Various types of faults and their effects that may occur in distribution systems are the following:

- *Short Circuits*: They may be generated inside or outside system equipment leading to equipment deterioration. A majority of short-circuit faults are breakdown in equipment insulation caused by switching surges or lightning strokes, degradation of insulating materials due to temperature rise or internal partial discharge, and accidental electric connection between two conductors through a foreign body (tools, animals, tree branches).
- *Faults on Motors*: Overheating and mechanical shocks on couplings due to several successive start-ups or excessive start-up time or rotor locking. When a short-circuit fault occurs in the system, motors act as generators and will contribute current to the fault. For synchronous motors, the fault occurrence in the system may cause a voltage drop, which yields the motor to receive less power from the system for driving its load. The inertia of the

Electric Distribution Systems, First Edition. Abdelhay A. Sallam, Om P. Malik.

Published 2011 by John Wiley & Sons, Inc.

motor and its load act as a prime mover, and with field excitation maintained, the motor acts as a generator. As the motor slows down, the fault current contribution diminishes and the motor field excitation decays [46].

Also, squirrel-cage induction motors contribute current to the fault. This is generated by inertia driving the motor in the presence of a field flux that is produced by induction from the stator. The current contribution to a fault at motor terminals drops off at a rapid exponential rate and disappears completely after a few cycles since the flux decays on removal of source voltage caused by the fault. There is no steady-state value of fault current as for synchronous motors since the field excitation is not maintained.

Wound-rotor induction motors normally operating with their rotor rings short circuited will contribute current to the fault in the same manner as a squirrel-cage induction motor. In some cases, large wound-rotor motors operated with some external resistance maintained in their rotor circuits may have sufficiently low short-circuit time constants that their fault current contribution is not significant and may be neglected.

- *Faults on Synchronous Generators*: Overload or malfunction of synchronous generators as frequency regulators may cause a variation of frequency. Also, a fault in the rotor circuit may cause a loss of excitation, which in turn leads to the overheating of rotor and stator, and loss of synchronism with the network.

  If a fault occurs at generator terminals, the short-circuit current starts out at a high value and decays exponentially to a lower steady-state value some time after the fault initiation. The steady-state short-circuit current will persist unless interrupted by switching means because the synchronous generator continues to be driven by its prime mover and to have its field excitation externally excited [52].
- *Overvoltages*: These are disturbances superimposed on circuit rated voltage. Abnormal voltages are caused most frequently by lightning, but some other disturbances can also cause damaging voltage surges such as the following:
  - Overvoltage due to insulation fault on a three-phase network when the neutral is unearthed or earthed through impedance.
  - Overvoltage on a long off-load line (Ferranti effect) may occur when a long line is energized at one of its ends and not connected at the other. This is due to resonance, which takes the form of a voltage wave increasing in linear fashion along the line.
  - Overvoltage by ferromagnetic resonance. This is a result of a special resonance that occurs when a circuit contains both a capacitor (voluntary or stray) and an inductance with saturable magnetic circuit (e.g., a transformer).
  - Switching overvoltage due to sudden changes in electric network structure. These changes give rise to transient phenomena frequently result-

ing in the creation of an overvoltage or of a high-frequency wave train of a periodic or oscillating type with rapid damping.

- *Overloads*: On cables, transformers, motors, or generators.
- *Voltage Variations*: Due to faulty operation of the on-load tap changers of a transformer, or network under- or overload. Types of voltage variations are explained in Chapter 8.
- *Overheating and Loss of Synchronism*: The presence of a negative-phase component due to a nonsymmetrical voltage source, a large single-phase consumer, a connection error, or phase disconnection leads to overheating of the motors or generators and a loss in generator synchronism.
- *Arcing Faults*: Distribution system faults may also be arcing in nature. Arcing faults can display a much lower level of short-circuit current than a bolted fault (fault connection has zero impedance) at the same location. These lower levels of current are due in part to the impedance of the arc. While system components should be capable of interrupting and withstanding the thermal and mechanical stresses of bolted short-circuit currents, arcing faults usually present different problems. Arcing faults may be difficult to detect because of the smaller currents. Sustained arcs can present safety hazards to people and also cause extensive damage because of the burning and welding effect of the arc as well as from the conductive products of ionization. More details are given in Chapter 6, Section 6.12.
- *Series Faults*: Distribution systems may be exposed to one of the following unbalances: One or more of the three phases are open or unbalanced line impedance discontinuity. These unbalances are associated with a redistribution of the prefault load current. Series faults are of interest when assessing the effects of situations that result in the flow of unbalanced currents such as failure of cable joints, failure of breakers to open all poles, inadvertent breaker energization across one or two poles, blown fuses, and snapped overhead phase wires [53].

Therefore, to minimize damage due to faults and other disturbances, the electric distribution network is equipped with a protection system to protect the equipment against overloads, short circuits, overvoltages, switching surges, and reversal of direction of energy flow. In addition, this necessitates the protection system to decide the correct action at the correct time, to minimize the part of network necessary to isolate the fault, to maintain the continuity of supply, and to achieve the cost of protective gear to be as low as possible.

4.2 SHORT-CIRCUIT ANALYSIS

It is necessary to perform a short-circuit analysis of distribution systems to calculate the current flow into the different elements such as transformers, cables, and circuit breakers when a fault occurs. This short-circuit current is required for defining the specifications of such elements and designing the earthing system. In relation to the protection system, short-circuit calculation is required to specify the setting of protective devices at the various locations in the distribution system and to select the appropriate switching devices.

The wave form of short-circuit current has two components: alternating current (AC) component (periodic component) and a decaying direct current (DC) component (nonperiodic component). The amplitude of DC component is high during the first few cycles after fault occurrence (transient state) and then decays to approximately zero value at stable conditions (steady state). The time of decay and the amplitude of short-circuit current characterize the fault and depend on the system impedance seen at the fault point.

Consequently, the root mean square (rms) value of short-circuit current under the worst fault conditions (symmetrical three-phase short circuit at a location nearest to the equipment) where the short-circuit current is maximum, its first peak of transient period (immediately after fault initiation), and the minimum short-circuit current must be calculated. The rms value of maximum short-circuit current at first cycle is required to determine the breaking capacity of switching devices (circuit breakers and fuses) and temperature stress that the equipment must withstand. The first peak value of transient period (commonly at one-half cycle after fault initiation) is required to determine the making capacity of such switching devices and electrodynamic withstand of switchgear. These two values, rms and first peak, are referred to as "first-cycle values." The first cycle values are also required for coordinating protective devices according to their time-current characteristics. Maximum values of short-circuit current after a few cycles (1.5–8 cycles) are required for comparison with the breaking ratings of medium-voltage (MV) circuit breakers. The reduced fault currents at about 30 cycles and the minimum short-circuit current in case of using sources with high internal impedance or long cables must be calculated to define the tripping characteristic of circuit breakers and fuses or to set the thresholds of overcurrent protection [52].

4.2.1 Nature of Short-Circuit Currents

To illustrate the general shape of short-circuit current wave and the main factors affecting the waveform, two cases are studied. In the first case, the short-circuit fault is located far enough from the generators (sources) where the effect of their parameters as rotating machines can be ignored (Fig. 4.1a). In the second case, the short-circuit fault is located at generator terminals and the effect of generator parameters is taken into account (Fig. 4.1b). This case is applied when the distribution network has distributed generation (DG) units used for feeding local loads and for emergency or backup operations in private industrial plants. More details about DG are given in Chapter 14.

4.2.1.1 Case 1 The fault location divides the network into two parts. One part is the network upstream of fault and includes the source. The other part

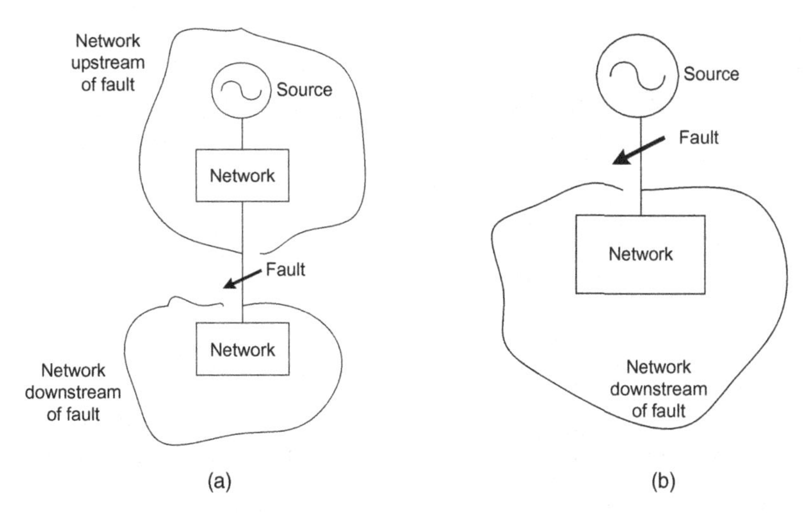

Figure 4.1 Schematic diagram of distribution network. (a) Fault location far from the source. (b) Fault location at generator terminals.

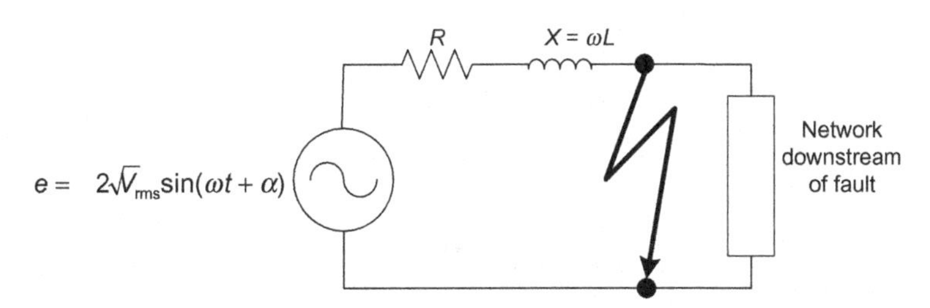

Figure 4.2 Equivalent circuit of case 1.

represents the network downstream of fault. To study the short-circuit current as a function of time indicating its transient nature, a simplified equivalent circuit as shown in Figure 4.2 is applied. In this circuit, the network upstream of fault is represented by a voltage source e applied to a series resistance R and a series reactance $X = \omega L$.

The voltage source is $e = \sqrt{2}V_{\text{rms}}\sin(\omega t + \alpha) = V_{\max}\sin(\omega t + \alpha)$, where V_{rms} is the rms value of the source voltage and α is the switching angle that defines the point on source sinusoidal voltage at the moment of fault occurrence.

Then, the behavior of short-circuit current I_{sc} can be obtained by solving the mathematical differential Equation written below:

$$e = iR + L\frac{di}{dt}. \tag{4.1}$$

Assume that

$$\begin{aligned} Z &= \text{impedance of network upstream of fault} \\ &= \sqrt{R^2 + \omega^2 L^2} \end{aligned}$$

and

$$\begin{aligned} \varphi &= \text{the network's phase displacement} \\ &= \tan^{-1}\frac{\omega L}{R}. \end{aligned}$$

Equation 4.1 has constant coefficients and can be solved by using Laplace transform method. Assuming zero current before the fault, it can be written as

$$E(s) = RI(s) + LsI(s).$$

From Laplace transform tables,

$$\begin{aligned} E(s) &= V_{\max}\frac{\omega\cos\alpha + s\sin\alpha}{s^2 + \omega^2} \\ &= I(s)[R + Ls]. \end{aligned}$$

Hence,

$$\begin{aligned} I(s) &= \frac{V_{\max}}{R + Ls}\frac{\omega\cos\alpha + s\sin\alpha}{s^2 + \omega^2} \\ &= \frac{V_{\max}}{L}\frac{\omega\cos\alpha + s\sin\alpha}{(s^2 + \omega^2)\left(s + \frac{R}{L}\right)}. \end{aligned} \tag{4.2}$$

By the method of partial fractions, Equation 4.2 can be written as

$$I(s) = \frac{As + B}{s^2 + \omega^2} + \frac{C}{s + \frac{R}{L}},$$

where

$$C = -\frac{V_{\max}}{Z}\sin(\alpha - \varphi),$$

$$A = -C = \frac{V_{\max}}{Z}\sin(\alpha - \varphi),$$

$$B = \frac{V_{\max}\omega}{Z}\cos(\alpha - \varphi),$$

that is,

$$I(s) = \frac{V_{\max}}{Z}\left[\frac{s\sin(\alpha-\varphi)+\omega\cos(\alpha-\varphi)}{s^2+\omega^2} - \frac{\sin(\alpha-\varphi)}{\left(s+\frac{R}{L}\right)}\right]. \tag{4.3}$$

By using the tables of Laplace transforms, Equation 4.3 can be written in time domain as below:

$$i(t) = \frac{V_{\max}}{Z}\left[\sin(\omega t+\alpha-\varphi) - \sin(\alpha-\varphi)e^{-(R/L)t}\right]. \tag{4.4}$$

Equation 4.4 shows that the current is the sum of a sinusoidal component and an exponential component. The sinusoidal component is the sinusoidal steady-state current resulting from the sinusoidal applied voltage, and the exponential component having the exponential form is characteristic of the circuit. The total current starts as an asymmetrical sinusoidal wave and becomes symmetrical as the transient exponential component of current dies away (Fig. 4.3).

From Equation 4.4, the fault current is maximum when $\alpha - \varphi = -(p + ½)\pi$ where p is an integer and its peak occurs at the time at which $\omega t + \alpha - \varphi = (p + ½)\pi$, that is, $\omega t = (2p + 1)\pi$. Thus, the peak value during the transient period is given as

$$\begin{aligned} I_{\text{peak}} &= \frac{V_{\max}}{Z}\left[1+e^{-(R/L)t}\right] \\ &= \frac{V_{\max}}{Z}\left[1+e^{-(R/X)(2p+1)\pi}\right]. \end{aligned} \tag{4.5}$$

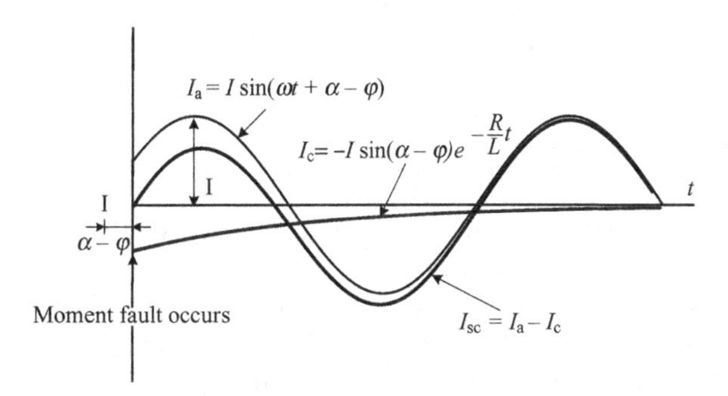

Figure 4.3 Short-circuit fault current versus time.

TABLE 4.1 Values of K in Relation to Time for R/X = 0.05, 0.1, 0.2, and 0.3 [t (ms) = $(p + ½)T$ and T = 20 ms for 50 Hz)]

	t (ms)				
R/X	10 (p = 0)	30 (p = 1)	50 (p = 2)	70 (p = 3)	∞
0.05	2.62	2.30	2.06	1.89	1.41
0.10	2.45	1.97	1.71	1.57	1.41
0.20	2.17	1.63	1.48	1.43	1.41
0.30	1.97	1.50	1.43	1.42	1.41

The steady-state rms value of fault current is given as

$$I_{rms} = \frac{V_{rms}}{Z}. \quad (4.6)$$

For short-circuit analysis, it is worth knowing how much the peak transient current is greater than the steady-state rms value. This can be defined by the ratio peak-to-rms (K). Dividing Equation 4.5 by Equation 4.6 gives

$$K = \sqrt{2}\left[1+e^{-(R/L)t}\right] = \sqrt{2}\left[1+e^{-(R/X)(2p+1)\pi}\right]. \quad (4.7)$$

It is seen from Equation 4.7 that K varies from 2.83 to 1.41 corresponding to (R/X) = 0 and ∞, respectively. It is interesting to define the coefficient K in relation to the ratio (R/X), which characterizes the network. This ratio ranges between 0.05 and 0.3 in MV and between 0.3 and 0.6 in low voltage (LV) (near transformers) [54].

As an example, the variation of K with time (from 10 to ∞ ms) and R/X (from 0.05 to 0.3) are given in Table 4.1 and plotted as in Figure 4.4.

It is common to use circuit breakers with lower tripping time than required for aperiodic exponential component to die away. So, the making capacity of the circuit breaker is necessary to be determined such that it will enable it to interrupt the peak value of the transient current including the two components, sinusoidal and exponential components, as calculated by Equation 4.5, and to determine electrodynamic withstand. American National Standards Institute (ANSI) stipulates a making capacity with a coefficient K = 2.7 corresponding to a ratio R/X = 0.03 [55].

4.2.1.2 Case 2 When the short circuit occurs close enough to generator terminals, the generator produces four components of short-circuit current decaying with different time constants: one in each of the three characteristic states (subtransient, transient, and steady states) and one aperiodic component similar to that obtained in case 1 when applying an AC voltage to a resistor-inductor (RL) circuit. The decaying pattern is produced as a result of non-

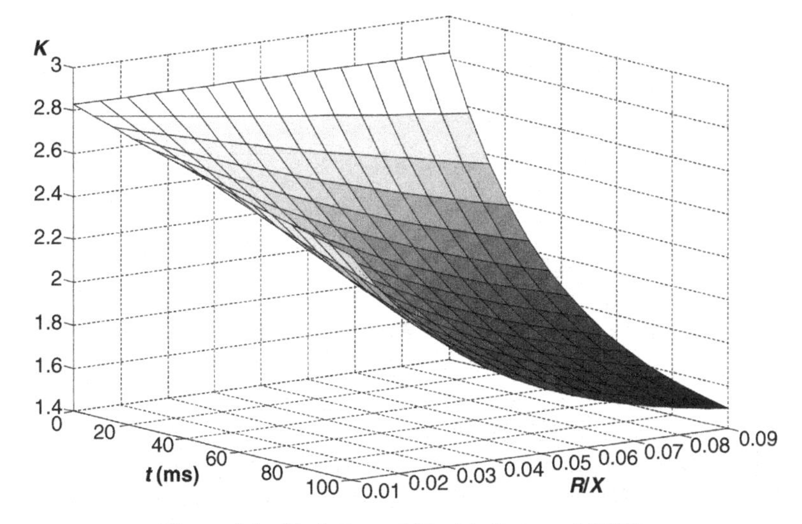

Figure 4.4 Variation of K with time and (R/X).

instantaneous change of magnetic flux in machine windings (armature reaction). Assuming the generator to be on no-load prior to the fault, the prefault values of generator internal electromotive forces (EMFs) in subtransient, transient, and steady states are equal to the terminal voltage "V" [56]. The four components can be calculated by the equations below:

Subtransient component:

$$i''(t) = V_{\max}\left[\frac{1}{X''_{\mathrm{d}}} - \frac{1}{X'_{\mathrm{d}}}\right] e^{-t/T''_{\mathrm{d}}} \cos(\omega t + \alpha),$$

Transient component:

$$i'(t) = V_{\max}\left[\frac{1}{X'_{\mathrm{d}}} - \frac{1}{X_{\mathrm{d}}}\right] e^{-t/T'_{\mathrm{d}}} \cos(\omega t + \alpha),$$

Steady-state component:

$$i(t) = \frac{V_{\max}}{X_{\mathrm{d}}} \cos(\omega t + \alpha),$$

Aperiodic component:

$$i_{\mathrm{a}}(t) = -\frac{V_{\max}}{X''_{\mathrm{d}}} e^{-t/T_{\mathrm{a}}} \cos\alpha.$$

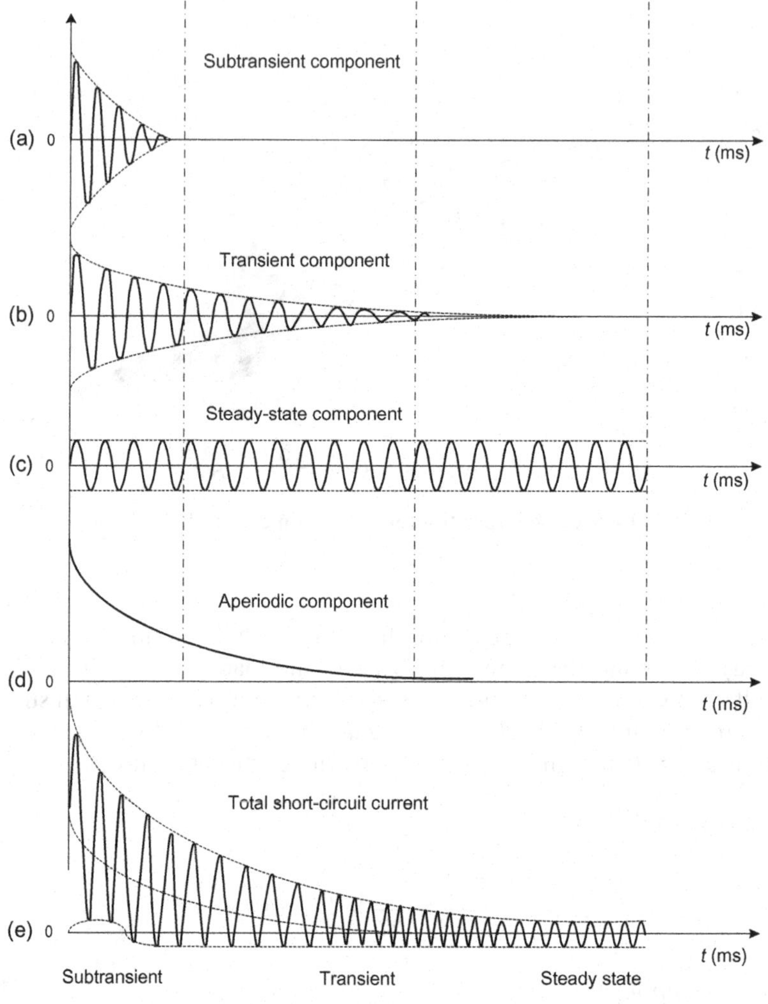

Figure 4.5 Components of generator short-circuit current.

The short-circuit current is the sum of four components: three decaying sinusoidal currents and one aperiodic decaying current as shown in Figure 4.5. It is given as

$$i_{\mathrm{sc}}(t) = i''(t) + i'(t) + i(t) + i_{\mathrm{a}}(t).$$

Hence,

$$i_{\mathrm{sc}}(t) = V_{\max}\left[\left(\frac{1}{X''_{\mathrm{d}}} - \frac{1}{X'_{\mathrm{d}}}\right)e^{-t/T''_{\mathrm{d}}} + \left(\frac{1}{X'_{\mathrm{d}}} - \frac{1}{X_{\mathrm{d}}}\right)e^{-t/T'_{\mathrm{d}}} + \frac{1}{X_{\mathrm{d}}}\right]\cos(\omega t + \alpha)$$
$$- \frac{V_{\max}}{X''_{\mathrm{d}}}e^{-t/T_{\mathrm{a}}}\cos\alpha, \qquad (4.8)$$

where

V_{max} = maximum value of phase voltage at generator terminals,
X''_d = direct-axis subtransient reactance,
X'_d = direct-axis transient reactance,
X_d = direct-axis synchronous reactance,
T''_d = direct-axis subtransient time constant,
T'_d = direct-axis transient time constant,
T_a = armature (aperiodic) time constant, and
α = switching angle that defines the point on voltage wave at which the fault occurs.

The short-circuit current is maximum for $\alpha = 0$ and its peak values occur at the times at which $\omega t = \pi(2p + 1)$, where n is an integer number.

Thus, the peak current is given as

$$I_{peak} = V_{max}\left[\left(\frac{1}{X''_d} - \frac{1}{X'_d}\right)e^{-t/T''_d} + \left(\frac{1}{X'_d} - \frac{1}{X_d}\right)e^{-t/T'_d} + \frac{1}{X_d}\right] + \frac{V_{max}}{X''_d}e^{-t/T_a}. \tag{4.9}$$

It is common to neglect the prefault load current assuming the system is at steady-state before a short circuit occurs. This is based on the premise that its magnitude is much smaller than the fault current. The importance of considering the prefault load current in the system depends on the rated system voltage and certain system loading patterns. That is the reason to assume a prefault voltage of 1 per-unit at every bus for typical industrial system studies. In the case that the prefault loading pattern is a concern, the fault simulation should be preceded by a prefault load flow analysis in order to ascertain a voltage profile that will be consistent with the existing system loads, shunts, and transformer tap settings. All static loads and capacitive line/cable shunts are essentially retained for the fault simulation in case of modeling the actual prefault system condition. Attempts to address this issue by using elevated prefault voltages and impedance correction factors for the synchronous generators have been provided [53, 57].

4.2.2 Calculation of Short-Circuit Current

As explained in Section 4.2.1, the short-circuit current is highly dependent on the equivalent impedance of the network seen at the fault point. The value of this impedance depends on the network configuration, type of system earthing, elements included in the network, type of fault, and fault location [58].

Two methods can be applied to calculate short-circuit current with the same degree of accuracy when using the same data. They are identified as the *direct* method and the *per-unit* method.

In the direct method, the system one-line diagram, system and equipment data in physical units (volts, amperes, and ohms), and basic electric equations and relationships are used directly without applying special diagrams or abstract units. It is adapted to a whole or portion of a system analysis starting at the source, and the short-circuit values at each location out to the end of the various circuits are determined.

In the per-unit method, the system one-line diagram is converted into an equivalent impedance diagram by using the parameters that will have an effect on short-circuit current. This diagram is reduced to a single impedance value by delta and wye conversion equations. The appropriate Equation is then applied to provide the short-circuit value. The per-unit method with its special mathematical technique is adapted to calculating short-circuit values at one or more specific points in the system and in particular when several voltage levels exist between the source and the short-circuit point. Each point has its own impedance diagram and consequently the reduced diagram and calculations.

Direct and per-unit methods are based on representing the three-phase electric system by one-line diagram. This representation is also successfully applied to other studies such as power flow and transient stability studies. For the purpose of short-circuit calculations, these two methods can only be applied to symmetrical three-phase short-circuit calculations since the fault condition is balanced involving all three phases. In contrast, the one-line diagram methods are inadequate to analyze faults involving asymmetry and unbalance because each of the three phases needs to be treated individually. Two approaches to handle this problem can be applied. First, the system is represented on three-phase basis and the identity of all three phases is explicitly retained. In this case, the analysis is difficult and may not be easily tractable as it can be data intensive. In the second approach, the system is analyzed by applying symmetrical components technique as in Section 4.2.2.2.

For complicated and large networks with many machines contributing to the fault current, more accurate methods based on computer programs are applied to calculate the short-circuit current as a function of time from the moment of fault inception (time domain fault analysis). This requires detailed and complex models of network components, for example, the machines and their controllers. Other programs simplify the calculations by considering linear models and analyze the fault at steady state (quasi-steady-state fault analysis). With the aid of linear algebra theory and numerical advances in matrix computation, powerful computer programs have been implemented [53]. They are able to form and process the bus-admittance and bus-impedance matrices by applying mathematical techniques to manipulate sparse matrices of high dimensions, and simulating the fault accurately. On the other hand, to use such programs, more effort should be paid to system data preparation.

In the next section, the direct method is explained where it gives reasonable results and more understanding to the reader, as well as the calculation can be made by hand. It is applied to symmetrical three-phase short-circuit faults. The method is based mainly on the impedance calculation of the different

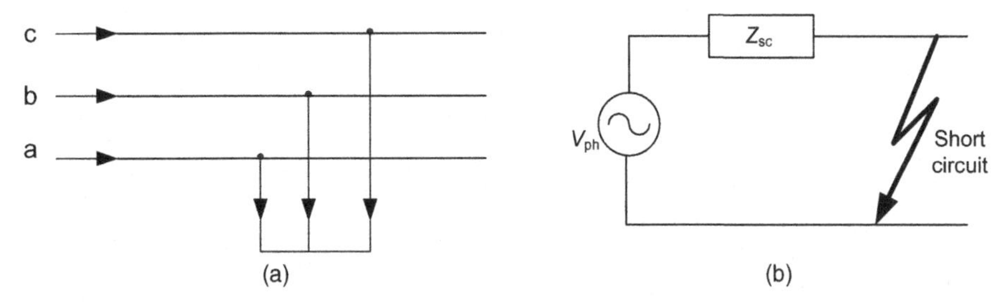

Figure 4.6 Symmetrical three-phase short circuit. (a) Three-phase symmetrical short circuit; (b) equivalent circuit.

elements in the network such as rotating machines, transformers, overhead lines (OHLs), and cables.

4.2.2.1 Symmetrical Three-Phase Short Circuit This type of fault is the most severe fault as it produces the highest short-circuit current with equal magnitudes in each of the three phases (Fig. 4.6a). So, it helps the protection engineers to determine the adequate specification of protective equipment at different locations in the network.

As shown in Figure 4.6b, the short-circuit current is $I_{sc} = V_{ph}/Z_{sc}$, where V_{ph} is the phase-to-neutral voltage and Z_{sc} is the equivalent impedance seen at the fault point. It is calculated by defining the impedance of each element in the path in which the short-circuit current flows. Therefore, the calculation of impedance of such elements is indicated below.

In MV networks, a coefficient of 1.1 is applied to the nominal voltage in order to calculate the maximum short-circuit current, thereby giving $I_{sc} = 1.1V_{ph}/Z_{sc}$ [57].

Transformers The specification of transformers given by manufacturers commonly implies the short-circuit voltage as a percentage of rated voltage (V_{sc} %) and the nominal apparent power S_n in kVA. The short-circuit voltage is defined as the primary voltage that, when applied to the transformer with its secondary short circuited, gives a current equal to the nominal current.

Thus,

$$V_{sc}(\%) \times V_{ph} = Z_{Tr} I_n \times 100$$

and

$$S_n = 3V_{ph} I_n.$$

Hence,

$$Z_{Tr} = \frac{V_{sc}(\%)}{100} \frac{3V_{ph}^2}{S_n},$$

that is,

$$Z_{Tr} = \frac{V_{sc}(\%)}{100}\frac{V_{line}^2}{S_n}, \tag{4.10}$$

where

Z_{Tr} = transformer impedance (Ω),
I_n = nominal current (A),
V_{ph} = nominal phase-to-neutral voltage (V),
V_{line} = nominal phase-to-phase voltage = $\sqrt{3}V_{ph}$, and
S_n = nominal rating (VA).

The transformer impedance, as seen from the primary, is given in Equation 4.10, if V_{line} is substituted by the nominal phase-to-phase primary voltage. Also, it gives the transformer impedance, as seen from the secondary, when V_{line} is substituted by the nominal phase-to-phase secondary voltage.

It is found that the transformer resistance is much less than its reactance [54], for example, for MV/LV transformers, $R_{Tr}/X_{Tr} \approx 0.3$; thus,

$$Z_{Tr}^2 = R_{Tr}^2 + X_{Tr}^2 = (0.3)^2 X_{Tr}^2 + X_{Tr}^2 = 1.09 X_{Tr}^2,$$

that is,

$$X_{Tr} \approx 0.96 Z_{Tr}.$$

For high-voltage (HV)/MV transformers, $R_{Tr}/X_{Tr} \approx 0.05$, which means that $X_{Tr} \approx 0.998 Z_{Tr}$. Therefore, it is an accepted approximation to consider the transformer impedance as a pure reactance: $Z_{Tr} \cong X_{Tr}$.

OHLs The resistance of OHLs, R, is calculated by

$$R = \frac{\rho l}{A} \quad \Omega, \tag{4.11}$$

where

ρ = conductor's resistivity (Ω mm²/m),
l = cable length (m), and
A = cross-sectional area of conductor (mm²).

To calculate the reactance of OHLs, the lines are assumed to be long with straight conductors; that is, the sag from pole to pole is neglected. The error introduced by this assumption is small and negligible [59]. For a long single-phase line, straight cylindrical conductor with a circular cross Section of radius

r, uniform current density, and a return conductor separated at distance d, its reactance is given as

$$X = 2\pi f\left(461\log\frac{d}{r} + 50\right)\times 10^{-6}\ \Omega/\text{km}. \qquad (4.12)$$

The inductance value $2\pi f \times 50 \times 10^{-6}\,\Omega$/km results from the flux within the conductor linking the current, assuming no skin effect. To make the Equation more general so that skin effect, nonsymmetrical conductors, conductor's configuration, and so on, can be dealt with, the value $2\pi f \times 50 \times 10^{-6}\,\Omega$/km can be replaced by 0.1447 $\log 10^k$ at 50 Hz or 0.1737 $\log 10^k$ at 60 Hz to give

$$X = 0.1447\left(\log\frac{d}{r} + \log 10^k\right)\Omega/\text{km at 50 Hz},$$

$$X = 0.1737\left(\log\frac{d}{r} + \log 10^k\right)\Omega/\text{km at 60 Hz}.$$

Also, $(r/10^k)$ is replaced by GMR and d is replaced by GMD. Thus, Equation 4.12 is in the form

$$\begin{aligned} X &= 0.1447 \log\frac{\text{GMD}}{\text{GMR}}\ \Omega/\text{km at 50 Hz}, \\ X &= 0.1737 \log\frac{\text{GMD}}{\text{GMR}}\ \Omega/\text{km at 50 Hz}, \end{aligned} \qquad (4.13)$$

where GMR is the geometric mean radius of the conductor. It is the radius of an infinitesimally thin circular tube that has the same internal reactance as the actual conductor. It can include skin effect, the effects of stranding, noncircular conductors, and multiple conductors for a phase line. GMR is equal to a conductor's radius r for homogeneous round single conductor neglecting skin effect. By including skin effect, GMR equals $0.799r$, and it ranges from 0.7 to 0.85 for stranded conductors and various combinations of different materials, such as aluminum cored steel reinforced (ACSR). The GMR values are usually provided as part of OHL characteristics and in conductor tables. The general form of GMR equation for parallel conductors 1, 2, … , n separated by different distances d is

$$\text{GMR}_n = \sqrt[n^2]{(\text{GMR}_1 d_{12} d_{13} \ldots d_{1n})(\text{GMR}_2 d_{21} d_{23})\ldots(\text{GMR}_n d_{n1} d_{n2} \ldots d_{n(n-1)})}. \qquad (4.14)$$

GMD is the geometric mean distance or separation between conductor(s) and the return conductor(s). The individual distances are measured between the conductor centers. For a single-phase line separated by a distance d from its return conductor, GMD equals d, and between three conductors 1, 2, and 3 separated by distances d_{12}, d_{13}, and d_{23}, the equivalent spacing or GMD is calculated by GMD $= \sqrt[3]{(d_{12}d_{13}d_{23})}$. The general Equation of the GMD between two areas or between two wires of a conductor, i, where $i = 1, 2, \ldots, n$ and its return or associated conductors, j, where $j = 1, 2, \ldots, m$, is

$$\mathrm{GMD}_{nm} = \sqrt[nm]{(d_{11}d_{12}d_{13}\ldots d_{1m})(d_{21}d_{22}d_{23}\ldots d_{2m})\ldots(d_{n1}d_{n2}d_{n3}\ldots d_{nm})}. \quad (4.15)$$

Adding the resistance per unit length, R Ω/km, provided by manufacturer, the impedance of OHLs can be calculated by

$$Z = R + j0.1447\log\frac{\mathrm{GMD}}{\mathrm{GMR}}\ \Omega/\mathrm{km\ at\ 50\ Hz} \quad (4.16)$$

and

$$Z = R + j0.1737\log\frac{\mathrm{GMD}}{\mathrm{GMR}}\ \Omega/\mathrm{km\ at\ 60\ Hz}.$$

Cables The most common cables used in distribution systems are either single-conductor cables or three-conductor cables. It is usual with cables to express the dimensions in inches, and this convention will be followed in deducing the formulas of impedances except otherwise stated. The resistance and reactance of each type is calculated as below [59].

SINGLE-CONDUCTOR CABLES As shown in Figure 4.7, the voltage drop, V_d, along the conductor is

$$V_d = IZ_c - jI_{sh}X_m. \quad (4.17)$$

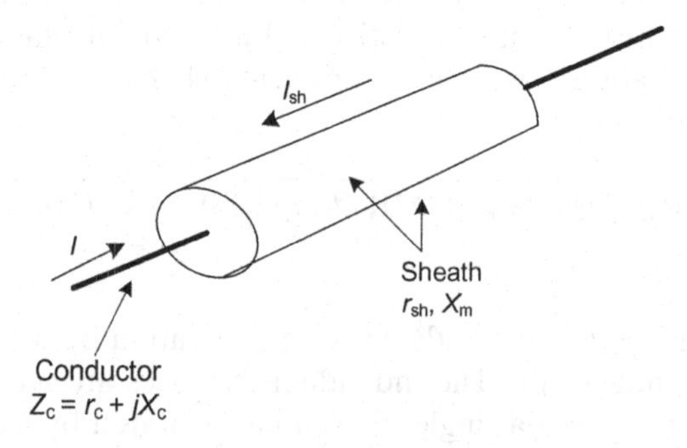

Figure 4.7 Single-conductor cable.

The drop along the sheath is zero as it is earthed, that is,

$$0 = (r_{sh} + jX_m)I_{sh} - jIX_m. \tag{4.18}$$

The sheath reactance, X_m, is based on the assumption that the conductor is concentric within the sheath. From Equations 4.17 and 4.18, the voltage drop along the conductor can be written as

$$V_d = \left[\left(r_c + \frac{r_s X_m^2}{r_{sh}^2 + X_m^2}\right) + j\left(X_c - \frac{X_m^3}{r_{sh}^2 + X_m^2}\right)\right] I. \tag{4.19}$$

It is indicated in Equation 4.19 that the equivalent AC conductor resistance, R, is

$$R = r_c + \frac{r_{sh} X_m^2}{r_{sh}^2 + X_m^2}, \tag{4.20}$$

where

r_c = AC resistance of the conductor including skin effect,
r_{sh} = AC resistance of the sheath, and
X_m = mutual reactance between the conductor and the sheath.

For the trefoil formation shown in Figure 4.8, where a group of three conductors are used, X_m and r_{sh} are given as

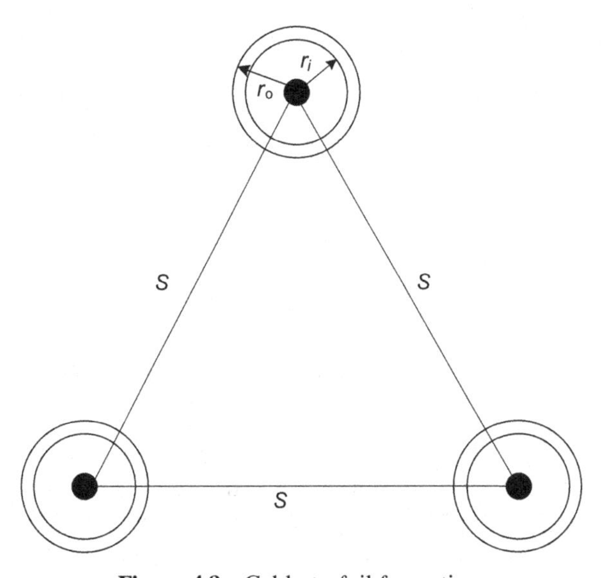

Figure 4.8 Cable trefoil formation.

$$X_m = 0.007495 f \log \frac{2S}{r_o + r_i} \ \Omega\text{/phase/km}, \tag{4.21}$$

$$r_s = \frac{0.32186}{(r_o + r_i)(r_o - r_i)} \ \Omega\text{/phase/km for lead}, \tag{4.22}$$

where

S = spacing between conductors (inches),
r_o = outer radius of the sheath (inches), and
r_i = inner radius of the lead sheath (inches).

Term $(r_o + r_i)/2$ in Equation 4.21 is the mean radius of the sheath and is approximately the GMR of the sheath. In addition, S can be replaced by DMD_{cond} to give an equivalent of unequal spacing between the phase cables. From Equation 4.19, the reactance of single-conductor cables is

$$X = 0.007459 f \log \frac{S}{GMR_{cond}} - \frac{X_m^3}{r_s^2 + X_m^2} \ \Omega\text{/phase/km}, \tag{4.23}$$

where X_m and r_{sh} are given in Equations 4.21 and 4.22, respectively.

THREE-CONDUCTOR CABLES The following formula is used to calculate the resistance R

$$R = r_c + \frac{71,066 S_1^2}{r_{sh}(r_o + r_i)^2} \times 10^{-6} \ \Omega\text{/phase/km}, \tag{4.24}$$

where S_1 is the effective center of each conductor and the cable center. For round cables, $S_1 = \frac{1}{\sqrt{3}}[D + 2(r_o - r_i)]$, where D is the conductor diameter in inches. The second term in Equation 4.23 represents the sheath loss. It is usually neglected except for large sizes.

So, the reactance is calculated by

$$X = 0.007495 f \log \frac{S}{GMR_{cond}} \ \Omega\text{/phase/km}, \tag{4.25}$$

where S = GMD, the geometric mean distance between three conductors.

Rotating Machines The values of machine impedance (motor or generator) in the characteristic states, subtransient X_d'', transient X_d', and steady-state X_d are required to calculate the short-circuit current as indicated in Equation 4.8. The manufacturer provides their values as a percentage based on the nominal apparent power S_n in kVA.

Therefore,

$$X(\Omega)=\frac{X(\%)}{100}\frac{V_{\mathrm{ph}}}{I_{\mathrm{n}}}=\frac{X(\%)}{100}\frac{V_{\mathrm{line}}^2}{S_{\mathrm{n}}}. \tag{4.26}$$

Thus,

$$\begin{bmatrix} X''_{\mathrm{d}} \\ X'_{\mathrm{d}} \\ X_{\mathrm{d}} \end{bmatrix}(\Omega)=\frac{V_{\mathrm{line}}^2}{100\,S_{\mathrm{n}}}\begin{bmatrix} X''_{\mathrm{d}}(\%) \\ X'_{\mathrm{d}}(\%) \\ X_{\mathrm{d}}(\%) \end{bmatrix} \tag{4.27}$$

The short-circuit current is developed according to the following three periods [54]:

- subtransient (X''_{d}): lasting half to one cycle after the fault inception,
- transient (X'_{d}): lasting up to 5–20 cycles, and
- steady state (X_{d}): synchronous reactance to be considered after the transient period.

Motors' Contribution to Short-Circuit Current The motors, found as loads in distribution systems, are either synchronous or asynchronous motors. In case of the occurrence of a fault, they feed the fault with a current, which is added to that supplied by the source to give the total short-circuit current.

The synchronous motor operates as a generator when a short-circuit occurs and is represented by an equivalent voltage source with internal impedance X''_{d} during subtransient period and X'_{d} during the transient period.

To check the making capacity of protective switching devices and electrodynamic stress withstand, the peak value of the current contribution of the motor is required. It is calculated on the basis of subtransient reactance where $X''_{\mathrm{d}} < X'_{\mathrm{d}} \ll X_{\mathrm{d}}$ and is given as

$$I_{\mathrm{peak}}=\frac{V_{\mathrm{ph}}}{X''_{\mathrm{d}}}.$$

On the other hand, the maximum short-circuit current supplied by the motor in transient state is required to check the breaking capacity of protective switching devices and temperature stress that the equipment must withstand. It is calculated by

$$I_{\max}=\frac{V_{\mathrm{ph}}}{X'_{\mathrm{d}}}.$$

For asynchronous motors, their contribution to symmetrical three-phase short-circuit current decreases rapidly with a time constant that depends on the motor type (squirrel-cage or slip-ring rotor) and motor parameters. For instance, it is 10 ms for single-cage motors up to 100 kW, 20 ms for double-cage motors over 100 kW, and from 30 to 50 ms for 1000-kW motors [54].

The asynchronous motor, when a short-circuit current occurs, is equivalent to a voltage source with an internal transient reactance X'_{M} equal to

$$X'_{\mathrm{M}} = \frac{V_{\mathrm{line}}^2}{P}\frac{I_{\mathrm{n}}}{I_{\mathrm{st}}}\eta\cos\phi \qquad (\Omega), \tag{4.28}$$

where

P = output mechanical power of motor = $\eta S_{\mathrm{n}} \cos\varphi$,
S_{n} = motor apparent power in kVA,
I_{n} = motor rated current,
I_{st} = motor starting-up current,
η = motor efficiency, and
$\cos\varphi$ = motor power factor.

As in Equation 4.26, the motor transient reactance as a percentage $X'_{\mathrm{M}}(\%)$ is given as

$$X'_{\mathrm{M}}(\%) = \frac{S_{\mathrm{n}}}{V_{\mathrm{line}}^2} X'_{\mathrm{M}}(\Omega) \times 100. \tag{4.29}$$

Equations 4.28 and 4.29 yield to

$$X'_{\mathrm{M}}(\%) = \frac{I_{\mathrm{n}}}{I_{\mathrm{st}}} \times 100. \tag{4.30}$$

It means that the motor transient reactance as a percentage is equal to the ratio of rated current to starting-up current. In practice, if a complete data is not available, the starting-up current may be taken as six times the rated current and $\eta\cos\varphi = 0.8$, that is, $X'_{\mathrm{M}} = 16.7\%$.

Example 4.1:

In a distribution network, assuming a low-voltage distribution point (LVDP) operating at 0.4 kV is supplied by a 22-kV utility network, through a 2.0-km OHL and MV/LV substation. The substation has one 1000-kVA transformer to feed a primary LVDP.

A secondary LVDP feeds four asynchronous motors, 100 kW each, through identical cables. The motors are working when a short circuit occurs. It is required to calculate the three-phase symmetrical short-circuit current at different locations F_1, F_2, F_3, and F_4 as shown in Figure 4.9.

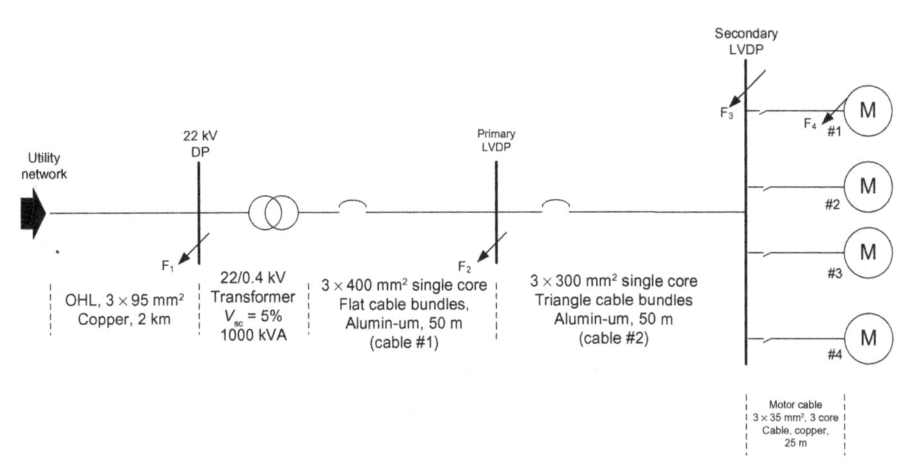

Figure 4.9 Distribution installation diagram.

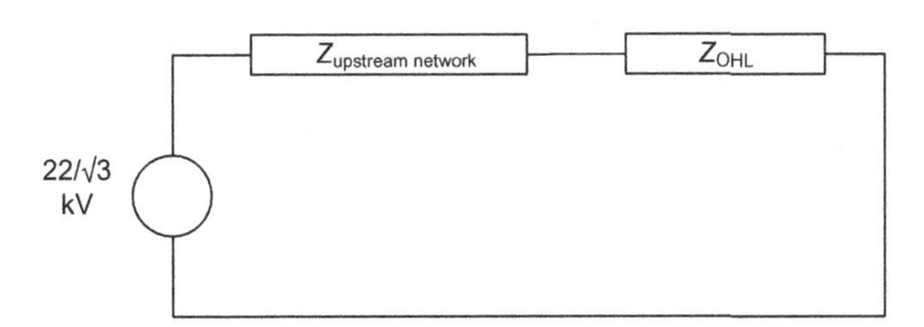

Figure 4.10 Equivalent circuit for fault at F_1.

The utility network which is the upstream network of the installation shown in Figure 4.9 has a short-circuit rating S_{sc} of 400 MVA, V_{line} of 22 kV, and (R/X) of 0.3. The reactance of OHL is taken as 0.3 Ω/km and the transformation ratio of transformer $n = 22/0.4 = 55$. The reactance of LV cables is 0.15 Ω/km. For motors, the four motors are identical, 100 kW each, (I_{st}/I_n) is 6, and $\eta \cos \varphi$ is 0.8. The bus-bar impedance at distribution points (DPs) is ignored. The resistivity is 0.0225 and 0.036 Ω mm²/m for copper and aluminum, respectively.

To solve the problem, the symmetrical three-phase short circuit produces a current of the same value in each phase. Therefore, a single-line diagram can be used to represent the equivalent network.

1. Calculation of short-circuit current at F_1:

The equivalent circuit includes a voltage source, $22/\sqrt{3}$ kV, and the impedance seen from the fault point Z_{F1}. This impedance is the sum of upstream network impedance ($Z_{upstream\ network}$) and the impedance of OHL (Z_{OHL}) (Fig. 4.10):

$$Z_{\text{upstream network}} = \frac{V_{\text{line}}^2}{S_{\text{sc}}} = \frac{(22\times 10^3)^2}{400\times 10^6} = 1.21\,\Omega.$$

$\frac{R}{X} = 0.3$, then, $R = 0.3 \times X$ and $X = 0.96Z$.

Thus, $X_{\text{upstream network}} = 0.96 \times 1.21 = 1.16\Omega$ and $R_{\text{upstream network}} = 0.3 \times 1.16 = 0.348\,\Omega$.

For OHL, the reactance is $0.3\,\Omega$/km and resistivity of copper conductor ρ is $0.0225\,\Omega\ \text{mm}^2/\text{m}$. Therefore,

$$X_{\text{OHL}} = 0.3\times 2 = 0.6\,\Omega,$$

$$R_{\text{OHL}} = \frac{\rho l}{A} = \frac{0.0225\times 2000}{95} = 0.47\,\Omega.$$

The impedance seen from the fault point $Z_{\text{F1}} = \sqrt{R_{\text{F1}}^2 + X_{\text{F1}}^2}$ where

$$R_{\text{F1}} = R_{\text{upstream network}} + R_{\text{OHL}} = 0.348 + 0.47 = 0.818\,\Omega \text{ and}$$

$$X_{\text{F1}} = X_{\text{upstream network}} + X_{\text{OHL}} = 1.16 + 0.6 = 1.76\,\Omega.$$

Thus, $Z_{\text{F1}} = \sqrt{(0.818)^2 + (1.76)^2} = 1.94\,\Omega$.

The short-circuit current $I_{\text{F1}} = \frac{V_{\text{ph}}}{Z_{\text{F1}}} = \frac{22000}{\sqrt{3}\times 1.94} = 6.555\,\text{kA}$.

Peak value of the short-circuit current I_{peak} can be determined by multiplying the K value (Eq.4.7) by the short-circuit current I_{F1}:

$$R_{\text{F1}}/X_{\text{F1}} = 0.818/1.76 = 0.46,$$

$$K = \sqrt{2}\left[1 + e^{-0.46\pi}\right] = 1.75.$$

Thus, $I_{\text{peak}} = 1.75 \times 6.555 = 11.47\,\text{kA}$.

2. Calculation of short-circuit current at F_2:

The equivalent circuit is shown in Figure 4.11. It includes the same voltage source and the impedance seen from the fault point F_2. This impedance is the

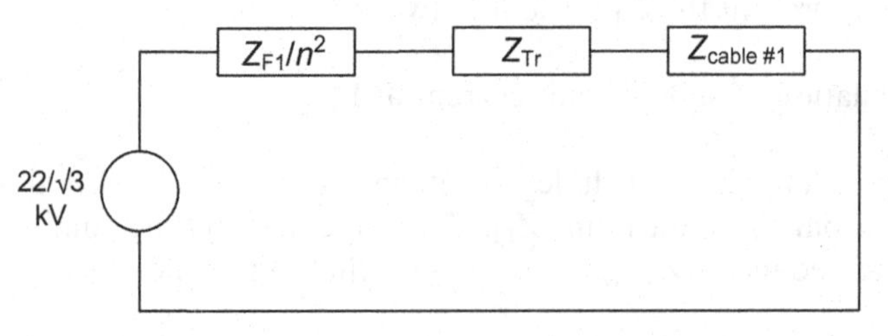

Figure 4.11 Equivalent circuit for fault at F_2.

sum of Z_{F1} referred to the LV side, transformer impedance Z_{Tr}, and the impedance of cable #1:

$$\frac{X_{F1}}{n^2} = \frac{1.76}{55^2} = 5.82 \times 10^{-4}\ \Omega$$

and

$$\frac{R_{F1}}{n^2} = \frac{0.818}{55^2} = 2.7 \times 10^{-4}\ \Omega.$$

Applying Equation 4.10 to calculate transformer impedance,

$$Z_{Tr} = \frac{5}{100}\frac{400 \times 400}{1000 \times 10^3} = 8 \times 10^{-3}\ \Omega.$$

Assuming $(R_{Tr}/X_{Tr}) \approx 0.3$,

$$\begin{aligned} X_{Tr} &= 0.96 Z_{Tr} = 0.96 \times 8 \times 10^{-3} \\ &= 7.7 \times 10^{-3}\ \Omega \\ R_{Tr} &= 0.3 \times X_{Tr} = 2.3 \times 10^{-3}\ \Omega \end{aligned}$$

$$\text{Cable \#1: } X_{\text{cable \#1}} = 0.15 \times \frac{50}{1000} = 7.5 \times 10^{-3}\ \Omega.$$

$$R_{\text{cable \#1}} = \frac{\rho l}{A} = \frac{0.036 \times 50}{400} = 4.5 \times 10^{-3}\ \Omega.$$

Therefore,

$$\begin{aligned} X_{F2} &= (0.582 + 7.7 + 7.5)\ 10^{-3} = 15.782 \times 10^{-3}\ \Omega, \\ R_{F2} &= (0.27 + 2.3 + 4.5)\ 10^{-3} = 7.07 \times 10^{-3}\ \Omega, \\ Z_{F2} &= \sqrt{(15.782)^2 + (7.07)^2} \times 10^{-3} = 17.29 \times 10^{-3}\ \Omega. \end{aligned}$$

Hence,

$$I_{F2} = \frac{400}{\sqrt{3} \times 17.29 \times 10^{-3}} = 13.36\ \text{kA}.$$

To determine the peak value of short-circuit current,

$$\frac{R_{F2}}{X_{F2}} = \frac{7.07}{15.782} = 0.45,\ K = \sqrt{2}\left[1 + e^{-0.45\pi}\right] = 1.76$$

$$I_{\text{peak}} = 13.36 \times 1.76 = 23.51\ \text{kA}.$$

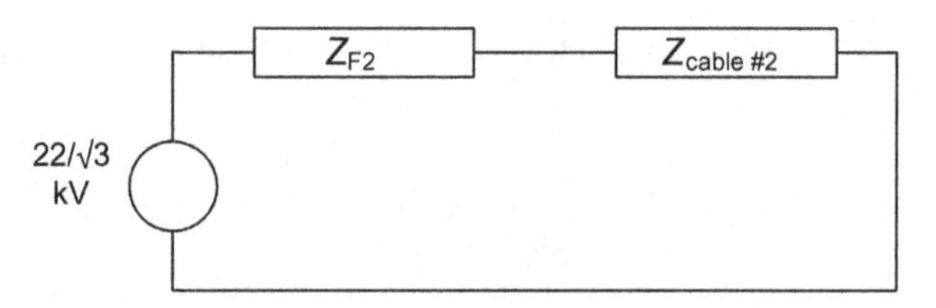

Figure 4.12 Equivalent circuit for fault at F_3.

3. Calculation of short-circuit current at F_3:

The short-circuit current flows into the equivalent impedance Z_{F3}, where $Z_{F3} = Z_{F2} + Z_{cable\#2}$ as shown in Figure 4.12:

$$X_{cable\ \#2} = 0.15 \times 50 \times 10^{-3} = 7.5 \times 10^{-3}\ \Omega,$$

$$R_{cable\ \#2} = \frac{\rho l}{A} = \frac{0.036 \times 50}{300} = 6 \times 10^{-3}\ \Omega.$$

Then,

$$X_{F3} = (15.782 + 7.5)\ 10^{-3} = 23.282 \times 10^{-3}\ \Omega,$$
$$R_{F3} = (7.07 + 6)\ 10^{-3} = 13.07 \times 10^{-3}\ \Omega,$$
$$Z_{F3} = \sqrt{(23.282)^2 + (13.07)^2} \times 10^{-3} = 26.7 \times 10^{-3}\ \Omega.$$

The short-circuit current I_{F3} is given as:

$$I_{F3} = \frac{400}{\sqrt{3} \times 26.7 \times 10^{-3}} = 8.65\ \text{kA}.$$

To determine the peak value,

$$\frac{R_{F3}}{X_{F3}} = \frac{13.07}{23.282} = 0.56,$$
$$K = \sqrt{2}\left[1 + e^{-0.56\pi}\right] = 1.65.$$

Hence,

$$I_{peak} = 1.65 \times 8.65 = 14.27\ \text{kA}.$$

4. Calculation of short-circuit current at F_4:

The impedance of equivalent circuit (Fig. 4.13) is the sum of Z_{F3} and the impedance of motor cable at which the fault occurs. The motor cable is a three-core copper cable. Taking its reactance as $0.08\,\Omega$/km:

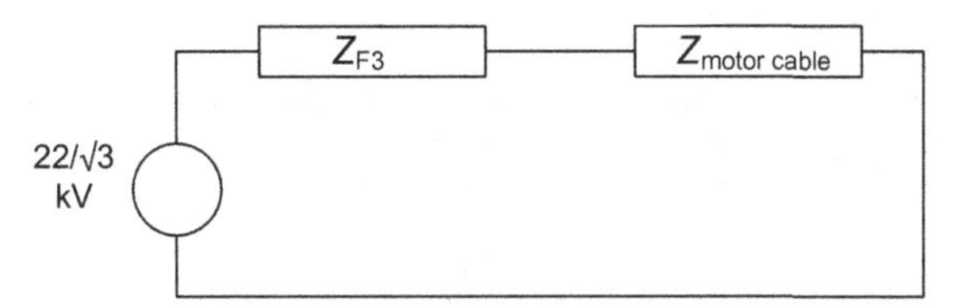

Figure 4.13 Equivalent circuit for fault at F_4.

$$X_{\text{motor cable}} = 0.08 \times 25 \times 10^{-3} = 2 \times 10^{-3}\ \Omega,$$

$$R_{\text{motor cable}} = \frac{\rho l}{A} = \frac{0.0225 \times 25}{35} = 16 \times 10^{-3}\ \Omega,$$

and

$$X_{F4} = (23.282 + 2)\ 10^{-3} = 25.282 \times 10^{-3}\ \Omega,$$

$$R_{F4} = (13.07 + 16)\ 10^{-3} = 29.07 \times 10^{-3}\ \Omega,$$

$$Z_{F4} = \sqrt{(25.282)^2 + (29.07)^2} \times 10^{-3} = 38.53 \times 10^{-3}\ \Omega.$$

Hence,

$$I_{F4} = \frac{400}{\sqrt{3} \times 38.53 \times 10^{-3}} = 5.99\ \text{kA}.$$

To determine the peak value,

$$\frac{R_{F4}}{X_{F4}} = \frac{29.07}{25.282} = 1.15,\ K = \sqrt{2}\left[1 + e^{-1.15\pi}\right] = 1.46,$$

$$I_{\text{peak}} = 1.46 \times 5.99 = 8.75\ \text{kA}.$$

It is found that the coefficient K, which characterizes I_{peak} to rms steady-state current, for an identical voltage level gradually decreases as the fault point moves away from the source, that is, $K_{F4} < K_{F3} < K_{F2}$. Thus, the aperiodic component is greater when the fault is near the transformer.

On the other hand, the contribution of motors to short-circuit current should be calculated. The installation has four motors; each one is an independent source when the fault occurs. So, the sum of their contributing current gives the total current supplied by all motors to the short circuit. Starting with the first fault location with respect to the source (F_4), the calculation of impedance seen from the fault is easier as indicated below. For the sake of simplicity, the motor resistance is ignored.

1. Short circuit at F_4:
   - For the motor on whose terminals the fault occurs, M_1, its impedance is calculated by applying Equation 4.28:

$$X_{M1} = X'_M = \frac{(400)^2}{100 \times 10^3} \frac{0.8}{6} = 0.213\,\Omega,$$

$$R_{M1} \cong 0.$$

Thus,

$$I_{M1} = \frac{400}{\sqrt{3} \times 0.213} = 1.084\,\text{kA}.$$

   - For each of the other three motors (M_2, M_3, M_4)

     The impedance between motor and fault location, $Z_{M,F4}$, is the sum of motor impedance and two times of motor cable, that is,

$$X_{M,F4} = X'_M + 2\,X_{\text{motor cable}} = 0.213 + 2 \times 2 \times 10^{-3}$$
$$= 217 \times 10^{-3}\,\Omega$$

$$R_{M,F4} = 2\,R_{\text{motor cable}} = 2 \times 16 \times 10^{-3} = 32 \times 10^{-3}\,\Omega$$

$$Z_{M,F4} = \sqrt{(217)^2 + (32)^2} \times 10^{-3} = 219.35 \times 10^{-3}\,\Omega.$$

   Hence,

$$I_{Mi} = \frac{400}{\sqrt{3} \times 0.219} = 1.054\,\text{kA}, i = 2, 3, 4.$$

   The current contribution of M_2, M_3, and $M_4 = 3 \times 1.054 = 3.162\,\text{kA}$. Therefore, the total current contributed by the four motors, $I_{\Sigma M}$, is given as

$$I_{\Sigma M,F3} = 1.084 + 3.162 = 4.246\,\text{kA}.$$

2. Short-circuit at F_3:
   For each motor,

$$X_{Mi,F3} = X'_M + X_{\text{motor cable}}, i = 1, 2, 3, 4$$

$$X_{Mi,F3} = 0.213 + 0.002 = 0.215\,\Omega.$$

   Similarly,

$$R_{Mi,F3} = 0 + 0.016 = 0.016\,\Omega.$$

Then,

$$Z_{\text{Mi,F3}} = \sqrt{(215)^2 + (16)^2} \times 10^{-3} = 215.6 \times 10^{-3}\ \Omega$$

$$I_{\text{Mi,F3}} = \frac{400}{\sqrt{3} * 0.2156} = 1.071\ \text{kA}.$$

The total return current of our motors, $I_{\Sigma\text{M,F3}} = 4 \times 1.071 = 4.284\ \text{kA}$.

3. Short circuit at F_2:

$$X_{\text{Mi,F2}} = X_{\text{Mi,F3}} + X_{\text{cable \#2}},\ i = 1, 2, 3, 4$$
$$= 0.215 + 0.0075 = 0.2225\ \Omega,$$
$$R_{\text{Mi,F2}} = R_{\text{Mi,F3}} + R_{\text{cable \#2}}$$
$$= 0.016 + 0.006 = 0.022\ \Omega,$$
$$Z_{\text{Mi,F2}} = \sqrt{(222.5)^2 + (22)^2} \times 10^{-3} = 223.6 \times 10^{-3}\ \Omega,$$
$$I_{\text{Mi,F2}} = \frac{400}{\sqrt{3} \times 0.2236} = 1.033\ \text{kA}.$$
$$I_{\Sigma\text{M,F2}} = 4 \times 1.033 = 4.132\ \text{kA}.$$

4. Short circuit at F_1:
The impedance is referred to MV side:

$$X_{\text{Mi,F1}} = n^2\ (X_{\text{Mi,F2}} + X_{\text{cable \#1}} + X_{\text{Tr}})$$
$$= (55)^2\ (222.5 + 7.5 + 8.0)\ 10^{-3} = 719.95\ \Omega,$$
$$R_{\text{Mi,F1}} = n^2\ (22 + 4.5 + 2.4)\ 10^{-3} = 87.42\ \Omega,$$
$$Z_{\text{Mi,F1}} = \sqrt{(719.95)^2 + (87.42)^2} = 725.29\ \Omega,$$
$$I_{\text{Mi,F1}} = \frac{22{,}000}{\sqrt{3} \times 725.29} = 17.5\ \text{A},$$
$$I_{\Sigma\text{M,F1}} = 4 \times 17.5 = 70\ \text{A}.$$

The results are summarized in Table 4.2. The ratio of current contribution by motors $I_{\Sigma\text{M,Fi}}$ to short-circuit current I_{Fi} (ratio of motor contribution [RMC]) and the total fault current are given in this table. The total fault current, I_{TFi}, is defined as the sum of the short-circuit current and the current contributed by the motors, that is, $I_{\text{TFi}} = I_{\text{Fi}} + I_{\Sigma\text{M,Fi}}$.

4.2.2.2 Unsymmetrical Short Circuits A majority of the faults in power systems are unsymmetrical faults. These faults occur between one line and earth (L-E) or between two lines (L-L) or two lines and earth (2L-E) (Fig. 4.14). Thus, the voltages or currents of three phases are different in magnitude

TABLE 4.2 Summary of Results

	MV	LV		
Fault Location	F_1	F_2	F_3	F_4
Z_{Fi} (Ω)	1.94	17.29×10^{-3}	26.7×10^{-3}	38.53×10^{-3}
I_{Fi} (kA)	6.555	13.36	8.65	5.99
I_{peak} (kA)	11.47	23.51	14.27	8.75
$Z_{M,Fi}$ (kA)	725.29	223.6×10^{-3}	215.6×10^{-3}	$M_{1,F}$: 0.213 $M_{2/3/4,F}$: 0.219
$I_{\Sigma M,Fi}$ (kA)	0.07	4.132	4.284	4.246
RMC (%)	1.06	30.9	49.5	70.9
I_{TFi} (kA)	6.625	17.492	12.934	10.236

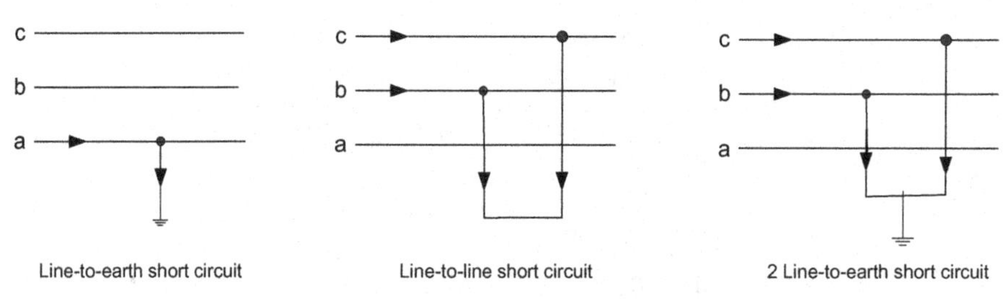

Figure 4.14 Three types of unsymmetrical faults.

and/or not displaced by equal angle (120°) providing an unbalanced three-phase system.

The unbalanced system cannot be solved by using an equivalent single-phase impedance network as in the case of the balanced system provided by symmetrical three-phase short circuit. It can be solved by using the theory of symmetrical components [59]. The method of symmetrical components breaks down an unbalanced system into the sum of three balanced systems: positive-sequence, negative-sequence, and zero-sequence systems (Fig. 4.15). This yields three single-phase networks. Only the positive-sequence network contains a driving voltage (V_{ph}) and each network has its own impedance: positive-sequence impedance $\mathbf{Z}_1$, negative-sequence impedance $\mathbf{Z}_2$, and zero-sequence impedance $\mathbf{Z}_o$ (Fig. 4.16).

By the superposition principle, the theory of symmetrical components states that any set of unbalanced three-phase system of sequence *abc*, for example, voltages V_a, V_b, V_c, can be replaced by a sum of three balanced sets of voltages:

i) a positive-sequence set of three-phase voltages having the same sequence as the original set, *abc*, and denoted by V_{a1}, V_{b1}, and V_{c1}.
ii) a negative-sequence set of three-phase voltages having a sequence opposite to that of the original set, *acb*, and denoted by V_{a2}, V_{b2}, and V_{c2}.

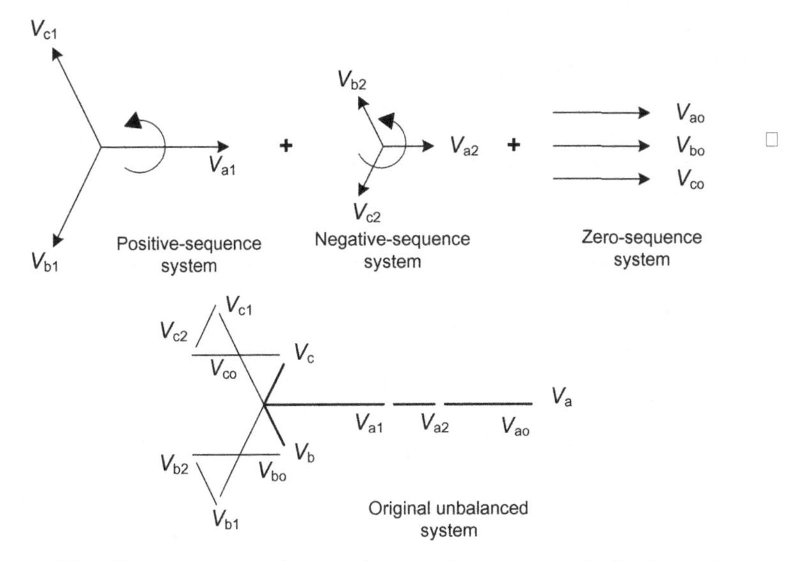

Figure 4.15 Original three-phase unbalanced system equivalents to three sets of three-phase balanced systems.

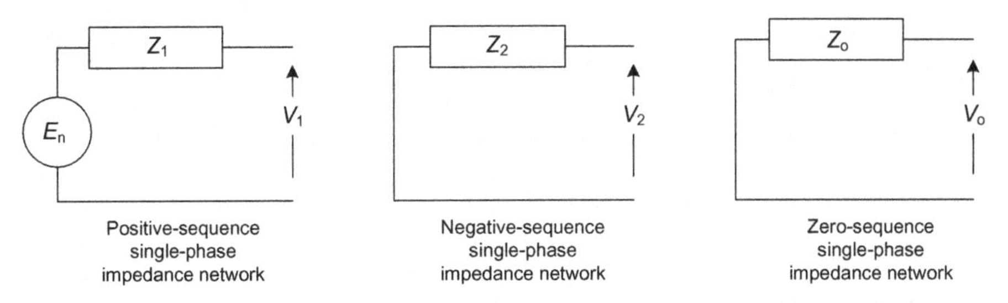

Figure 4.16 Equivalent single-phase impedance networks.

iii) a zero-sequence set of three voltages equal in magnitude and in phase, denoted by V_{ao}, V_{bo}, and V_{co}, each one of them equal to V_o.

An operator "a" is defined as an operator that rotates a vector by 120° in the positive anticlockwise direction without any change in magnitude. Thus,

$$a = -\frac{1}{2} + j\frac{\sqrt{3}}{2},\ a^2 = -\frac{1}{2} - j\frac{\sqrt{3}}{2},\ a^3 = 1$$

and

$$1 + a + a^2 = 0.$$

From the principles of symmetrical components as indicated in Figure 4.15, it is seen that

$$\begin{aligned} V_a &= V_{a1} + V_{a2} + V_{ao}, \\ V_b &= V_{b1} + V_{b2} + V_{bo}, \\ V_c &= V_{c1} + V_{c2} + V_{co}. \end{aligned} \tag{4.31}$$

By applying the operator "a" to Equation 4.31 and taking phase "a" as a reference, they can be rewritten as

$$\begin{aligned} V_a &= V_{a1} + V_{a2} + V_o, \\ V_b &= a^2 V_{a1} + a V_{a2} + V_o, \\ V_c &= a V_{a1} + a^2 V_{a2} + V_o. \end{aligned}$$

Therefore, the following matrix relationship can be established:

$$\begin{bmatrix} V_a \\ V_b \\ V_c \end{bmatrix} = \begin{bmatrix} 1 & 1 & 1 \\ a^2 & a & 1 \\ a & a^2 & 1 \end{bmatrix} \begin{bmatrix} V_{a1} \\ V_{a2} \\ V_o \end{bmatrix}. \tag{4.32}$$

This system of equations can be solved for the three components, V_{a1}, V_{a2}, and V_o, where the determinant

$$\begin{vmatrix} 1 & 1 & 1 \\ a^2 & a & 1 \\ a & a^2 & 1 \end{vmatrix} \neq 0.$$

Hence,

$$V_{a1} = \frac{1}{3}(V_a + aV_b + a^2 V_c),$$

$$V_{a2} = \frac{1}{3}(V_a + a^2 V_b + aV_c),$$

$$V_o = \frac{1}{3}(V_a + V_b + V_c),$$

that is,

$$\begin{bmatrix} V_{a1} \\ V_{a2} \\ V_o \end{bmatrix} = \frac{1}{3} \begin{bmatrix} 1 & a & a^2 \\ 1 & a^2 & a \\ 1 & 1 & 1 \end{bmatrix} \begin{bmatrix} V_a \\ V_b \\ V_c \end{bmatrix}. \tag{4.33}$$

4.2.2.3 Sequence-Impedance Networks As shown in Figure 4.16, the three-phase system is represented by three equivalent single-phase sequence impedance networks: Z_1, Z_2, and Z_o. The combination of these three networks depends on the type of asymmetrical short circuit. Each sequence impedance is obtained by calculating the equivalent impedance of interconnected elements from the upstream utility network to the fault point, which is the system impedance as viewed from the short-circuit location. All synchronous and asynchronous machines are replaced by their internal impedances. The system zero-sequence impedance includes three times the neutral-to-earth impedance Z_N.

The sequence impedances of electric elements can be calculated as below and can also be easily measured [57].

Positive-sequence short-circuit impedance Z_1 of electric element is the ratio of line-to-neutral voltage to the short-circuit current of the corresponding line conductor of electric element when fed by a symmetrical positive-sequence system of voltages (Fig. 4.17a).

Negative-sequence short-circuit impedance Z_2 of electric element is the ratio of the line-to-neutral voltage to the short-circuit current of the corresponding line conductor of electric element when fed by a symmetrical negative-sequence system of voltages (Fig. 4.17b).

Zero-sequence short-circuit impedance Z_o of electric element is the ratio of the line-to-earth voltage to the short-circuit current of one line conductor of electric element when fed by an AC voltage source, if the three paralleled line conductors are used for the outgoing current and the fourth line and/or earth as a joint return (Fig. 4.17c).

Values of Sequence Impedances of Static Electric Elements [54] The positive sequence of static elements, such as transformers, OHLs, and cables, is the same as impedance Z_{sc} of the element when a symmetrical three-phase short-circuit occurs. Thus,

$$Z_1 = Z_{sc}.$$

Because of the symmetrical nature of these passive elements, the negative-sequence impedance is equal to their positive-sequence impedance, that is,

$$Z_2 = Z_1 = Z_{sc}.$$

The zero-sequence impedance differs from one element to another and is given as below.

When calculating unbalanced short-circuit currents in MV systems, the zero-sequence capacitances of lines (OHLs or cables) and the zero-sequence shunt admittances are to be considered. The capacitances of LV lines may be neglected in the positive-, negative-, and zero-sequence systems. In the case of earthed neutral system and depending on network configuration, the results

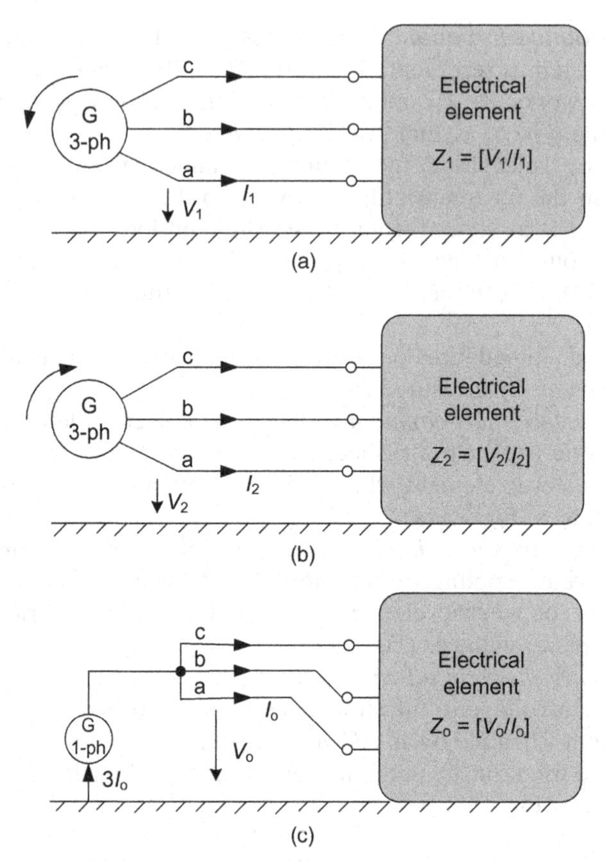

Figure 4.17 Sequence short-circuit impedances of electric element. (a) Positive-sequence short-circuit impedance (Z_1); (b) negative-sequence short-circuit impedance (Z_2); (c) zero-sequence short-circuit impedance (Z_o).

may be slightly higher than the real values of the short-circuit currents when line capacitance is neglected.

Transformers The zero-sequence impedance depends on transformer connection and whether the neutral point is earthed. Its value as seen from secondary for different transformer connections is given in Table 4.3.

OHLs The zero-sequence impedances of OHLs are referred to an earth resistivity of ρ is 100 Ωm and therefore to an equivalent depth of earth return d_e is 930 m (50 Hz) and d_e is 850 m (60 Hz) [60].

The zero-sequence impedance, Z_o, of a conductor or group of conductors with earth return of equivalent depth d_e can be calculated by

TABLE 4.3 Zero-Sequence Impedance of Transformers Seen from the Secondary [54]

Transformer Connection	Zero-Sequence Impedance Seen from the Secondary	Equivalent Diagram
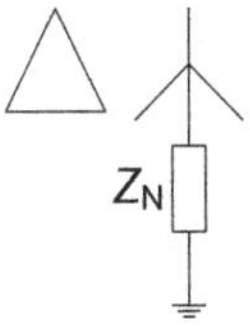 Z_N	$Z_o = Z_1 + 3Z_N$, where $Z_N = 0$ for directly earthed neutral	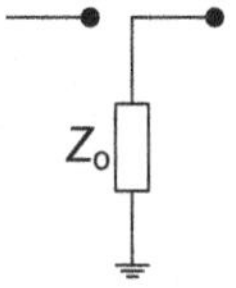 Z_o
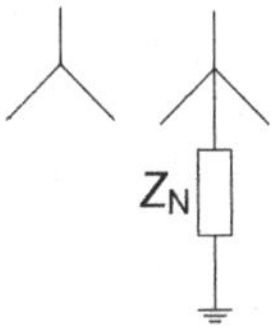 Z_N	$Z_o = 10Z_1 + 3Z_N$ for a driven flux transformer that is generally the case. $Z_N = 0$ for directly earthed neutral.	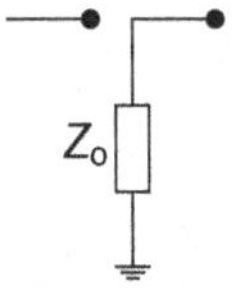 Z_o
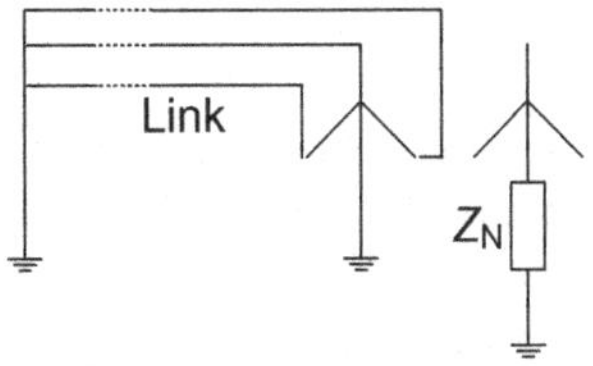 Link Z_N	$Z_o = Z_1 + 3Z_N$ in the case where the network upstream of the transformer has an earthed neutral allowing the primary earth current to be reclosed. It is noted that where the primary and secondary neutrals are earthed, there is no galvanic insulation between the two windings. The overvoltages thus spread through the transformer.	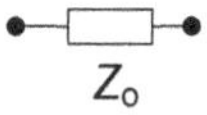 Z_o
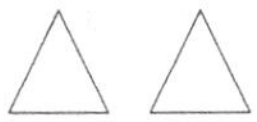	$Z_o = \infty$	
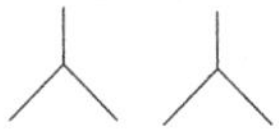	$Z_o = \infty$	
	$Z_o = \infty$	

$$Z_{oc} = r_c + 0.00296f + j0.00868f \log \frac{d_e}{\text{GMR}} \ \Omega/\text{km} \tag{4.34}$$

where r_c is the conductor resistance per kilometer and $d_e = 2160\sqrt{\rho/f}$.

The mutual coupling in zero sequence is of importance when parallel circuits share the same right of way and geometric arrangement is such that current flow in one circuit causes a voltage drop in the other. A typical example is exposed OHLs sharing the same support structure. Neglecting the mutual coupling causes an error in the calculation of earth-fault currents. Although relatively infrequent for industrial power system analysis, it should be borne in mind and treated accordingly [59]. The zero-sequence mutual impedance, Z_{om}, can be calculated by

$$Z_{om} = r_c + 0.00296f + j0.00868f \log \frac{d_e}{\text{GMR}} \ \Omega/\text{km}. \tag{4.35}$$

The zero-sequence reactance is generally equal to three times the positive-sequence reactance (Section 4.2.2.1), that is, $X_0 \approx 3X_1$ and the zero-sequence capacitance is taken approximately as 5 nF/km.

Cables The flow of zero-sequence currents in three-phase conductors depends on the return path. If the cable is solidly bonded and unearthed, or if the resistance of the earth is relatively high, no appreciable current flows in the earth and all return current is in the sheath. In the case where the sheath is broken by insulating sleeves, or highly resistive sheath, all return current is in the earth and none in the sheath. Otherwise, the return current is in sheath and earth in parallel.

Equation 4.34 is also applicable to cable circuits to calculate the zero-sequence impedance, Z_{oc}, of a group of three parallel conductors with earth return and no sheath return.

The impedance of the sheath with earth return and no conductor groups is

$$Z_{osh} = N_{sh} r_{sh} + 0.00296f + j0.00868f \log \frac{d_e}{\text{GMR}_{sh}} \ \Omega/\text{phase/km}. \tag{4.36}$$

The mutual impedance between conductors and sheath with common earth return is

$$Z_{om} = 0.00296f + j0.00868f \log \frac{d_e}{\text{GMD}_{\text{sheath to cond.}}} \ \Omega/\text{phase/km}, \tag{4.37}$$

where GMR_c is the geometric mean radius of the three conductors as a group (inches), GMR_{sh} is the geometric mean radius of sheath return, and $\text{GMD}_{\text{sheath to cond.}}$ is the geometric mean distance between sheaths and conductors.

The three impedances calculated by Equations 4.34, 4.36, and 4.37 can form the equivalent zero-sequence impedance as below.

- Return current in the sheath only:

$$Z_o = Z_{oc} + Z_{osh} - 2Z_{om} \ \Omega/\text{phase/km}, \tag{4.38}$$

- Return current in the earth only:

$$Z_o = Z_{oc} \ \Omega/\text{phase/km}, \tag{4.39}$$

- Return current in sheath and earth in parallel:

$$Z_o = Z_{oc} - \frac{Z_{om}^2}{Z_{osh}} \ \Omega/\text{phase/km}. \tag{4.40}$$

These formulas are applicable to either three-phase cable or three single-phase cables (Fig. 4.18).

Typical data of sequence impedances for cables, lines, and transformers are given in Reference 61.

Rotating Machines The sequence impedances are considered as pure reactance where $R << X$ for either generators or motors.

Synchronous Generators The generator has a time-varying reactance to positive-sequence current (X_d'', X_d', X_d). The subtransient reactance is used in the positive-sequence network ($Z_1 = X_d''$) when the maximum mechanical forces during a fault are required. The transient reactance is commonly used during the period the protective system is operating and a circuit breaker is opening ($Z_1 = X_d'$).

The negative-sequence impedance is the actual reactance of the inductive circuit when the magnetic field produced by a negative-sequence three-phase

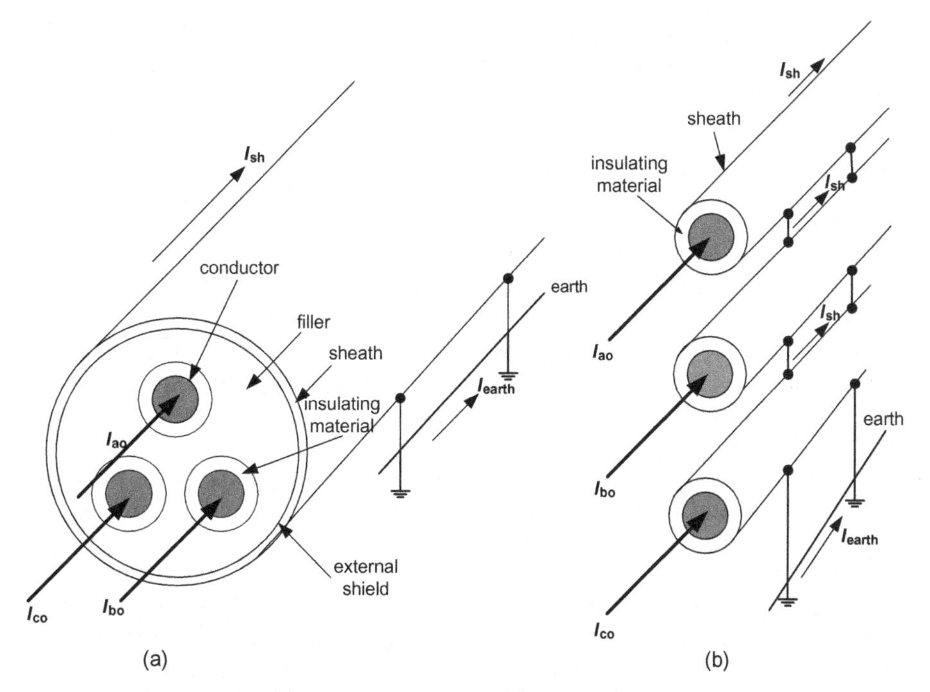

Figure 4.18 (a) Three-phase cable. (b) Three single-phase cables.

system of currents rotates in the opposite direction to that of the machine. Thus, $Z_2 \approx X''_d$, where its difference from positive-sequence impedance may be small during the first few cycles after a fault.

The zero-sequence impedance is obtained when a zero-sequence three-phase voltage system is applied to the stator. Consequently, there is no rotating field and the flux induced on the rotor is zero. This means that the impedance only depends on stator winding and therefore its value is low. When the generator neutral is earthed with neutral earthing impedance Z_N, its zero-sequence impedance $Z_o = X_o + 3Z_N$ if a phase-to-earth fault occurs. Then, $Z_o \cong 3Z_N$ since $X_o << 3Z_N$. It is obvious to see that $Z_o = \infty$ when generator neutral is unearthed.

Asynchronous Motors The positive-sequence impedance is equal to the short-circuit impedance X'_M (transient impedance), which is calculated by Equation 4.28. Thus,

$$Z_1 = X'_M.$$

The negative-sequence impedance is assumed to be slightly different from the positive-sequence impedance and it is taken as $X_{(1)}$, that is,

$$Z_2 = Z_1 = X'_M.$$

The zero-sequence impedance depends only on the stator winding and is low. Thus, it is ignored because of the winding connection being delta or unearthed wye.

Example:

A distribution system has one transformer (Δ/Y) and a generator. The neutral point of the secondary winding of the transformer is earthed through impedance Z_N and generator neutral point is not earthed. The transformer and generator are connected to a DP that feeds three loads through a ring system. Asymmetrical short-circuit fault occurs at point F as shown in Figure 4.19.

Impedances of transformer, generator, bus bar of DP and loads are denoted by Z_T, Z_G, Z_{BB}, and Z_L, respectively. All of them are subscripted by corresponding sequence notation.

The layout of both positive- and negative-sequence impedance networks is identical. They only differ in values of generator reactance and the absence of negative-sequence EMF. The zero-sequence impedance network does not include generator zero-sequence impedance since its neutral point is isolated from earth and there is no path to earth for zero-sequence current. Accordingly, the positive-, negative-, and zero-sequence impedance networks can be drawn as shown in Figure 4.20. The system positive-, negative-, and zero-sequence equivalent impedances (Z_1, Z_2, and Z_o, respectively) can thus be found by reducing each sequence network into single impedance by combining

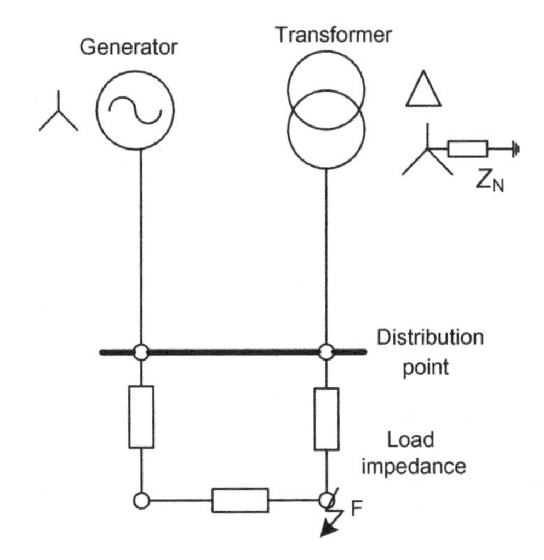

Figure 4.19 Layout of a distribution system.

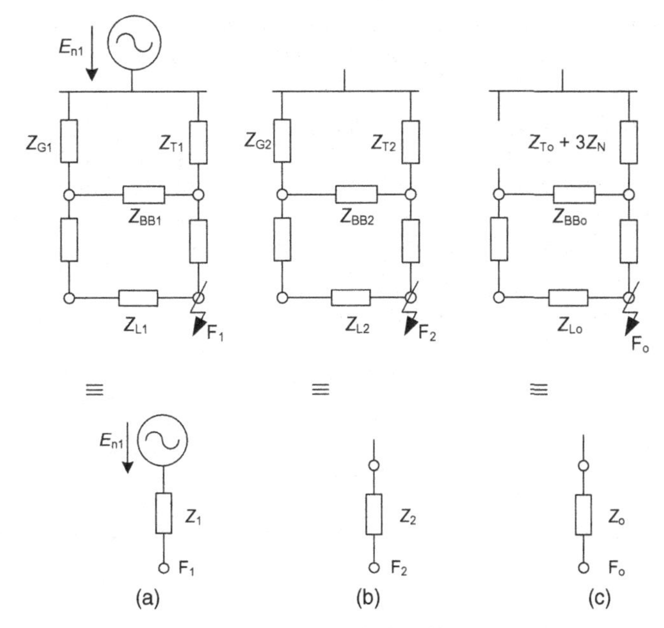

Figure 4.20 (a)Positive-, (b) negative-, and (c) zero-sequence networks.

elements in series or parallel and using delta-star transformations [62]. The fault current is then calculated by defining the interconnection of sequence networks needed to satisfy the conditions of asymmetrical fault that occurred in the original network.

This type of fault is the most common one that may occur in distribution systems. The circuit diagram that represents a system with an earth fault on phase "a" at point F is shown in Figure 4.21. The system impedances between

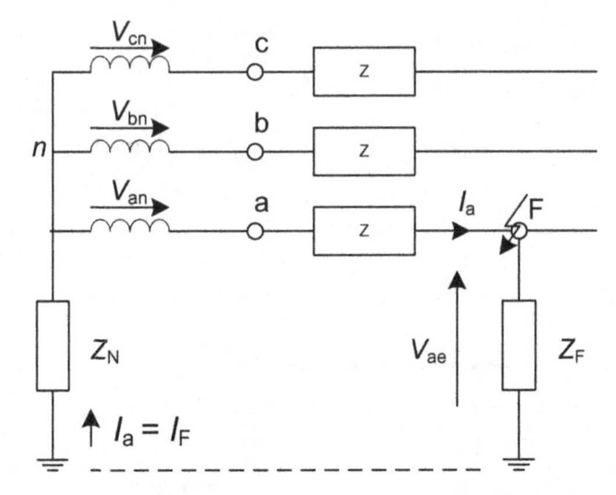

Figure 4.21 Circuit diagram of a system with L-E fault.

upstream network Z and the three-phase source voltages (V_{na}, V_{nb}, V_{nc}) are assumed to be balanced, so that there is no mutual coupling between the sequence networks. As a general case, the star point of the source is earthed via impedance Z_N and the fault path has impedance Z_F.

4.2.2.4 Line-to-Earth Fault (L-E Fault) The first step to calculate the fault current I_F is a deduction of three equations for phases a, b, and c in terms of phase currents and voltages to define the fault conditions. The next step is to apply the symmetrical components method to provide the relation by which the fault current is calculated and to determine the interconnection of three sequence networks necessary to satisfy this relation.

For this particular fault, it is found that

$$V_{ae} = Z_F I_a, \tag{4.41}$$

$$I_b = 0, \tag{4.42}$$

$$I_c = 0. \tag{4.43}$$

Applying symmetrical components, Equations 4.41–4.43 can be rewritten as

$$V_{a1} + V_{a2} + V_{ao} = Z_F(I_{a1} + I_{a2} + I_{ao}), \tag{4.44}$$

$$a^2 I_{a1} + aI_{a2} + I_{ao} = 0, \tag{4.45}$$

$$aI_{a1} + a^2 I_{a2} + I_{ao} = 0. \tag{4.46}$$

Subtracting Equation 4.46 from Equation 4.45,

$$(a^2 - a)I_{a1} + (a - a^2)I_{a2} = 0,$$

$$I_{a1} = I_{a2}.$$

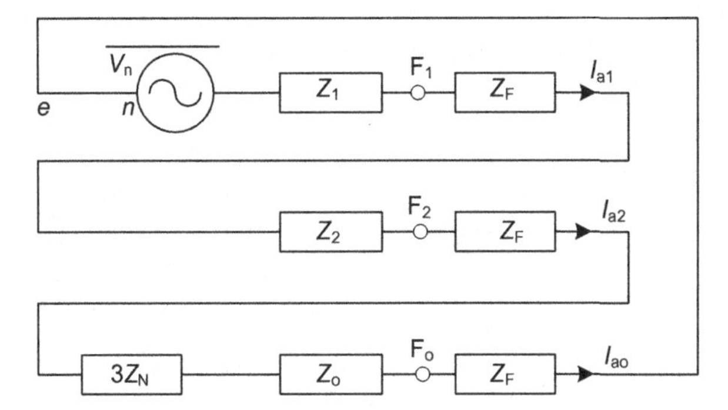

Figure 4.22 Interconnection of sequence networks for L-E fault.

Substituting in Equation 4.45,

$$(a^2 - a)I_{a1} + I_{ao} = 0,$$
$$I_{a1} = I_{ao}.$$

Thus,

$$I_{a1} = I_{a2} = I_{ao} = I_a/3 \tag{4.47}$$

Equation 4.44 can be replaced by

$$[V_{an} - I_{a1}Z_1] + [-I_{a2}Z_2] + [-I_{ao}(Z_o + 3Z_N)] = 3Z_F I_{a1},$$
$$V_{an} = I_{a1}(Z_1 + Z_2 + Z_o + 3Z_N + 3Z_F),$$

and

$$I_F = I_a = 3I_{a1} = \frac{3V_n}{Z_1 + Z_2 + Z_o + 3Z_N + 3Z_F}, \tag{4.48}$$

where $V_{an} = V_{bn} = V_{cn} = V_n$ (balanced system).

The interconnection of three sequence networks that satisfies Equations 4.47 and 4.48 is shown in Figure 4.22.

It is seen that the short-circuit fault current depends on the type of earthing system and the neutral earthing impedance Z_N. The value of Z_N is zero when the neutral is solidly grounded, infinite for isolated neutral, and a specific value for limiting the current. There is another factor that may affect the fault current value. It is the network capacitances between healthy phases and earth (Fig. 4.23a). The fault current is a sum of current passing through the neutral I_N and that passing through the phase-earth capacitance of healthy phases I_C, that is,

$$I_F = I_N + I_C.$$

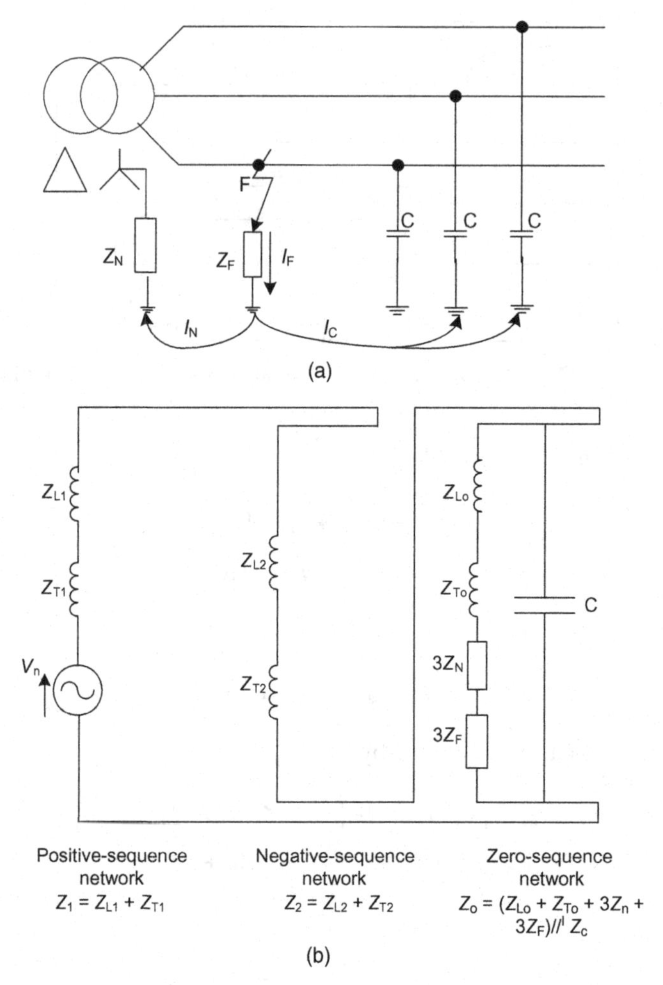

Figure 4.23 (a) L-E fault with network capacitance between phases and earth. (b) Sequence networks.

The sequence networks as shown in Figure 4.23b give the impedances

$$
\begin{aligned}
Z_1 &= Z_{L1} + Z_{T1}, \\
Z_2 &= Z_{L2} + Z_{T2}, \\
Z_o &= Z_{\text{neto}} //^{l} Z_C \\
&= \frac{Z_{\text{net(o)}} \frac{1}{j\omega C}}{Z_{\text{net(o)}} + \frac{1}{j\omega C}} = \frac{1}{\frac{1}{Z_{\text{net(o)}}} + j\omega C},
\end{aligned}
\tag{4.49}
$$

where

Z_{Li} = line impedance of sequence i,
Z_{Ti} = transformer impedance of sequence i,
Z_C = healthy phase zero-sequence capacitive reactance,
Z_{neto} = sum of zero-sequence impedance of network elements
= $Z_{Lo} + Z_{To} + 3Z_N + 3Z_F$, and
C = healthy phase zero-sequence capacitance.

Thus,

$$I_F = \frac{3V_n}{Z_1 + Z_2 + Z_o} = \frac{3V_n}{2Z_1 + Z_o} \ (Z_1 = Z_2 \text{ for static elements}). \qquad (4.50)$$

When using Z_N, usually its value is much higher than Z_1. Therefore, $2Z_1$ can be neglected compared with Z_o. The fault current becomes

$$I_F = \frac{3V_n}{Z_o}.$$

From the relations above, the effect of earthing a system can be illustrated as below.

1. For isolated neutral or using high impedance Z_N, Equation 4.49 yields

$$Z_o = 1/j\omega C.$$

Then,

$$I_F = 3j\omega C V_n.$$

The value of C is small for LV and MV distribution networks. So, I_F is negligible where its value is low and not dangerous for cables and transformers.
2. Using $Z_N >> Z_{neto}$ and Z_F is usually small. In this case, the system zero-sequence impedance and fault current are given by

$$Z_{(o)} = \frac{1}{\frac{1}{3Z_N} + j\omega C},$$

$$I_F = 3V_n\left(\frac{1}{3Z_N} + j\omega C\right) = \frac{V_n}{Z_N} + 3j\omega C V_n$$
$$= I_N + I_C, \qquad (4.51)$$

where

$$I_N = \frac{V_n}{Z_N} \text{ and } I_C = 3j\omega CV_n.$$

Therefore, Z_N is used as line-to-ground limiting impedance. It is noted that for HV networks, the capacitive fault current I_C is approaching the neutral current I_N. This makes the protection system more complex.

If Z_N is a pure reactance of inductance L_N, the fault current value is

$$I_F = V_n\left[\frac{1}{j\omega L_N} + 3j\omega C\right] = V_n\left[3j\omega C - j\frac{1}{\omega L_N}\right]$$

if $3\omega C = \frac{1}{\omega L_N}$, that is, $\omega = \frac{1}{\sqrt{3L_N C}}$.

Then, the fault current cannot be seen because its value is zero. This is the reason to use variable reactance that is easy to tune (Paterson coil).

3. For solidly grounded neutral, $Z_N = 0$ and Z_F is not negligible. Thus,
4.

$$I_F = \frac{3V_n}{2Z_1 + Z_{neto}} = \frac{3V_n}{2Z_1 + Z_{Lo} + Z_{To} + 3Z_F},$$

where $Z_{neto} = Z_{Lo} + Z_{To} + 3Z_F$ and the capacitive fault current is neglected.

If $Z_{Lo} + Z_{To}$ is close to Z_1, which is the case of Δ-Y transformer and lines, then

$$I_F = \frac{3V_n}{3Z_1 + 3Z_F} = \frac{V_n}{Z + Z_F}. \tag{4.52}$$

It is seen from Equation 4.52 that the fault current is very high and close to the three-phase short-circuit value.

As a practical application to indicate the importance of the capacitive current during short-circuit occurrence, assume a transformer connected to a DP, which feeds several outgoing feeders (Fig. 4.24). A residual current measuring device is installed on each outgoing feeder. The value of current seen by measuring devices at different locations during fault must be calculated as accurately as possible to be able to choose and set the adequate protective devices.

Assume an L-E fault is located on one of the outgoing feeders (say, the first feeder). The measuring devices measure the residual current, which is the

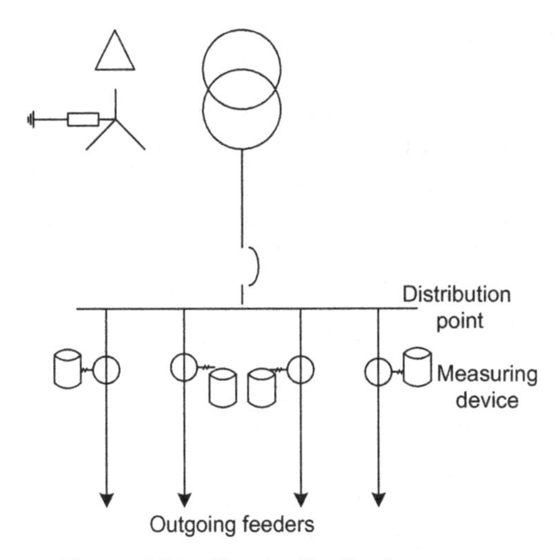

Figure 4.24 Simple distribution system.

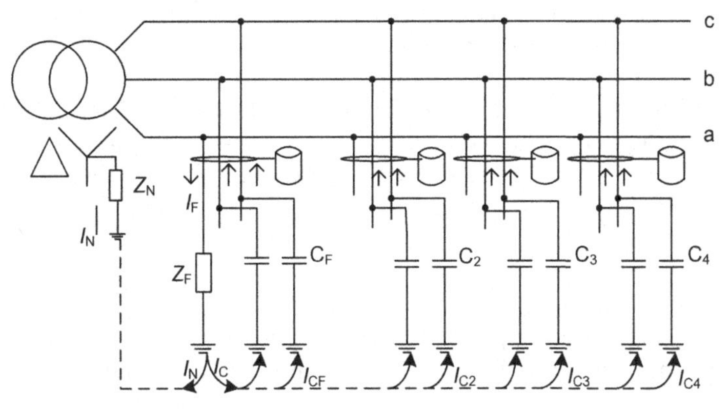

Figure 4.25 Three-phase diagram of a distribution system with four outgoing feeders (one faulty and three healthy).

current flowing to earth through phase capacitance. The corresponding three-phase diagram is shown in Figure 4.25.

The capacitances of phase "a" of all feeders have no effect because they are short circuited. As it is seen in Figure 4.25, the fault current is divided into two paths; the first flows through Z_N and the second flows through the capacitance of both the faulty feeder and the healthy feeders. Apply Equation 4.51 by substituting C as the equivalent network capacitance per phase. Then,

$$I_F = \frac{V_n}{Z_N} + 3j\omega C V_n,$$

where $C = C_F + C_2 + C_3 + C_4$ and C_F is the capacitance of faulty feeder per phase.

The current detected by measuring devices of faulty and healthy feeders can be calculated. For faulty feeder, the measuring device detects the vectorial sum of currents in phases a, b, and c. The current in phase "a" is the fault current and the currents in phases b and c are I_{cFb} and I_{cFc}, respectively. These two currents are in opposite direction to the current in phase a. Consequently, the current measured by the device I_{dev} is equal to the fault current reduced by the capacitive current produced by the capacitance per phase C_F:

$$I_{dev} = \frac{V_n}{Z_N} + 3j\omega(C - C_F)V_n.$$

For healthy feeders, the current detected by measuring device installed on each feeder is given as

$$I_{dev} = 3j\omega C_i V_n, i = 2, 3, 4.$$

Therefore, it is essential to set the protection threshold above the detected capacitive current to avoid the tripping of healthy feeders.

4.2.2.5 Line-to-Line Fault (L-L Fault) As shown in Figure 4.26, there is no path for the zero-sequence current $I_{a(o)}$ to pass through and no voltage across Z_N, that is, $I_{ao} = 0$. The three equations that represent the conditions of this type of fault are

$$I_a = 0, I_b = -I_c, V_{be} - V_{ce} = Z_F I_b,$$

where V_{be} and V_{ce} are the phase voltages of phases a and b, respectively.

By applying symmetrical components method and using the relations above, it is found that

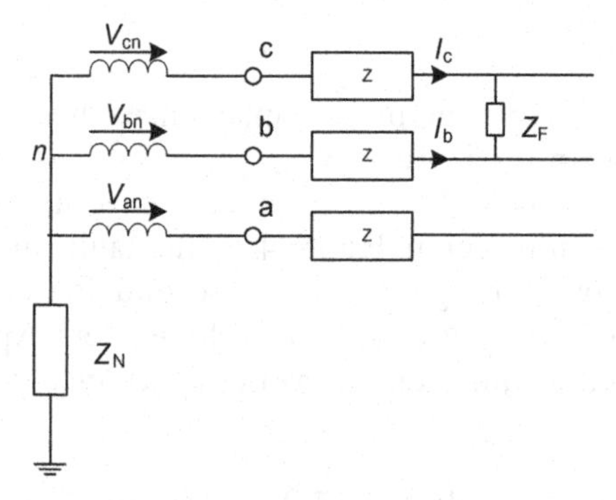

Figure 4.26 Circuit diagram of a system with L-L fault.

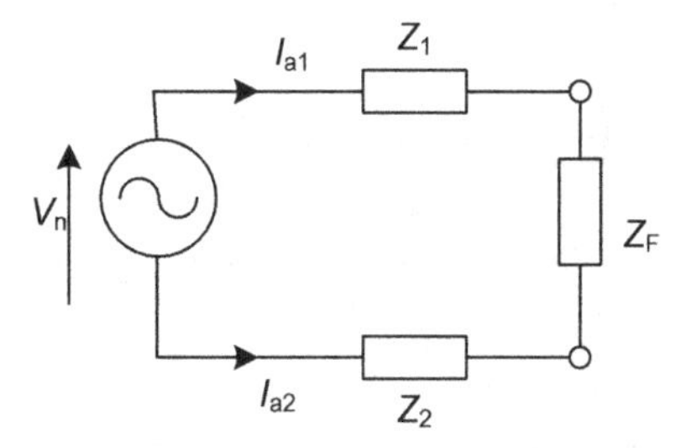

Figure 4.27 Interconnection of sequence networks for L-L fault.

$$I_{a1} + I_{a2} + I_{ao} = 0 \text{ and } I_{a1} = -I_{a2} \tag{4.53}$$

$$\begin{aligned} V_{be} - V_{be} &= Z_F I_b \\ V_{be} &= a^2 V_{a1} + a V_{a2} + V_{ao} \\ V_{ce} &= a V_{a1} + a^2 V_{a2} + V_{ao} \\ V_{be} - V_{be} &= (a^2 - a)V_{a1} + (a - a^2)V_{a2} \\ &= Z_F(a^2 I_{a1} + a I_{a2}) \\ (a^2 - a)V_{a1} + (a - a^2)V_{a2} &= Z_F I_{a1}(a^2 - a) \\ V_{a1} - V_{a2} &= Z_F I_{a1}. \end{aligned} \tag{4.54}$$

Equation 4.51 can be rewritten as

$$\begin{aligned} (V_n - I_{a(1)}Z_{(1)}) - (-I_{a(2)}Z_{(2)}) &= Z_F I_{a(1)} \\ V_n &= I_{a(1)}(Z_{(1)} + Z_{(2)} + Z_F). \end{aligned} \tag{4.55}$$

Equations 4.53 and 4.55 specify the interconnection of sequence networks as shown in Figure 4.27. The fault current $I_F = I_b$ is given as

$$I_F = a^2 I_{a1} + a I_{a2} = (a^2 - a)I_{a1} = -j\sqrt{3} I_{a1}.$$

4.2.2.6 Double Line-to-Earth Fault (2L-E Fault) The system with 2L-E fault is represented by the circuit diagram shown in Figure 4.28.

The three equations describing the fault conditions are $I_a = 0$, $V_b = 0$, and $V_c = 0$.

Their symmetrical components are

$$\begin{aligned} &I_{a1} + I_{a2} + I_{ao} = 0, \\ &a^2 V_{a1} + a V_{a2} + V_{ao} = 0, \\ &a V_{a1} + a^2 V_{a2} + V_{ao} = 0. \end{aligned} \tag{4.56}$$

Subtracting the last two equations gives

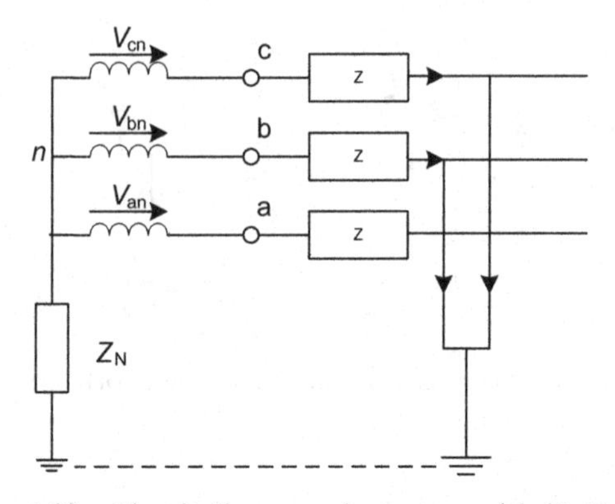

Figure 4.28 Circuit diagram of a system with 2L-E fault.

$$(a^2 - a)V_{a1} + (a - a^2)V_{a2} = 0$$
$$V_{a1} = V_{a2}.$$

Substituting into one of them

$$(a^2 + a)V_{a1} + V_{ao} = 0$$
$$V_{a1} = V_{a2} = V_{ao}. \tag{4.57}$$

Replace Equation 4.57 by equivalent quantities of sequence voltages to give

$$V_n - I_{a1}Z_1 = -I_{a2}\ Z_2 = -Z_o I_{ao}.$$

Thus,

$$I_{ao} = \frac{Z_2}{Z_o} I_{a2}.$$

Substituting into Equation 4.56 gives

$$I_{a1} + \left(1 + \frac{Z_2}{Z_o}\right) I_{a2} = 0,$$

$$I_{a(2)} = -\frac{Z_{(o)}}{Z_{(2)} + Z_{(o)}} I_{a(1)}.$$

Therefore,

$$V_n = I_{a(1)}Z_{(1)} - I_{a(2)}Z_{(2)},$$

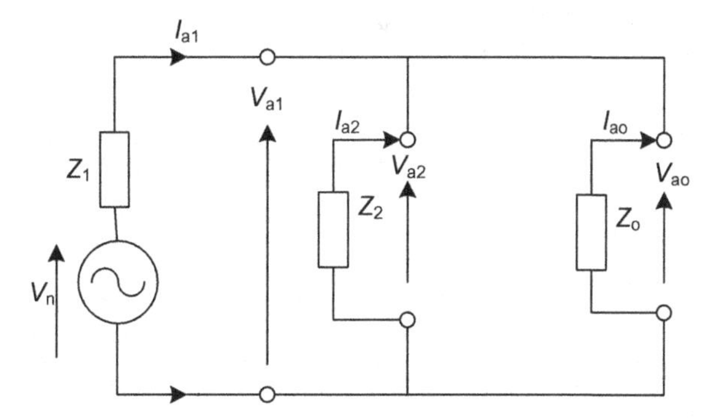

Figure 4.29 Interconnection of sequence networks for 2L-E fault.

that is,

$$V_n = I_{a1}\left(Z_1 + \frac{Z_2 Z_o}{Z_2 + Z_o}\right). \tag{4.58}$$

The interconnection of sequence networks that satisfies Equations 4.56 and 4.58 is shown in Figure 4.29.

The currents in the two faulty phases (b and c) can be calculated in terms of I_{a1}, I_{a2}, and I_{ao}. The fault current in the earth path equals

$$I_b + I_c = 3I_{ao}.$$

It must be noted that in the case of existence of a fault impedance Z_F, three times its values is added to the network zero sequence impedance to obtain system zero-sequence impedance ($Z_o = Z_{neto} + 3Z_F$).

Example 4.2:

A 22-kV upstream network with a short-circuit rating of 300 MVA and R/X of 0.3 is connected to an 800 kVA, 22/0.4 kV, delta-star transformer as shown in Figure 4.30. The star neutral point is directly connected to ground. The transformer resistance R_T and reactance X_T are 2.68 and 9.07 mΩ, respectively, and the short-circuit voltage ratio is 4.5%. The transformer feeds a load via an aluminum cable of 100 m length, 95-mm^2 cross-sectional area, and reactance 0.08 mΩ/m. It is required to calculate the fault current when a fault (three-phase short circuit, L-E, L-L, and 2L-E) occurs at the load point.

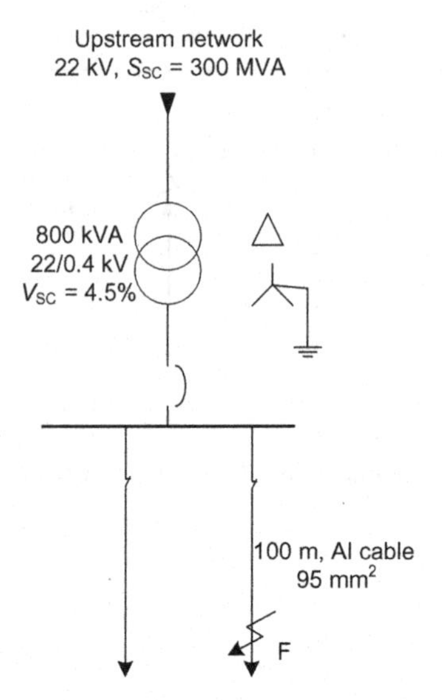

Figure 4.30 A simple distribution network.

As a first step, the network sequence impedances Z_1, Z_2, and Z_o are calculated as below,

$$Z_1 = Z_{\text{upstream}} + Z_T + Z_{\text{cable}}$$
$$= \sqrt{[R_1]^2 + [X_1]^2},$$

where

$$Z_{\text{upstream}} = \frac{(22\times10^3)^2}{300\times10^6} = 1.613\,\Omega \text{ and } (R/X)_{\text{upstream}} = 0.3,$$

$$X_{\text{upstream}} = 0.96 Z_{\text{upstream}} = 0.96\times1.613 = 1.548\,\Omega,$$

$$R_{\text{upstream}} = 0.3\times1.548 = 0.516\,\Omega,$$

$$R_T = 2.68\times10^{-3}\,\Omega,$$

$$X_T = 9.07\times10^{-3}\,\Omega,$$

$$R_{\text{cable}} = (\rho L)/A = 0.043\times100/95 = 0.045\,\Omega,$$

$$X_{\text{cable}} = 0.08\times100\times10^{-3} = 0.008\,\Omega,$$

$$R_1 = 0.516 + 0.00268 + 0.045 = 0.564\,\Omega,$$

$$X_1 = 1.548 + 0.0091 + 0.008 = 1.565\,\Omega, \text{ and}$$

$$Z_1 = \sqrt{(0.564)^2 + (1.565)^2} = 1.663\,\Omega.$$

The network does not include rotating machines and, therefore, $Z_1 = Z_2$. Considering the neutral impedance $Z_N = 0$ and assuming $Z_o \approx Z_1$, the fault current of different types of short circuit can be calculated as follows:

$$\textit{Three-phase short circuit: } I_F = \frac{V_n}{Z} = \frac{22 \times 10^3}{\sqrt{3} \times 1.663} = 7.64\ \text{kA},$$

$$\textit{L-E fault: } I_F = \frac{3V_n}{Z_1 + Z_2 + Z_o} = \frac{3V_n}{3Z_1} = \frac{V_n}{Z_1} = 7.64\ \text{kA},$$

$$\textit{L-L fault: } I_F = \frac{\sqrt{3}V_n}{Z_1 + Z_2} = \frac{22 \times 10^3}{2 \times 1.663} = 6.615\ \text{kA},$$

$$\textit{2L-E fault: } I_{a1} = \text{kA},$$

$$I_{a2} = -I_{a1}\frac{Z_o}{Z_2 + Z_o} = -5.09 \times 0.5 = -2.545\ \text{kA},$$

$$I_{ao} = -I_{a1}\frac{Z_2}{Z_2 + Z_o} = -2.545\ \text{kA}.$$

The fault current in the earth is $3I_{ao} = 7.635\,\text{kA}$.

It is noted that the return path of zero-sequence current is the earth with negligible impedance. If the network has a neutral line as it is on the LV side, its impedance must be included according to the grounding system used when calculating fault current.

To set the protective devices accurately and clearly, the fault at appropriate time to avoid equipment damage and injury to living beings, the minimum short-circuit current regardless of its type and location must be calculated. The value of minimum short-circuit current is mainly dependent on the upstream network configuration and the method of grounding. Accordingly, the type of fault at the farthest point from the source, which gives minimum short-circuit current, can be defined.

4.2.2.7 Calculation of Minimum Short-Circuit Current in LV Distribution Networks As explained in Chapter 3, three earthing systems are commonly used in LV distribution networks: TN, IT, and TT systems. The distribution network shown in Figure 4.31 is taken as an example to illustrate the method of minimum short-circuit current calculation for different earthing systems. It is noted that the impedance of bus bar at the DP is neglected.

1. *LV Distribution Network Using TN Grounding System*: The three-phase circuit diagram shown in Figure 4.32 indicates the route of the fault current when L-E fault occurs on phase "a" at a load (load #2) farthest from the source. This type of fault and its location give the lowest value of short-circuit current because its route is the longest with highest impedance.

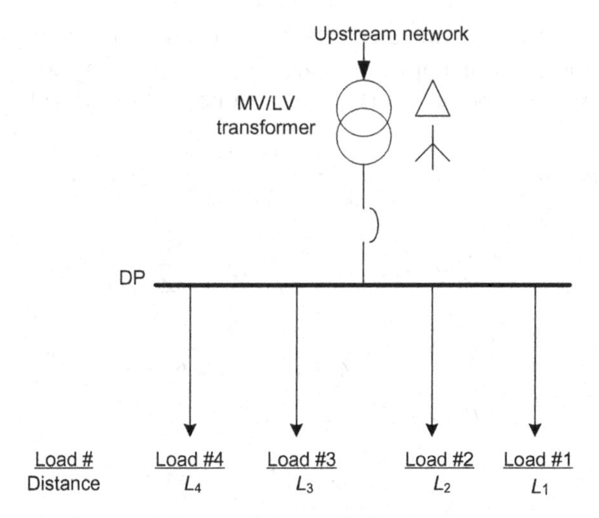

Figure 4.31 A sample of distribution network.

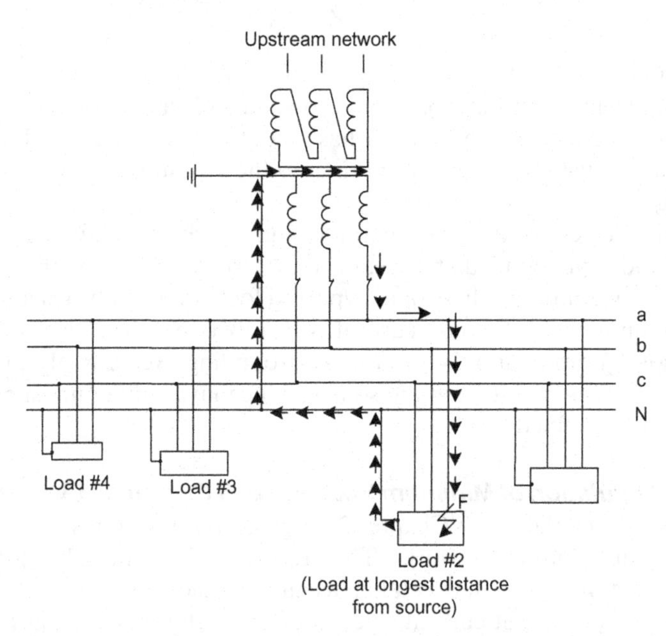

Figure 4.32 Three-phase circuit diagram of TN system with L-E fault on phase a.

In the TN earthing systems, the neutral is directly connected to earth, and the exposed conductive part of the loads is connected to the neutral line. The fault current is given in Equation 4.50,

$$I_{\mathrm{F}} = \frac{3V_{\mathrm{n}}}{Z_1 + Z_2 + Z_o},$$

where

$Z_1 = \sqrt{R_1^2 + X_1^2}$ = positive-sequence impedance of fault circuit,

$R_1 = R_{\text{upstream}} + \frac{\rho L_2}{A_{\text{ph}}} + \frac{\rho L_2}{A_{\text{neutral}}}$,

R_{upstream} = resistance of network upstream of switching device,

ρ = resistivity of conductor material,

A_{ph} = cross-sectional area of phase conductor,

A_{neutral} = cross-sectional area of neutral conductor,

$X_1 = X_{\text{upstream}} + X_{\text{ph}} + X_{\text{neutral}}$,

X_{upstream} = reactance of upstream network of switching device,

X_{ph} = phase reactance,

X_{neutral} = neutral line reactance,

$Z_1 = Z_2$ (because of the absence of rotating machines), and

$Z_o = Z_{\text{upstreamo}} + Z_{\text{pho}} + Z_{\text{neutralo}}$.

2. *LV Distribution Network Using IT Earthing System*: In the IT earthing system without a distributed neutral, the neutral is isolated from earth and the exposed conductive parts of loads are connected to an earthed conductor as shown in Figure 4.33. This Figure illustrates the fault current flow during the fault, which gives minimum short-circuit current. The fault is L-L fault, between phase "b" at load #1 and phase "c" at load #2. The loads #1 and #2 are located at distances L_1 and L_2, respectively, from the source.

   In case of L-L fault, the fault current is calculated as below,

$$I_F = \frac{\sqrt{3}V_n}{Z_1 + Z_2},$$

where

$Z_1 = \sqrt{(R_1^2 + X_1^2)}$,

$R_1 = R_{\text{upstream}} + \frac{\rho(L_1 + L_2)}{A_{\text{ph}}} + \frac{\rho(L_1 + L_2)}{A_{\text{neutral}}}$,

$X_1 = X_{\text{upstream}} + X_{L1} + X_{L2}$, and

X_{Li} = reactance of phase conductor connected to load #i.

The network does not include rotating machines, so, $Z_1 = Z_2$. If a rotating machine exists in the network, it is found that $Z_2 < Z_1$ and using their equality as an approximation, even in the case of the presence of rotating machine, will minimize the short-circuit current and yield safer setting of switching devices.

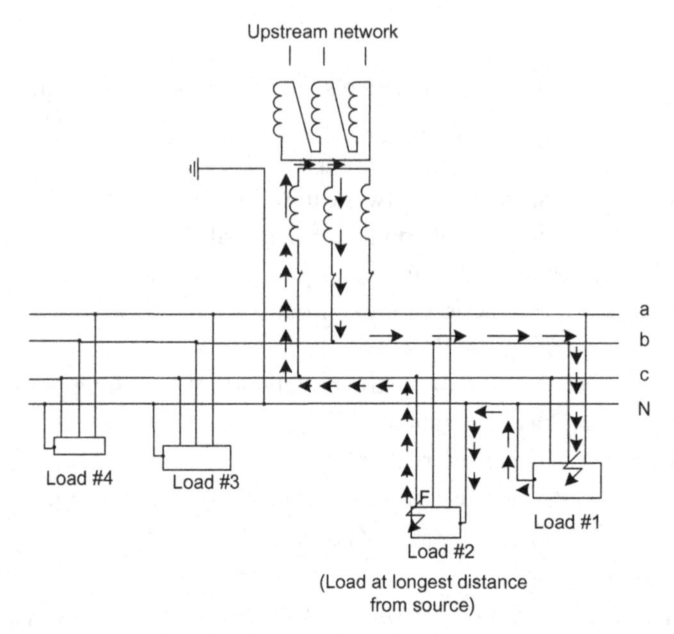

Figure 4.33 Three-phase circuit diagram of IT system with L-L fault (phase b at load #1 and c at load #2).

3. *LV Distribution Network Using TT Earthing System*: In the TT earthing system, the neutral is directly earthed and the exposed conductive parts of loads are grouped and connected to earth. The fault, which gives the lowest value of short-circuit current is L-L fault (phase b to phase c). It is not seen by residual current devices where the circulation of fault current is as shown in Figure 4.34.

   It is noted that in a TT system, the residual current devices must be used to protect the network against L-E fault since both neutral and exposed conductive parts of the loads are earthed.

   The fault current is calculated by applying the relation

$$I_{\mathrm{F}} = \frac{\sqrt{3}V_{\mathrm{n}}}{Z_1 + Z_2} = \frac{\sqrt{3}V_{\mathrm{n}}}{2Z_1},$$

where

$$Z_1 = \sqrt{\left[R_{\mathrm{upstream}} + \frac{\rho L_2}{A_{\mathrm{ph}}}\right]^2 + \left[X_{\mathrm{upstream}} + X_{\mathrm{ph}}\right]^2}.$$

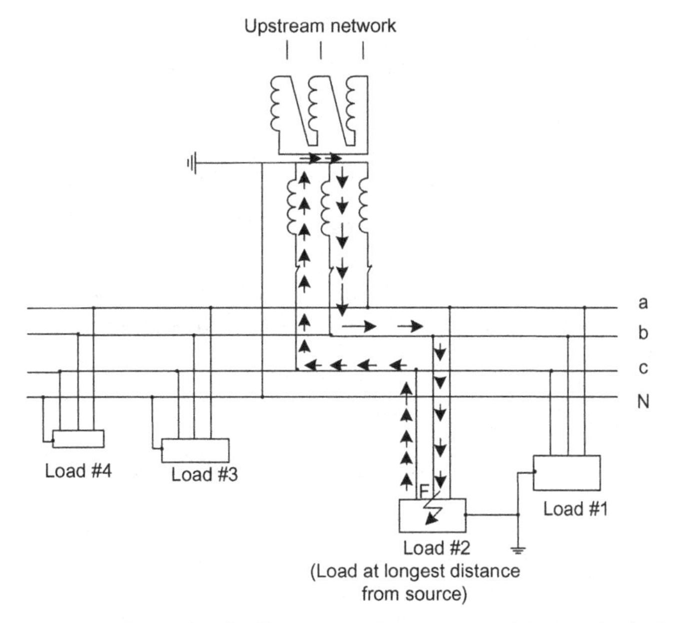

Figure 4.34 Three-phase circuit diagram of TT system with L-L fault (phase b to phase c).

Example 4.3:

The data of the sample system shown in Figure 4.31 are given below. It is required to calculate the minimum short-circuit current $I_{min\text{-}sc}$ in case of using (1) TN system, (2) IT system without distributed neutral, and (3) TT system.

Data:

Upstream MV network: 22 kV, $S_{sc} = 300\,\text{MVA}$, $R/X = 0.3$

Transformer: 22/0.4 kV, $V_{sc} = 4.5\%$, $R_T = 2.68\,\text{m}\Omega$, $X_T = 9.07\,\text{m}\Omega$

Load feeders: aluminum cable $3 \times 95 + 1 \times 50\,\text{mm}^2$, length 80 m for load #1 and 100 m for load#2, $\rho = 0.043\,\Omega\,\text{mm}^2/\text{m}$, $X_{cable} = 0.08\,\text{m}\Omega/\text{m}$. $L_2 > L_1 > L_3 > L_4$.

Solution:

The upstream network impedance of switching device $Z_{up\text{-}dv}$ is the sum of MV upstream network impedance $Z_{upstream}$ and transformer impedance Z_T:

$$Z_{\text{upstream}} = (400)^2/(300 \times 106) = 0.533\ \text{m}\Omega$$
$$X_{\text{upstream}} = 0.96 \times Z_{\text{upstream}} = 0.512\ \text{m}\Omega$$
$$R_{\text{upstream}} = 0.3 \times 0.512 = 0.1536\ \text{m}\Omega$$
$$R_{\text{up-dv}} = R_{\text{upstream}} + R_{\text{T}} = 2.834\ \text{m}\Omega$$
$$X_{\text{up-dv}} = X_{\text{upstream}} + X_{\text{T}} = 9.582\ \text{m}\Omega$$

a) *TN System*: The type of fault that gives minimum short-circuit current is L-E fault at farthest load from source, that is, phase "a" to earth at load #2:

$$\begin{aligned} R_1 &= R_{\text{up-dv}} + R_{\text{cable}} + R_{\text{neutral}} \\ &= 2.834 \times 10^{-3} + (0.043 \times 100/95) + (0.43 \times 100/50) \\ &= 0.1338\ \Omega, \end{aligned}$$

$$\begin{aligned} X_1 &= X_{\text{up-dv}} + X_{\text{cable}} + X_{\text{neutral}} \\ &= 9.582 \times 10^{-3} + 0.08 \times 100 \times 10^{-3} + 0.08 \times 100 \times 10^{-3} \\ &= 25.582\ \text{m}\Omega, \end{aligned}$$

$$Z_1 = 0.136\ \Omega,$$

$$Z_1 = Z_2 \text{ and assuming } Z_{\text{o}} \approx \text{Z}_1.$$

Thus,

$$I_{\text{min-sc}} = (400/\sqrt{3})/Z_1 = 1698\ \text{A}.$$

b) *TT System without Distributed Neutral*: The minimum short-circuit current is obtained when L-L fault occurs on two different circuits, for example, phase "b" at load #1 to phase "c" at load #2:

$$\begin{aligned} R_1 &= R_{\text{up-dv}} + R_{\text{cable}} + R_{\text{neutral}} \\ &= 2.834 \times 10^{-3} + (0.043 \times 180/95) + (0.043 \times 180/50) \\ &= 0.2388\ \Omega, \end{aligned}$$

$$\begin{aligned} X_1 &= X_{\text{up-dv}} + X_{\text{cable}} + X_{\text{neutral}} \\ &= 9.582 + 0.08 \times 180 + 0.08 \times 180 = 38.38\ \text{m}\Omega, \end{aligned}$$

$$Z_1 = 0.242\ \Omega.$$

Thus,

$$I_{\text{min-sc}} = (\sqrt{3} \times V_{\text{n}})/[Z_1 + Z_2] = (400)/(2 \times 0.242) = 826.45\ \text{A}.$$

c) *TT System*: The fault type that gives minimum short-circuit current is L-L fault, for example, phase "b" to phase "c" at load #2:

$$R_1 = R_{\text{up-dv}} + R_{\text{cable}}$$
$$= 2.834 \times 10^{-3} + (0.043 \times 100/95) = 47.833\ \text{m}\Omega,$$
$$X_1 = X_{\text{up-dv}} + X_{\text{cable}}$$
$$= 9.582 \times 10^{-3} + 0.08 \times 100 \times 10^{-3} = 17.582\ \text{m}\Omega,$$
$$Z_1 = 50.96\ \text{m}\Omega.$$

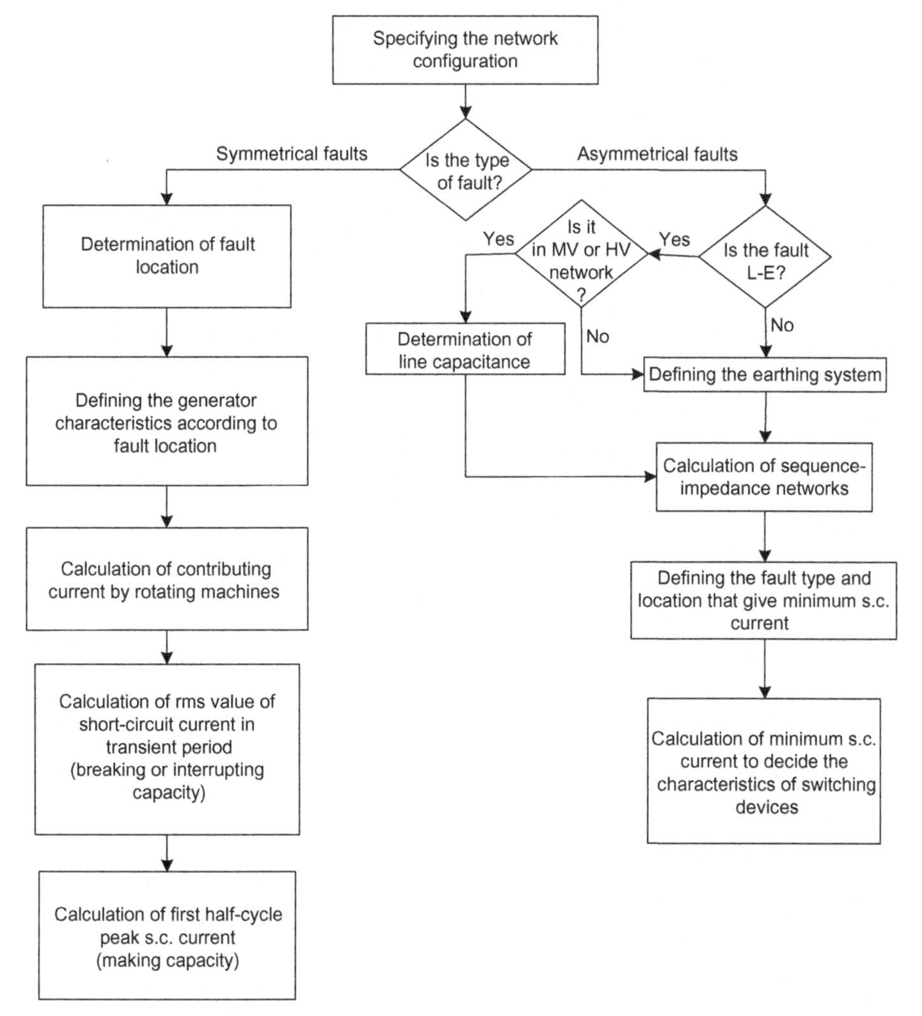

Figure 4.35 Main steps of peak, rms, and minimum short-circuit current calculations. s.c. = short-circuit.

Thus,

$$I_{\text{min-sc}} = (\sqrt{3} \times V_n)/[Z_1 + Z_2] = (400)/(2 \times 50.96 \times 10^{-3})$$
$$= 3925 \text{ A}.$$

As mentioned at the beginning of this chapter, the main objectives of short-circuit calculations are to calculate the peak, rms, and minimum values of fault current. The main steps to satisfy these objectives, as explained in the previous sections, can be summarized as depicted in the chart shown in Figure 4.35.

CHAPTER 5

PROTECTION OF ELECTRIC DISTRIBUTION SYSTEMS

5.1 INTRODUCTION

The distribution system is expected to experience different types of faults: faults due to internal or external causes and temporary or permanent. The faults due to internal causes (failure in system components and mainly short circuits) as described in Chapter 4 have some of the consequences below:

- Thermal effect where a high fault current flow results in system equipment (cables, lines, machines, and transformers windings) overheating, which may damage the insulation material and melt conductors. For instance, a three-phase short-circuit on medium-voltage (MV) bus bars can melt up to 50 kg of copper in 1 s and the temperature at the center of the arc can exceed 10,000°C. The current must, therefore, be switched by a circuit breaker (CB) or a fuse in a short enough time for the equipment temperature not to reach a critical value.
- Electrodynamic effect: Electromagnetic effects of current cause electrodynamic forces observed in the network equipment. The CBs or in general the switching devices must be able to withstand the high dynamic force proportional to the peak fault current. Otherwise, the separable contacts of these devices tend to open under the effect of what is called repulsive electrodynamic force.

Electric Distribution Systems, First Edition. Abdelhay A. Sallam, Om P. Malik.

Published 2011 by John Wiley & Sons, Inc.

- The voltages sag in the healthy network elements, which is often lower than the allowable load voltage. It may lead to motor stopping if the fault lasts too long.
- Switching surges: The switching of a current in an inductive circuit and critical situations such as resonance can cause overvoltages (switching surges) where the maximum peak value may reach two or three times the root mean square (rms) value of the nominal voltage.
- Overvoltages: An overvoltage occurs between the healthy phases and earth when a phase-to-earth fault occurs as explained in Chapter 3. Also, it may result from an earth fault itself and its elimination.

The faults due to external causes (surges from outside the system) are mainly due to lightning strokes that largely affect the overhead lines and outdoor equipment by producing overvoltage. This overvoltage may cause flashovers and damage the equipment insulation.

Therefore, the distribution system must be provided with protection devices to continuously monitor the electric status of system components and de-energize them (e.g., by tripping a CB) when they are the site of a serious disturbance (e.g., short circuit and insulation failure).

5.1.1 Protection System Concepts

The protection system consists of a string of devices. Among their multiple purposes, these devices aim at

- limiting the thermal, dielectric, and mechanical stresses on the equipment;
- maintaining stability and service continuity of the network; and
- protecting living beings against electric hazards.

To attain these objectives, the protection system must be fast enough to clear the fault at minimum operating time, operating correctly (reliable), disconnecting the minimum section of the network necessary to isolate the fault (selectivity), and ensuring coordination at cost as low as possible. Protection has its limits because the fault must occur before the protection system can react. It, therefore, cannot prevent disturbances; it can only limit their effects and their duration.

The general philosophy of applying protection to distribution systems is to divide the system into separate zones (e.g., motors, cables, lines, bus bars, transformers, generators). Each zone is individually protected (primary protection) to be isolated from the rest of the system when a fault occurs within it and keeping the other zones in-service. The zones may overlap at some points. In this case, these points will be protected by more than one set of protection relays. In addition, backup protection is used as a second line of

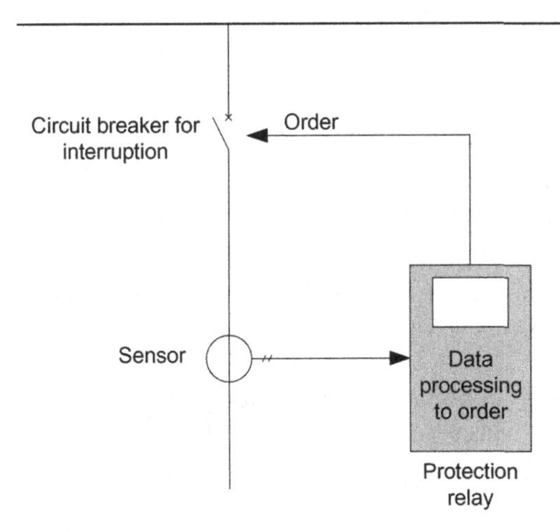

Figure 5.1 Protection system components.

defense in order to operate when the primary protection does not work. This necessitates that the time delay of the backup relay be more than that of the primary protection to allow time for the primary protection to operate first.

Two distinct steps are taken to design the protection of a distribution system.

Step 1: Protection system study that includes the choice of protection components and a consistent, overall structure suited to the distribution system. The choice of protection components is not the result of an isolated study, but rather one of the most important steps in the design of the distribution system. The main components of a protection system, as shown in Figure 5.1, include the following:

- measurement sensors (current transformers [Cts] and voltage transformers [vts]) supplying the data required for fault detection at lower levels (conventionally, currents are reduced either to 5 or 1 A and voltages are reduced to 110 or 120 V);
- protection relays for continuous monitoring and processing the data received from sensors (currents and voltages) to formulate and deliver the orders to the trip circuit to clear the faulty parts; and
- CBs or combination of switches or contactors and fuses to isolate the faulty parts.

To determine such components based on fault type, the parameters below must be considered:

- the structure of distribution system (as described in Chapter 1) as well as the different operating modes,
- neutral-earthing systems (as explained in Chapter 3),

- characteristics of current sources and their contribution in the event of faults,
- types of loads, and
- the need for continuity of service.

Step 2: Protection coordination that deals with determination of setting for each protection unit based on protection function to ensure the best possible operation in all operating modes of distribution system. The best settings are the result of complete calculations based on the detailed characteristics of the various elements in the installation.

The forthcoming sections introduce the types of relay construction, and protection systems; overcurrent protection (different types and faults); and the protection devices, autoreclosers, sectionalizers, and fuses. Overvoltage protection is presented as well.

5.2 TYPES OF RELAY CONSTRUCTION

With regard to relay construction, the most common types of relays used in industrial power system for fault protection are classified into three types: electromagnetic, static, and digital relays [46].

5.2.1 Electromagnetic Relays

Electromagnetic relays are constructed with electric, magnetic, and mechanical components. They have an operating coil and various contacts. These relays are very robust and reliable. Some of the varieties of construction of such relays are described below.

i) *Electromagnetic Attraction Relays*: Many relays are based on electromagnetic attraction principle and most of them are solenoid relays. The main element of these relays is a solenoid wound around an iron core and steel plunger or armature that moves inside the solenoid and supports the contacts. These relays operate without any intentional time delay, usually within one-half cycle. They are called "instantaneous overcurrent relays" and operate with a definite-current characteristic.

ii) *Induction Disk Relays*: This type of relay has a provision for the variation of the time adjustment and permits change of operating time for a given current. This adjustment is called "time dial setting" and the relay operates with a definite-time characteristic. Construction of an induction disk relay is similar to a watt-hour meter since it consists of an electromagnet and a movable armature (metal disk) on a vertical shaft restrained by coiled spring. The relay contacts are operated by a movable armature. This type of relay is used when the system experiences overcurrent of a transient nature (e.g., caused by motor starting or sudden overload of a short duration) since it has a time delay that permits a current of several times relay setting to persist for a limited period without tripping.

5.2.2 Static Relays

Static relays are based on analog electronic components such as diodes, transistors, integrated circuits, and capacitors. The technology applied to this type of relays overcomes the restrictions of the electromagnetic relays and has some advantages such as

- very fast reset times since it is not limited by disk inertia,
- the shape of time-current characteristic (TCC) can be controlled,
- accurate predetermined operating set points,
- low burden level instrument transformers are needed,
- multifunction availability where a single static relay can perform the functions of many electromagnetic relays,
- equivalent functions can often be obtained at lower cost, particularly if a multifunction relay is used, and
- capability of communications that is not available in electromagnetic relays.

The drawbacks of static relays are mainly the possibility of failure in adverse environment, both physical and electric as well as the possibility of total failure of the protective system due to failure of one component such as a common power supply.

5.2.3 Digital Relays

The main parts of the digital relay are analog input system, digital processor, digital output system, and independent power supply [63]. A simplified block diagram of a digital relay is given in Figure 5.2. The analog input signals received from the Cts or vts are converted into digital form within the analog input system before being analyzed by the processor (this is the main difference between digital relays and other types).

The analog input system includes the following:

- A surge filter to suppress the large inrush in the input signals for relay safety. It typically consists of capacitors and isolating transformers. Zener diodes are also used to protect electronic circuits against surges. Their placement depends on the exact physical circuit arrangements used.

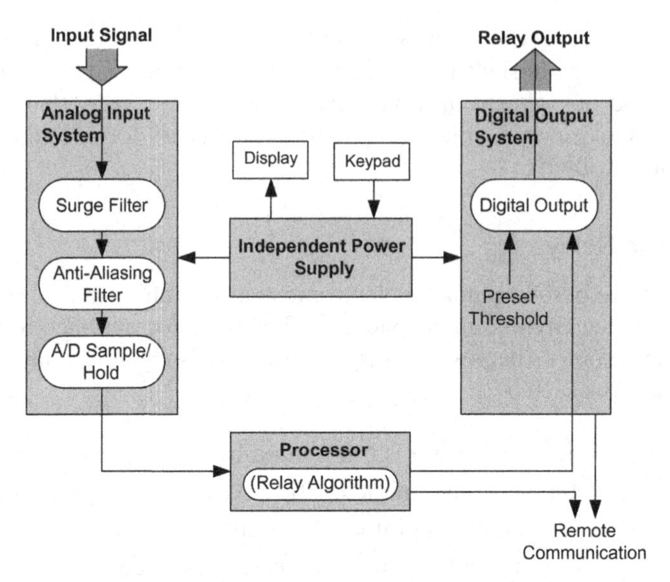

Figure 5.2 Block diagram of a digital relay [63].

- An anti-aliasing filter used to avoid possible errors in reconstructing the input signal.

  The analog signals received from the Cts and/or vts have to be filtered to remove the undesired high frequencies before sampling. This is commonly performed using a low-pass filter.

  The sampling theorem states that a band-limited signal can be uniquely specified by its samples if and only if the sampling frequency, f_s, is at least twice the maximum frequency component, f_m, contained within the original signal ($f_s \geq 2f_m$) [64]. If this condition is not satisfied, in time domain a low-frequency component, dotted curve in Figure 5.3a, that does not exist in the original signal will appear. In frequency domain and due to the actual characteristics of the low-pass filter, this error appears as an overlap of adjacent parts of the sampling function (Fig. 5.3b). It causes an error in the analysis as a result of the difficulty in distinguishing between low- and high-frequency components. This error is commonly known as an "aliasing error." It can be removed by filtering high-frequency components (components with frequency greater than one-half the sampling frequency) from the input by what is called "anti-aliasing filter."
- A/D sample/hold circuit that is adopted to convert the input signal from analog to digital. The signal is scanned by a sliding data window of limited length. Within the window at a specific time (moment of sampling), the wave shape is intermittent in the form of a certain number of recorded samples. So, more samples are obtained at various snapshots of time with window moving forward (Fig. 5.4). The length and shape of the sampling

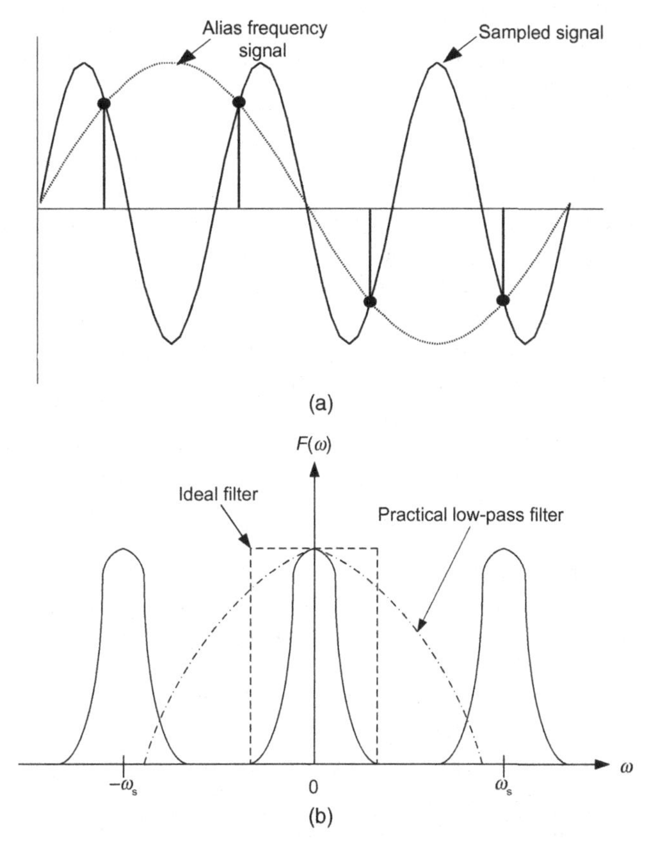

Figure 5.3 An illustration of aliasing phenomenon. (a) In time domain. (b) In frequency domain.

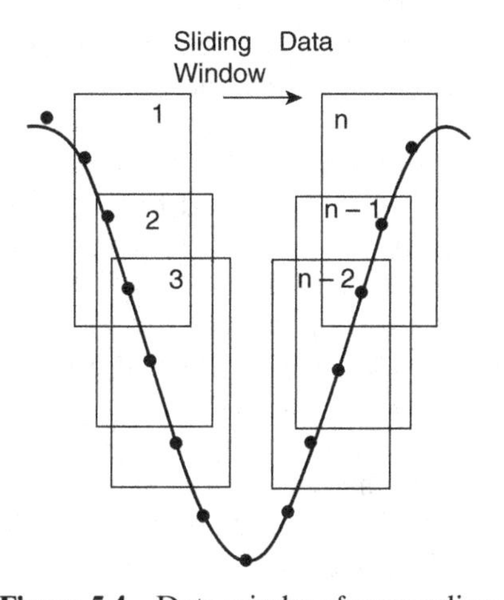

Figure 5.4 Data window for sampling.

window and the number of samples in the window depend on the relay algorithm (e.g., discrete Fourier transform [DFT], least squares [LS], Kalman algorithms).

A digital processor includes a read-only memory (ROM) and a random access memory (RAM) in which the relay algorithm is stored. It determines the way to reconstruct the input signal based on the digital samples from the A/D converter. As the input signal may contain unwanted components such as harmonics, interharmonics, and direct current (DC), the algorithm is designed to remove them as much as possible. The algorithm acts as a digital filter to extract the fundamental component of the input signal on which the relay operation is based. The shape of the sliding data window varies with different algorithm principles. The rectangular window shown in Figure 5.4 is only one option. The length of the data window depends on the required decision speed of the algorithm and on the tolerance against disturbances. The number of sample points in the data window is determined by the sampling frequency.

A digital output system provides signals for the operation of the relay according to a comparison between the preset threshold and a detection function. The detection function is calculated based on the reconstructed signal.

Digital protective relays are fast becoming the norm in distribution systems throughout the world. They typically consist of one or more of the following basic protection components: undervoltage and overvoltage units, overcurrent units, directional/nondirectional overcurrent units, negative-sequence overcurrent unit, directional power unit, differential protection, and breaker monitoring and automatic reclosing.

Digital relays can easily be used for measurement functions where the input signals (three-phase currents, voltages) are sampled and the fundamental is extracted by the relay algorithm. So, the voltage, current, frequency, active power, reactive power, and apparent power can be metered and viewed locally at the front panel human machine interface, or remotely through any communication port.

For control purposes, the digital relays have several virtual switches, which can be used to trip and close switches and breakers, or enable and disable certain functions. By communication ports, stored information, and using ASCII command interface, the distribution system information can be retrieved from a remote location, control operations can be performed, and metering information and reports can be retrieved. In addition, communication ports can be connected to computers to access and analyze the transmitted binary data, which are supported by relay communication protocols.

Digital relays can provide many advanced fault reporting features such as fault summary reports, sequence of events recorder reports, and oscillographic records, as well as visual flags that indicate the type of fault.

Their main advantages and drawbacks are listed below:

Advantages:

- high level of functionality integration,
- additional monitoring functions,
- functional flexibility,
- capable of working under a wide range of temperatures,
- implementation of more complex functions and generally more accurate,
- self-checking and self-adaptability, and
- ability to communicate with other digital equipment (pear to pear).

Drawbacks:

- short lifetime due to the continuous development of new technologies (the devices become obsolete rapidly),
- susceptibility to power system transients, and
- as digital systems become increasingly more complex, they require specially trained staff for proper maintenance of the settings and monitoring data.

5.3 OVERCURRENT PROTECTION

Distribution systems are exposed to overcurrent flow into their elements. The reasons for overcurrent may be due to abnormal system conditions such as overload and short-circuit faults or due to normal system conditions such as transformer inrush current and motor starting. Therefore, for normal system conditions, some tools such as demand-side management, load shedding, and soft motor starting can be applied to avoid overloads. In addition, distribution systems are equipped with protective relays that initiate action to enable switching equipment to respond only to abnormal system conditions. The relay is connected to the circuit to be protected via Cts and vts according to the required protection function. For instance, the directional overcurrent protection uses both Cts and vts (Section 5.5) while earth-fault protection only uses Cts. The relay operates when the received signals (current and voltage) exceed a predetermined value. It sends a tripping signal to the CB to isolate the circuit experiencing overcurrent on any phase or to disconnect defective equipment from the remainder of the system. The protective relay and switching element may be combined together in one element such as molded case circuit breakers (MCCBs) and fuses.

Relays need to be energized to operate. This energy can be provided by the monitored circuit itself or by using energy-storage system such as capacitor trip devices (for small low-voltage [LV] systems) or battery sets (for large switchgear). In case of providing energy directly from monitored circuit, there is a risk of relay malfunction because of the system voltage drop that may occur suddenly.

5.3.1 Overcurrent Relays

The operating current for all overcurrent relays is either fixed or adjustable. The relay contacts close and initiate the CB tripping operation when the current flowing from the secondary of Ct to the relay exceeds a given setting. There are three types of operating characteristics of overcurrent relays:

i) *Definite-Current Protection*: When the current reaches or exceeds the setting threshold, the relay operates instantaneously as shown in the characteristic curve of Figure 5.5a. The relay setting is adjusted based on its location in the network. Its value is getting lower as the relay location is furthest away from the source. For instance, the overcurrent relay connected to the receiving end of a distribution feeder will operate for a current lower than that connected to the sending end, especially, when the feeder impedance is large. If the feeder impedance is small compared with upstream network impedance, distinguishing between the fault currents at both ends is difficult and leads to poor discrimination in addition to little selectivity at high levels of short-circuit current.

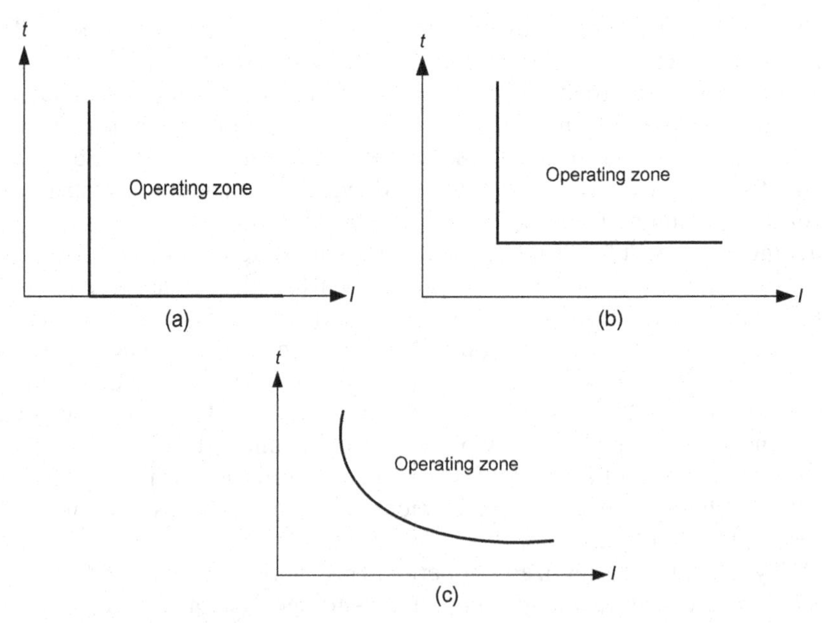

Figure 5.5 Time-current characteristics of overcurrent protection. (a) Definite-current characteristic; (b) definite-time characteristic; (c) inverse-time characteristic.

So, a poor discrimination and little selectivity are considered as disadvantages of this type of protection. On the other hand, when the feeder impedance or the impedance of the element to be protected is high, instantaneous protection has advantages of reducing the relay's operating time for severe faults and avoiding loss of selectivity when the system uses relays with different characteristics.

The adjustment of instantaneous overcurrent relay setting depends on relay location and type of element to be protected. For instance, setting of instantaneous overcurrent relays for protecting lines between substations is taken at a minimum of 1.25 times the rms current for the maximum symmetrical fault level at the next substation. The adjustment of setting starts from the furthest substation and moving back toward the source. For distribution lines ended by MV/LV transformers, the setting is taken as 50% of maximum short-circuit current at the point of Ct connection or in the range from 6 to 10 times the maximum circuit rating. When the instantaneous overcurrent relay is installed on the primary side of the transformer, its setting lies between 1.25 and 1.50 times the short-circuit current flowing into the bus bar in LV side, referred to the high-voltage (HV) side. This value is sufficient to keep the coordination with the higher current encountered due to magnetic inrush current when energizing the transformer [65].

ii) *Definite-Time Protection*: The overcurrent relays of this type of protection, definite-time overcurrent relays, have an operating time that is independent of the current magnitude after a certain current value is reached (Fig. 5.5b). The settings of overcurrent relays at different locations in the network can be adjusted in such a way that the breaker close to the fault is tripped in the shortest time and then the other breakers in the direction toward the upstream network are tripped successively with longer time delay. This type of relay has a current setting (sometimes called pickup or plug or tap setting) to select the current value at which the relay will start and a time dial setting to define the actual time of relay operation.

The disadvantage of this type of protection is that the short-circuit fault close to the source may be cleared in a relatively long time in spite of its highest current value.

iii) *Inverse-Time Protection*: The inverse-time overcurrent relays operate faster as current increases. Their characteristic is as shown in Figure 5.5c. They are available with inverse, very inverse, and extremely inverse-time characteristics to fit the requirements of the particular application (Fig. 5.6).

The operating time of both overcurrent definite-time relays and overcurrent inverse-time relays must be adjusted in such a way that the relay closer to the fault trips before any other protection. This is known as time grading. The difference in operating time of two successive

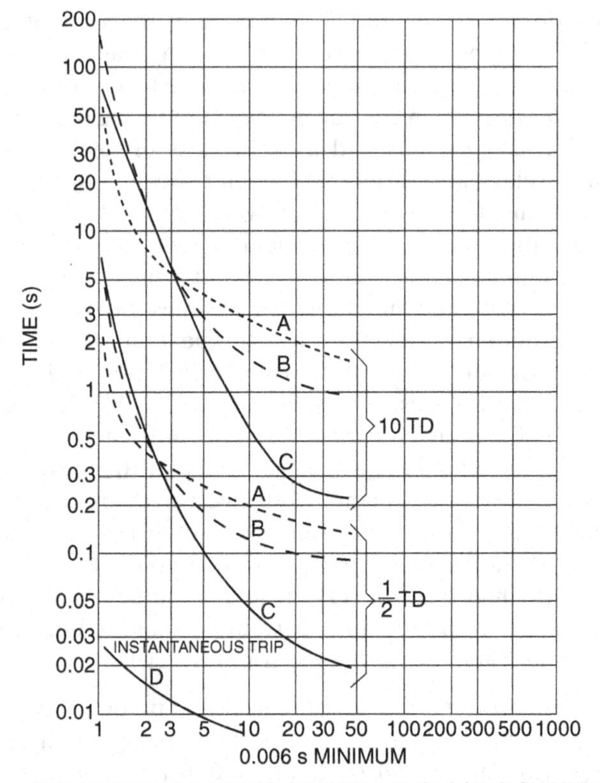

Figure 5.6 Typical relay time-current characteristics [46]. (A) Inverse; (B) very inverse; (C) extremely inverse; (D) instantaneous. TD = relay time dial setting.

relays at the same fault level is defined as "discrimination margin." It is typically taken in the order of 0.25–0.4 s for electromagnetic and static relays, and 0.2 s for the digital relays.

The adjustment of definite-time and inverse-time relays can be carried out by determining two settings: time dial setting and pickup setting. Time dial setting, sometimes referred to as time multiplier, is determined for the different relays installed on the system by defining the time delay of each relay keeping the time grading and discrimination margin. Pickup setting is used to define the pickup current of the relay by which the fault current exceeds its value. It is determined by allowing a margin for overload above the nominal current as in the relation below,

$$\text{Pickup setting} = \frac{K_{\text{ld}} \times I_{\text{nom}}}{\text{CTR}}. \tag{5.1}$$

where

K_{ld} = overload factor,
I_{nom} = nominal rated current, and
CTR = current transformer ratio.

The value of K_{ld} is recommended to be taken as 1.05 for motors, in the range of 1.25–1.5 for lines, transformers, and generators, and 2 for distribution feeders under emergency conditions where it is possible to increase the loading.

iv) *Overcurrent Voltage-Controlled Relays (OC-VCRs) and Overcurrent Voltage-Restrained Relays (OC-VRRs)*: Some faults may result in voltage drop and fault current reduction such as that occurs close to isolated generators. So, to ensure proper operation and coordination of generator protection, the overcurrent time-delay relays should be equipped with voltage control devices. Consequently, relay reliability is enhanced by its operation before extra reduction of generator current. OC-VCRs and OC-VRRs are the two types having this feature.

The current setting of OC-VCR is below rated current and the relay does not operate until the voltage drops with a certain margin below nominal preset voltage. The pickup of this type of relay is fixed, and it can be coordinated with other relays easily. On the other hand, the pickup of OC-VRR varies as it decreases with reducing voltage. So, its coordination with other overcurrent relays in the system is difficult.

5.3.2 Coordination of Overcurrent Relays

Coordination of overcurrent protection relays is based on isolating the fault part of the distribution system and only that part as quickly as possible, leaving all the fault-free parts in service. Some of the different means to implement coordination in distribution system protection are time-based coordination, current-based coordination, and logic coordination.

5.3.2.1 Time-Based Coordination Overcurrent relays at different locations in the distribution system are assigned by various time delays. The closer the relay is to the source, the longer the time delay. As shown in Figure 5.7,

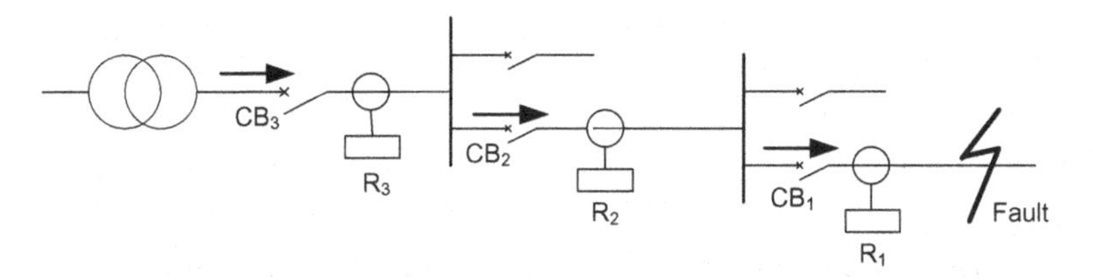

Figure 5.7 Fault detection by three protection units.

the fault is detected by protection units R_1, R_2, and R_3. The relay R_1 operates faster than R_2, which in turn operates faster than R_3. When the CB_1 (associated with relay R_1) is tripped and the fault is cleared, protection units R_2 and R_3 return to the standby position.

The difference in operating time Δt between two successive protection units is the coordination interval. It is given by the summation $t_c + t_r + 2Dt + m$, where t_c is the breaking time of downstream CB, which consists of the breaker response time and the arcing time. Dt is the time delay tolerance, t_r is the upstream protection unit overshoot time, and m is a safety margin. The safety margin takes a value that avoids losing coordination due to breaker opening time, relay overrun time after fault clearance (it does not exist when using digital relays), variation in fault levels, errors in Cts, and deviations from characteristic curves of relays. A typical value of safety margin is 110ms for a coordination interval of 0.3s.

This coordination system has the advantages of simplicity and providing its own backup, for example, if R_1 fails, R_2 is activated Δt later. On the other hand, in the presence of a large number of cascading relays, since the furthest upstream relay has the largest time delay, the fault clearing time becomes incompatible with short-circuit current withstand.

The principle of time-based coordination is applied to radial distribution systems. The time delays set are activated when the current exceeds the relay settings. Three types of time delay are applied according to protection characteristics: definite-time, inverse-time, and combined inverse-time and instantaneous protection systems. They are illustrated by using the system shown in Figure 5.7 as below:

- *Definite-time protection* where the time delay is constant and independent of the current. Therefore, the protection tripping curves are plotted in Figure 5.8a. The current threshold setting, $I_{set,Ri}$, of relays R_1, R_2, and R_3 must be such that $I_{set,R3} > I_{set,R2} > I_{set,R1}$.
- *Inverse-time protection* (it is also referred to as inverse definite minimum time [IDMT]) where the higher the current is, the shorter the time delay. The protection tripping curves are shown in Figure 5.8b. If the current thresholds are set at a value close to the nominal currents, $I_{n,Ri}$ (e.g., $I_{set,Ri} = 1.2I_{n,Ri}$), protection against both overloads and short circuits is ensured. The time delays are set to obtain the coordination interval Δt for the maximum current seen by the downstream protection relay.
- *Combined inverse-time and instantaneous protection*: It has the same principle of IDMT as well as the instantaneous tripping at maximum downstream short-circuit current, max $I_{sc,Ri}$. The tripping curves are shown in Figure 5.8c.

5.3.2.2 Current-Based Coordination

This coordination uses the principle that within a distribution system, as the fault is far from the source, the

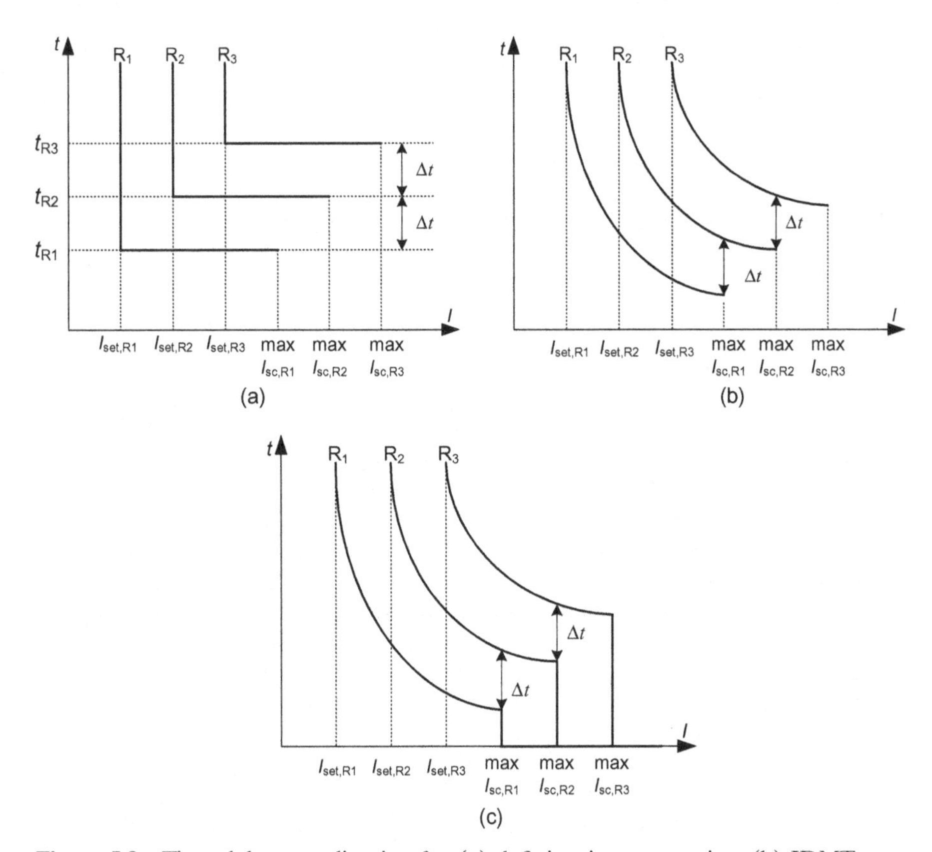

Figure 5.8 Time-delay coordination for (a) definite-time protection, (b) IDMT protection, and (c) combined inverse-time and instantaneous protection.

fault current is less. It is installed at the starting point of each section. The threshold is set at a value lower than the minimum short-circuit current caused by a fault downstream (outside the monitored area). This system can be used advantageously for two line sections separated by a transformer (Fig. 5.9a) since it is simple, economical, and tripping without time delay. To ensure coordination between the two protection units R_1 and R_2, the current setting of R_2, $I_{set,R2}$, must satisfy the relation 1.25 (max $I_{sc,R1}$) < $I_{set,R2}$ < 0.8 (min $I_{sc,R2}$), where max $I_{sc,R1}$ is the maximum short-circuit current at R_1 referred to the upstream voltage level (generally, symmetrical three-phase short circuit), and min $I_{sc,R2}$ is the minimum short-circuit current at R_2 (generally, phase-to-phase short circuit clear of earth). The tripping curves are shown in Figure 5.9b. Time delays are independent and t_{R2} may be less than t_{R1}.

On the other hand, current-based coordination has drawbacks where the upstream protection unit R_2 does not provide backup for the downstream protection unit R_1. In addition, practically, in the case of MV systems except for sections with transformers, there is no notable decrease in current between

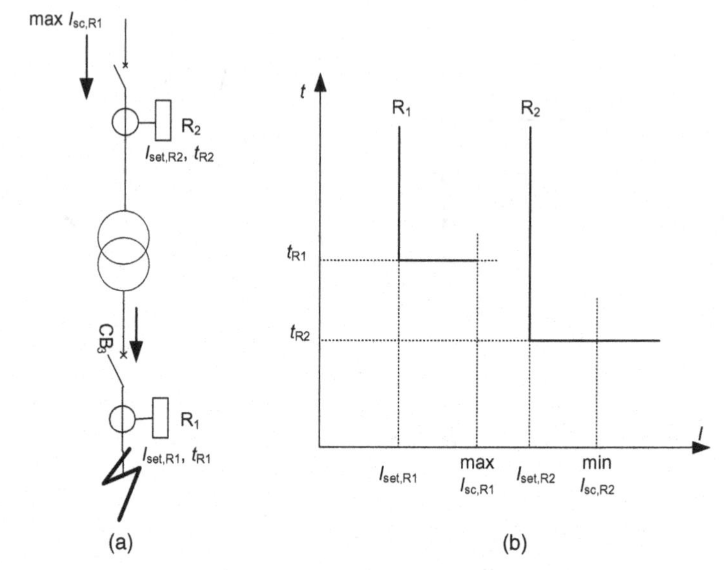

Figure 5.9 Current-based coordination. (a) Line diagram; (b) tripping curves.

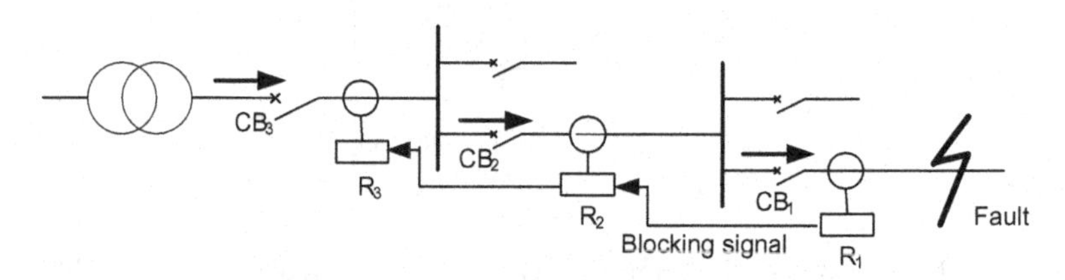

Figure 5.10 Logic coordination principles.

two adjacent areas. Therefore, to define the settings for two cascading protection units and ensuring the coordination is difficult.

5.3.2.3 Logic Coordination It is designed and developed to solve the drawbacks of both time-based and current-based coordination. With this system, coordination intervals between two successive protection units are not needed. Furthermore, the tripping time delay of the CB closest to the source is considerably reduced.

When a fault occurs in a radial system (Fig. 5.10), the protection units upstream from the fault (R_1, R_2, R_3) are activated and those downstream are not. Each protection unit activated by the fault sends a blocking signal to the upstream level as an order to increase the upstream relay time delay. Only the first CB, CB_1, directly upstream of the fault is tripped since the protection unit R_1 has not received a blocking signal from the downstream level.

Therefore, protection unit R_1 gives a tripping order to CB_1 and sends a blocking signal to R_2, which in turn sends a blocking signal to R_3. The tripping order given by R_1 is provided after a time delay t_{R1} and the duration of blocking signal to R_2 is limited to $t_{R1} + t_1$, where t_1 is the sum of opening and arc extinction time of CB_1. Thus, if CB_1 fails to trip, protection unit R_2 gives a tripping order at $t_{R1} + t_1$ as a backup protection. Of course, if a fault occurs between CB_1 and CB_2, the protection unit R_2 triggers tripping after a time delay t_{R2}.

To implement the logic coordination, extra wiring is needed for transmitting logic signals between the different protection units. This causes difficulty in the case of long links as the protection units are far apart from each other. This problem may be solved by using logic coordination in the nearby switchboards and time-based coordination between zones that are far apart.

Example 5.1:

The inverse-time characteristics of all relays used for the protection of the system in Figure 5.11 are shown in Figure 5.12. The relays applied in this example may have a combination of instantaneous and inverse-time characteristics as shown in Figure 5.8c. Pickup setting of relays is in the range of 1–10 A in steps of 1 A. Instantaneous setting is in the order of 1–100 A in steps of 1 A. For CBs at locations 1, 2, 3, and 4 in Figure 5.11, it is required to find the following:

- nominal rated current and short-circuit level;
- current transformation ratios (CTRs);
- pickup, time dial, and instantaneous settings of all phase relays keeping a proper coordinated protection scheme; and
- coverage of full length of the load feeder by overcurrent instantaneous unit of relay.

Solution:

Nominal Current Calculation for CB #i "$I_{nom\#i}$":

$$I_{\text{nom}\#1} = \frac{2\times10^6}{\sqrt{3}(11\times10^3)} = 105\text{ A},$$

$$I_{\text{nom}\#2} = \frac{(1.5+2+1)\times10^6}{\sqrt{3}(11\times10^3)} = 236.2\text{ A},$$

$$I_{\text{nom}\#3} = \frac{15\times10^6}{\sqrt{3}(11\times10^3)} = 787.3\text{ A},$$

$$I_{\text{nom}\#4} = I_{\text{nom}\#3}\times\frac{11}{66} = 131.2\text{ A}.$$

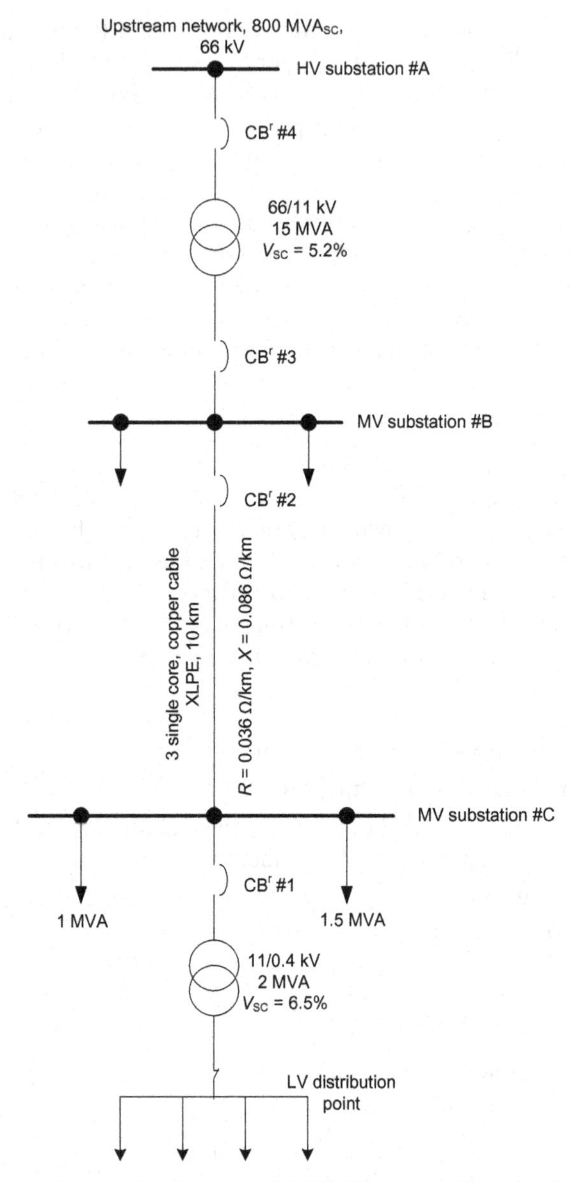

Figure 5.11 System for Example 5.1. XLPE = cross-linked polyethylene.

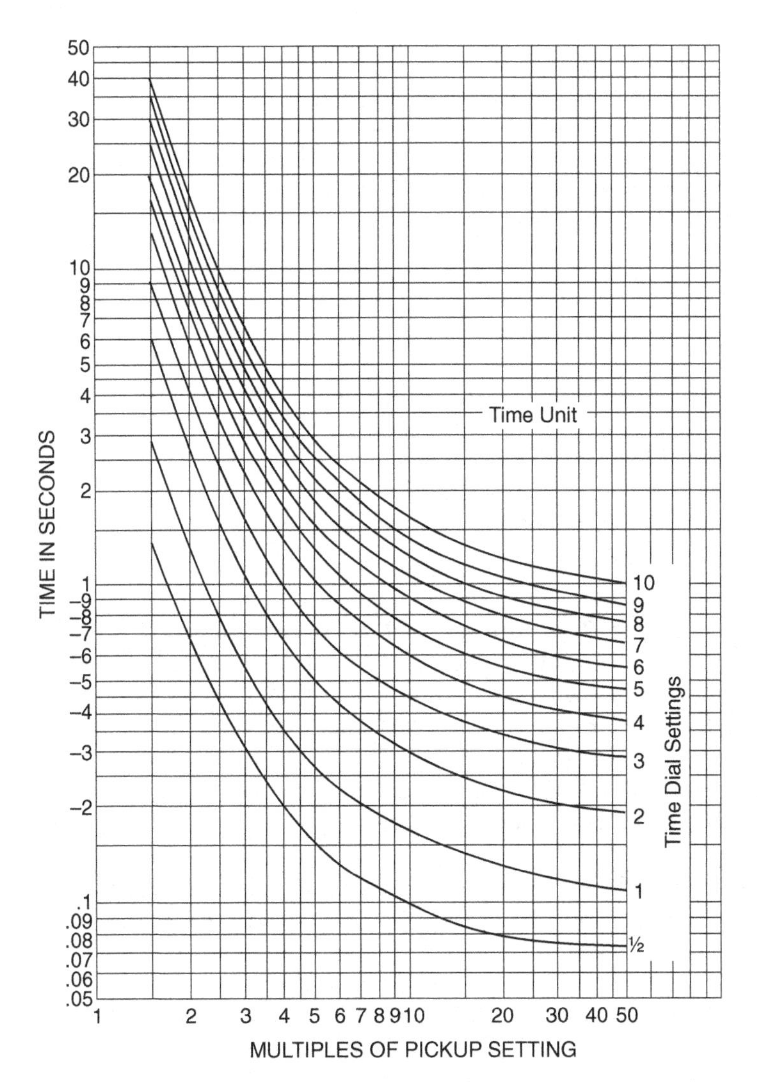

Figure 5.12 Typical time-current characteristics for overcurrent inverse-time relays [46].

Short Circuit Current at Each CB "I_{Fi}" (Referred to HV Side):

$$Z_{\text{upstream}} = \frac{(66\times 10^3)^2}{800\times 10^6} = 5.44\ \Omega,$$

$$Z_{\text{T}} = \frac{5.2}{100}\times\frac{(66\times 10^3)^2}{15\times 10^6} = 15.1\ \Omega,$$

$$Z_{\text{BC}} = 0.086\times 10\ (66/11)^2 = 30.96\ \Omega,$$

Then,

$$I_{FC} = \frac{66 \times 10^3}{\sqrt{3}(30.96 + 15.1 + 5.44)} = 739.9 \text{ A},$$

$$I_{FB} = \frac{66 \times 10^3}{\sqrt{3}(15.1 + 5.44)} = 1855.2 \text{ A},$$

$$I_{FA} = \frac{66 \times 10^3}{\sqrt{3}(5.44)} = 7004.6 \text{ A}.$$

The short circuit level at CB #3 is the same as that at CB #2.

The fault currents at substations #B and #C are needed to be referred to the MV side when calculating CTR. Their values are

$$I_{FC} = 739.9\ (66/11) = 4439.4 \text{ A},$$

$$I_{FB} = 1855\ (66/11) = 11131.2 \text{ A}.$$

Ratio of Current Transformers (CTR) Connected at the Locations of CBs: Selecting a Ct with 100 A core and total burden of 1 Ω, the short-circuit current must be below the saturation level of the Ct. This can be satisfied by checking the relation $I_{SC}(5/I_{PCt}) \leq 100$ A, where I_{PCt} is the upper limit of the Ct primary current and $(I_{PCt}/5)$ is the CTR.

The value of I_{PCt} is, therefore, determined by the larger of the two values, I_{nom} and $[I_{SC}(5/100)]$, as given in Table 5.1.

Pickup Setting: Pickup setting of all relays is calculated by applying Equation 5.1 and taking the overload factor $K_{ld} = 1.5$, and determining the corresponding adequate setting step.

Relay #1: Pickup setting = 1.5 × 105 (5/300) = 2.625, thus, set at 3 A.
Relay #2: = 1.5 × 236.2 (5/600) = 2.95, thus, set at 3 A.
Relay #3: = 1.5 × 787.3 (5/800) = 7.38, thus, set at 8 A.
Relay #4: = 1.5 × 131.2 (5/400) = 2.46, thus, set at 3 A.

TABLE 5.1 Determination of CTR

Relay #	I_{nom} (A)	I_{SC} (A)	(5/100) I_{SC}	CTR	Pickup (A)
1	105	4,439.4	221.97	300/5	3
2	236.2	11,131.2	556.56	600/5	3
3	787.3	11,131.2	556.56	800/5	8
4	131.2	7,004.6	350.23	400/5	3

Time Dial and Instantaneous Settings: The inverse-time characteristics shown in Figure 5.12 illustrate the variation of relay tripping time versus multiples of pickup setting (MPS) at different time dial settings. MPS is defined as the ratio of short-circuit current in secondary of Ct to pickup setting, that is,

$$\text{MPS} = \frac{\text{Short-circuit current in Ct secondary}}{\text{Pickup setting}}. \tag{5.2}$$

Relay #1:

$$\begin{aligned} I_{\text{pickup-prim}} &= \text{pickup setting} \times \text{CTR} \\ &= 3\ (300/5) = 180\ \text{A} \end{aligned}$$

Selecting time-dial setting = 1.0.

This relay is located on a line ended by an MV/LV transformer. Therefore, its instantaneous current setting is 50% of the short-circuit current in the Ct secondary. Thus, instantaneous setting = 0.5 × 4439.4 (5/300) = 36.995 A. The steps are in 1 A. Accordingly, the relay instantaneous setting is 37 A.

The primary instantaneous tripping current = 37 (300/5) = 2220 A.

From Equation 5.2,

$$\text{MPS} = \frac{2220\ (5/300)}{3} = 12.33 \text{ times.}$$

From Figure 5.12, at MPS of 12.33 times and time dial setting of 1, the tripping time delay of relay #1, t_1 is 0.15 s.

Relay #2:
The tripping current of relay #1 should make relay #2 to operate in at least 0.15 + 0.4 = 0.55 s (where, discrimination time = 0.4 s). The value of MPS corresponding to this current is 2220 (5/600) (1/3) equals 6.17 times.

Therefore, time dial setting can be chosen 3, which corresponds to MPS of 6.17 and t of 0.55 s in Figure 5.12.

From Equation 5.1 and taking K_{ld} = 1.25, instantaneous setting = 1.25 × 4440 (5/600) = 46.25 A, set at 47 A.

The primary instantaneous tripping current = 47 (600/5) = 5640 A:

$$\text{MPS} = 5640\ (5/600)\ (1/3) = 15.66 \text{ times.}$$

From Figure 5.12, at MPS of 15.66 times and time dial setting of 3, the tripping time delay of relay #2, t_2 = 0.35 s.

Relay #3:
The tripping current of relay #2 (5640 A) should make relay #3 to operate in at least 0.35 + 0.4 = 0.75 s, which yields

$$\text{MPS} = 5640\ (5/800)\ (1/8) = 4.4 \text{ times,}$$

and from Figure 5.12, the time dial setting can be chosen as 3.

Bus-bar B is downstream of relay #3 and feeds other feeders in addition to feeder BC. To keep the coordination between relays installed on those feeders and their upstream relay, relay #3 is an inverse-time relay without an instantaneous unit. Its characteristic is based on short-circuit current at bus-bar B. Therefore, the short-circuit current is 11,131.2A and

$$\text{MPS} = 11{,}131.2\ (5/800)\ (1/8) = 8.7 \text{ times.}$$

With time dial setting of 3 and MPS of 8.7 times, the tripping time delay for relay #3 is 0.45 s.

Relay #4:
For I_{SC} = 11,131.2A, MPS = 11,131.2 (11/66) (5/400) (1/3) = 7.73 times.

At the same fault current, relay #4 operates in at least 0.45 + 0.4 = 0.85 s.

This yields a time dial setting of 5.

Instantaneous setting = 1.25 × 1855.2 (5/400) = 28.98A, set at 29A.

$$\text{The instantaneous tripping current referred to 66 kV} = 29\ (400/5) = 2320 \text{ A.}$$

$$\text{The instantaneous tripping current referred to 11 kV} = 2320\ (66/11). = 13{,}920 \text{ A.}$$

Thus, MPS = 2320 (5/400) (1/3) = 9.66 times and time dial setting is chosen as 5, which yields the operating time t_4 = 0.73 s.

The results are summarized in Table 5.2 and relay coordination curves are plotted in Figure 5.13.

Percentage of the BC Line Length Covered by the Instantaneous Unit of Relay #2: The current at the end of line BC "$I_{BC\text{-}end}$" is given as

TABLE 5.2 Summary of Results

Relay #	Time Dial Setting	Instantaneous Setting (A)	Relay Operating Time (s)
1	1	37	0.15
2	3	47	0.35
3	3	Disable	0.45
4	5	29	0.37

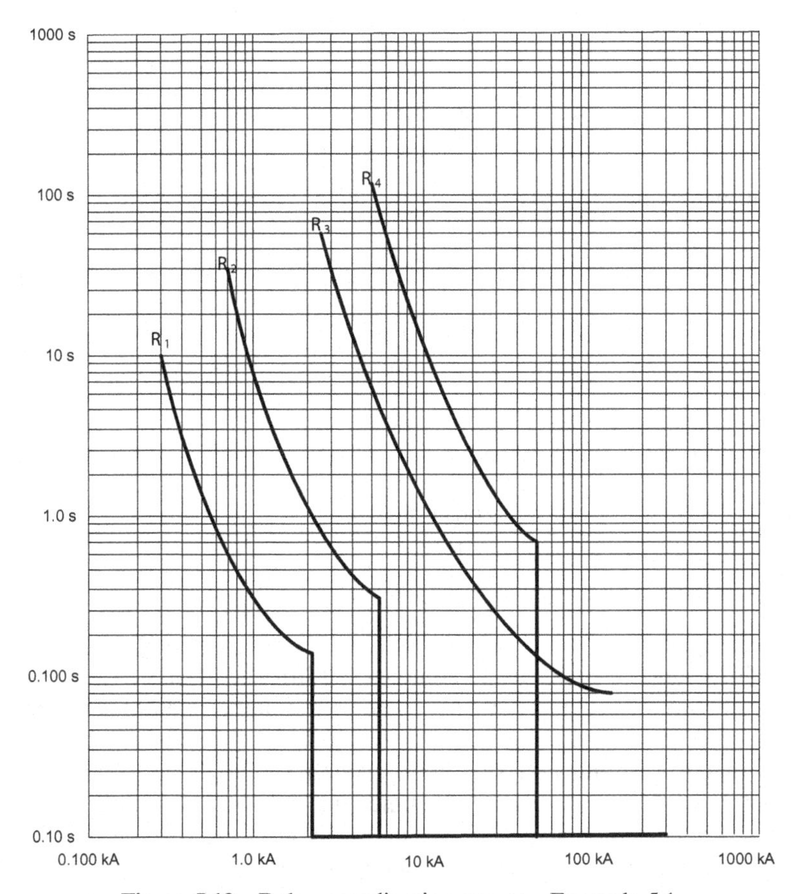

Figure 5.13 Relay coordination curves—Example 5.1.

$$I_{\text{BC-end}} = \frac{V_s}{Z_{\text{upstream}} + Z_T + Z_{BC}}, \tag{5.3}$$

where

V_s = source phase voltage,
Z_{upstream} = upstream network impedance,
Z_T = transformer impedance, and
Z_{BC} = line BC impedance.

Assuming L is the percentage of the protected line length. Thus, the minimum current of relay pickup is given as

$$I_{\text{pickup}} = \frac{V_s}{Z_{\text{upstream}} + Z_T + LZ_{BC}}. \tag{5.4}$$

Assuming

$$Z_S = Z_{upstream} + Z_T, K_1 = \frac{I_{pickup}}{I_{BC\text{-}end}}, K_2 = \frac{Z_S}{Z_{BC}}.$$

Equations 5.3 and 5.4 give

$$L = \frac{1 + K_2(1 - K_1)}{K_1} \tag{5.5}$$

$$K_1 = \frac{5640}{4439.4} = 1.27 \qquad K_2 = \frac{5.44 + 15.1}{30.96} = 0.66$$

Applying Equation 5.5,

$$L = \frac{1 + 0.66\,(1 - 1.27)}{1.27} = 64.57\%.$$

As seen from Example 5.1, determination of relay's settings can be easily obtained by hand calculations, to some extent, for radial or simple interconnected distribution systems. For large systems, manual calculations will be too hard. Therefore, it is more expedient to develop computer algorithms to perform this process and analyze different system topologies as well. Computer algorithms are designed to (i) determine the current for setting the relays when a fault occurs at a specific location, (ii) identify the pairs of relays to be set to determine which one acts as a backup for the other, and (iii) ensure the adequacy of relay coordination. To obtain these outputs, it necessitates entry data that mainly includes short-circuit currents for faults at system bus bars, system margins and constraints, available settings of relays to be coordinated, and the location of relays with respect to the loads and boundaries of other protected zones in the system. If the results obtained are inadequate, the algorithm repeats the process using lower coordination margins and new relays that have different characteristics. The process is terminated when the algorithm results in reasonable coordination.

5.3.3 Earth-Fault Protection

Earth faults are the most common types of faults that may occur in distribution networks. So, the protection system must have a scheme to protect the network against such faults. Earth-fault current flows through the fault to earth with a value that depends on the network configuration, type of fault (e.g., L-E, 2L-to-E), fault location, and type of earthing system. This value is the summation of phase currents and is not equal to zero as in normal balanced operating system conditions. It is known as residual current I_{res} and thus,

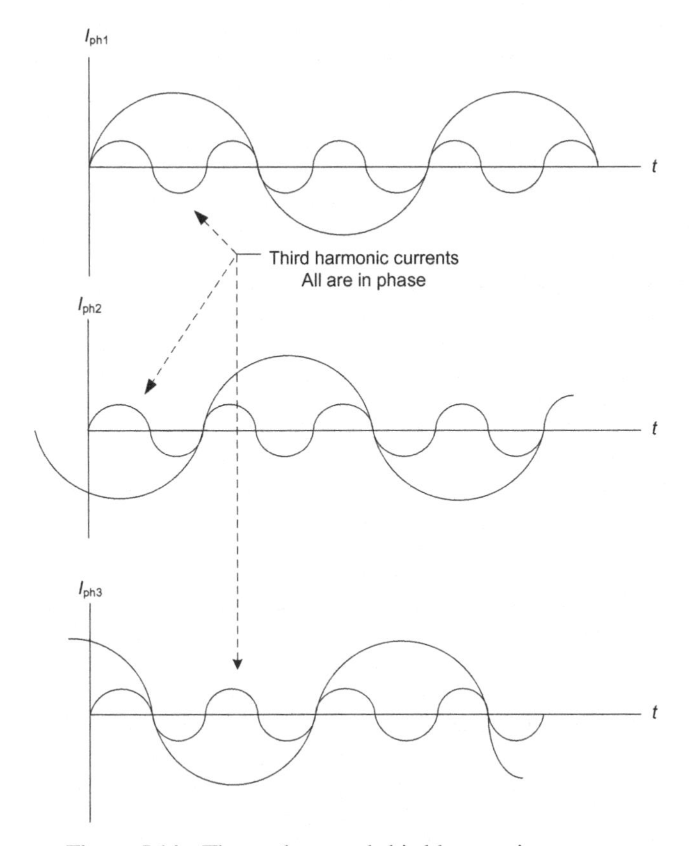

Figure 5.14 Three-phase and third harmonic currents.

$$I_{res} = \sum I_{ph} = I_{ph1} + I_{ph2} + I_{ph3} > 0.$$

Earth-fault protection scheme operates in a way similar to that applied to overcurrent protection as far as the characteristics are concerned. It is required to be highly sensitive to fault currents that might have low values, in particular, when the fault impedance or neutral earthing impedance is high. On the other hand, the current setting threshold of earth fault protective relays should be above the neutral current resulting from permitted unbalance operating conditions to avoid a risk of false tripping. Furthermore, the protection system should be insensitive to third harmonic and its multiples. Third harmonic component of three-phase currents are equal in magnitude and have the same direction (Fig. 5.14). This yields, in absence of fault, the residual current to be three times the third harmonic of a single-phase current. The same is applied to multiples of third harmonics as in the analysis below.

The three-phase balanced currents are

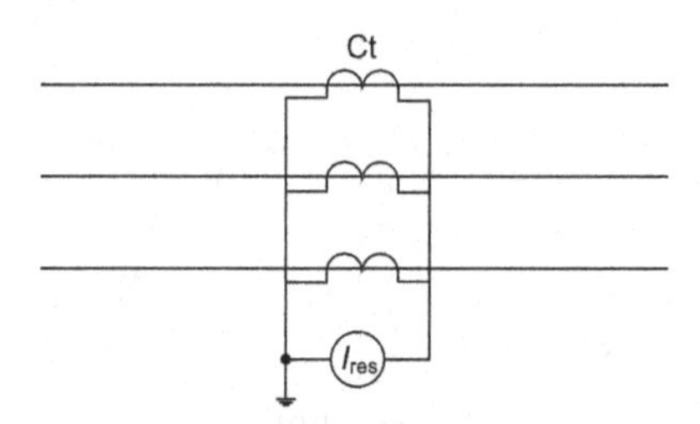

Figure 5.15 Residual current measurements.

$$i_{ph1}(t) = I_{max} \cos \omega t,$$

$$i_{ph2}(t) = I_{max} \cos\left(\omega t + \frac{2\pi}{3}\right),$$

$$i_{ph3}(t) = I_{max} \cos\left(\omega t + \frac{4\pi}{3}\right).$$

And their third harmonics are

$$i_{ph1\text{-}h3}(t) = I_{max\text{-}h3} \cos 3\omega t,$$

$$i_{ph2\text{-}h3}(t) = I_{max\text{-}h3} \cos 3\left(\omega t + \frac{2\pi}{3}\right) = I_{max\text{-}h3} \cos(3\omega t + 2\pi)$$
$$= I_{max\text{-}h3} \cos 3\omega t,$$

$$i_{ph3\text{-}h3}(t) = I_{max\text{-}h3} \cos 3\left(\omega t + \frac{4\pi}{3}\right) = I_{max\text{-}h3} \cos(3\omega t + 4\pi)$$
$$= I_{max\text{-}h3} \cos 3\omega t.$$

Thus,

$$i_{ph1\text{-}h3}(t) + i_{ph2\text{-}h3}(t) + i_{ph3\text{-}h3}(t) = 3I_{max\text{-}h3} \cos 3\omega t.$$

The residual current can be measured by using three Cts connected to the distribution element as shown in Figure 5.15, making the sum $I_{res} = I_{ph1} + I_{ph2} + I_{ph3}$.

5.4 RECLOSERS, SECTIONALIZERS, AND FUSES

Most devices used in the protection of the distribution systems, rather than the relays described in the previous sections, are autoreclosers, sectionalizers, and fuses.

5.4.1 Reclosers

Studies of faults on overhead distribution lines have shown that most are transient and can be cleared without interrupting customer supply. This is

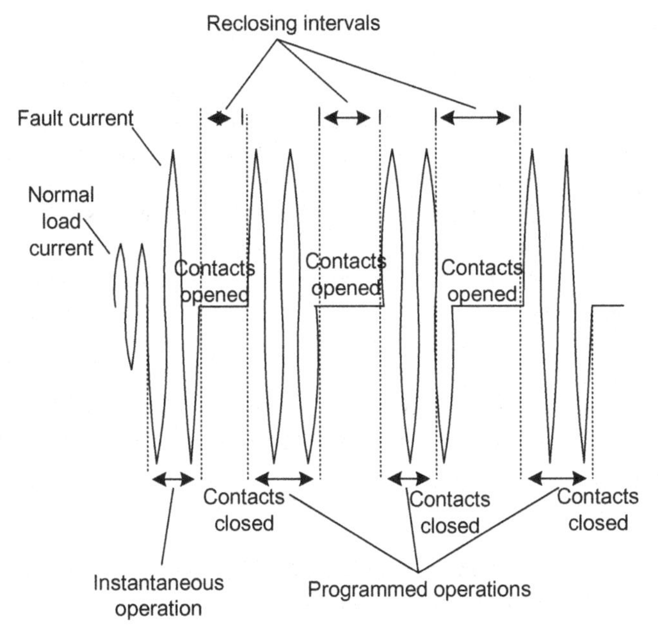

Figure 5.16 Sequence of recloser operation.

especially true of areas that experience lightning activity or mix power lines with trees.

At its most basic, a recloser will detect a fault and then open for a preprogrammed time before closing automatically. This automatic close is referred to as an autoreclose, and multiple open and close operations can be utilized to clear transient faults. If the fault is transient and cleared when the line is de-energized, the next autoreclose will restore the supply. If the fault is permanent, the recloser will eventually open and not attempt to close until instructed by an operator. This state is referred to as recloser lockout. Manufacturers have standardized on a maximum of four protection trip operations before lockout occurs. Usually, the first trip is instantaneous while the next trips are programmed in time (Fig. 5.16).

Early reclosers simply detected a level of current, opened after a fixed time, closed and repeated the sequence for a preprogrammed number of times. These devices were graded by a number of operations; that is, downstream reclosers had fewer autoreclose operations than upstream reclosers. This method of grading subjects all the customers on the affected feeder to outages, but only the first recloser upstream of the fault locks out and isolates downstream supply.

The next advance was the introduction of current dependent time to trip. Terminology widely used to describe current-dependent operating speed is IDMT curves. IDMT application in the first reclosers used electric and hydraulic mechanisms to provide a variable period from detection of fault current to opening the recloser. This time was inversely proportional to the magnitude of the fault current and allowed devices in series to be graded on current through the use of Cts with different turns ratios. These reclosers also incorporated mechanical means to provide a time multiplier to the curve, allowing better coordination of series devices with different operating characteristics. Grading on current relies on downstream reclosers opening more quickly at a given fault level than upstream reclosers. Only the recloser closest to the fault operates; upstream supply is not interrupted at all. Hydraulic operating mechanisms are limited by a TCC governed by the operating mechanism and the inaccuracy of the time to trip on any given fault current. This inhibits accurate grading resulting in compromises having to be made in the accommodation of all fault conditions.

Subsequent generations of reclosers benefited from the introduction of electronic relays to refine flexibility and accuracy of time to trip. The advent of microprocessor control has led to the current generation of reclosers with increasingly sophisticated protection capabilities.

5.4.1.1 *Locations of Reclosers Installation* Reclosers can be installed at different locations in distribution systems such as the following:

- *High Lightning Exposed Overhead Lines*: An electric storm can generate extremely large fault currents with enormous potential for damage. A fast tripping time reduces equipment risk and danger to public safety. A time of 75 ms between the initiation and clearing of a fault can be achieved by utilizing an instantaneous element. Use of instantaneous operation can allow fuse-saving strategies to be employed to prevent fuses on an upstream spur or transformer from blowing. This strategy is generally only successful with large fuses.
- *Remote Site*: Remotely controlled reclosers allow fast switching and network reconfiguration to minimize customer time lost. This can provide immediate benefits in customer satisfaction and reduce call out rates when travel times are significant.
- *Radial Feeders*: Long radial feeders require multiple protective devices to provide overlapping zones of protection. This is achieved through coordination of multiple reclosers in series to minimize the possibility of low-level faults not being seen by the nearest upstream device.
- *Substations*: To provide primary protection of a particular circuit.
- *Branches*: Reclosers are used to prevent the tripping of the main feeder when faults occur in branches.
- *Cogeneration Interchange Sites*: Cogeneration plants can use reclosers at the interchange sites. This allows connection of separate supplies once they are synchronized and prevents closing onto unsynchronized supplies. Under and over frequency tripping allows load to be progressively shed and restored without operator intervention.

5.4.1.2 Series Reclosers Coordination To achieve coordination on series reclosers, the operating time of a recloser must be faster than any upstream device and slower than any downstream device. A safe margin between operating times of successive devices must be maintained for all fault levels on the segment of the network being protected.

Preprogrammed IDMT curves or definite time to trip can be utilized for phase and earth overcurrent protection. Provision of curves to International Electromechanical Commission (IEC) and Institute of Electrical and Electronics Engineers (IEEE) standards allows close grading with substation protection relays. Therefore, the principle of coordination between two successive reclosers is mainly based on time separation between the operating characteristics.

A typical autorecloser mounted on a wood pole and the associated control and communication cabinet (CCC) are shown in Figure 5.17. The CCC houses

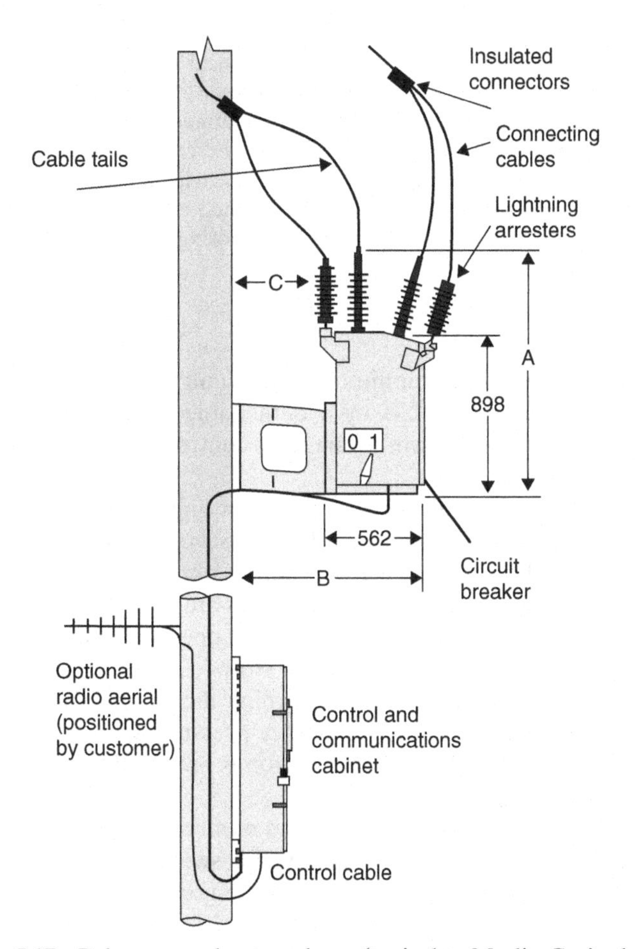

Figure 5.17 Pole-mounted autorecloser (typical to Merlin Gerin design).

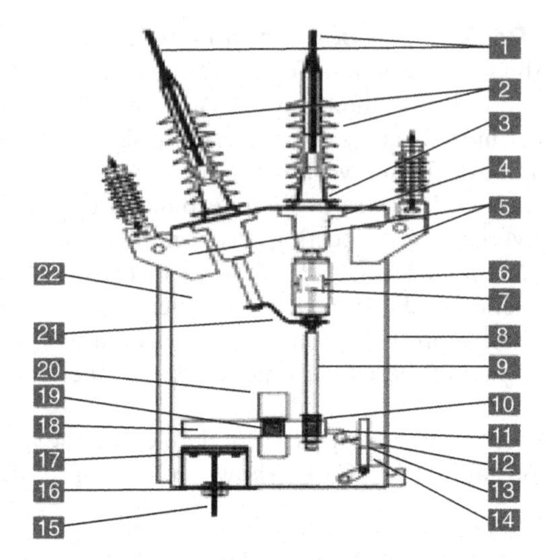

Figure 5.18 Cross section of a circuit breaker (typical to Merlin Gerin three-phase circuit recloser). 1 = Insulated cable tails; 2 = polymeric brushing boot; 3 = bushings; 4 = current transformer; 5 = lightning arresters (optional) and mountings; 6=vacuum interrupter; 7 = contacts; 8 = stainless steel tank; 9 = contact pushrod; 10 = contact pressure spring; 11 = latch; 12 = trip bar armature; 13 = trip bar; 14 = trip coil; 15 = cable to control cabinet; 16 = cable entry cover; 17 = switch cable entry module (SCEM); 18 = mechanism plate; 19 = opening spring; 20 = closing solenoid; 21 = flexible connecting arm; 22 = gas-filled tank.

an electronic controller that monitors the CB and provides protection functions. It is connected to the CB by a detachable control cable. Combined together, the CB and CCC form a remotely controlled and monitored automatic circuit recloser.

The cross section of a typical CB is shown in Figure 5.18. It is operated by sending a controlled pulse of current from a storage capacitor in the CCC through a solenoid. This attracts the mechanism plate, which in turn closes the contacts in the vacuum interrupter. The contacts are held in the closed position by latch tongues resting on the trip bar. Contact opening is achieved by releasing a controlled pulse of current from the storage capacitor through the trip coil. This attracts the trip bar armature, turning the trip bar and releasing the latch. The opening spring and the contact pressure springs accelerate the opening of the contacts. A flexible connection is provided to allow movement of the contacts to occur.

All relevant calibration data, ratings, and number of operations conducted are stored in a memory that is used by the switch control entry module (SCEM). This switch interfaces the CCC to the recloser.

To specify the autorecloser, some major points should be considered such as single-phase or three-phase, control mechanisms (hydraulic, electronic, etc.), insulation used (air, vacuum, or SF_6), installation location (indoor, outdoor), system voltage, short-circuit level (maximum and minimum), maximum load current, coordination required with other devices (upstream and downstream), and sensitivity of operation for earth faults.

5.4.2 Sectionalizers

To improve open-ring security, one or more sectionalizing CBs with time graded overcurrent and earth-fault protection can be used between the primary feeder and the normal open point (NOP). Such schemes have not proven popular due to two limitations:

1. lack of remote indication of the sectionalizing breaker status which means that in case of fault, customers could be off supply for a long time if nobody alerts the utility; and
2. long fault clearance times at the primary feeder breaker (>1 s) cause stresses on the MV cable sheaths and extend the duration of the voltage dip.

Motorized sectionalizing CBs equipped with remote terminal units (RTUs) and telecommunications equipment have been introduced to overcome the first limitation. Supply restoration can take place within minutes, gaining a significant improvement in security and availability. Telecontrol [66] can also be used in conjunction with modern numerical relays to alter the IDMT protection settings remotely when the MV network is reconfigured. An open ring with a single IDMT graded sectionalizing point, which effectively creates two zones, is shown in Figure 5.19. The improvement of security resulting from this scheme is expected to be 25%.

For further improvement on the open-ring performance, more than two zones are needed. To avoid the penalty of long tripping times, overcurrent with logic selectivity could be used. In this scheme, the IDMT relays at the primary feeder and at the sectionalizing breakers need to have a "start" output contact and a "block" input signal. IDMT relays in the sectionalizing points have the same setting as the primary feeder to allow the clearance of any MV/LV transformer fault. Upon detection of a fault, all relays will instantaneously operate their "start" contact to send a signal to the relay upstream to prevent it from tripping. The relay nearest to the fault does not receive a blocking signal and, therefore, trips. The blocking signal is automatically removed 200 ms after a trip has been issued to provide backup in case the downstream breaker fails to clear the fault. As the relays are a distance apart, a communication channel is required to send the blocking signal. Three sectionalizing CBs with IDMT protection and logic discrimination, effectively creating four zones are shown in Figure 5.20. This gives probably the highest cost/benefit ratio, with an expected 37% improvement in security.

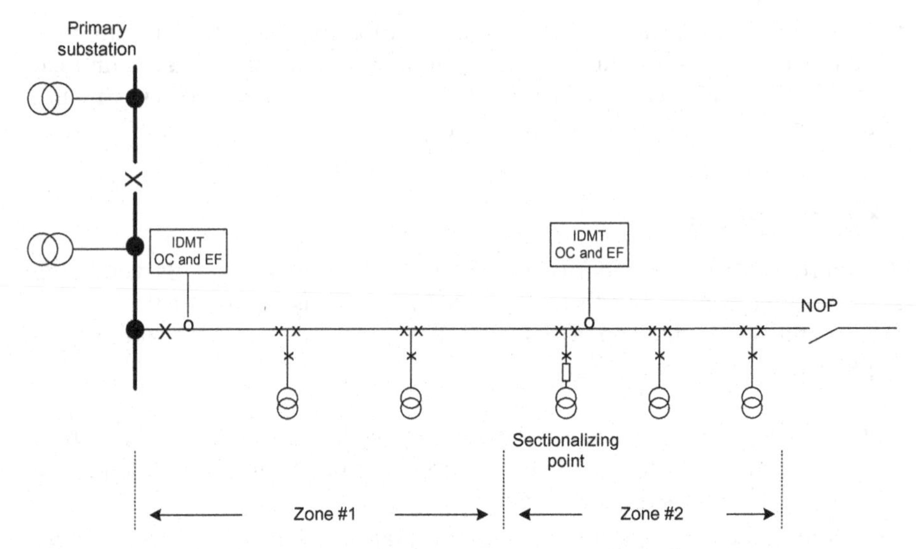

Figure 5.19 Open ring with single sectionalizing point [67]. OC = over current; EF = earth fault.

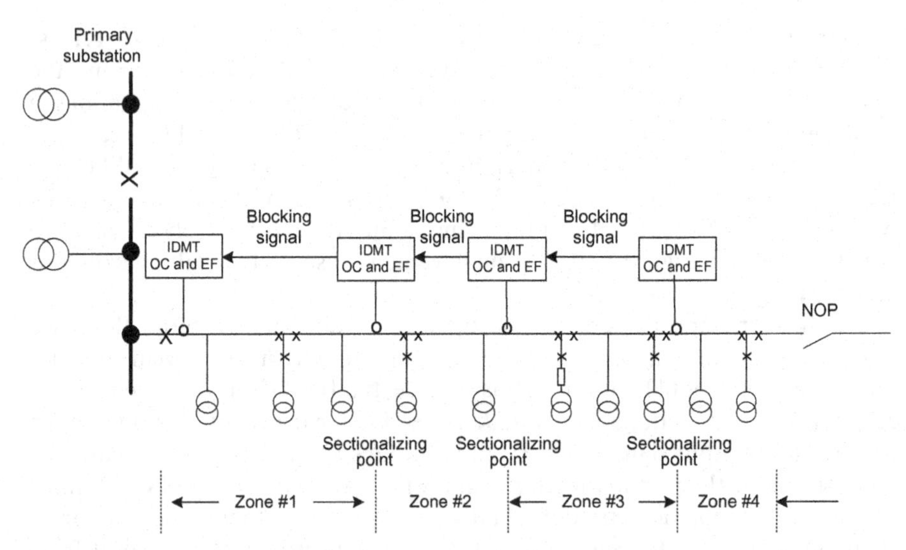

Figure 5.20 Open ring with three sectionalizing points, IDMT, and logic selectivity [67].

Recent advances in technology have changed the alternatives that are available to utilities to improve the performance of their MV distribution networks. The main developments are

- outdoor SF_6 CBs available as single units or as part of ring main units (RMUs),
- microprocessor-based protection relays,

- micro-RTUs and dedicated control cabinets for the remote control of secondary MV switchgear, and
- reliable radio and distribution power line carrier (DPLC) telecommunication systems.

A block diagram illustrating the main components of automatic controlled sectionalizer is shown in Figure 5.21a, and its installation as pole-mounted sectionalizer is shown in Figure 5.21b.

In Figure 5.21a, a motor is used in feeder automation schemes to facilitate remote control. The motor is fitted to the switch and its power is supplied from batteries installed in the pole-mounted CCCs. Cts and capacitive voltage transformers (CVTs) are built into the tank. These are connected to the control electronics in the control cabinet to provide fault detection, line sensing, and measurement. The control cubicle houses the operator control panel, microelectronics, and the electronic controller that monitors the load-break switch and provides sectionalizer functions. Combined together, the load-break switch and control cabinet form a remotely controlled and monitored pole-mounted load-break switch/sectionalizer.

5.4.3 Fuses

A fuse is an overcurrent protection device that includes both sensing and interrupting elements. It responds to a combination of magnitude and duration of circuit current flowing through it. A fuse can carry continuous current, and, therefore, when this current exceeds its rated continuous current due to any fault, it responds to de-energize and interrupt the affected phase or phases of the circuit or equipment that is faulty. The mechanism of interruption is performed by two processes: thermal process and interrupting process.

Thermal Process: The current-responsive element in the fuse is sensitive to the current flowing through it. Its temperature rises when heat is generated as a result of the current flow. For a current less than or equal to the fuse rated continuous current, the fuse is in a stable condition where the heat generated equals the heat dissipated. When a current higher than rated continuous current flows through the fuse with sufficient magnitude, it causes the current-responsive element to melt before other steady-state temperature conditions are achieved. After melting, the interruption of current must be carried out. The melting TCCs of the fuses are provided by manufacturers to illustrate the relationship between the current, which causes melting, and the time needed to melt (Fig. 5.22).

Interrupting Process: The current continues to flow through an arc after the current-responsive element melts. The arcing takes some time until the current is interrupted. It is called "arcing time" and is defined as "the time elapsing from the melting of the current-responsive element to the final interruption of the circuit." After interruption, the fuse may experience immediate

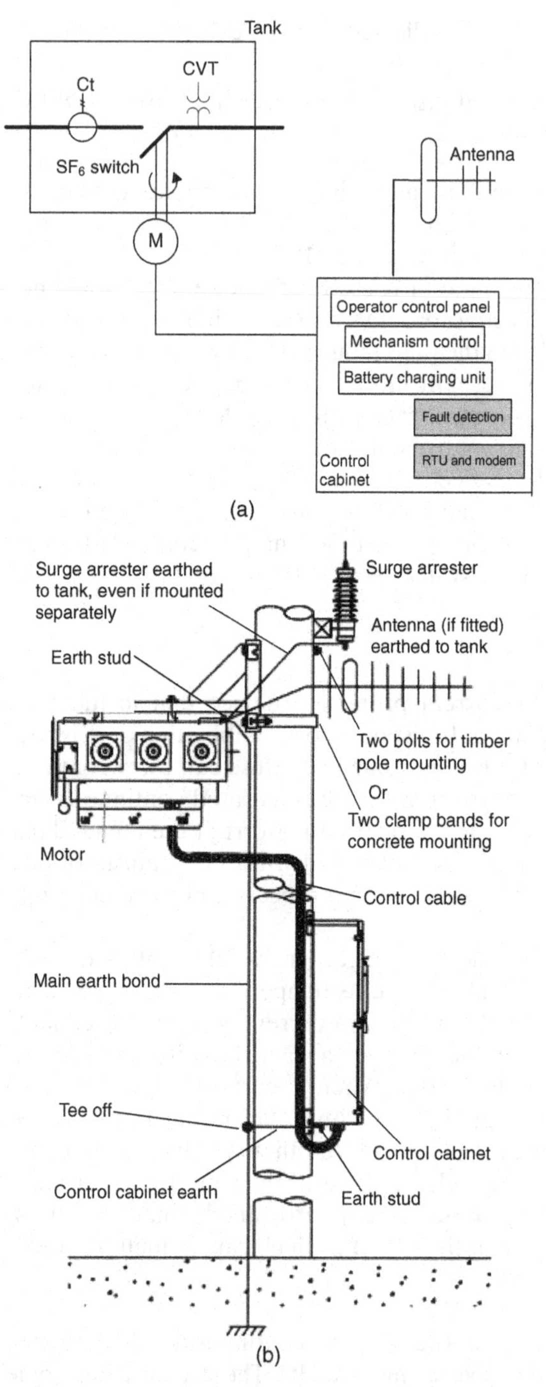

Figure 5.21 Typical sectionalizer and its controller. (a) Components of automatic controller sectionalizer (typical to Merlin Gerin manufacturing). (b) Typical pole-mounted sectionalizer.

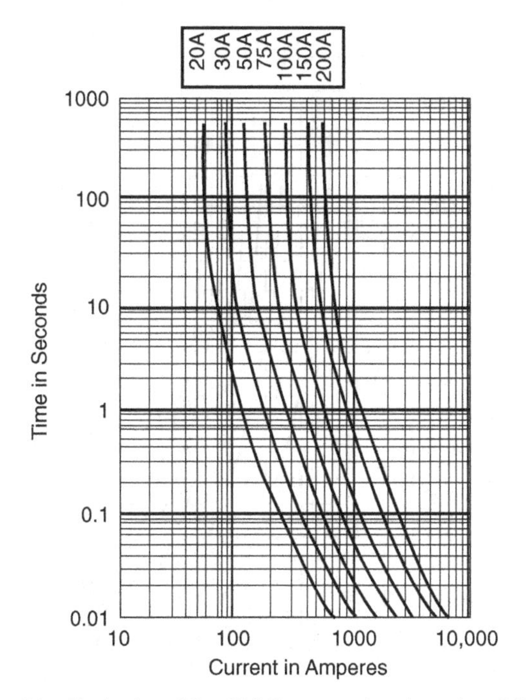

Figure 5.22 Typical melting TCC curves developed at 25°C [68].

transient recovery voltage condition and subsequent steady-state recovery voltage. It must be designed to withstand such conditions.

The arcing period in the interrupting process is added to the melting time in thermal process to give the total clearing time. Typical total clearing TCC curves are shown in Figure 5.23.

The melting and total clearing TCCs are used to coordinate with comparable characteristics of other protective devices for appropriate selectivity to protect system equipment and isolate faulted circuits.

The fuses are classified into power-class and distribution-class fuses. Power-class fuses have specifications based on particular requirements for generating sources and substations. The specifications of distribution-class fuses are more closely matched with requirements of distribution systems. Some of these specifications in accordance with American National Standards Institute (ANSI) Std. for the two classes are given in Table 5.3.

Another classification of fuses is based on interrupting characteristics. The fuses are classified into expulsion and current-limiting fuses.

Expulsion Fuses: The fuse allows the full peak of fault current to pass through it until interruption at the circuit's current zero. On the other hand, it limits the fault duration, but transient recovery voltage may occur because there is no guarantee for normal current zero to be close enough to voltage zero.

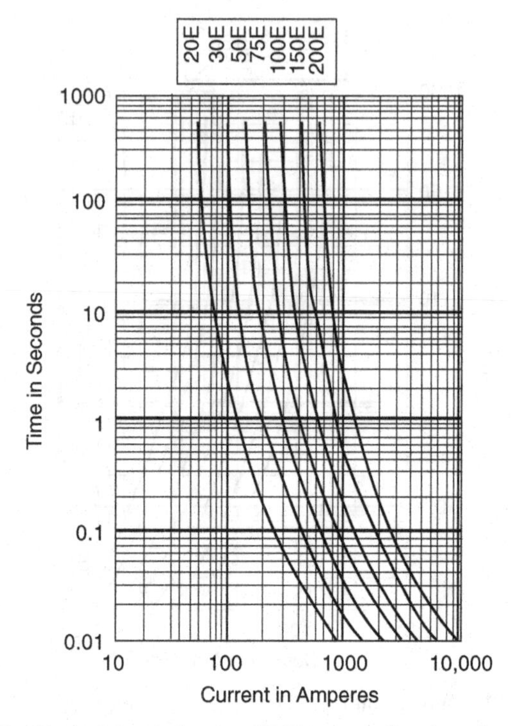

Figure 5.23 Typical total clearing TCC curves developed at 25°C [68].

TABLE 5.3 Comparison of ANSI Fuses Specifications [68]

Requirements	ANSI C37.42 Distribution Cutouts and Fuse Links	ANSI C37.44 Cutouts and Fuse Links	ANSI C37.46 Power Fuses	ANSI C37.47 Distribution Current-Limiting Fuses
Rated kV	2.6–38	2.5–15	2.8–169	2.8–27
Rated continuous current (A)	To 200	To 200	To 700	To 200
Rated interrupted current (kA)	2.6–16	2.2–7.1	1.25–80	12.5–50
(X/R) ratio at maximum interrupting current	1.33–15	2.3–12	≥15	≥10

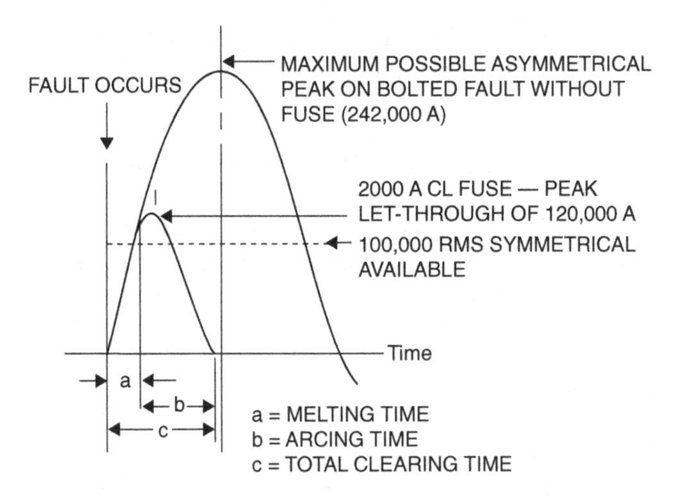

Figure 5.24 Typical current limitation showing peak let-through current and total clearing time [69].

Current-Limiting Fuses: The fuse limits both magnitude and duration of fault current. The current limitation is based on introducing quick increase of fuse equivalent resistance into the fault circuit during current-responsive element melting. This can be achieved by using long fuse element interacting with constraining and cooling medium (e.g., sand). Therefore, the fuse prevents the current from reaching its peak value. In addition, the fuse changes a high current, low power factor circuit into a lower-current, higher-power factor circuit. This results in forcing the current to near zero before the normal current zero of the circuit and close to the voltage zero.

A typical example of current-limiting fuse is shown in Figure 5.24. It indicates the maximum instantaneous current through the fuse during the total clearing time (peak let-through current) which is much less than the possible peak passing without fuse.

5.4.3.1 Fuse–Fuse Coordination The principle of fuse–fuse coordination is that the maximum clearing time for a main fuse does not exceed 75% of the minimum melting time of the backup fuse. The percent 75% is used to compensate for the effects of load current and ambient temperature, fatigue in the fuse element caused by fault currents flow through the fuse to fault downstream, which is not large enough to melt the fuse.

5.4.4 Coordination of Reclosers, Sectionalizers, and Fuses

Coordination of series of reclosers and fuse–fuse coordination have been described in Sections 5.4.1.2 and 5.4.3.1, respectively. Different combinations of devices such as reclosers, sectionalizers, and fuses may be required to be coordinated as explained below.

Recloser–Fuse Coordination: It depends on the location of both devices with respect to the source and load; that is, the recloser on the source side backs up the fuse on the load side, or vice versa.

- *Fuses on the Load Side*: The coordination must satisfy that the maximum clearing time of the fuse must be less than the recloser delayed curve. The recloser should have at least two delayed operations to prevent loss of service in case the recloser trips when the fuse operates. In addition, the minimum melting time of the fuse must be greater than the recloser fast curve.
- *Fuses at Source Side*: All recloser operations should be faster than the minimum melting time of the fuse.

Recloser–Sectionalizer Coordination: The principle of this coordination is that the number of operations of the backup recloser can be any combination of instantaneous and timed shots. The sectionalizer is set for one shot less than those of reclosers so that the sectionalizer opens and isolates the fault, and enables the recloser to reenergize the section and restore the circuit.

Relay–Recloser Coordination: This coordination is described as first, the interrupter opens the circuit some cycles after the associated relay trips and, second, the relay integrates the clearance time of the recloser. The relay reset time is long and if the fault current is reapplied before complete relay reset, the relay moves toward its operating point from this partially reset position.

5.5 DIRECTIONAL PROTECTION

This type of protection is applied to distribution systems that contain multiple sources, parallel circuits, and loops. It is necessary for such systems to be protected against fault currents that could circulate in both directions avoiding disconnection of unnecessary circuits. As normal overcurrent relays cannot provide this function, a directional unit is added to activate the relay when the fault current flow is in a predetermined direction.

A bus bar supplied by two parallel sources via two CBs and one transformer for each, and the current circulation when a fault occurs at point F where two fault currents are established is shown in Figure 5.25. The fault current I_{F1} flows from source #1 to F via CB #1, and I_{F2} flows from source #2 to F via CBs #2, 3, and 4. Using normal overcurrent relays, CB #3 may operate if I_{F2} exceeds the setting threshold of the relay associated with it providing unnecessary disconnection of source #2.

Therefore, relays associated with CBs #2 and 3 should be directional relays, which operate when the current flow is in the direction indicated in the figure. Thus, relay #2 will be active, while relay #3 will be inactive. This causes CB #2 to trip and I_{F2} is interrupted. The overcurrent relay associated with CB #1 will

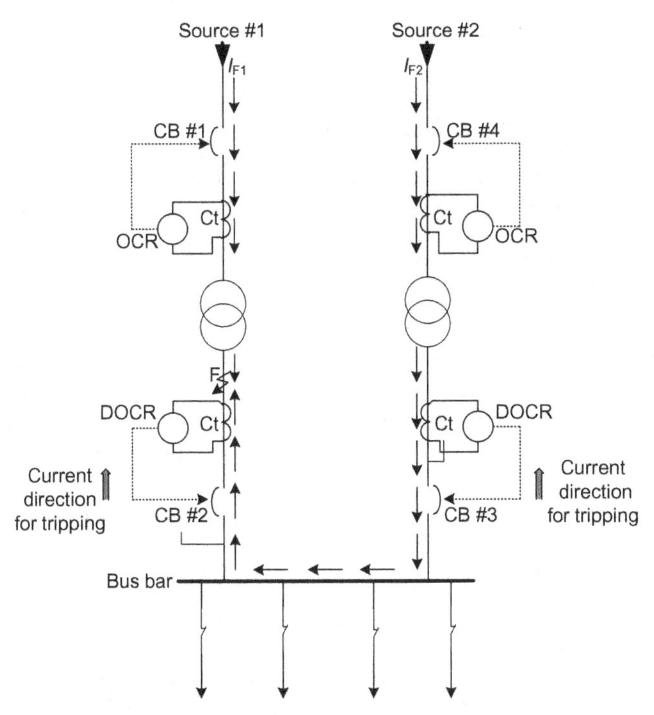

Figure 5.25 Bus bar supplied by two sources. CB = circuit breaker; Ct = current transformer; OCR = overcurrent relay; DOCR = directional overcurrent relay.

trip and I_{F1} is interrupted, and a complete isolation of faulty section is achieved. It is noted that the time delay of directional relays should be less than that of the overcurrent relays.

5.5.1 Directional Overcurrent Relays

Directional overcurrent relays consist of two units: overcurrent unit and directional unit. The overcurrent unit is actually the same as for normal overcurrent relays and has similar characteristics (definite, inverse, very inverse, and extremely inverse-time characteristics).

The directional unit determines the direction of current flow in its operating element by comparing the phase angle with a reference signal in another element called "polarizing element." The overcurrent and directional units operate jointly for a predetermined current magnitude and direction. The relay is activated only for current flow to a fault in one direction and is insensitive to the current flow in the opposite direction. Such relays can be supplied with voltage restraint on the overcurrent unit, in particular when faults result in LV and current.

Directional relays are usually directionally controlled, which means that the directional element detects, first, the direction of current and then it activates the overcurrent unit if the current direction is the same as the predetermined direction.

5.5.2 Directional Relays Operation

The phase displacement between voltage and short-circuit current must be determined to detect the current direction. The active power seen by the directional relay is positive when $-\frac{\pi}{2} \leq \varphi \leq \frac{\pi}{2}$ ($\cos\varphi \geq 0$) and is negative when $\frac{\pi}{2} \leq \varphi \leq \frac{3\pi}{2}$ ($\cos\varphi \leq 0$).

The displacement φ can be determined by comparing the phase current with a polarizing voltage as a reference. If the current is in phase "a," the polarizing voltage will be V_{bc}, which is perpendicular to the current I_a at $\varphi = 0$ (Fig. 5.26a). In the same way, if the current is in phase "c," the polarizing voltage will be V_{ab} and is perpendicular to I_c at $\varphi = 0$ (Fig. 5.26b). The angle between phase current and chosen polarizing voltage, defined as the connection angle, equals 90° at $\varphi = 0$.

The polarizing voltage is chosen as phase-to-phase voltage between the two phases rather than the phase at which the current is detected to guarantee sufficient voltage amplitude.

For directional protection, two relays, for example, on phases "a" and "c," are sufficient to protect the distribution element against symmetrical and unsymmetrical faults since at least one of the phases "a" or "c" is involved. The directional relay is activated if the three conditions below are satisfied:

- The fault current is higher than the setting threshold.
- The fault current flows for a time equal to the relay time delay.
- Current displacement with respect to the polarizing voltage is in the tripping zone.

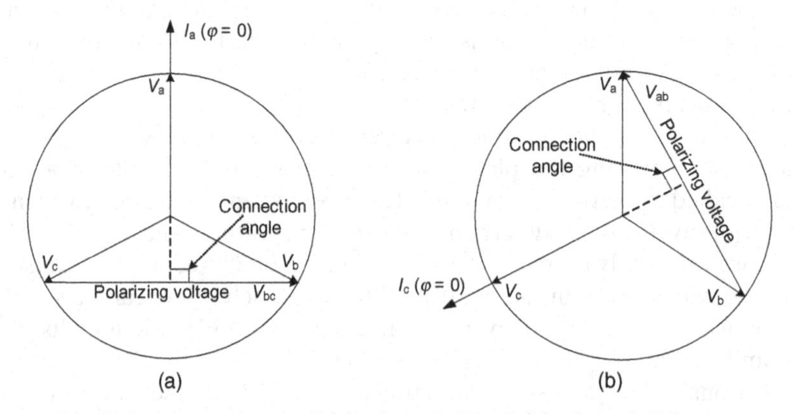

Figure 5.26 Polarizing voltage of (a) phase "a" and (b) phase "c" for a current in phase "a."

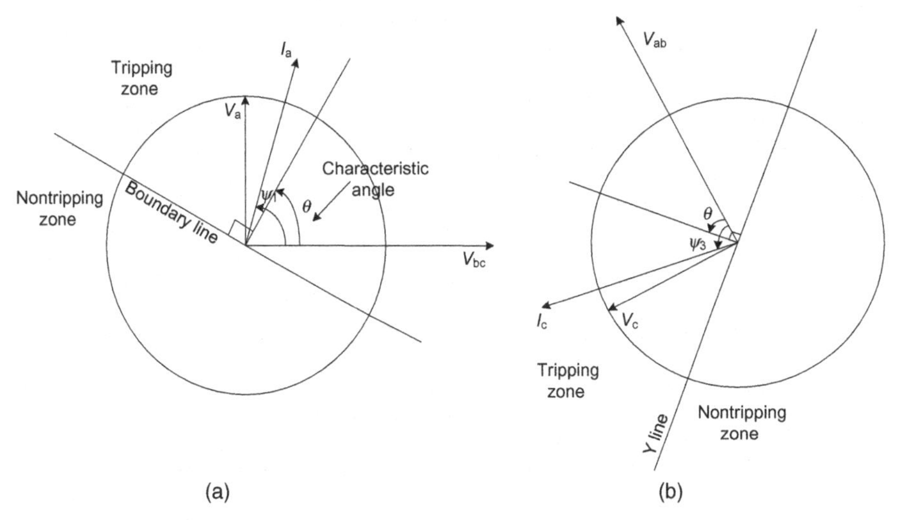

Figure 5.27 Directional protection tripping zones on (a) phase "a"; (b) phase "c" for a characteristic angle $\theta = 45°$.

The tripping zone is a half plane defined by characteristic angle θ. This angle is the angle between polarizing vector and a line perpendicular to the boundary line (Fig. 5.27). The usual values of θ are 30° or 45° or 60°. It is seen that the phase current I_a is in the tripping zone when $\theta - \pi/2 < \psi_1 < \theta + \pi/2$ and in the nontripping zone when $\theta + \pi/2 < \psi_1 < \theta + 3\pi/2$. Similarly, the current I_c is in the tripping zone when $\theta - \pi/2 < \psi_3 < \theta + \pi/2$ and in the nontripping zone when $\theta + \pi/2 < \psi_3 < \theta + 3\pi/2$. Angles ψ_1 and ψ_3 correspond to φ_1 and φ_2, respectively ($\varphi_i = \psi_i + 90°$).

Example 5.2:

A symmetrical three-phase short circuit occurs at a point with upstream impedance $R + jX$, where (R/X) is in the range 0.05–0.3 in an MV distribution system. Find the phase displacement ψ with connection angle 90°:

$$\tan\psi_1 = X/R \text{ and } 3.3 < (X/R) < 20.$$

Hence, $73° < \varphi < 87°$ and ψ_1 is negative ($3° < |\psi_1| < 17°$).
The phase current I_a is leading polarizing voltage V_{bc}.

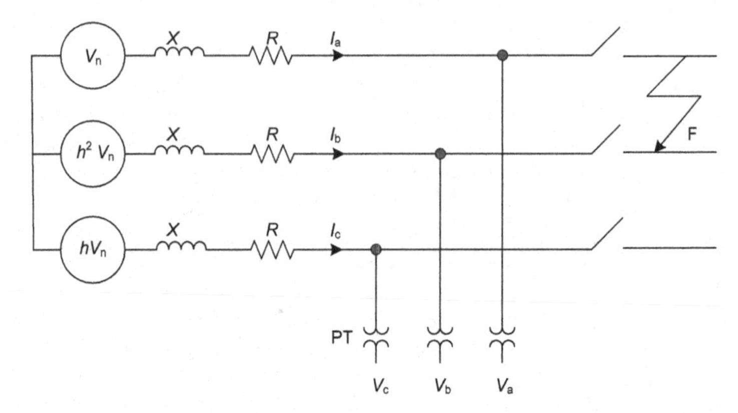

Figure 5.28 Network with L-L fault at F.

Example 5.3:

For the problem in Example 5.2, find the phase displacement ψ when a line-to-line short circuit occurs at point F as shown in Figure 5.28.

The measured voltages by vts when L-L fault occurs are

$$V_a = V_b = \frac{V_n + h^2 V_n}{2} = -h\frac{V_n}{2} \text{ and } V_c = hV_n,$$

$$I_a = -I_b = -\frac{V_n - h^2 V_n}{2(R + jX)} = -h\frac{V_n(1-h^2)}{2(R+jX)}.$$

Thus,

$$V_{bc} = V_b - V_c = -\frac{3}{2}hV_n.$$

Hence,

$$\frac{V_{bc}}{I_a} = \frac{-3h}{1-h^2}(R + jX) = \sqrt{3}(-jR + X).$$

The phase displacement ψ_1 between V_{bc} and I_a is given as

$$\psi_1 = \tan^{-1}(R/X),$$

where $0.05 < (R/X) < 0.3$.

Then, ψ_1 is negative ($3° < |\psi_1| < 17°$) and I_a is leading V_{bc}.

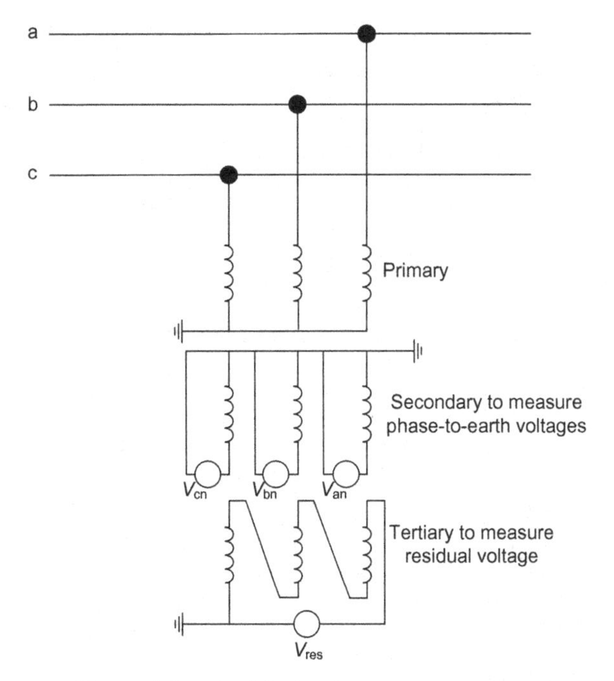

Figure 5.29 Residual voltage measurements.

5.5.3 Directional Earth-Fault Protection

Directional earth-fault protection is based on the same principles as the directional overcurrent protection. The protection is activated when the residual current exceeds the setting threshold and its phase displacement with respect to a polarizing voltage is inside a predefined tripping zone. The residual current can be measured as illustrated in Section 5.3.3. The residual voltage is taken as polarizing voltage. It is the vectorial sum of the three phase-to-earth voltages: V_{an}, V_{bn}, and V_{cn}. It can be measured by one of the two methods:

- using a device that performs the vectorial sum of the voltages V_{an}, V_{bn}, and V_{cn};
- using three-winding potential transformers as indicated in Figure 5.29.

5.6 DIFFERENTIAL PROTECTION

Differential protection is based on comparing phase current entering the protected distribution element "I_{in}" with current leaving this element "I_{out}" on the same phase. At normal operating conditions or fault occurrence outside the element, the two currents are approximately equal. If there is a difference

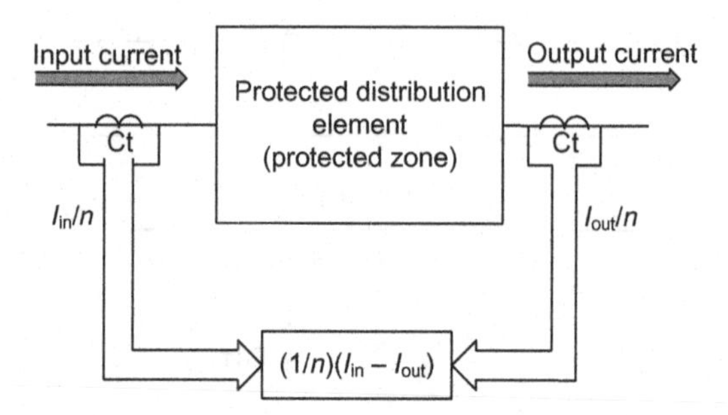

Figure 5.30 Block diagram representing differential protection scheme.

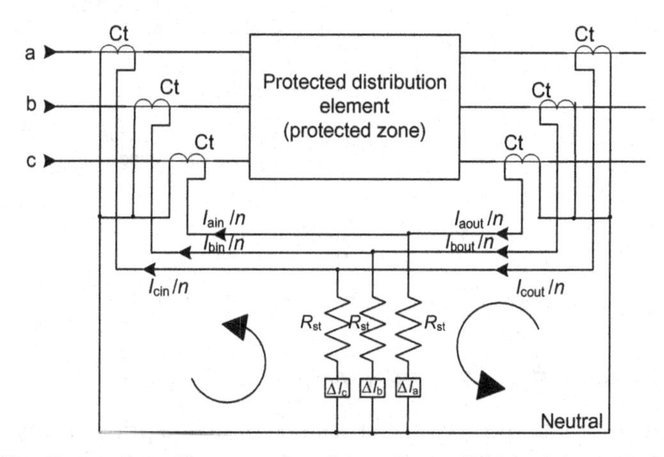

Figure 5.31 Connection diagram of a three-phase differential protection against three-phase or L-L faults.

between them, then a fault inside the element is detected and the fault current is proportional to this difference. A sample of both I_{in} and I_{out} is supplied to differential protection through Ct with turns ratio n and thus the two currents measured by the differential protection are I_{in}/n and I_{out}/n (Fig. 5.30). Each phase of differential protection includes a differential relay in series with a stabilizing resistor R_{st}. The connection diagram of a three-phase distribution element protected by differential protection against symmetrical and unsymmetrical faults is shown in Figure 5.31. It is shown that the current in differential branch $\Delta I_s = (1/n)(I_{in}-I_{out})$.

When differential protection is needed to protect a distribution element only against L-E fault, the connection diagram will be as shown in Figure 5.32. Summation of the Cts' secondary currents, on both sides of the distribution element, gives the residual currents $I_{res\text{-}in}$ and $I_{res\text{-}out}$. Thus, the current in differential branch $\Delta I_{res} = I_{res\text{-}in} - I_{res\text{-}out}$, where $I_{res\text{-}in} = (1/n)(I_{ain} + I_{bin} + I_{cin})$ and $I_{res\text{-}out} = (1/n)(I_{aout} + I_{bout} + I_{cout})$.

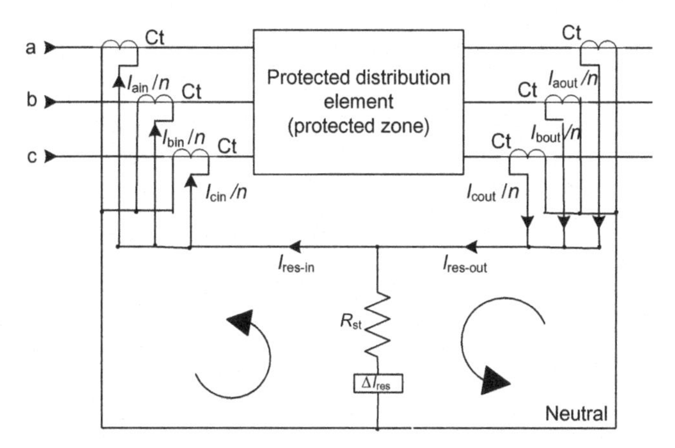

Figure 5.32 Connection diagram of differential protection against L-to-G fault.

To emphasize a proper operation of differential protection, the main aspects to be taken into account are (1) protection is insensitive to external faults, (2) presence of sufficient voltage at differential branch to activate the relay for protection against internal faults, (3) limiting the voltage at secondary terminals to avoid Ct damage, and (4) protection can detect a minimum fault current. These aspects can be achieved as below:

1. *Protection Is Insensitive to External Faults*: It is important to guarantee that the current difference flowing in differential branch is actually due to fault occurrence inside the protected element, not outside it. Because of the saturation of Cts, the secondary currents at both ends of the element may not be equal when short circuit occurs outside the protected zone. The nonperiodic component of the short-circuit current may reach 80% of alternating current (AC) component peak value for the duration of a few cycles (two to five cycles). So, this nonperiodic component may cause Cts to saturate regardless of the short-circuit current value (if it is less than the accuracy limit) and secondary voltage produced by this current (if it is less than the knee-point voltage). In addition, Cts are not saturated at the same level, that is, one may saturate more than the other, because their manufacturing is not identical. Consequently, the current difference in the differential branch is not equal to zero and may cause spurious activation of protection.

   The risk of Ct saturation problem can be avoided by connecting the stabilizing resistor R_{st} in the differential branch to limit the current difference due to Ct saturation to be lower than setting threshold when short circuit occurs outside the protected zone. Connection diagram including the equivalent circuit of Cts is shown in Figure 5.33. R_{st} value can be calculated as below.

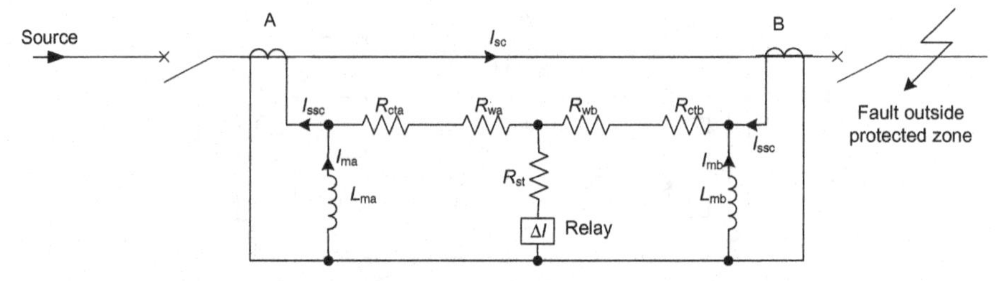

Figure 5.33 Connection diagram indicating Ct equivalent circuits.

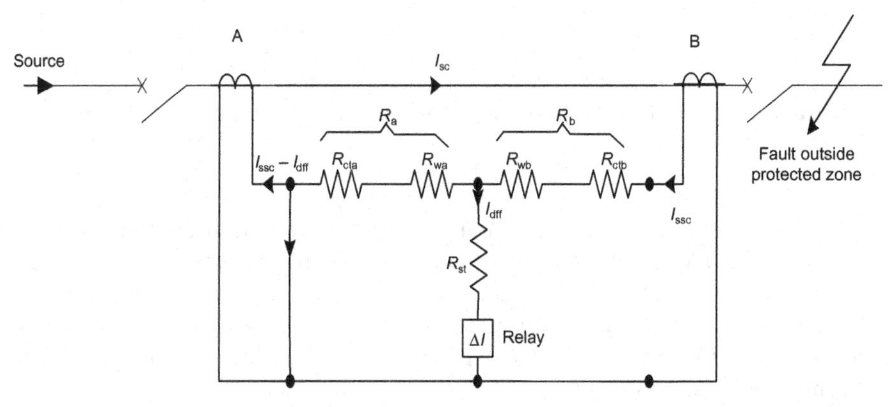

Figure 5.34 Simplified connection diagram.

TABLE 5.4 Ct Operating Conditions

Ct location	Status	Magnetizing Current	Magnetizing Inductance
End (A)	Saturated	High value	$L_{ma} \approx 0$, can be neglected
End (B)	Unsaturated	Very low value	Very high value, $L_{mb} \approx \infty$

Assume that R_{st} includes relay resistance and the Ct at end A is saturated while the Ct at end B is not saturated. In this, the most problematic case, Ct at end B supplies the actual short-circuit current in its secondary, $I_{ssc} = I_{sc}/n$. The operating conditions of Cts are given in Table 5.4.

The wires connecting Ct secondaries at both ends A and B to differential relay have a small cross-sectional area. Thus, their inductances, ωL_{wa} and ωL_{wb}, are much less than their resistances R_{wa} and R_{wb} and can be neglected. Therefore, the connection diagram in Figure 5.33 can be simplified as shown in Figure 5.34. From this Figure it is found that

$$I_{dff} = I_{ssc} \frac{R_{cta} + R_{wa}}{R_{cta} + R_{wa} + R_{st}}.$$

To avoid the risk of spurious tripping upon the occurrence of a fault outside the protected zone, the relay setting threshold should be higher than the current in differential branch, $I_{set} > I_{dff}$.

Thus,

$$I_{set} > I_{ssc} \frac{R_{cta} + R_{wa}}{R_{cta} + R_{wa} + R_{st}},$$

$$(R_{cta} + R_{wa} + R_{st}) > (R_{cta} + R_{wa}) \frac{I_{ssc}}{I_{set}},$$

$$R_{st} >> R_{cta} + R_{wa}.$$

Therefore,

$$R_{st} > (R_{cta} + R_{wa}) \frac{I_{ssc}}{I_{set}}, \qquad R_{st} > R_a \frac{I_{ssc}}{I_{set}}, \tag{5.6}$$

where $R_a = R_{cta} + R_{wa}$.

Applying the same procedure considering Ct at end (A) is not saturated and Ct at end B is saturated, R_{st} should satisfy the relation

$$R_{st} > (R_{ctb} + R_{wb}) \frac{I_{ssc}}{I_{set}}, \qquad R_{st} > R_b \frac{I_{ssc}}{I_{set}}, \tag{5.7}$$

where $R_b = R_{ctb} + R_{wb}$.

Assuming $R_{max} = \max[R_a, R_b]$, Equations 5.6 and 5.7 give

$$R_{st} > R_{max} \frac{I_{ssc}}{I_{set}}. \tag{5.8}$$

2. *Sufficient Voltage at Differential Branch Terminals*: Ct parameters must be accurately chosen so that the voltage at differential branch terminals is sufficient to activate the protection when a fault occurs inside the protected zone. For instance, the connection diagram for solid fault inside the protected zone is shown in Figure 5.35.

   The knee-point voltage V_k of a Ct is defined in accordance with Std. BS 3938 class X as the applied voltage at rated frequency to secondary terminals, which when increased by 10% causes a maximum increase of 50% in the magnetizing current.

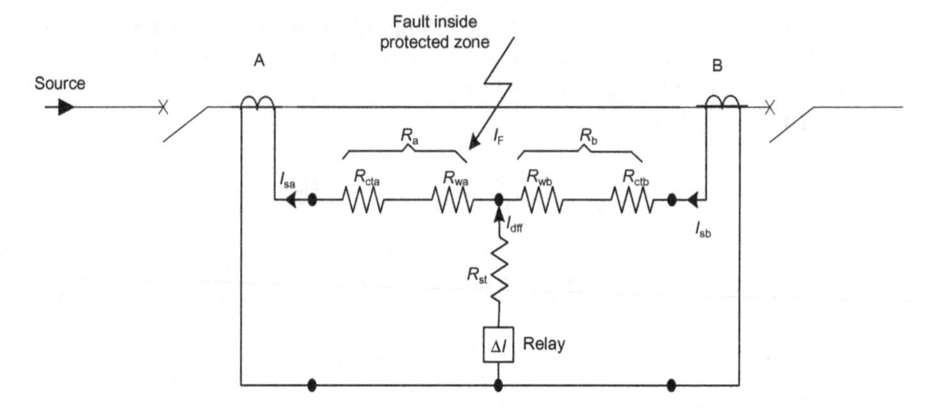

Figure 5.35 Connection diagram for internal solid fault.

The value of V_k is chosen such that when $I_{dff} = I_{set}$, $V_k \geq kV_{ter}$, where

$$1.5 \leq k \leq 2,$$

$$V_{ter} = \text{voltage at Ct secondary terminals}$$
$$= (R_{max} + R_{st})I_{set}, \text{ and}$$

$$R_{st} >> R_{max}.$$

Thus,

$$V_k \geq kR_{st}I_{set}.$$

Substituting the value of $R_{st}I_{set}$ from Equation 5.8 gives

$$V_k \geq kR_{max}I_{ssc}. \tag{5.9}$$

Satisfying this relation is essential to ensure protection activation for a fault current higher than I_{set}.

3. *Voltage Peaks Limitation*: Upon occurrence of severe fault that has maximum short-circuit current inside the protected zone, Ct is expected to be highly saturated and HV peaks occur. These voltage peaks may damage the Ct and differential relay that are not dimensioned to withstand such voltage peaks. Therefore, a nonlinear resistor R_{non} is connected in parallel with the differential branch to limit the voltage at Ct terminals to be within a permissible value.
4. *Detection of Minimum Fault Current*: The connection diagram including nonlinear resistor and magnetizing inductance of both Cts at the two ends A and B is shown in Figure 5.36.

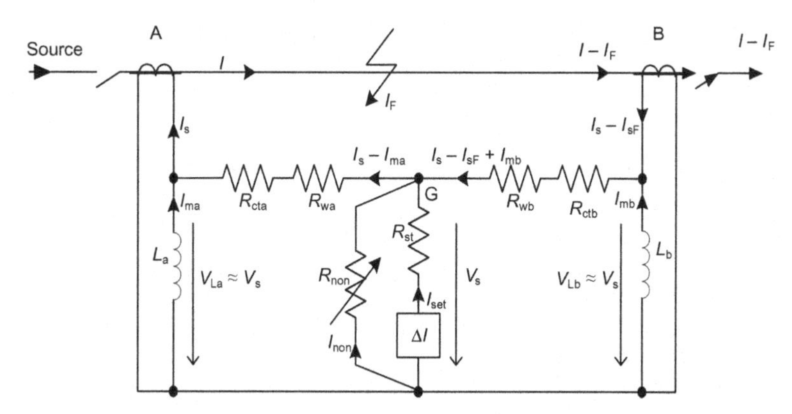

Figure 5.36 Connection diagram with nonlinear resistor.

The voltage applied to nonlinear resistor V_s is the voltage created by current flow I_{set} in stabilizing resistance R_{st} assuming that the fault current I_F causes a current I_{set} in the differential branch.

From the current flow shown in Figure 5.36 and equivalent circuit of Ct at end A, it is found that

I_{mi} = secondary magnetizing current of ith Ct,

I_s = secondary current of I and $= I/n$ (n is Ct turns ratio),

I_{sF} = secondary fault current $= I_F/n$, and

$$\begin{aligned} V_s &= I_{set}R_{st} \\ &= (I_s - I_{ma})R_{max} + V_{La} \\ &= (I_s - I_{ma})(R_{cta} + R_{wa}) + V_{La}, \end{aligned}$$

where

V_{La} = voltage across magnetizing inductance L_a,
$R_{max} = R_{cta} + R_{wa}$ assuming $R_{cta} + R_{wa} > R_{ctab} + R_{wb}$, and
$I_{ma} << I_s$.

Thus,

$$V_s = I_s R_{max} + V_{La} \text{ since } I_{ma} << I_s$$
$$V_s - I_s R_{max} = V_{La}.$$

From Equation 5.8,

$$I_{set}R_{st} > I_{ssc}R_{max}$$
$$V_s > I_{ssc}R_{max} >> I_s R_{max} \text{ since } I_{ssc} >> I_s.$$

That is I_s R_{max} can be neglected with respect to V_s and hence,

$$V_s \approx V_{La}.$$

Applying the same procedure to Ct at end B, it is found that

$$V_s \approx V_{Lb}.$$

Therefore, the voltage applied to the nonlinear resistor is approximately the same as that applied to the magnetizing inductance of both Cts at the two ends.

At point G, the Equation below can be written as

$$I_s - I_{ma} = I_{non} + I_{set} + I_s - I_{sF} + I_{mb}.$$

If the two Cts are identical, then $I_{ma} = I_{mb} = I_m$ and

$$I_{sF} = I_{set} + I_{non} + 2I_m.$$

Thus, the minimum fault current referred to Ct primary is

$$I_F = n(I_{set} + I_{non} + 2I_m). \tag{5.10}$$

To apply differential protection to distribution elements (motors, generators, transformers, etc.), Equations 5.8–5.10 must be satisfied to determine stabilizing resistance, minimum Ct knee-point voltage, and minimum detected fault current.

5.6.1 Motor Differential Protection

The main point to be taken into account is that the starting-up current of motor ($I_{st\text{-}up}$) may reach five to seven times its nominal current. The differential protection does not operate for this current by satisfying the condition

$$R_{st} \geq R_{max} \frac{I_{st\text{-}up}}{I_{set}}.$$

Ct is dimensioned to meet the condition $V_k \geq kR_{max}I_{st\text{-}up}$.

The minimum fault current that can be detected is

$$I_F = n(I_{set} + I_{non} + 2I_m).$$

The connection diagram of motor differential protection is shown in Figure 5.37.

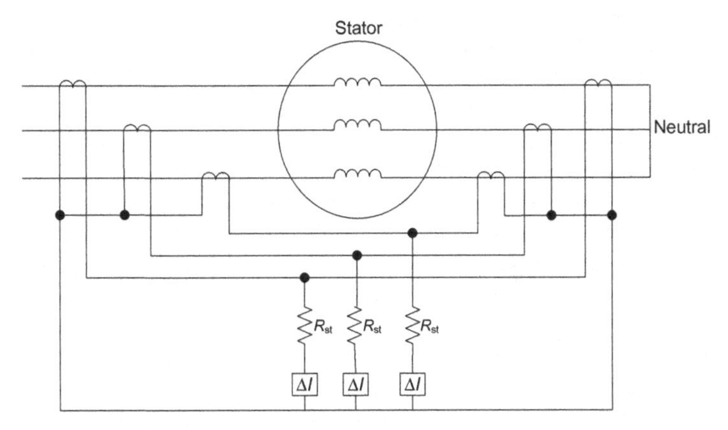

Figure 5.37 Motor or generator differential protection.

5.6.2 Generator Differential Protection

The protection of generators usually has no time delay and operates instantaneously. Generator short-circuit current I_{Gsc} is the short-circuit current during subtransient period. It may reach 5–10 times the nominal generator current. To stabilize the differential protection and avoid spurious tripping when faults occur downstream of the generator, the values of stabilizing resistance and knee-point voltage are

$$R_{st} \geq R_{max} \frac{I_{Gsc}}{I_{set}},$$

$$V_k \geq kR_{max} I_{Gsc}.$$

The detected minimum fault current is

$$I_F = n(I_{set} + I_{non} + 2I_m).$$

The connection diagram of generator protection is the same as shown in Figure 5.37.

5.6.3 Transformer Differential Protection

The connection of differential protection with transformer depends on whether its neutral is earthed or not.

1. *Differential Protection of Transformer with Earthed Neutral*: In this case, the transformer neutral is either directly earthed or earthed through limiting impedance.

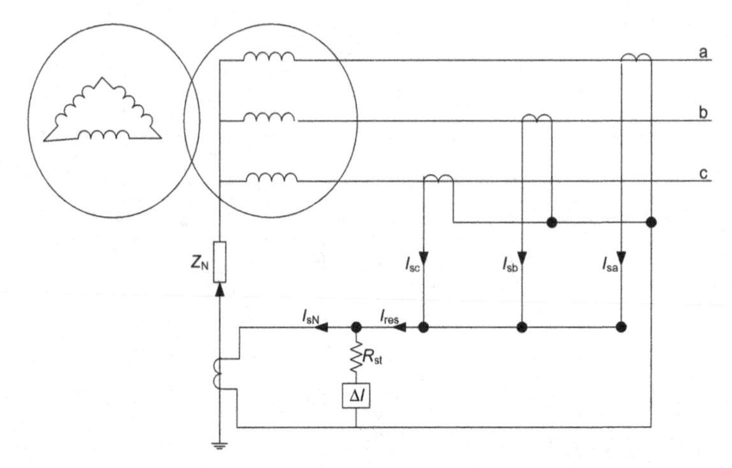

Figure 5.38 Differential protection of transformer with earthed neutral through limiting impedance.

As shown in Figure 5.38, the summation of secondary phase currents I_{sa}, I_{sb}, and I_{sc} is the residual current, which is equal to the earth-fault current. When a fault occurs inside the protected zone, the residual current flows in a direction opposite to that circulating in the earth connection and produces a current in the differential branch. Thus, the protection senses the fault.

If the earth fault occurs outside the transformer, the residual current is equal in magnitude and direction to the current circulating in the earth connection, and there is no current in the differential branch. Consequently, the differential protection is insensitive to faults outside transformer.

Differential protection must remain stable when maximum short-circuit current flows in the protected circuit. This maximum short-circuit current is when there is a three-phase short circuit at the secondary terminals and the three conditions to be met are the following:

- avoiding protection operation upon fault occurrence downstream of transformer by providing

$$R_{st} \geq R_{max} \frac{I_{sc}}{I_{set}},$$

- Ct rating is $V_k \geq kR_{max}I_{sc}$, and
- minimum fault current to be detected is $I_F = n(I_{set} + I_{non} + 4I_m)$.

The Ct magnetizing current is multiplied by 4 because there are four Cts connected to the differential branch.

2. *Differential Protection of a Transformer with Unearthed Neutral*: The maximum current at which the protection remains stable is the

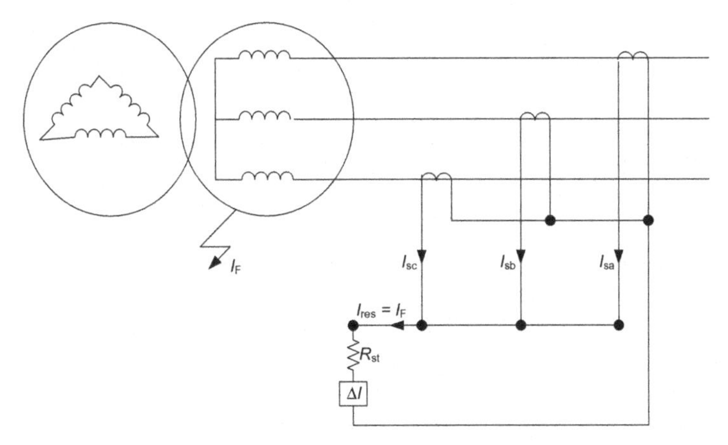

Figure 5.39 Differential protection of unearthed neutral transformer.

secondary three-phase short circuit. As shown in Figure 5.39, if the earth fault occurs downstream of the transformer, no current circulates in the differential branch where the residual current is zero. On the other hand, if the earth fault occurs inside the transformer, the residual current is equal to the fault current producing a current in the differential branch. The three conditions to be satisfied are the following:

- $R_{st} \geq R_{max} \dfrac{I_{sc}}{I_{set}}$;
- $V_k \geq kR_{max}I_{sc}$; and
- $I_F = n(I_{set} + I_{non} + 3I_m)$, where three Cts are connected to the differential branch.

5.6.4 Differential Protection of Buses

Based on differential protection principles, the application of differential protection to buses means that a comparison between currents entering the bus and currents leaving it should be provided. Normally, the bus is supplied with more than one incoming feeder to feed a number of outgoing feeders, and in some cases, the bus is sectionalized into two sections or more to improve system reliability. Therefore, differential protection is not just to compare the currents at the two ends of the protected zone because it has a number of ends. This necessitates having multicircuits to provide the sum of currents flowing in the incoming feeders and others to provide the sum of current flowing in the outgoing feeders.

The connection diagram of differential protection applied to a bus bar with two incoming and three outgoing feeders is shown in Figure 5.40 as an example to protect it against an L-G fault. The summation of currents in the three phases of each incoming and outgoing feeder is required. Then, the current

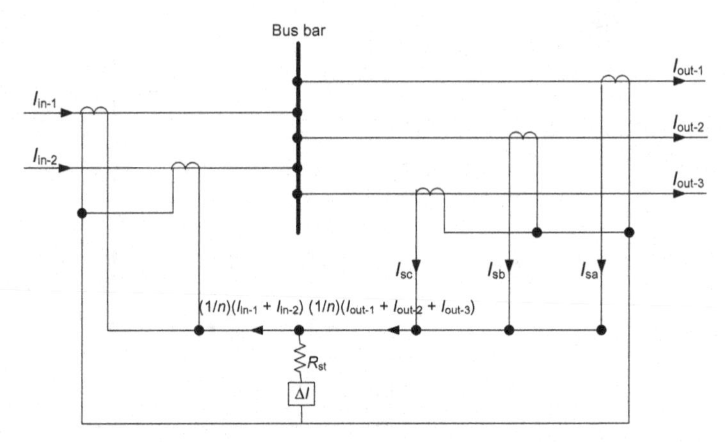

Figure 5.40 Bus-bar differential protection.

sums of the three phases of the outgoing feeders are compared with that of the incoming feeders.

The maximum short-circuit current for which the protection is stable is the sum of maximum short-circuit currents of each incoming feeder. The conditions to be satisfied for proper operation of the protection are the following:

- $R_{st} \geq R_{max} \dfrac{I_{sc}}{I_{set}}$;
- $V_k \geq kR_{max}I_{sc}$; and
- $I_F = n(I_{set} + I_{non} + HI_m)$ where, H is the number of incoming sources plus the number of outgoing feeders.

5.6.5 Differential Protection of Cables and Lines

Differential protection can be applied to protect cables or lines against short-circuit faults by using pilot wire along the cable, which is terminated at the two ends with a Ct for each. The input current to the cable is approximately equal to the output current in normal conditions. The main idea of this type of protection is how to get a current flow in the relay, which is related to the current difference to enable the protection to operate properly.

This can be done by using what is known as percentage differential protection since the current flow in the relay is proportional to the difference between the input and output currents. Because of the long length of cables, two relays are used, one at each end. The main elements necessary for constructing percentage differential protection of the cables are shown in Figure 5.41.

The pilot wire resistance R_{plt} in addition to operating resistances is assumed to be much greater than the restraining resistance R_{rest}:

$$R_{plt} + 2R_{op} >> R_{rest}.$$

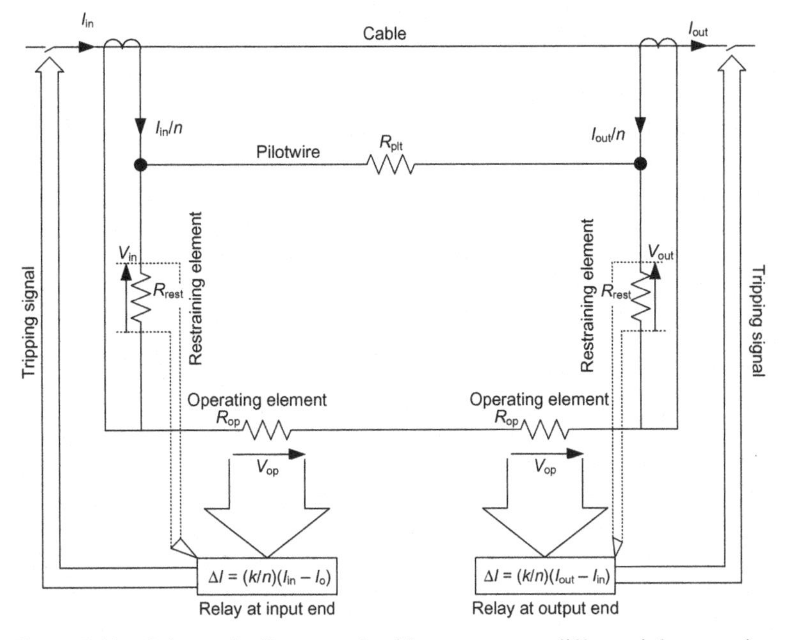

Figure 5.41 Schematic diagram of cable percentage differential protection.

The secondary current I_{in}/n is flowing approximately in R_{rest} creating the voltage V_{in}. Thus,

$$V_{in} = I_{in} R_{rest}/n$$

and similarly

$$V_{out} = I_{out} R_{rest}/n.$$

The voltage V_{op} created by current flow in R_{op} is applied to relay terminals. Its value is given as

$$\begin{aligned} V_{op} &= (V_{in} - V_{out}) \frac{R_{op}}{2R_{op} + R_{plt}} \\ &= (1/n)(I_{in} - I_{out}) \frac{R_{rest} R_{op}}{2R_{op} + R_{plt}}. \end{aligned}$$

Hence,

$$\begin{aligned} \frac{V_{op}}{V_{in}} &= \frac{I_{in} - I_{out}}{I_{in}} \times \frac{R_{op}}{2R_{op} + R_{plt}} \\ &= \Delta I \frac{R_{op}}{2R_{op} + R_{plt}} = H \Delta I, \end{aligned}$$

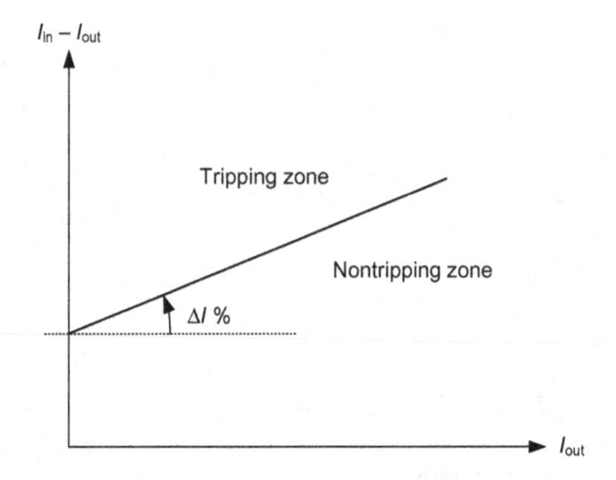

Figure 5.42 Current difference versus input current.

where

$$H = \frac{R_{op}}{2R_{op} + R_{plt}}$$

and

$$\Delta I = \frac{I_{in} - I_{out}}{I_{in}}.$$

At no-load, both the current difference $(1/n)(I_{in} - I_{out})$ and input current I_{in} are approximately equal to zero, which in turn leads to ΔI being undetermined and a false tripping may occur. To avoid this situation, the relay operates when the input current is higher than the minimum load value I_{Lo}.

Accordingly, the protection is activated when

$$I_{in} - I_{out} > I_{in}\Delta I + I_{Lo}.$$

This is depicted graphically in Figure 5.42 showing the relation between current difference and input current, which is a straight line acting as a boundary between tripping and nontripping zones.

5.7 THERMAL PROTECTION

Thermal protection is used to protect the machines (motors, generators, transformers, etc.) in distribution systems against overloads by determining the heating level of the machines.

The thermal model is defined by the differential equation:

$$\tau \frac{dH}{dt} + H = \left(\frac{I}{I_n}\right)^2, \tag{5.11}$$

where

τ = machine thermal time constant,
H = heat rise,
I_n = nominal current, and
I = rms current.

The heat energy supplied to the machine due to current I for time dt

$= RI^2 dt$
= machine heat dissipation via convection with external environment ($K\theta dt$, where K is the characteristic constant of heat exchange; $\theta = T_i - T_e$; T_i is the machine internal temperature; and T_e is the machine external temperature)
+ heat stored by machine due to its temperature rise ($mCd\theta$, m is the machine mass and C is the machine average heat capacity).

Thus,

$$RI^2 dt = K\theta dt + mCd\theta.$$

Dividing by dt gives

$$mC\frac{d\theta}{dt} + K\theta = RI^2. \tag{5.12}$$

By definition, $H = \dfrac{\theta}{\theta_n}$,
where θ_n = machine temperature—outside environment temperature, at normal conditions (when the machine is operating at its nominal current for a sufficient period with stabilized temperature):

$$\frac{d\theta}{dt} = 0 \text{ at } I = I_n \text{ and } \theta = \theta_n.$$

From Equation 5.12, it is found that $\theta_n = \dfrac{R}{K} I_n^2$.

Using the definition of H and substituting R in terms of θ_n into Equation 5.12, it can be rewritten as

$$\frac{mC}{K}\frac{dH}{dt} + H = \left(\frac{I}{I_n}\right)^2. \tag{5.13}$$

From Equations 5.11 and 5.13, it is seen that the machine thermal time constant is given as

$$\tau = \frac{mC}{K}. \tag{5.14}$$

Equation 5.14 is used by thermal overload protection together with the load current to determine the heating level of the machine and how the heat rise develops.

The initial condition is needed to solve the differential Equation 5.13. Therefore, if the machine is loaded with a constant current I_L and $H = 0$, that is, the machine temperature is the same as temperature of the environment (cold state), the solution becomes

$$H = \left(\frac{I_L}{I_n}\right)^2 \left(1 - e^{-\frac{t}{\tau}}\right), \tag{5.15}$$

and the variation of H versus time is as shown in Figure 5.43. It indicates that the heat rise reaches 63% at time equal to the thermal time constant of the machine as given by the manufacturer.

From Equation 5.15, at a specific setting threshold H_{set}, the activation time of protection t_{act} is calculated by

$$t_{act} = \tau \,\mathrm{Ln}\left[\frac{1}{1 - H_{set}\left\{\frac{I_n}{I_L}\right\}^2}\right]. \tag{5.16}$$

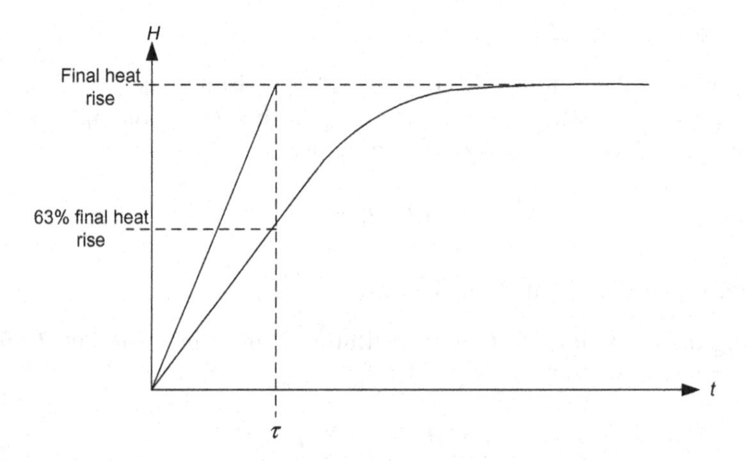

Figure 5.43 Heat rise versus time for a cold machine.

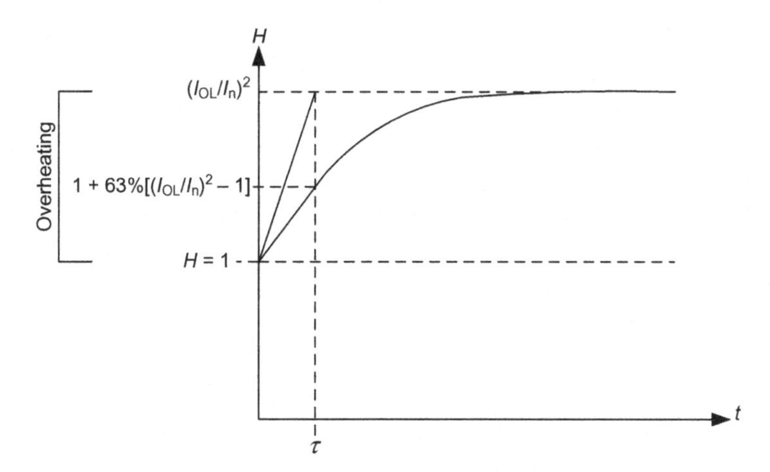

Figure 5.44 Heat rise versus time for a hot machine.

If the machine is operated at its nominal current I_n for a long enough time for its temperature to be stable, then H is equal to unity ($H = 100\%$). The initial condition in this case to solve Equation 5.13 and calculate the heat rise when the machine operates at overload current I_{OL} is $I = I_{OL}$ and $H = 1$. This condition is defined as "hot machine."

The solution becomes

$$H = \left(\frac{I_{OL}}{I_n}\right)^2 - \left[\left(\frac{I_{OL}}{I_n}\right)^2 - 1\right] e^{-\frac{t}{\tau}}, \tag{5.17}$$

which is plotted graphically to illustrate the change of H versus time as in Figure 5.44.

Therefore, at a specific setting threshold H_{set}, the activation time of protection t_{act} is given as

$$t_{act} = \tau \operatorname{Ln}\left[\frac{\left(\frac{I_{OL}}{I_n}\right)^2 - 1}{\left(\frac{I_{OL}}{I_n}\right)^2 - H_{set}}\right]. \tag{5.18}$$

5.8 OVERVOLTAGE PROTECTION

The overvoltage disturbances (or voltage surges) are voltage impulses or waves superimposed on the rated network voltages (Fig. 5.45a). Their variation and random nature makes them hard to characterize, allowing only a statistical approach to their duration, amplitudes, and effects. Overvoltage disturbances are therefore characterized by (Fig. 5.45b) the following:

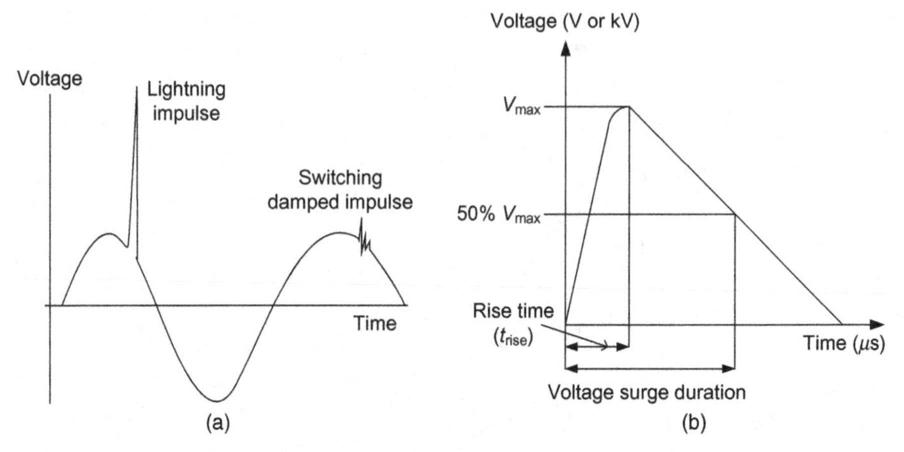

Figure 5.45 (a) Example of voltage surges. (b) Main overvoltage characteristics.

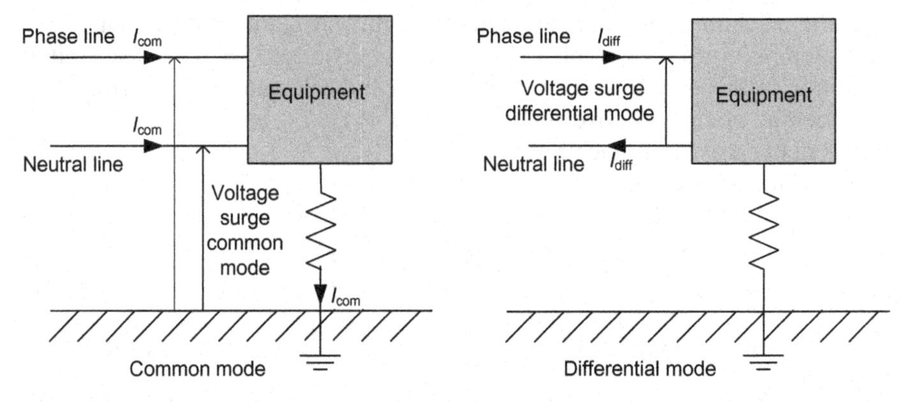

Figure 5.46 Surge propagation modes (I_{com} and I_{diff} are the common and differential mode currents, respectively).

- the rise time (t_{rise}) in μs and
- the gradient S in kV/μs.

This type of voltage disturbance may cause risk of malfunctions, electromagnetic radiation, destruction of equipment, and, consequently, lack of service continuity. These effects may occur on the installations of both utilities and consumers.

These disturbances can occur in two modes (Fig. 5.46):

- Common mode surges occur between the live conductors and the earth. They are especially dangerous for devices whose frame is earthed due to the risk of dielectric breakdown.

- Differential mode surges circulate between live conductors. They are especially dangerous for electronic equipment, sensitive computer equipment, and so on.

5.8.1 Types of Overvoltages

5.8.1.1 Switching Overvoltages Sudden changes in an electric network cause transient phenomena to occur. These transients result in the creation of an overvoltage or of a high-frequency wave train of a periodic or oscillating type with rapid damping. They have a slow gradient and have frequency that varies from several tens to several hundred kilohertz. Switching overvoltages may be created by the following [38]:

i) *Breaking of Small Inductive Currents*: Overvoltages in MV networks are caused by current pinch-off, re-arcing, and pre-arcing phenomena. The circuit shown in Figure 5.47 represents the utility network upstream as a main source of sinusoidal voltage with equivalent inductance L_1 and capacitance C_1 supplying an inductive load L_2 with distributed capacitance C_2 through a CB. The CB has its stray elements L_{CB} and C_{CB}.

Current Pinch-Off Phenomenon: The arc produced when the CB breaks a current much less than its rated current takes up little space and undergoes considerable cooling. So, it becomes unstable and its voltage (remains less than network voltage) may present high relative variations. These variations may generate oscillating currents of high frequency in the adjacent stray and voluntary capacitances and with non-negligible amplitude.

In the CB, the high-frequency current is superimposed on the fundamental current (50- or 60-Hz wave) resulting in a current passing through zero several times around the zero of the fundamental wave (Fig. 5.48). The CB is capable of breaking at the first current zero (point A) since the current is much less than the CB rated current. At this

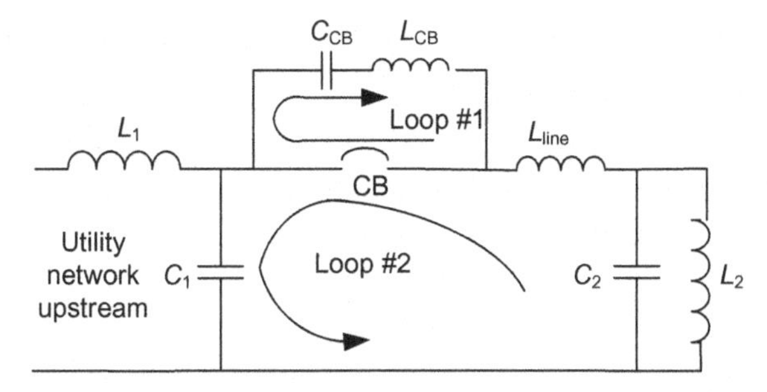

Figure 5.47 Equivalent circuit for overvoltage study when breaking small inductive current [38].

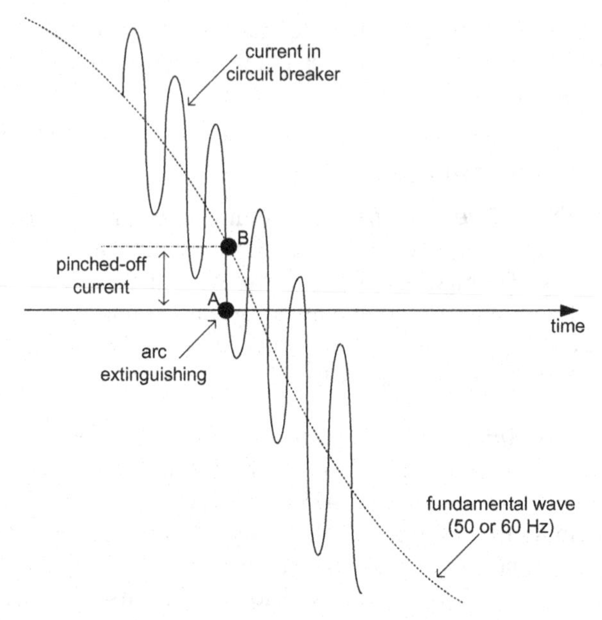

Figure 5.48 High-frequency oscillating current superimposed on fundamental current wave [38].

point, the circuit current is not zero (point B). This nonzero value represents the instantaneous value of fundamental wave on arc extinguishing and is known as the "pinched-off current."

If the load inductance is high as the load current is low, the load at the instant of arc extinguishing will have energy $\frac{1}{2}L_2I^2$. By energy conservation, the L_2C_2 circuit is exposed to a voltage at the C_2 terminals with a peak value V approximately given by $\frac{1}{2}L_2I^2 = \frac{1}{2}C_2V^2$. The value V may lead to a risk for the CB or load installation, particularly if C_2 is only made up of stray capacitances with respect to frames.

Re-Arcing Phenomenon: The pinching-off phenomenon causes an overvoltage at CB terminals. Then, an arc can occur if the CB cannot withstand this overvoltage. Actually, three oscillating phenomena at respective frequencies f_1, f_2, and f_3 occur simultaneously as consequences of current breaking and re-arcing.

First oscillating phenomenon at frequency f_1 around a few megahertz through loop #1 (Fig. 5.48) is given as

$$f_1 = \frac{1}{2\pi\sqrt{L_1C_1}}. \tag{5.19}$$

Second oscillating phenomenon at frequency f_2 around 100–500 kHz through loop #2 (Fig. 5.47) is given as

$$f_2 = \frac{1}{2\pi}\sqrt{\frac{C_1 + C_2}{L_{ine}C_1C_2}}. \tag{5.20}$$

Third oscillating phenomenon at frequency f_3 around 5–20 kHz through the equivalent circuit (Fig. 5.47) is given as

$$f_3 = \frac{1}{2\pi}\sqrt{\frac{L_1 + L_2}{L_1L_2(C_1 + C_2)}}. \tag{5.21}$$

It will be noticed from Equation 5.21 that line inductance plays no role in the value of f_3.

Thus, multiple re-arcing occurs until it is stopped by increasing contact clearance. This re-arcing is characterized by high-frequency wave trains of increasing amplitude. These overvoltage trains upstream and downstream from the CB can present a high risk for equipment containing windings.

Pre-Arcing Phenomenon: Pre-arcing overvoltages occur when closing CB contacts since there is a moment at which the dielectrics withstand between contacts is less than the applied voltage. This phenomenon is extremely complex as the resulting overvoltages depend on factors such as

- closing speed, rapid or slow with respect to 50 or 60 Hz;
- CB characteristics (dielectric properties, capacity to break high-frequency currents, etc.);
- cable characteristic impedance; and
- natural frequencies of load circuit.

ii) *Breaking of Strong Current*: The breaking of a short-circuit current generates surges if breaking is very quick and without energy consumption by the arc. Surges may be great when certain fuses are blown.

iii) *Switching on Capacitive Circuits:* Capacitive circuits are the circuits that are made up of capacitor banks and off-load lines. For instance, when capacitor banks are connected to the network, they are energized normally without initial load. In the case of slow operating CB, arcing occurs between the contacts around the 50- or 60-Hz wave peak. Damped oscillations of the equivalent inductive-capacitive (LC) system then occur. The frequency of this oscillation is generally much higher than power frequency, and voltage oscillation is mainly around the fundamental wave peak value. In the case of faster operating CB, arcing does not systematically occur around the peak value, and the overvoltage, if any, is thus lower.

A statistical study of switching surges has resulted in standardization of the waves shown in Figure 5.49 [70].

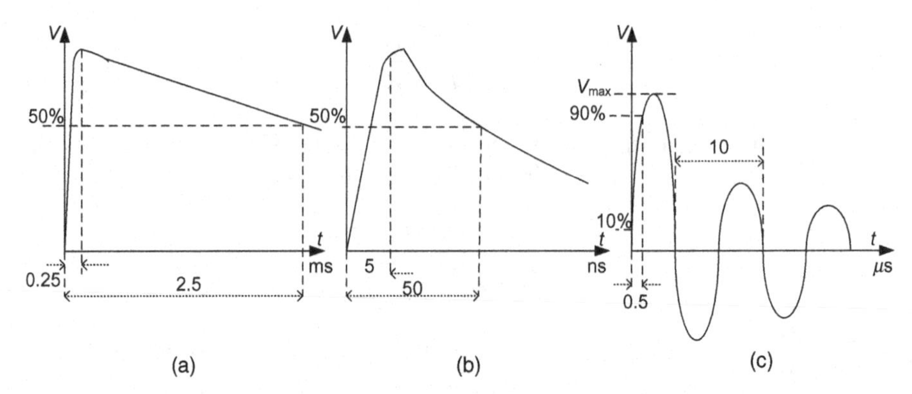

Figure 5.49 Switching surges representation by standardized waveforms: (a) 250/2500 μs long damped wave; (b) 5/50 ns recurrent pulse wave; (c) 0.5 μs/100 kHz damped sinusoidal wave [66].

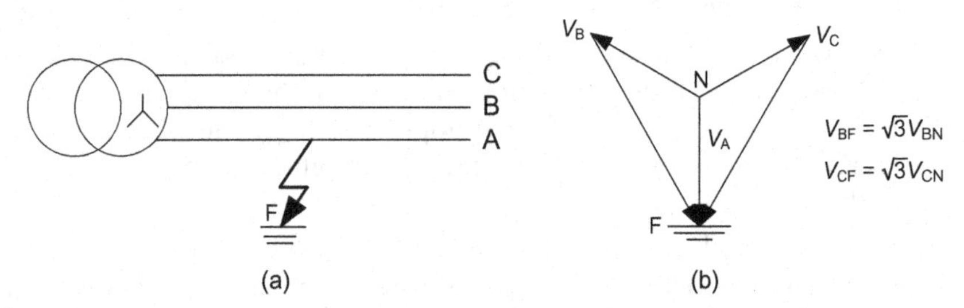

Figure 5.50 Voltages of healthy phases B and C on the occurrence of phase A-to-earth fault [38].

5.8.1.2 Power-Frequency Overvoltages Power-frequency overvoltages include all overvoltages that have the same frequency as the network (50 or 60 Hz). In distribution systems, they can mainly be caused by phase-to-frame or phase-to-earth faults on a network with an insulated or impedance-earthed neutral, or by breakdown of the neutral conductor. In occurrence of such faults, the faulty phase is at earth potential, and the other two healthy phases are then subjected, with respect to earth, to the phase-to-phase voltage. As shown in Figure 5.50a,b, when the phase A-to-earth fault occurs at F, the voltages of phases B and C are raised to the value of phase-to-phase voltage, that is, $V_{BF} = \sqrt{3}V_{BN}$ and $V_{CF} = \sqrt{3}V_{CN}$.

The ratio of the voltage of healthy phases B and C with respect to the network phase-to-neutral voltage is defined as the earth-fault factor (EFF). This factor is given as

$$\mathrm{EFF} = \frac{\sqrt{3}(k^2 + k + 1)}{k + 2}, \tag{5.22}$$

where

$$k = \frac{X_0}{X},$$

X_0 = zero-sequence reactance, and
X = the equivalent network reactance seen from the fault point.

It is noted that

- if the neutral is unearthed, $X_0 = \infty$ and EFF $= \sqrt{3}$;
- if the neutral is earthed, $X_0 = X$ and EFF = 1; and
- if $X_0 \leq 3X$ which is the common case, EFF$\leq$1.25.

5.8.1.3 Lightning Overvoltages Lightning is a dangerous and destructive natural phenomenon. It may cause fires, death of people and animals, and disruption of both electric and telephone lines. Lightning is linked to the formation of storm, which combines with the earth to form a genuine dipole. The electric field on the earth may then reach 20 kV/m. A leader develops between the cloud and the earth in a series of leaps, creating the ionized channel in which the return arc or lightning stroke flows (Fig. 5.51). According to the polarity of the cloud with respect to the earth, the stroke is either negative

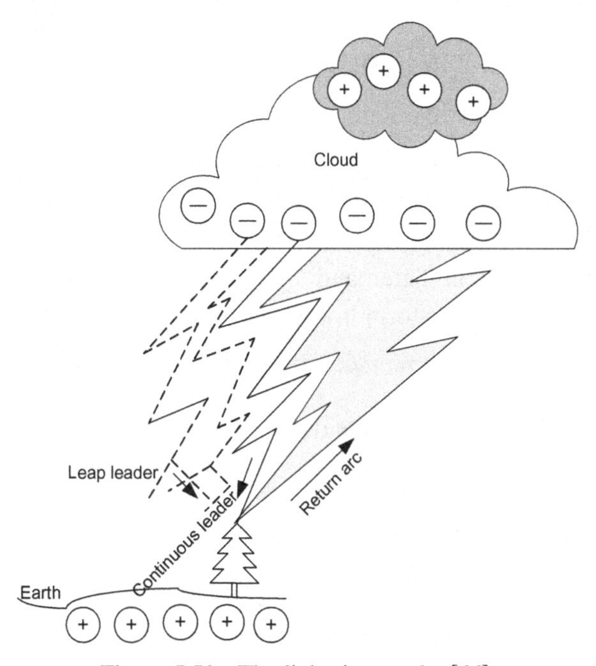

Figure 5.51 The lightning stroke [66].

(negative cloud) or positive (positive cloud), and according to the origin of the leader, the stroke is either ascending or descending.

Lightning overvoltages are described as severe transient overvoltages as a result of lightning striking on, or near the distribution system equipment. This may cause flashovers. Lightning strokes terminating on distribution equipment are defined as "direct strokes." Strokes to other objects surrounding the distribution equipment may induce substantial overvoltage when objects and distribution equipment are sufficiently close to each other. In this case, the strokes are defined as "induced strokes" or "indirect stroke." The distribution equipment that is commonly protected against these strokes is the insulation of transformers, regulators, overhead line, and cables.

Direct Stroke It affects mainly overhead lines by producing insulation flashover. Its crest current is of the order of tens of kiloamperes, while crest line voltage is of the order of megavolts. This voltage, to a large extent, exceeds insulation levels of overhead lines and, therefore, causes flashover and circuit interruption.

In absence of significant trees or tall building near the lines, that is, the overhead line pole is the taller structure, flash collection rate is estimated by [71]

$$N = N_g\left(\frac{28h^{0.6}+b}{10}\right)$$

and

$$N_g = 0.054T_h^{1.1},$$

where

N = flashes/100 km/year,

N_g = earth density (flashes/km^2/year),

h = pole height in meter,

b = structure width in meter, and

T_h = number of thunderstorm hours/year.

This relation indicates how much the flash rate increases by pole height increase.

The voltage on an overhead line due to direct stroke is given as [72]

$$V_F = I_{stroke}Z_{line},$$

where

V_F = the voltage to earth of struck overhead line conductor,

I_{stroke} = magnitude of direct stroke current, and

Z_{line} = surge impedance of overhead line.

Induced Stroke In the presence of taller objects such as trees and buildings surrounding the distribution line, the lightning that hits them may cause flashes, which in turn induce voltages on the line. So, the line performance is affected by the height of the taller object and its distance from the line. In addition, flashes may be collected by taller objects.

The maximum voltage induced in a line (single conductor and infinitely long) at the point closest to strike can be estimated by [73, 74]

$$V_{max} = 38.8\frac{I_{peak}h_{av}}{y} \text{ (assuming the ground is perfectly conducting),}$$

where

I_{peak} = lightning peak current,
h_{av} = average height of the line over earth level, and
y = the closest distance between line and lightning stroke.

Also, when lightning strikes, the earth near the earthing connections of installations, a rise in earthing potential occurs when the lightning current flows off through the earth. This variation in earthing potential affects installations. Thus, at a given distance D from the point of lightning impact, the potential V is expressed by [70]

$$V = 0.2I\rho_s/D,$$

where

I = lightning current and
ρ_s = earth resistivity.

As a schematic diagram, Figure 5.52 shows the voltage distribution along the earths of neutral point, installation, and point of lightning impact. It shows the resulting potential difference between neutral earth and installation earth, ΔV.

The main properties of the three types of overvoltages described above are summarized in Table 5.5. When the network is exposed to some of these disturbances, an interruption may occur. This interruption is either short disconnection (e.g., when using automatic reclosers on MV public networks) or long disconnection to change the damaged insulators or to replace the equipment. Therefore, the network must be protected to limit these risks.

5.8.2 Methods of Overvoltage Protection

5.8.2.1 Insulation Coordination To ensure safe distribution of electric power, the insulation coordination study is provided to determine the necessary and sufficient insulation characteristics of the various network components in order to obtain uniform withstand to normal voltages and to

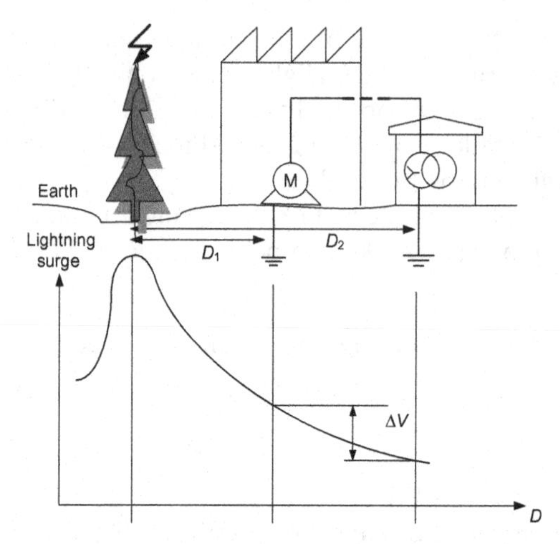

Figure 5.52 Potential profile along the earth of an electric installation [66].

TABLE 5.5 Main Properties of Different Overvoltage Types

Type of Overvoltage	Voltage Surge Coefficient[a]	Duration	Front Frequency	Damping
Switching	2–4	Short 1–100 ms	Average 1–200 kHz	Medium
Power frequency	$\leq\sqrt{3}$	Long 30 ms to 1 s	Rated frequency (50 or 60 Hz)	Low
Lightning	>4	Very short 1–100 μs	Very high 1–1000 kV/μs	High

[a] Voltage surge coefficient is the ratio of surge voltage to the rated voltage.

overvoltages of different origins [75, 76]. Two parameters must be defined in addition to insulation characteristics: the clearance and withstand voltage.

The clearance includes gas (e.g., air) clearance, which is the shortest path between two conductive parts, and creepage distance, which is the shortest path on the outer surface of a solid insulator between two conductors (Fig. 5.53). The creepage distances may be subjected to deterioration of their characteristics due to aging of insulators. This deterioration as well as environmental conditions (e.g., humidity, pollution), electric stresses, and gas pressure affect the voltage value that can be withstood.

The withstand voltage of insulating gases is a highly nonlinear function of clearance. For solid insulators, creepage distances of bus-bar supports, transformer bushings, and insulator strings are determined to obtain a withstand similar to direct air clearance between two end electrodes when they are dry and clean.

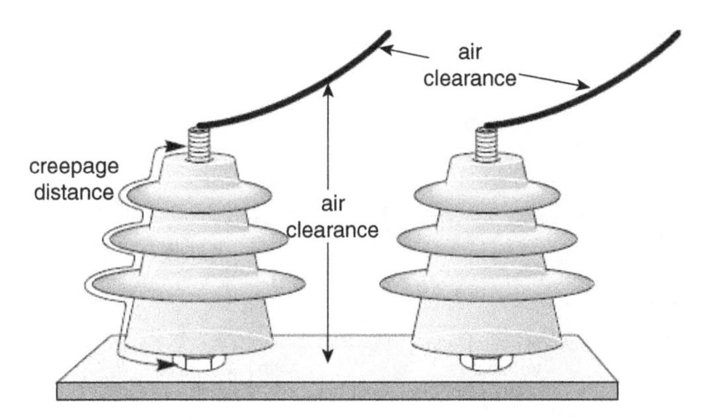

Figure 5.53 Air clearance and creepage distance [38].

Insulation characteristics and clearances are determined in order to withstand power frequency, switching, and lightning overvoltages in compliance with standard tests. Power-frequency withstand checked by 1-min dielectric tests is normally sufficient. However, withstand to an impulse voltage wave 250/2500 and 1.2/50 μs for switching and lightning overvoltages, respectively, must be tested [77].

Therefore, the overvoltage to which the equipment could be subjected during use must be defined to decide the right insulation level that will never be overshot, at least, when power frequency and switching impulses occur. As regards lightning, a compromise must generally be found between acceptable failure risk, insulation level, and the use of protective devices (e.g., arresters).

5.8.2.2 Surge Arresters Arresters act as high impedance at normal operating voltage and as low impedance during lightning-surge conditions. The arrester limits the voltage appearing across the equipment being protected. The surge current flows to the earth through its low impedance producing discharge voltage across arrester terminals [78]. There are different types of arresters such as gapped silicon carbide and gapped and nongapped metal oxide. The most common type used in MV and LV networks is the gapped/nongapped metal oxide [79]. The arresters are made of zinc oxide (ZnO) as nonlinear resistance (varistor) on which the operating principle is based. This nonlinearity varies widely depending on voltage rating and arrester class. Typical values of resistance decreases from 1.5 MΩ to 15 Ω. A schematic diagram of an arrester installation is shown in Figure 5.54a and a typical structure of porcelain-housed arrester is shown in Figure 5.54b.

The arrester is connected in parallel with equipment to be protected via connection leads. Therefore, the surge voltage produced by arrester across the device to be protected is the sum of discharge voltage of arrester and the voltage drop of the connection leads (VD_{leads}):

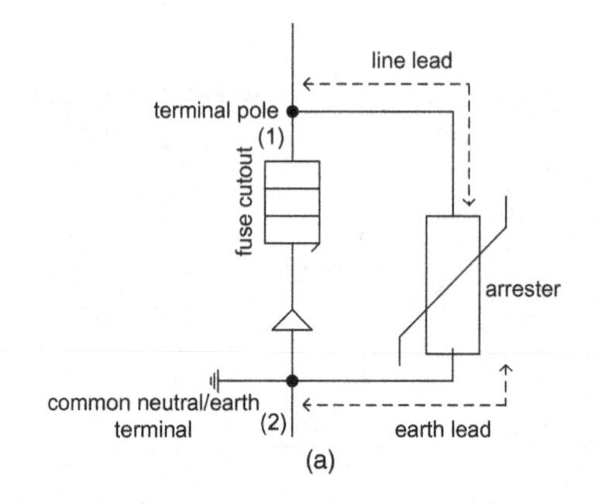

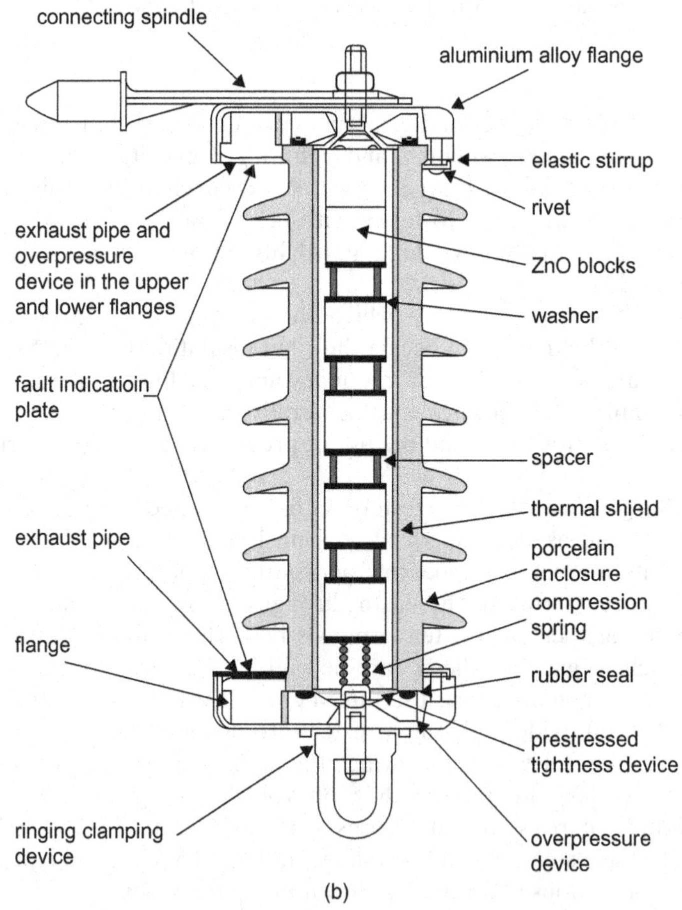

Figure 5.54 (a) Arrester installation. (b) Typical structure of MV porcelain-housed arrester [70].

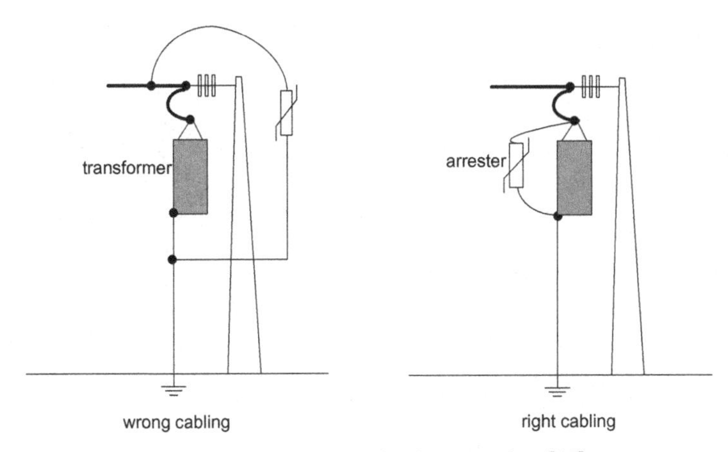

Figure 5.55 Arrester lead connection [38].

$$(VD_{\text{leads}}) = (L_{\text{L}} + L_{\text{G}})\frac{di}{dt},$$

where L_{L} is the inductance of line lead, which connects the phase conductor with line terminal of the arrester, and L_{G} is the inductance of earth lead, which connects the common neutral/earth with both arrester earth terminal and line earth terminal.

The connection leads are kept as short as possible to minimize the total impressed voltage developed by the arrester installation. This voltage is a function of the magnitude and waveform of the current, the arrester class, design, and voltage rating. Typical wrong and right lead connections of arrester installation to protect an MV/LV transformer connected to an overhead line are shown in Figure 5.55.

5.8.2.3 Primary and Secondary Protection of LV Network Two major types of protection devices are used to suppress or limit lightning surges. They are referred to as primary protection devices and secondary protection devices.

Primary Protection Devices Sensors, specific conductors, and earth are the devices used as primary protection devices. Their functions are intercepting lightning strokes, flowing these strokes off to earth, and dissipating them in the earth.

The sensors, which act as interception devices, are lightning rods. They are available in different forms such as a tapered rod placed on top of the building and earthed by one or more copper strips (Fig. 5.56a) or meshed cage (Fig. 5.56b) or overhead earth wires stretched over the structure to be protected (Fig. 5.56c).

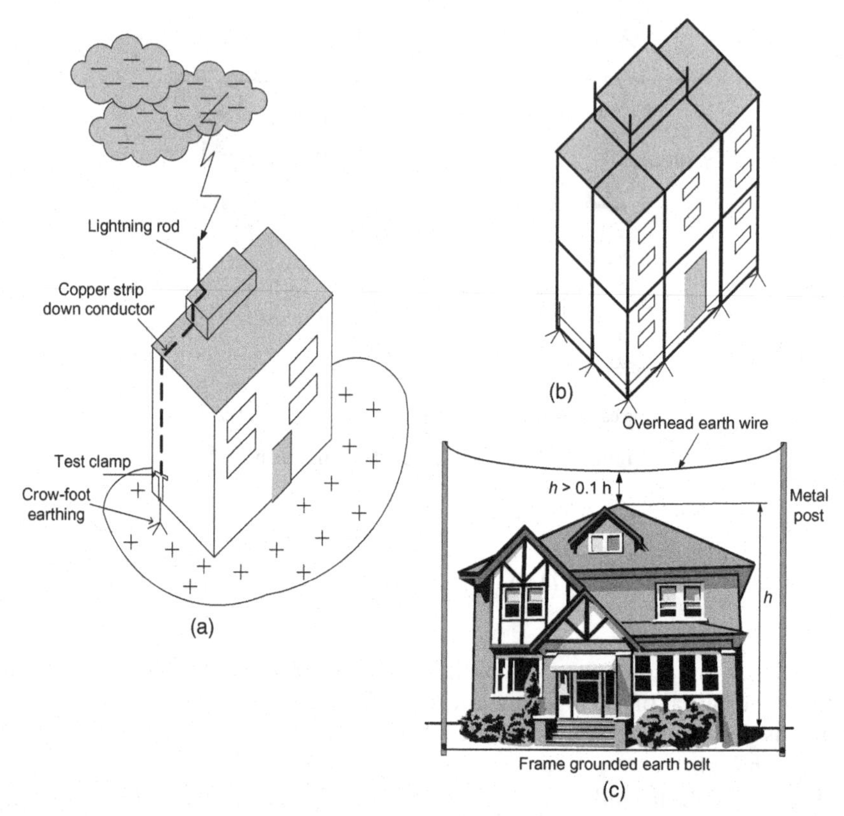

Figure 5.56 LV primary protection against lightning: (a) using lightning rod, (b) using meshed cage, and (c) using overhead earth wire.

When using lightning rods, attention must be paid to the copper strip paths, test clamps, crow-foot earthing to help high-frequency lightning currents run to the earth, and distances in relation to the wiring system (gas, water, etc.). Furthermore, it is preferred to use several copper strips in different paths to split the lightning current in two, four, or more paths. Thus, the effect of induced voltage surges, by electromagnetic radiation, in the buildings to be protected is minimized.

The same approach of splitting the lightning current flowing into earth to reduce electromagnetic fields is applied when using meshed cage. The number of down strips is multiplied and horizontal links are added, particularly if the building is high and housing computers or electronic equipment [80].

Secondary Protection Devices These devices provide protection against the indirect effects of lightning and/or switching and power frequency surges. The devices include the following:

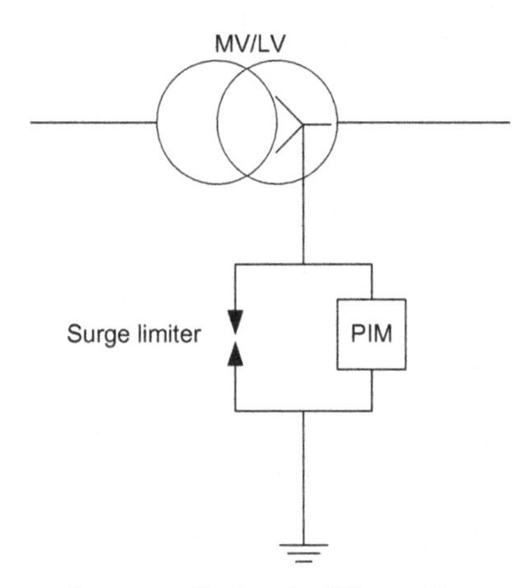

Figure 5.57 Installation of a surge limiter in IT earthing system. PIM = permanent insulation monitor.

- Surge arresters for LV networks as explained in Section 5.8.2.2.
- Filters that use the resistive-inductive-capacitive (RLC) circuit principle and are easily designed once the disturbance to be filtered has been properly identified. Filters are used particularly to attenuate switching surges.
- Wave absorbers that consist of arresters/filters. They are used to protect electronic equipment.
- Surge limiters to enable the flow off to earth of high energy surges and withstand the earth-fault current of the MV network. They are used on unearthed or earthed through impedance networks. Installation of a surge limiter is shown in Figure 5.57.
- Network conditioners and static uninterruptible power supplies (UPSs) sometimes contain several of the devices described above, and as such are part of the secondary protection devices.

CHAPTER 6

DISTRIBUTION SWITCHGEAR

6.1 NEED FOR SWITCHGEAR

Distribution systems not only include power lines that receive power from the transmission system or any other energy sources to be distributed to customers, but also include associated equipment for switching, measurement, monitoring, control, and protection.

Electric components, such as circuit breakers (CBs), load-break switches (LBSs), isolating switches, instruments and instrument transformers, and relays, are required to carry out these functions. These components are interconnected electrically and installed physically in a given enclosure known as "switchboard." The switchboard consists of a number of panels determined by the way of interconnection to achieve reliable and secure operations (Fig. 6.1).

For instance, to distribute incoming power to four outgoing feeders (customers), it necessitates a switchboard of five panels as shown in Figure 6.2a. For higher reliability, the incoming power is received from two sources and delivered to two sections coupled to each other through a CB (bus coupler). The switchgear used in this case consists of eight panels (Fig. 6.2b). It is noted that the coupling of the bus bar in one section with the other constitutes two panels, one for installing the CB and another called "bus-riser" panel. The bus-riser panel is necessary to adopt the termination of the bus bars installed at the top of the switchgear with the down terminals of the tie CB. Also, for connecting different loads through an open ring, each load needs a switchgear

Electric Distribution Systems, First Edition. Abdelhay A. Sallam, Om P. Malik.

Published 2011 by John Wiley & Sons, Inc.

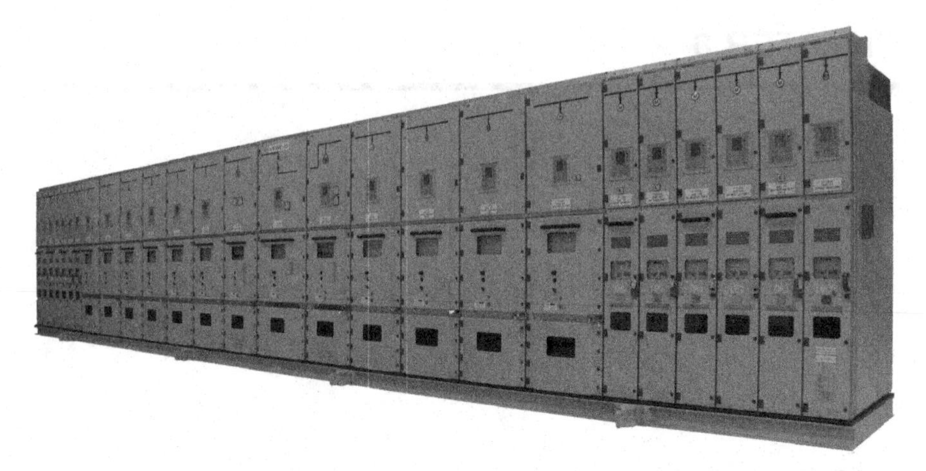

Figure 6.1 A typical ABB UniGear type ZS1® (courtesy of ABB). MV, arc-proof, air-insulated, metal-clad switchgear11.

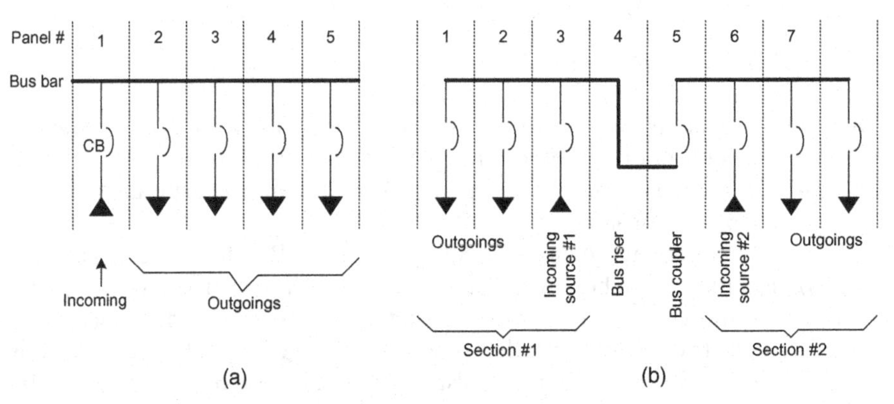

Figure 6.2 Switchgear layout. (a) Switchgear with one incoming feeder. (b) Switchgear with two incoming feeders.

of three panels (each panel usually includes LBS), which is called ring main unit (RMU) (Fig. 6.3).

The main functions of the switchgear are the following:

- Switching in or out of variable sections of the system (safe isolation from or connection to live parts). The aim of this function is to isolate or disconnect a circuit, apparatus, or an item of the system from the remainder energized parts, so that the personnel may carry out work on the isolated part in perfect safety. In principle, all parts of an installation shall have

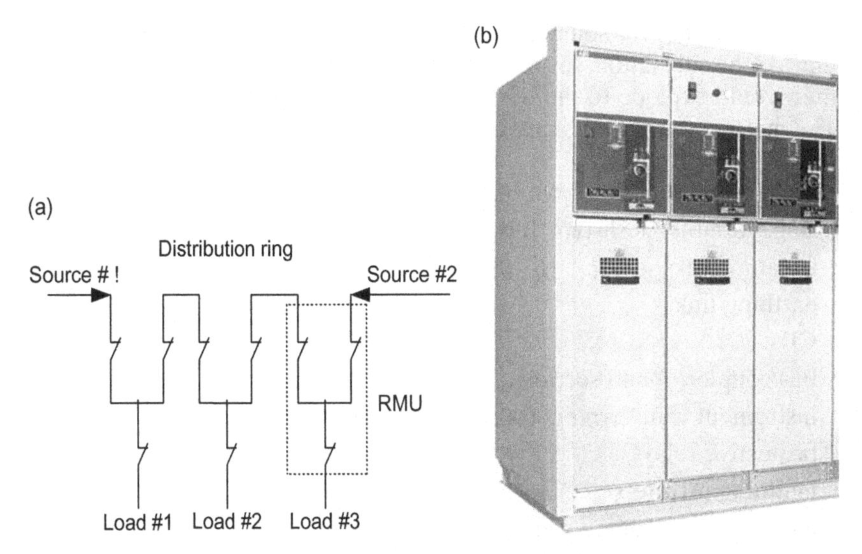

Figure 6.3 Ring main unit (RMU), ABB UniSwitch®. (a) Symbol of RMU in a distribution ring; (b) typical type of RMU (courtesy of ABB).

means to be isolated. In practice, in order to maintain optimum continuity of service, it is preferred to provide a means of isolation at the origin of each circuit.

- Allowing segregation of the faulty sections of the system, by local or remote switching control, to preserve continuity of supply via alternative circuits. This control also relates to all switching operations in normal service conditions for energizing and de-energizing a part of a system or installation, or an individual piece of equipment, and so on.
- Measurement and protection. System parameters being measured are selected by designers at each incoming and outgoing feeder in the switchgear (voltage, current, power, kWh, kvarh, etc.) as well as protection components.

6.2 SWITCHGEAR LAYOUT

There are mainly two types of switchgear, the outdoor and the indoor type. It is common to build the higher-voltage switchgear of the outdoor type, while the lower voltage could be of the indoor type. Irrespective of the type of switchgear, the substation auxiliaries, such as direct current (DC) batteries and their charging units or air compressors, must be indoors. With some exceptions, the power transformers could be of the outdoor type for safety reasons though this is not a rule of thumb. Nowadays, there are transformers of the dry insulated type, and for some applications, it is more convenient to have them just inside the substation, say at a side entrance. During any hazards, the transformer would explode to the outside taking the separating wall with it.

The main features and elements of a switchgear are the following:

- higher- and lower-voltage bus bars
- single/double/mesh-type bus bars
- isolating links
- earthing links
- CBs
- bus coupler(s) and sectionalizers
- instrument transformers (current and voltage)
- protective relays
- lightning arresters
- measuring instruments and integrating meters
- signaling apparatus
- DC auxiliary supplies
- compressed air/gas circuits and their associated compressors.

6.2.1 Environmental Requirements

Switchgear is governed by the following factors:

- the physical size
- general safety regulations
- degree of atmospheric pollution
- noise level
- vulnerability to damage
- ease of maintenance
- economic factors.

Clearly, these factors interact with each other but should be borne in mind when considering the types of installation in the forthcoming sections

6.2.2 Types of Switchgear Installations

Switchgear installations can be classified into three main types:

- metal-enclosed switchgear, cubical type/metal-clad type
- insulated—enclosed switchgear
- open-type switchgear.

The above three categories make the predefined classes of outdoor and indoor substations. These can be installed inside a building. When built weatherproof and with proper insulation class, they can be installed outdoors.

6.2.2.1 Metal-Enclosed Switchgear

The Compartmental Type The complete installation is enclosed by earthed metal and hence sometimes called the "package type." The circuits are usually segregated into separate metal compartments but the components of the circuit (bus bars, isolators, CBs, etc.) are not necessarily in different compartments. The main insulation is air and the CBs could be of any type. CBs are isolated either by withdrawal or by separate isolators.

The characteristics of this type of switchgear are the following:

- low space requirement
- interlocking may be intrinsic
- easy maintenance
- factory built (site erection reduced)
- less expensive than metal clad.

Installation of this type is most common up to 36kV, although it has been installed at 72kV and occasionally at 145kV. The main advantage of this type is the low space requirement, although the protection from pollution and malicious damage may also rank high.

Metal-Clad Type In this form of metal-enclosed switchgear, separate components are in earthed metal compartments. The insulation may be air (compressed or at atmospheric pressure) or sulfur hexafluoride. The CBs may be of the oil, vacuum, or gas blast type.

The main characteristics are the following:

- minimum space
- intrinsic interlocking giving high degree of safety
- factory built
- expensive.

The main advantage offered by this type of gear, that of minimum space requirements, has often been the main factor in its choice, particularly with very high-voltage installations.

6.2.2.2 Insulation-Enclosed Switchgear This type is similar to metal-enclosed switchgear but the enclosure is of insulating material, which must be capable of withstanding the voltages to earth on the outside surface of the insulating enclosure. This type has been manufactured and used in Europe since 1980s.

6.2.2.3 Open-Type Switchgear In this type of installation, the various components are manufactured as individual items and can be arranged in different ways with respect to each other. The main characteristics are the following:

- high space requirement,
- with adequate segregation service continuity can be high,
- high maintenance if in polluted area,
- easy replacement of items when required,
- very adaptable to specific requirements, and
- cheapest overall layout unless cost of land is predominant.

Layouts can vary from "low" designs where individual structures of concrete or steel are used as a support for the equipment, to "high" layouts where one large structure, usually of steel, supports all equipment except the CBs. There are many intermediate stages between the two extremes, but in general, one can assign the following characteristics to them:

- *Low Layouts:* Good segregation between equipment giving increased reliability and ease of maintenance. The layout is clear, circuits are easy to identify, and hence safer. The area required is large but the structure costs low. It is not obstructive in the landscape.
- *High Layouts:* Segregation of equipment may be poor and hence reliability may be down and maintenance is more difficult. Circuits may be difficult to identify. Structure cost is high.

6.3 DIMENSIONING OF SWITCHGEAR INSTALLATIONS

6.3.1 Dimensioning of Insulation

Insulation is the sum of the measures taken to isolate galvanically electrically conducting parts, which when in operation have voltage with respect to each other or with respect to earth (International Electromechanical Commission [IEC] publ. 71 and 664). According to the IEC publications, the requirement of sufficient insulating capability with respect to the operational voltage stress is satisfied when the insulation of the individual components of the installation withstands the insulation levels in Table 6.1. For cables, the test voltages and insulation wall thicknesses are in compliance with IEEE or IEC standards or any other international standards. Dimensioning of insulation is based on the anticipated stresses. The continuous stress due to the power-frequency voltage should be compromised with the stress due to overvoltages usually of short

TABLE 6.1 Standardized Insulation Levels for Voltage Range I: 1 kV < U_m < 52 kV [81]

Maximum Voltage for Apparatus V_m (rms Value) (kV)	Rated Lightning Impulse Withstand Voltage V_{rB} (Peak Value)		Rated Power-Frequency Withstand Voltage V_{rw} (rms Value) (kV)
	List #1 (kV)	List #2 (kV)	
3.6	20	40	10
7.2	40	60	20
12	60	75–95	28
17.5	75	95	38
24	95	125–145	50
36	145	170	70

Note: Standardized lightning surge voltage has 1.2/50 μs (front time/tail half time).

duration. The type of insulation is selected according to the maximum operating voltage of the device and taking into account that insulating capability is reduced by aging and degraded by contamination.

6.3.2 Insulation Coordination

Insulation coordination is a crucial measure taken to restrict flashover or breakdown of the insulation caused by overvoltages making the resulting damage as slight as possible. This can be achieved by insulation grading and using lightning arresters if it is appropriate as explained in Chapter 5. The standardized insulation levels for voltages below 52 kV are given in Table 6.1. In this table, the rated lightning impulse withstand voltage is classified into two lists: list #1 and list #2. The selection between the two lists depends on the risk of lightning overvoltages, method of neutral point earthing, and the nature of overvoltage protection (if any).

6.3.3 Dimensioning of Bar Conductors for Mechanical Short-Circuit Strength

Three-phase bus-bar arrangement with three main conductors H, each having three conductor elements T, with spacers Z is shown in Figure 6.4. The center-line spacing between main conductors is "a," and mean geometric spacing between conductor elements is "a_{ij}" (e.g., a_{12} between conductor elements 1 and 2). Distance between the two supports is "l," while l_s is the maximum distance from the spacer to the support point or to an adjacent spacer. F_d is the stress at support point that is at a distance h from the top edge of the support insulator.

The parallel conductors are subjected to uniform distributed forces along their length when carrying current ($l >> a$). When a short circuit occurs, the forces are large and cause bending stresses on the conductor, and cantilever

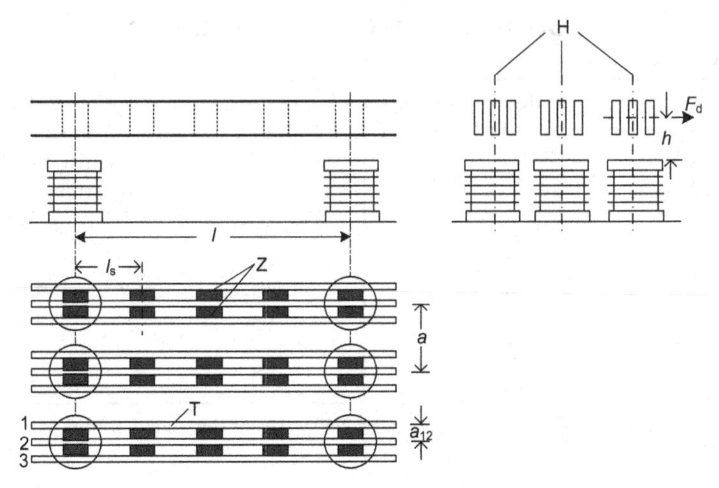

Figure 6.4 The bus-bar arrangement [81].

stresses and compressive stresses on the means of fixing. Therefore, the bus bars must be designed not only for the operating current, but also for the maximum short-circuit current. To produce this design, the calculation of stresses on the bus bars and their supports must be provided.

The electromagnetic force F between the main conductors, through which the same current flows, when three-phase fault occurs is given as

$$F = \frac{\mu_o}{2\pi} i_p^2 \frac{l}{a} \text{ Newton,} \tag{6.1}$$

where i_p is the peak short-circuit current (kA), μ_o is the magnetic field constant and equals $4\pi \times 10^{-7}$ (H/m), and l is in the same unit as a.

If the main conductor has t parallel-separate elements, the electromagnetic force F_s between the conductor elements is

$$F_s = \frac{\mu_o}{2\pi} \left(\frac{i_p}{t} \right)^2 \frac{l}{a} \text{ Newton.} \tag{6.2}$$

The peak short-circuit current i_p in Equations 6.1 and 6.2 is substituted by the peak short-circuit current of the middle element i_{p2}. These equations apply as a first approximation to conductors of any cross section, provided the distance between them is much greater than the largest dimension of the conductors. If this condition is not satisfied (e.g., the case of composite buses assembled from rectangular sections), the individual bars must be split into their component conductors. The actual effective distance between conductor elements a_s is given as

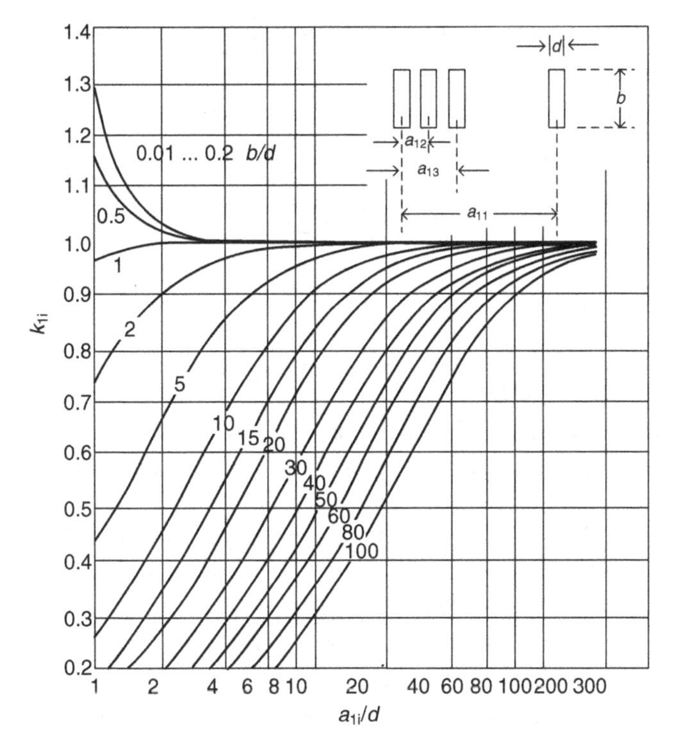

Figure 6.5 k_{1i} factors for the mean effective distance between conductor elements with $i = 1, 2, \ldots, t$ [81].

$$a_s = \frac{k_{12}}{a_{12}} + \frac{k_{13}}{a_{13}} + \ldots + \frac{k_{1t}}{a_{1t}},$$

where the factors k_{1i}, $i = 1, 2, \ldots, t$ can be calculated from Figure 6.5.

The electromagnetic force between the conductor elements is calculated by

$$F_s = \frac{\mu_o}{2\pi}\left(\frac{i_p}{t}\right)^2 \frac{l_s}{a_s} \text{ Newton.} \tag{6.3}$$

Stresses on conductors and forces on supports must be calculated. The bending stress σ applied to a bus bar under short-circuit conditions must not exceed a certain limit; otherwise, the material will be overstressed. When determining this limit, it was assumed that a permanent deformation of up to 1% of the supported length is acceptable. The strength of bus bars and the forces on the supports are influenced by the conductor's vibration behavior. This in turn depends on how the conductors are fastened, their permissible plastic deformation, and their natural frequency. Regarding plastic deformation, bending stress on the three main conductors, σ_m, and the bending stress on conductor element, σ_s, are calculated by the relations

$$\sigma_m = V_\sigma V_r \beta \frac{Fl}{8Z}, \tag{6.4}$$

$$\sigma_s = V_{\sigma s} V_r \frac{F_s l_s}{16 Z_s}, \tag{6.5}$$

where

V_σ, $V_{\sigma s}$ = factors for main conductor and conductor elements stress, respectively;
V_r = factor for failed three-phase autoreclosure;
β = factor for main conductor stress dependent on the nature of support and fixture;
Z, Z_s = section modulus of main conductor and conductor elements, respectively; and
$V_\sigma V_r$ = $V_{\sigma s} V_r = 2$ in DC installations;
= 1 in alternating current (AC) installations
= 1 in three-phase installations without autoreclosure
= 1.8 in three-phase installations with autoreclosure.

The resultant conductor stress is made up of the stresses on the main conductors and conductor elements, σ_{total}:

$$\sigma_{total} = \sigma_m + \sigma_s. \tag{6.6}$$

The force on each support

$$F_d = V_F V_r \alpha F \tag{6.7}$$

where

V_F = ratio of dynamic force to static force on support,
α = factor for force on support dependent on nature of support and fixing,
$V_F V_r$ = 2 in DC installations, otherwise,
$V_F V_r$ = 1 for $\sigma_{total} \geq 0.8 R'_{p0.2}$ and
$= \dfrac{0.8 R'_{p0.2}}{\sigma_{total}}$ for $\sigma_{total} < 0.8 R'_{p0.2}$,
$R'_{p0.2}$ = maximum yield strength.

In AC installations, $V_F V_r$ can be assumed to be not more than 2, and in three-phase installations not more than 2.7. If the condition $\sigma_{total} \geq 0.8 R'_{p0.2}$ is satisfied, the bus cannot transfer any forces greater than the static forces to the supports because it will deform beforehand ($V_F V_r = 1$). If σ_{total} is well below $0.8R'_{p0.2}$, the frequency response of the bus should be checked in accordance with the formula for conductor vibration [81].

6.3.4 Mechanical Short-Circuit Stresses on Cables and Cable Fittings

Cables and cable fittings are exposed to high forces occurring under short-circuit conditions. The forces are absorbed because of their action in the radial direction. In addition, the correct thermal size of cable for short-circuit conditions enables the cable to withstand the produced mechanical short-circuit stresses.

Parallel single-core cables are subjected to particularly high mechanical stresses under fault conditions. For example, the electromagnetic force, F, on two parallel single-core cables in relation to cables centerline spacing, a, and peak short-circuit current, i_p, is shown in Figure 6.6.

6.3.5 Dimensioning for Thermal Short-Circuit Strength

Under short-circuit conditions, bus bars, their T-off connections, and associated devices (e.g., switching devices, current transformers, and bushings) are

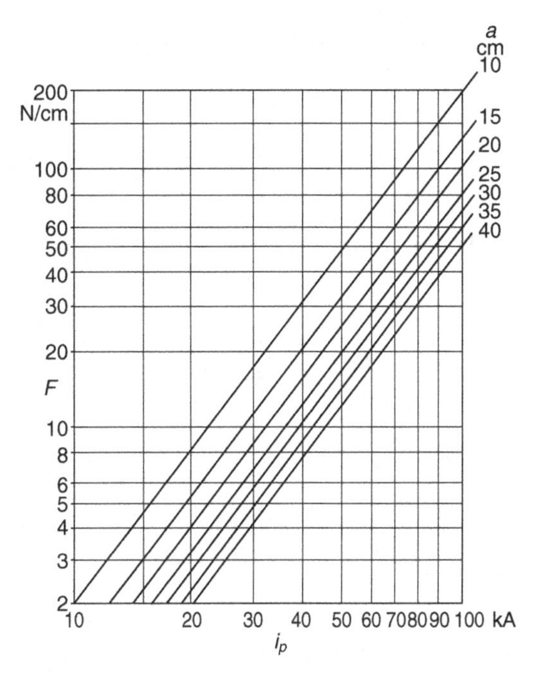

Figure 6.6 Electromagnetic force F versus peak short-circuit current i_p at different centerline spacing $a12$ [81].

designed to withstand not only mechanical stresses but also thermal stresses. Thermal stresses depend mainly on the short-circuit current as magnitude, duration, and characteristic with time.

The thermal equivalent short-time current, I_{th}, is the root mean square (rms) value, which generates the same quantity of heat as that generated by the short-circuit current (AC and DC components) for duration T_k. I_{th} is called the thermal effective mean value. Its value for a single short circuit of duration T_k is given as

$$I_{th} = I_k''\sqrt{m+n}, \tag{6.8}$$

where I_k'' is the initial symmetrical short-circuit current and m, n are factors obtained from Figure 6.7

A set of curves representing the factor m in relation to short-circuit duration T_k and the decay factor k are shown in Figure 6.7a. The decay factor takes into account the decay of DC component and can be calculated by

$$k = 1.02 + 0.98e^{-3R/X}, \tag{6.9}$$

where R and X are the resulting resistance and reactance, respectively, at fault location.

The factor n in relation to short-circuit duration and surge factor (SF) is depicted in Figure 6.7b. The SF is defined by the ratio of initial (I_k'') to steady-state (I_k) values of symmetrical short-circuit current, that is,

$$\text{SF} = I_k''/I_k.$$

When using, for example, rapid reclosure, a number of successive switching operations, n, occur. The individual values of thermal equivalent short-time current are found with Equation 6.8. Then, the resultant equivalent short-time current is

$$I_{th} = \sqrt{\frac{1}{T_k}\sum_{i=1}^{n} I_{thi}T_{ki}}, \tag{6.10}$$

where

$$T_k = \sum_{i=1}^{n} T_{ki}.$$

The rms value of the current, which the device will withstand for the rated short-circuit duration T_{kr}, represents the maximum permissible rated short-time current I_{thr} for the duration T_{kr} and it is provided by the manufacturer. Therefore, the electric device has relevant thermal strength when

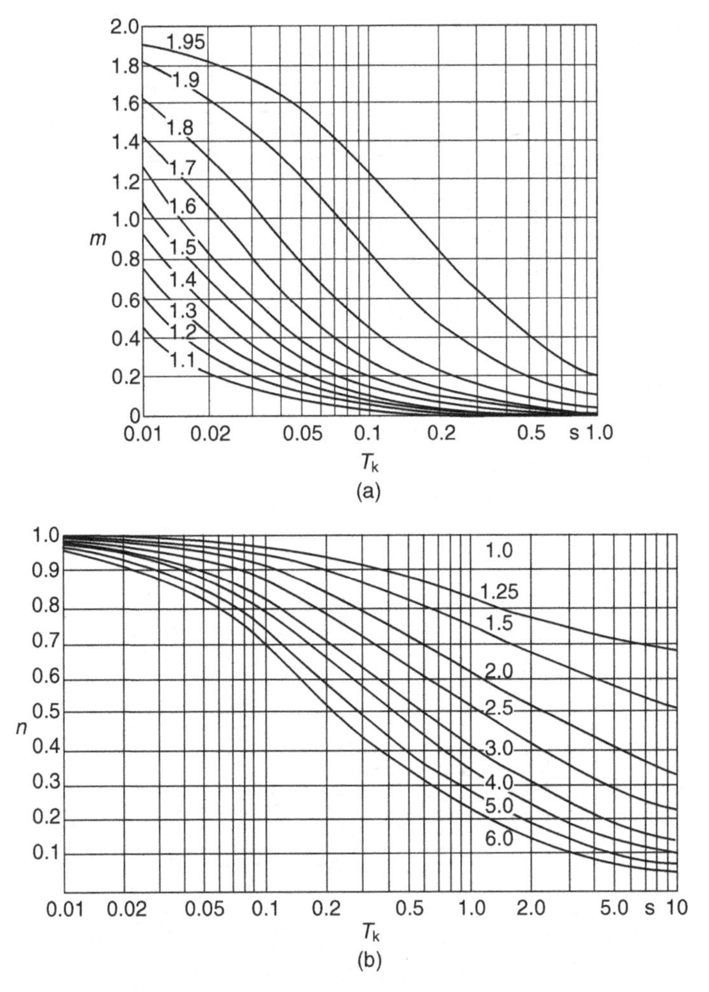

Figure 6.7 (a) The factor m versus T_k for different values of decay factor. (b) The factor n versus T_k at different surge factor values [81].

$$I_{th} \leq I_{thr} \text{ for } T_k \leq T_{kr}$$

and

$$I_{th} \leq I_{thr} \sqrt{\frac{T_{kr}}{T_k}} \text{ for } T_k \geq T_{kr}, \tag{6.11}$$

where T_k is the sum of relay response times and total break time of the CB.

For conductors, to determine the adequate conditions of thermal strength, the thermal equivalent short-time current density S_{th} is used. S_{th} must be

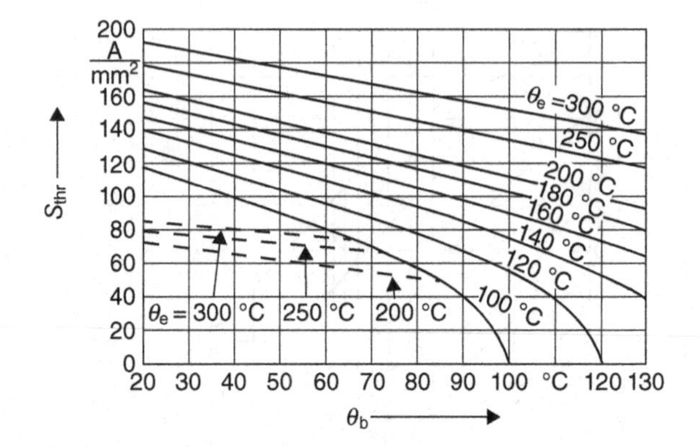

Figure 6.8 Rated short-time current density S_{thr} in relation to temperature (solid lines for copper and dotted lines for steel) [81].

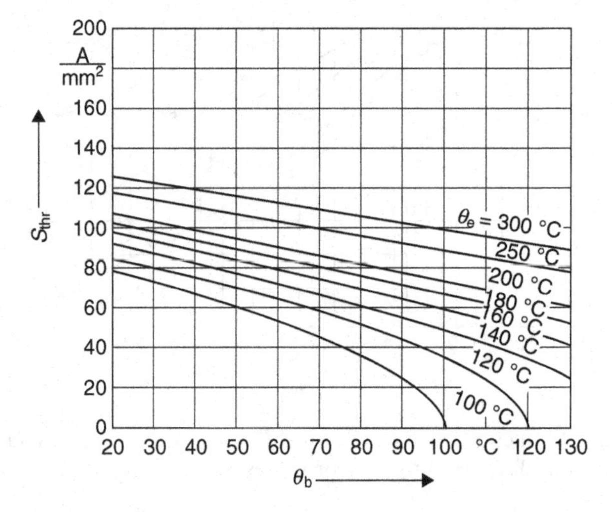

Figure 6.9 Rated short-time current density S_{thr} in relation to temperature (aluminum, aldrey, and Al/St) [81].

smaller than the rated short-time current density S_{thr}, which can be obtained from Figures 6.8 and 6.9. In these figures, θ_b is the initial temperature of the conductor and is considered as the maximum continuous operating temperature. The ultimate temperature of the conductor θ_e is its maximum temperature under short-time conditions. Therefore, the condition of thermal strength adequacy to be satisfied is given as

$$S_{th} \leq S_{thr} \frac{1}{\sqrt{\eta}} \sqrt{\frac{T_{kr}}{T_k}} \text{ for all } T_k, \tag{6.12}$$

where η is a factor taking account of the heat transfer to the insulation. Its value equals 1 for bare conductors.

Example 6.1:

A T-off connection of generator bus bar to a service transformer with cross-sectional area (CSA) 80 × 15 mm, copper, and a bushing-type current transformer are required to be checked for their size adequacy for the thermal stress under short-circuit event conditions. The total break time is 0.5 s. The design data of installation are given as follows:

- initial short-circuit current = 180 kA,
- steady-state short-circuit current = 55 kA,
- the decay factor $k = 1.9$,
- initial temperature of conductor = 90°C,
- ultimate temperature of conductor = 250°C, and
- transformer manufacturer provides that the thermal-current rating of current transformer = 170 kA for rated short-circuit duration of 0.5 s.

Solution:

From the design data: $I_k'' = 180$ kA, $I_k = 55$ kA, $k = 1.9$, $T_k = 0.5$ s, $\theta_b = 90$°C, and $\theta_e = 250$°C.

Thus, the SF = $I_k''/I_k = 180/55 = 3.27$.

It is found from Figure 8.7, $m = 0.2$ and $n = 0.5$.

Applying Equation 8.8, $I_{th} = 180\sqrt{0.2 + 0.5} = 150.6$ kA.

Bushing-type current transformer:

$$I_{thr} = 170 \text{ kA for } T_{kr} = 0.5 \text{ s.}$$

Therefore, $I_{th} < I_{thr}$ and the current transformer has sufficient thermal strength.

Bus-bar connection:

$$\text{CSA} = 80 \times 15 = 1200 \text{ mm}^2$$

$$S_{th} = \frac{I_{th}}{\text{CSA}} = \frac{150.6 \times 10^3}{1200} = 125.5 \text{ A/mm}^2$$

From Figure 8.8, at $\theta_b = 90°C$ and $\theta_e = 250°C$, it is found that

$$S_{thr} = 140\ A/mm^2.$$

Thus, $S_{th} < S_{thr}$, which means that the T-off bus-bar connection has sufficient thermal strength.

6.3.6 Dimensioning for Continuous Current Rating

Current flow through the equipment contained in switchgear panels produces losses, which in turn are dissipated to the surrounding air causing temperature rise. So, the specified temperature limits inside the panel must not exceed that the equipment can withstand for continuous correct operation.

It is experimentally found that the temperature distribution inside the panel depends on panel height and how the equipment is arranged. The heated air rises to the upper part of the panel because of differences in heated air density. For this reason, it is desirable to locate the equipment with relatively high power losses in the lower portion of the panel to allow the heat to dissipate from the enclosure as much as possible. Also, if the panel is fixed on a wall, a gap of air between the back of the panel and the wall should be left to help remove the heat.

The average temperature rise $\Delta\theta$ of the air inside the panel can be calculated by using the relation

$$\Delta\theta = \frac{P_{av\text{-}eff}}{\alpha A_M}, \tag{6.13}$$

where

$P_{av\text{-}eff}$ = average effective power loss with allowance for load factor (LF) as in IEC 439-1 and equals LF^2P_v (P_v is the average power loss per equipment);

A_M = heat dissipating surface area of panel;

α = hear transfer coefficient

= $6\,W/(m^2K)$ when heat sources are mainly in the lower half of the panel

= $4.5\,W/(m^2K)$ when heat sources are uniformly distributed over the panel height

= $3\,W/(m^2K)$ when heat sources are mainly in the upper half of the panel.

Allowing heat transfer from the air inside the panel to the surrounding air is more effective than with fully enclosed construction. Heat transfer is affected by

- size of panel,
- location and distribution of heat source equipment inside the panel,
- temperature difference,
- position of inlets and outlets, and
- positions of inlet openings with respect to outlet openings and their relation to surface area of total heat dissipation.

Example 6.2:

A panel includes 10 pieces of equipment, such as fuses, air-break contactors, and overcurrent relays (OCRs). Each has a power loss of 40 W with allowance for LF of 0.65. The panel heat dissipation area is $4.5\,m^2$. It is required to calculate the temperature rise given that the maximum allowable temperature for equipment is 45°C and the ambient temperature θ_o is 30°C. The heat sources are uniformly distributed over the panel height.

Solution:

Heat coefficient factor α is taken as $4.5\,W/(m^2 K)$

$$P_{\text{av-eff}} = (0.65)^2(10 \times 40) = 169 \text{ W},$$

$$\Delta\theta = \frac{P_{\text{av-eff}}}{\alpha A_M} = \frac{169\,(\text{W}\cdot\text{m}^2\text{K})}{4.5\,(\text{W})\cdot 4.5\,(\text{m}^2)} = 8.34 \text{ K}.$$

Thus, $\theta = \theta_o + \Delta\theta = 30 + 8.34 = 38.34°C < 45°C$.

6.4 CIVIL CONSTRUCTION REQUIREMENTS

Detailed information and layout drawings are required for civil engineers to prepare the civil construction drawings, such as plans for foundations, formwork and reinforcement, and services. The information required includes

- physical arrangement of equipment;
- dimensions of equipment;
- normal building services;
- gangways for transport, operation, and servicing;
- locations of cables and ducts penetration in ceiling and walls;
- loadings;

- type of doors and windows (e.g., fire resistant or fireproof) indicating how they open;
- details of services;
- ventilation and air-conditioning;
- steel work and cable trenches required in the floor;
- earthing for foundations and buildings;
- lightning protection (if any); and
- drainage.

6.4.1 Indoor Installations

The design of buildings and rooms for electric equipment must satisfy the operational requirements and other aspects as well. These aspects include the following:

- Selected rooms are far from groundwater and flooding.
- Easy access for operation, transport, and fire-fighting services.
- Walls, ceilings, and floors must be dry.
- In, over, and under the rooms containing switchgear, the piping for liquids, steam, and combustible gases should be avoided.
- Room dimensions must be appropriate to the nature, size, and arrangement of equipment.
- The escape route from exit doors must not be longer than 40 m.
- Inside surfaces of switchgear room walls should be as smooth as possible to avoid dust deposits.
- The floor surface must be easy to clean, pressure resistant, nonslip, and water resistant.
- Steps or sloping floor surfaces in switchgear rooms must be avoided in all circumstances.
- Windows can be easily opened or closed without personnel coming dangerously close to any live parts.
- Windows and doors must be secured.
- The rooms must be sufficiently ventilated to prevent condensation. Condensation and high relative humidity may lead to corrosion and reduced creepage distances.
- In areas with heavily polluted air, the rooms should be kept at slight positive pressure using filtered air. The necessary air vents must be protected against rain, water splashes, and small animals.
- In conventional switchgear installations, the arcing faults can create considerable pressure in the equipment room. Using pressure relief vents can avoid any probable damage to walls and ceiling resulting from excessive high pressure.

6.4.2 Outdoor Installations

Switchgear, transformers, and associated electric equipment are placed in an open area (switchyard). Account must be taken of foundations and access roads. Foundations are made of normal concrete and laid according to switchgear layout and steel structures. Foundations must be designed to withstand not only static loads but also the stresses occurring in operation due to short circuits, temperature rise, wind, and ice loads. The penetration of ground wires and cables through foundations should be considered during foundation laying process.

For transporting switchgear and transformers, the size and construction of access roads must be determined in such a way that the transportation is easy.

6.4.3 Transformer Installation

Rooms dedicated to the transformer and switchgear must be so arranged that they are easily accessible. The room dimensions are determined, taking noise, fire hazards, solid-borne sound, ability to replace equipment, operation and maintenance, and temperature rise into account. Oil-filled transformers do not require any special additional protection against climatic conditions while dry transformers require dry rooms for installation and special protection against moisture [82].

For self-cooling transformers installed indoors, ventilation openings in the room above and below the transformer must be provided so that the heat loss can be removed. If this is not sufficient, the forced-flow ventilation is essentially needed (Fig. 6.10).

When installing transformers, the requirements given below need attention:

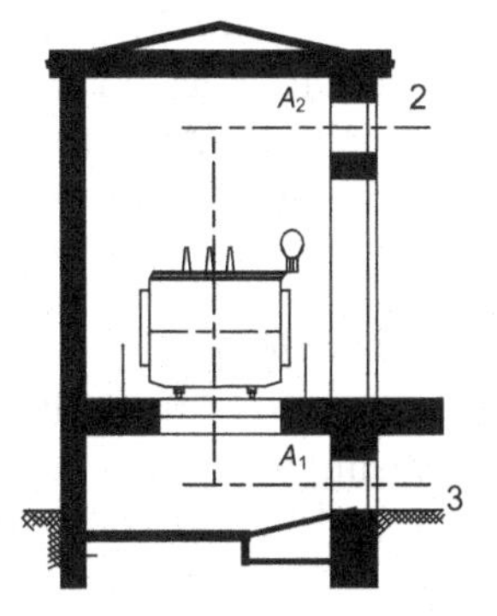

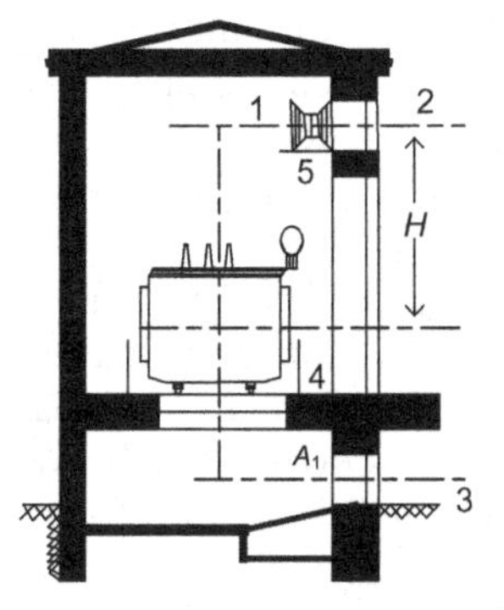

Figure 6.10 Transformer room with natural ventilation (on left side) and forced ventilation (on right side). A_1 = air intake area; A_2 = air exhaust area; H = height of fan shaft; 1 = fan; 2 = exhaust louvers; 3 = inlet grille; 4 = skirting; 5 = baffle plate.

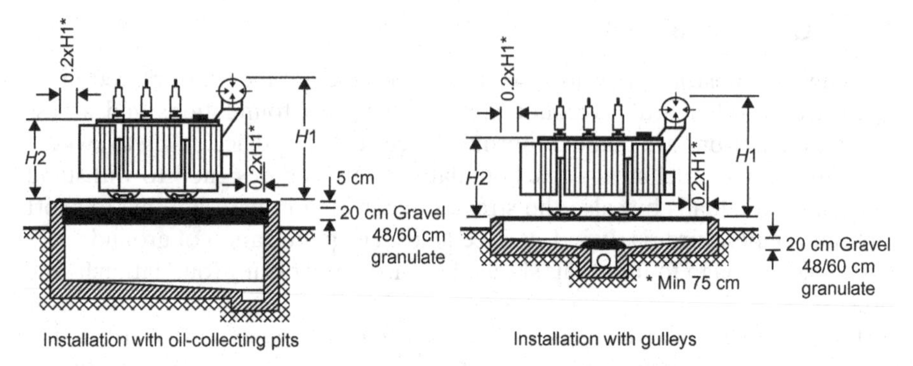

Figure 6.11 Transformer arrangements [81].

- safety distances;
- access for transport;
- cooling/ventilation;
- ease of transformer replacement;
- easy to operate and maintain;
- locations of auxiliary equipment;
- installation of fire protection system; and
- galleys, pits, and sumps must be provided under transformers containing insulating liquids to protect against fire and water pollution. They must be built so that the insulating fluid cannot escape into the soil (Fig. 6.11).

6.4.4 Ventilation of Switchgear Installations

Dimensioning for continuous current rating has been discussed in Section 6.3.6. Accordingly, the thermal loads due to heat emission from the equipment and climate conditions must be considered when designing switchgear installations. Heating, ventilation, cooling, and air-conditioning (HVAC) systems can be used to reduce the thermal loads. Selection of HVAC type depends on the value of maximum permitted ambient temperature $T_{\text{amb-max}}$ with respect to the outside temperature T_{out}.

When $T_{\text{amb-max}} > T_{\text{out}}$, ventilation equipment can be used to force cooling air into the switchgear and in turn the hot air is expelled through exhaust opening. As shown in Figure 6.12a, a fan is installed at the top of the switchgear. The forced air comes down into the switchgear causing hot air in the upper area to escape through an exhaust opening in the top at the opposite side.

If $T_{\text{amb-max}} < T_{\text{out}}$ cooling equipment and systems are used purely for removing heat. Interior and exterior heat loads are removed to keep the space cool (Fig. 6.12b).

In addition to removing heat, if it is required to maintain the indoor air at specific climate conditions, air-conditioning systems can be used (Fig. 6.12c).

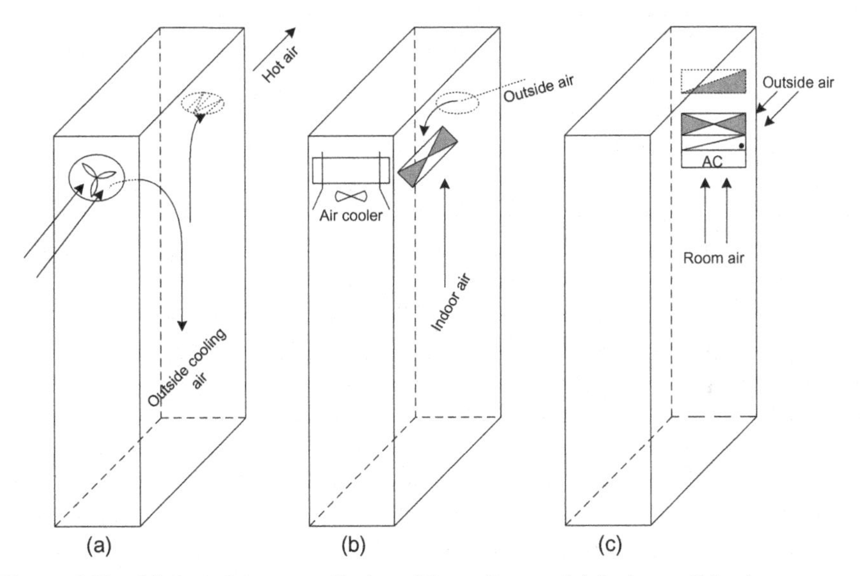

Figure 6.12 (a) Switchgear ventilation, (b) cooling, and (c) air-conditioning systems.

TABLE 6.2 2 Associated Notices with Drawings in Figure 6.13 (S = Rated Power, Cables up to 50 mm^2)

Symbol	Specified Conditions		Distance (mm)
A	With/without rear cabling, dry cables, $S < 1000$ kVA		220
	With/without rear cabling, dry cables, $S > 1000$ kVA		585
B	Without lateral cabling (dry cables) or earthing connection		25
	With earthing connection or lateral cabling (dry cables)		100
C	Minimum distance for operation, no CB/no transformer		1100
	Minimum distance for operation, with CB/no transformer		1500
H	MV switchboard with/without top cabling, $S < 1000$ kVA		2350
	MV switchboard with/without top cabling, $S > 1000$ kVA		2800
L	Varies with transformer	50, 100 kVA	1200
	rating kVA output,	200, 250 kVA	1600
	with/without	400–1000 kVA	2000
	transformer cubicle	1250 kVA	2400

Example 6.3:

An installation has an RMU, oil-immersed medium-voltage (MV)/low-voltage (LV) transformer, and main LV compartment. To prepare the civil drawings and construction of the room containing this equipment, the civil engineer must provide the layout of electric equipment arrangement. This layout should include the dimensions of the equipment, particular requirements (cable trenches, transformer pit, locations of windows and doors, ventilation, etc.), and the clearances (distance between equipment side and walls, distance between equipment height and ceiling). Figure 6.13 shows such a layout with some notices in Table 6.2 concerning minimum clearances.

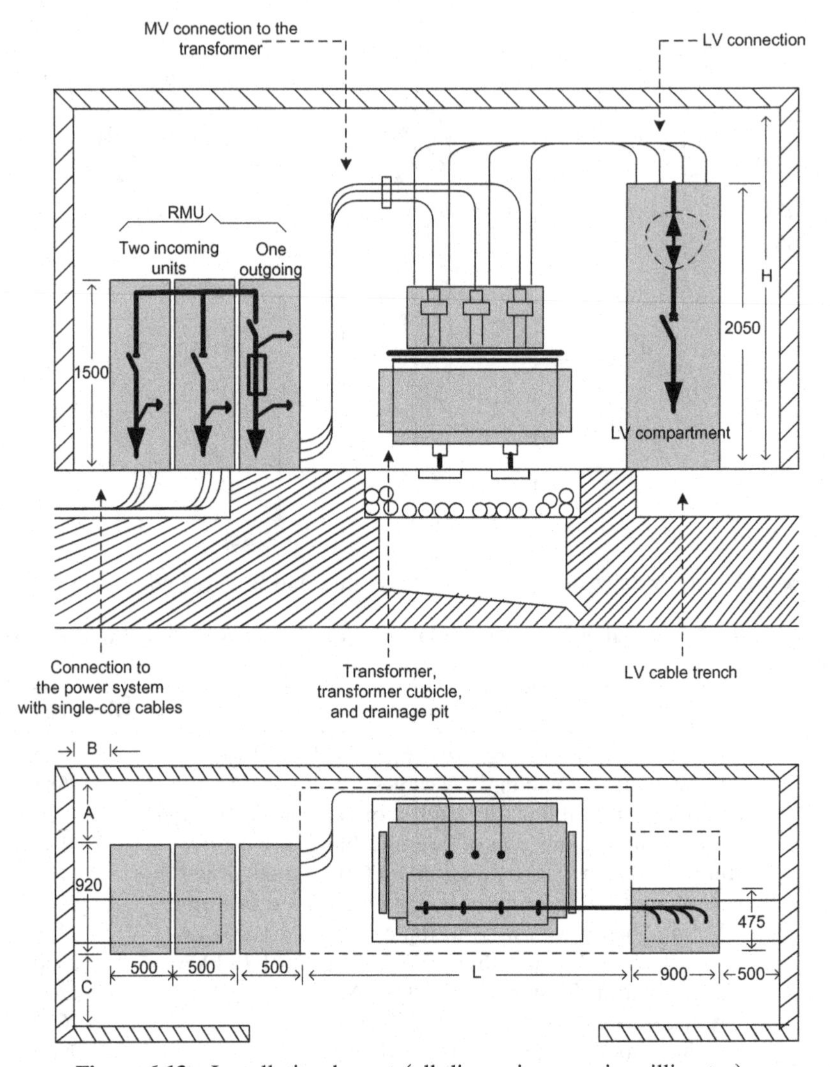

Figure 6.13 Installation layout (all dimensions are in millimeter).

Example 6.4:

An installation layout of distribution switchboard is required for room construction.

Figure 6.14 shows the layout of that switchboard. The switchboard contains six panels: two incoming feeders with LBSs, one for measurements, one for protection including CB, and two outgoing feeders with LBSs. The clearances are given as follows:

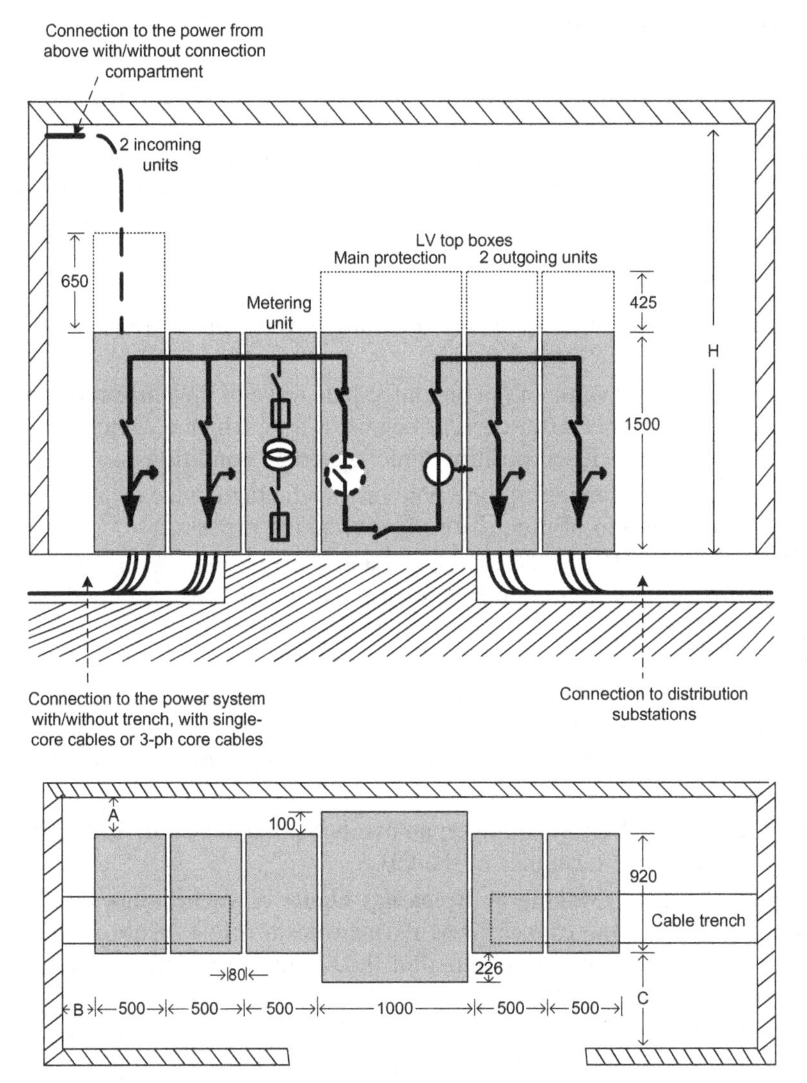

Figure 6.14 Installation layout for distribution switchboard (all dimensions are in millimeter; cables up to 50 mm^2).

- A without rear cabling and breaker = 100 mm, without rear cabling and with breaker = 180 mm;
- B without side cabling or earthing connection = 25 mm, with earthing connection or with side cabling = 100 mm;
- C minimum distance for operation and handling, switchboard without transformer or CB unit = 1100 mm, switchboard without transformer, with CB = 1500 mm; and
- H without top cabling or LV top box = 1900 mm, with top box = 2325 mm.

6.5 MV SWITCHGEAR DEVICES

MV switchgear contains electric equipment such as disconnectors, LBSs, earthing switches, CBs, relays, fuses, current limiting devices, and measuring instruments. Relays and fuses are explained in Chapter 5 as well as protection schemes. This chapter is concerned with the various types of switches and CBs.

6.5.1 Definitions

Making Current: Peak value of the first half-wave of the current in one pole of switching device during closing.

Peak Current: Peak value of the first large half-wave of the current during the transient occurrence after current begins to flow, which a switching device withstands in the closed position under specified conditions.

Breaking Current: Current in one pole of a switching device or a fuse at the instant of initiation of the arc during a breaking process.

Making Capacity: The value of the prospective making current which at a given voltage a switching device can make under specified conditions of application and performance; for switches, the value of the prospective service current.

Breaking Capacity: The value of the prospective breaking current which at a given voltage a switching device can interrupt under specified conditions of application and performance; for switches, the value of the prospective current.

Short-Line Fault: A short circuit on an overhead line at a short, not negligible distance from the terminals of the CB.

Switching Capacity (Making or Breaking) Under Asynchronous Conditions: Making or breaking capacity when synchronism is lost or absent between the network sections before and after the CB under specified conditions of application and performance.

Normal Current: The current that the main current path of a switching device can carry continuously under specified conditions of application and performance.

Short-Time Withstand Current: The rms value of the current which a switching device in the closed position can carry for a specified short time under specified conditions of application and performance.

Rated Voltage: The upper limit of the highest voltage of a network for which a switching device is designed.

Applied Voltage: The voltage between the terminals of a pole of a switching device or a fuse after the current is interrupted.

Opening Time: The time interval between the specified instant of initiation of the opening operation and the instant of separation of the arcing contacts in all poles.

Closing Time: The time interval between initiation of the closing operation and the instant when the contacts touch in all poles.

Rated Value: The value of a characteristic quantity used to define the operating conditions for which a switching device is designed and built, and which the manufacturer must guarantee.

Withstand Value: The maximum value of a characteristic quantity that a switching device will tolerate with no impairment of function. The withstand value must be at least equal to the rated value.

Standard Value: A value defined in official specifications on which the design of a device is to be based.

Rated Power-Frequency Withstand Voltage: The rms value of the sine-wave voltage alternating at system frequency which the insulation of a device must withstand for 1 min under the specified test conditions.

Rated Lightning Impulse Withstand Voltage: The peak value of the standard 1.2/50-μs voltage surge that the insulation of a device must withstand.

6.5.2 Knife Switches

Knife switches are used in MV switchgear to isolate specific equipment or feeder for maintenance or other purposes such as earthing (Fig. 6.15). They operate at no-load conditions by hand, or in remote-controlled installations, they are actuated by motor or compressed air. Blades of knife switches, mounted standing or suspended, must be prevented from moving spontaneously under their own weight. The physical size of this type of switches must be taken into consideration when deciding the dimensions of the switchgear. Usually, the switchgear requires greater depth.

6.5.3 LBSs

LBSs are increasingly being used in MV distribution systems. For instance, RMUs use LBSs in the two incoming feeders that connect the consumer's substation to the network. LBSs can be operated on load conditions. They have full making capacity and can handle all fault-free routine switching operations. Two mechanisms can be used for LBS operation:

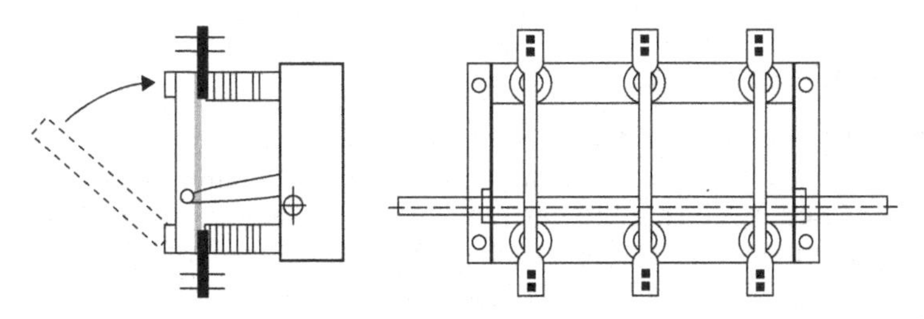

Figure 6.15 MV knife switch [81].

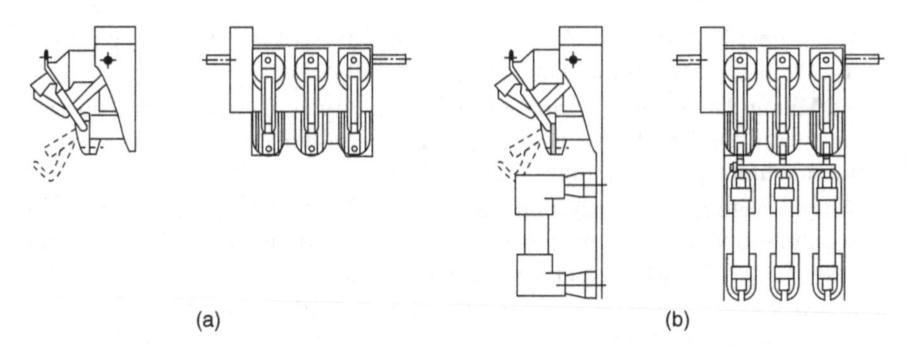

Figure 6.16 Knife-contact LBS type [81]. (a) Without fuses. (b) With fuses.

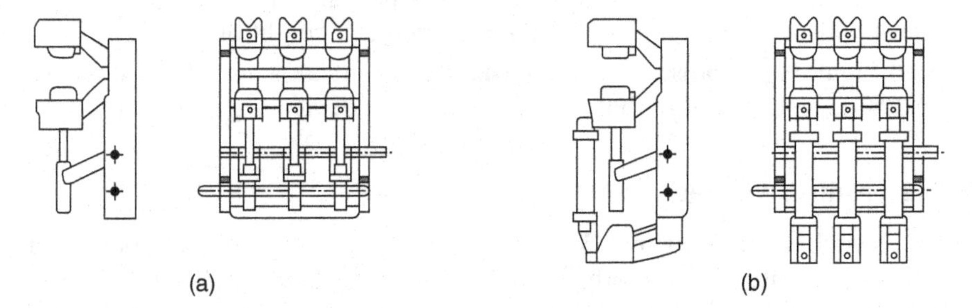

Figure 6.17 Slide-in LBS type [81]. (a) Without fuses. (b) With fuses.

- *Snap-Action Mechanism:* A spring is tensioned, which is released shortly before the switching angle is completed. Its force is used to move the contacts. The procedure is employed for both closing and opening.
- *Stored-Energy Mechanism:* It has one spring for closing and another for opening. During closing operation, the opening spring is tensioned and latched. The stored energy for opening operation is released by means of a magnetic trip or fuse.

Mainly two types of LBSs are used: knife-contact type with/without fuses (Fig. 6.16) and slide-in type with/without fuses (Fig. 6.17).

The switch blades are isolated by air (air LBS) or by using sulfur hexafluoride (SF_6 LBS).

6.5.4 Earthing Switches

Earthing switches are commonly used and installed in switchgear. When isolating any of the feeders (incoming or outgoing) for maintenance, the feeder must be earthed by closing the earthing switch to discharge any static charge carried by the feeder. Earthing switches are mounted separately ahead of the switchgear (Fig. 6.18a) or in the base of LBS (Fig. 6.18b) or just underneath the CB (Fig. 6.18c). The switchgear manufacturer must mechanically interlock

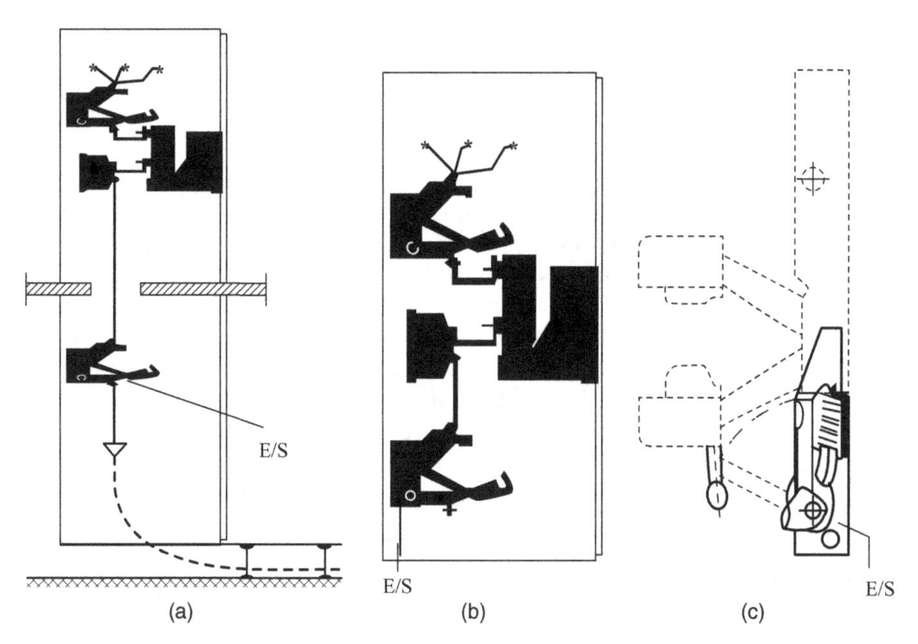

Figure 6.18 Installation of earthing switch (E/S) [81]. (a) In the cable basement; (b) underneath the circuit breaker; (c) in the base of LBS.

the earthing switch with the CB or LBS to avoid severe symmetrical short circuit if they are closed simultaneously.

6.5.5 CBs

The basic function of a CB is to break/make the continuity of the circuit. It is the consideration of the effect on the circuit of doing this, which principally dictates the choice of breaker. Therefore, CBs are mechanical switching devices able to make, continuously carry, and interrupt currents under normal circuit conditions and also for a limited time under abnormal conditions, such as short circuits. Compromise is needed both on economic grounds, taking into account probability of certain conditions, and technical grounds involving consideration of lower or higher speeds, of heavy or low current operation, and many other opposing influences (e.g., maximum operating voltage and current at location, system frequency, duration of short circuit, switching cycle, and climatic conditions). The basic elements of CBs are operating mechanism, insulators, interrupting chamber(s), capacitor, and resistor [83].

The main types of CBs include

- bulk oil
- minimum oil
- air

- air blast
- sulfur hexafluoride (SF_6)
- vacuum
- explosive.

Bulk Oil CBs: This type ranges from the lowest voltage with one break per phase to 330kV with six breaks per phase. It has the characteristics of simplicity, robustness, and quiet operation.

The operation principally depends on the formation of gases in the oil by the arc, which creates a high pressure in the restricted space. At high currents with the very high pressure built up, the interruption takes place in hydrogen. At lower currents, there is a lower pressure but a high cooling effect, which extinguishes the arc. Although the bulk oil CB is falling out of use at higher voltages due to its cost and weight, it has the great inherent advantage of internally mounted current transformers.

Minimum Oil CBs: These start from the lowest voltage with one break per phase, and have been designed for the highest voltages with 20 breaks per phase at 735kV. The characteristics of this type are simplicity, quiet operation, and cheapness.

The operating principles are as that of the bulk oil but in a smaller enclosed volume to utilize the arcing energy to extinguish itself instead of the bulk oil volume pressure. It is becoming more and more acceptable at all voltages because of its characteristics and the large amount of development testing that has gone into the small oil CB design. It does, however, require more frequent maintenance than other types of CBs.

Air CBs: These CBs are in general use up to about 22kV. For operation, they depend on lengthening and cooling the arc chutes. The features and characteristics are high ratings and low maintenance for a large number of operations. This type of CB is inherently expensive and is usually only used where a large number of operations are required, such as in auxiliary systems for power stations or substations.

Air Blast CB: This is used over a wide voltage range but, due to cost considerations, tend to be restricted to the high-voltage range (132kV and above) where very high current ratings are required. They are traditionally used for arc furnace switching at 33kV. The arc is drawn in high pressure air, and extinction is obtained by deionization and cooling.

The performance and characteristics of this type are highest rating, low maintenance, and noise level is low when fitted with silencers; it can be fitted with closing resistor to improve voltage recovery. The aforementioned factors make this CB most suitable for transmission systems of the highest capacity.

SF_6 (Sulfur Hexafluoride): This type is the most commonly used at present and its applications are from 11 kV upward. Its commercial use started in the early 1980s and it was then thought to be the principal type of CB at such voltages. Some problems have lately been discovered mainly with steep front voltage switching waveforms, which could produce ferroresonance overvoltages in voltage transformers. Newer designs of the gas type are being developed, one of which is Freon gas, but further development was stopped as a result of the recent regulations.

The interruption mechanism depends on the electronegative gas SF_6, which has a strong affinity for free electrons. When an arc is drawn in this gas, the conducting electrons are captured to form relatively immobile negative ions. This causes the arc to become unstable and easily extinguishable when the electric strength of the gap recovers very quickly.

The performance and characteristics of this type are highest ratings, the noise level is low due to the closed circuit gas system, and very suitable for metal-clad installations.

Vacuum CBs: These CBs are based on the use of vacuum interrupters that are now being produced up to 33 kV and current ratings above 1600 A. The breakers are virtually maintenance free and where the number of interrupters in series is kept down, they are very simple to maintain. They are used largely in distribution and railway electrification systems.

Figure 6.19 shows the two types of CBs that are commonly used in distribution systems.

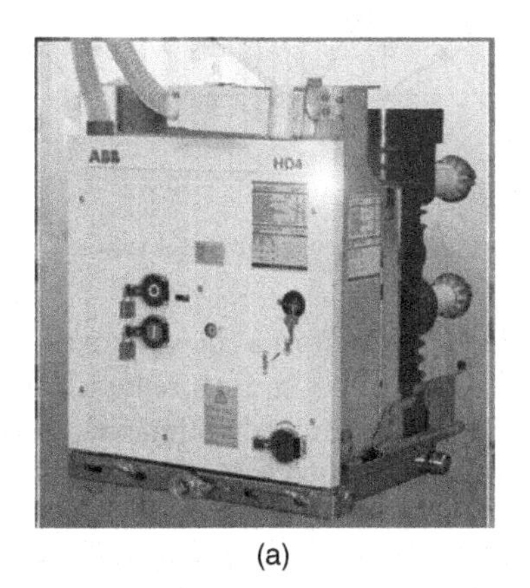

(a)

(b)

Figure 6.19 Typical examples of the most commonly used circuit breakers in MV primary distribution (courtesy of ABB). (a) SF_6 circuit breaker; (b) vacuum circuit breaker.

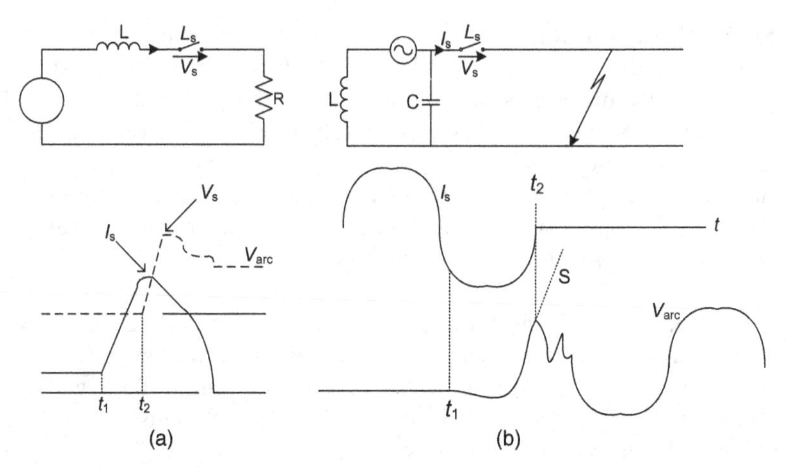

Figure 6.20 Forms of arc extinction process. (a) Equivalent circuit and waveforms of voltage and current for DC extinction. (b) Equivalent circuit and waveforms of voltage and current for AC extinction. t_1 = contact separation; t_2 = arc extinction; S = rate of recovery voltage rise.

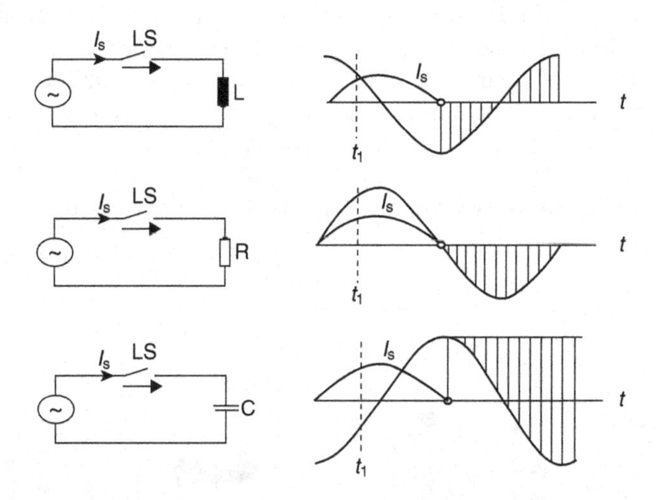

Figure 6.21 Breaker voltage when connecting to various load types.

6.5.5.1 Principles of Interruption Two basic forms of arc extinction process can take place; DC extinction and AC extinction. Enforcing a current zero is the only way to extinguish the DC arc. So, the arc voltage V_{arc} has to be higher than the voltage applied to the breaker contacts. A sufficiently high arc voltage can be built up only in LV and MV DC circuits (Fig. 6.20a). For AC arcs, they may extinguish at each current zero. In high-voltage circuits and without extra measures, the arc reignites after passing current zero, and so continues to burn.

One of the factors affecting breaker voltage is the type of load (inductive, resistive, and capacitive). In case of inductive loads, the breaker must be able to cope with the recovery voltage's rate of rise and its peak value. Once the arc is quenched, the dielectric strength between contacts must build up faster than the recovery voltage rise, if restriking is to be prevented. When interrupting a purely resistive load, current zero and voltage zero coincide. The recovery voltage at the breaker raises sinusoidal V_{arc} with the service frequency. The gap between the contacts has sufficient time to recover. If the load is capacitive, the supply voltage oscillates at system frequency after interrupting the current and the terminals of the breaker on the load side remain charged (Fig. 6.21).

6.6 LV SWITCHGEAR DEVICES

6.6.1 Isolators

The isolator is a switch manually operated (some types are equipped with automatic close/open mechanism), lockable, two-position (open/closed) device. It provides safe isolation of a circuit when locked in the open position. It is not designed to make or to interrupt current and no rated values for these functions are given in the standards.

The isolator must be capable of withstanding the flow of short-circuit currents for a limited time (short-time withstand capability) usually 1 s. For operational overcurrent, the time is longer. Therefore, LV isolator is essentially a dead system switching device to be operated with no voltage on either side of it, particularly, when closing. This is because of the possibility of an unsuspected short circuit on the downstream side. Interlocking with an upstream switch or CB is frequently used. The symbol shown in Figure 6.22a is used to represent the isolator in the drawings.

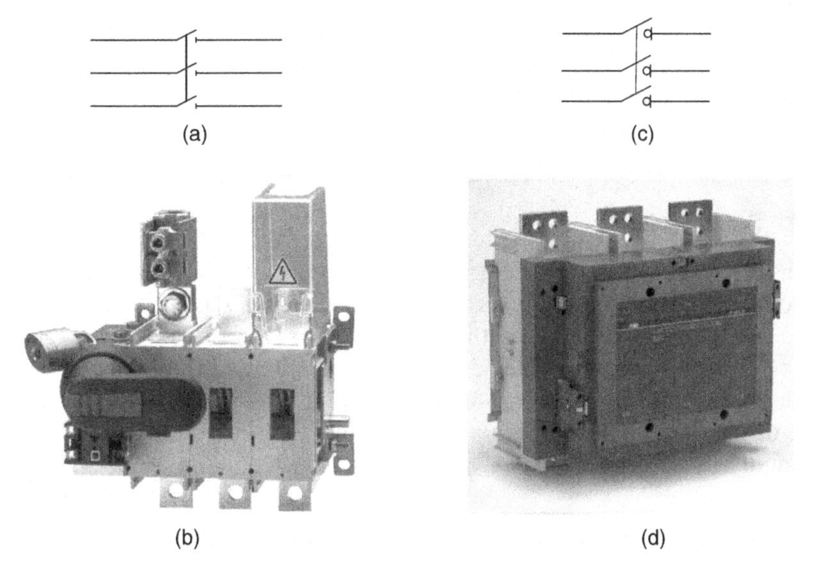

Figure 6.22 (a) Symbol of isolators; (b) typical type of isolator; (c) symbol of load-break switches; (d) typical type of load-break switch (courtesy of ABB).

6.6.2 LBS

LBS is a control switch, nonautomatic, two-position (open/close), operated manually, and sometimes provided with electric tripping for operator convenience. LBS is used to close and open loaded circuits under normal conditions. It does not provide any protection for the circuit it controls. Its characteristics are determined by the frequency of switch operation (600 close/open cycles per hour maximum), mechanical and electric endurance, and current making and breaking capacity for normal and infrequent situations. The symbol used for the LBS is shown in Figure 6.22b.

6.6.3 Contactors

The contactor is a solenoid-operated switching device that is generally held closed by reduced current through the closing solenoid. Different mechanically latched types can be used for specific applications (e.g., motor starting, switching capacitors). Contactors are designed to carry out numerous close/open cycles and commonly controlled by on/off push buttons. They may have auxiliary contacts, normal-close (N.C.) and normal-open (N.O.) contacts, to be used for control functions.

The characteristics of contactors are specified by the operating duration, the application in which to be used, the number of start/stop cycles per hour, and mechanical and electric endurance. The contactor is represented by the symbol shown in Figure 6.23a. In some applications, for example, motor controller and remote push-button control of lighting circuits, the contactor is equipped with a thermal-type OCR for protecting the circuit against overloading (Fig. 6.23b).

The contactor equipped or not equipped with OCR is not equivalent to a CB, since its short-circuit current breaking capacity is limited. For short-circuit protection, therefore, it is necessary to include either fuses or a CB in series with, and upstream of, the contactor.

6.6.4 Fuse Switch

It consists of three switch blades each constituting a double break per phase. These blades are not continuous throughout their length, but each has a gap in the center that is bridged by the fuse cartridge (Fig. 6.23c).

Single units of switching devices do not cope with all the requirements of the three basic functions: isolation, control, and protection. Combinations of these devices specifically designed for such functions are employed. The most commonly used combinations are

- fuse switch + contactor equipped with OCR (Fig. 6.24a) and
- automatic isolator + fuse + OCR (Fig. 6.24b).

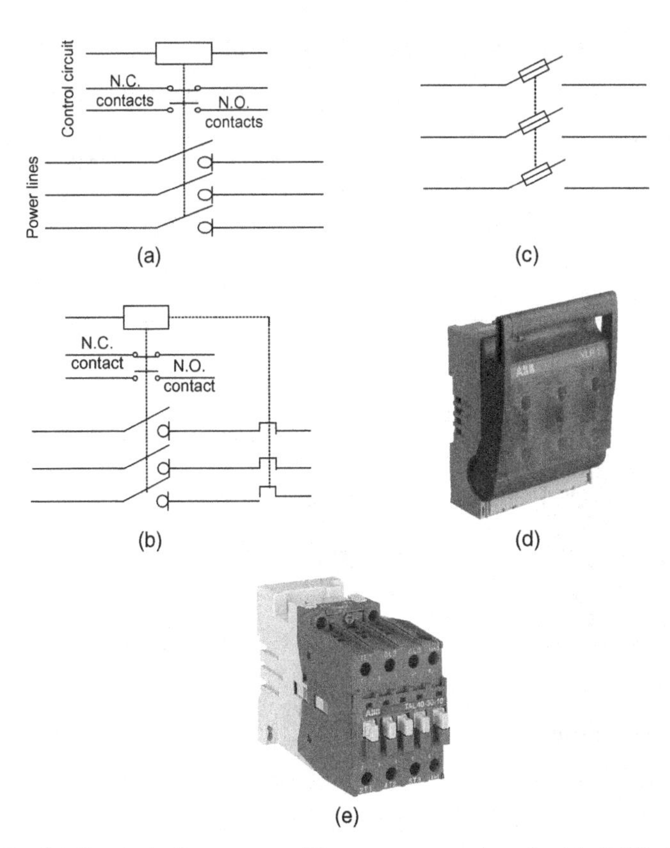

Figure 6.23 (a) Symbol of contactor, (b) contactor equipped with OCR, (c) symbol of nonautomatic fuse switch, and (d,e) typical types of fuse switch and contactor, respectively (courtesy of ABB).

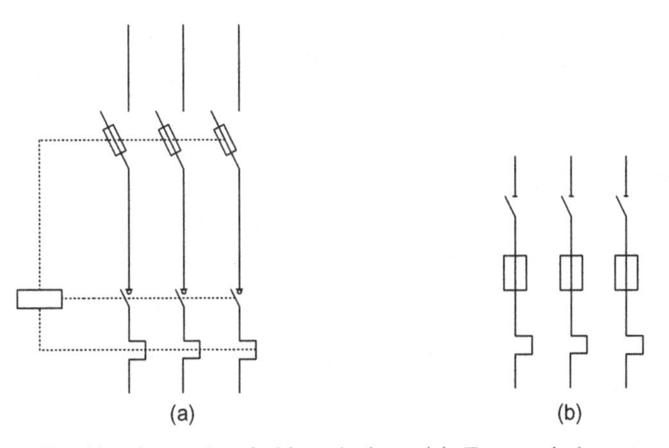

Figure 6.24 Combinations of switching devices. (a) Fuse switch + contactor with OCR; (b) automatic isolator + fuse + OCR.

TABLE 6.3 Summary of Switching Devices Capability

	Basic Switchgear Functions		
Switching Device	Isolation	Control	Protection
Isolators	x		
LBS	x	x	
Contactors		x	
Contactors with OCR		x	x
Fuse switch	x		x

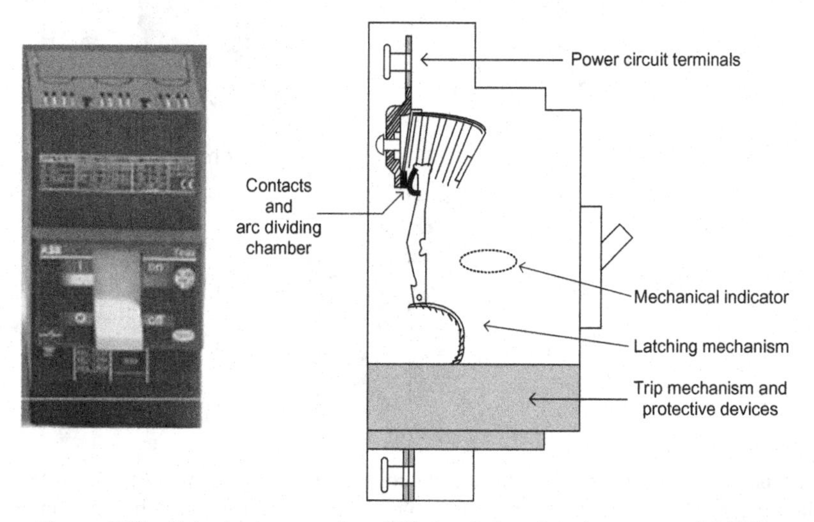

Figure 6.25 Principal parts of an LV circuit breaker (courtesy of ABB).

Therefore, referring to the three basic functions of switchgear, the capabilities of switching components to perform these functions are summarized in Table 6.3.

6.6.5 LV CBs

The CB is the only item of switchgear capable of simultaneously satisfying all the basic functions necessary in an electric installation (isolation, control, and protection).

6.6.5.1 Description LV CB consists of principal parts (Fig. 6.25) to carry out four essential functions:

- Circuit-breaking components, comprising the fixed and moving contacts and the arc-dividing chamber.

- Latching mechanism that becomes unlatched by the tripping device on detection of abnormal current conditions. This mechanism is also linked to the operation handle of the breaker.
- Trip-mechanism actuating device. It is either
  - a thermal-magnetic device in which a thermal-operated bimetal strip detects an overload conditions, while an electromagnetic striker pin operates at current levels reached in short-circuit conditions, or
  - an electronic relay operated from current transformers, one of which is installed on each phase.
- Space allocated to the several types of terminals currently used for the main power-circuit conductors.

Additional modules can be added to the CB to be adapted to provide further features such as sensitive detection (30 mA) of earth leakage current with CB tripping, remote control, and indication (on/off fault), and heavy-duty industrial CBs of large current ratings, which have numerous built-in communication and electronic functions (e.g., molded case circuit breaker [MCCB]).

6.6.5.2 Fundamental Characteristics

Rated Operational Voltage (V_n) This is the voltage at which the CB has been designed to operate, in normal conditions. Other values of voltage are also assigned to the CB corresponding to abnormal conditions, for example, rated insulation voltage and rated impulse-withstand voltage.

Rated Current (I_n) This is the maximum value of current that a CB, fitted with a specific overcurrent tripping relay, can carry indefinitely at an ambient temperature stated by the manufacturer, without exceeding the specified temperature limits of the current carrying parts. For example, a CB at $I_n = 125\,A$ for an ambient temperature 40°C will be equipped with a suitably calibrated overcurrent tripping relay (set at 125 A). The same CB can be used at higher values of ambient temperature, however, if suitably "derated." Thus, the CB in an ambient temperature of 50°C could carry only 117 A indefinitely, or again, only 109 A at 60°C, while complying with the specified temperature limit.

Derating of CB is achieved by reducing the trip-current setting of its overload relay. The use of an electronic type of tripping unit, designed to withstand high temperatures, allows CBs to operate at 60°C or even at 60°C ambient.

Overload Relay Trip-Current Setting (I_{rth} or I_r) Industrial CBs differ from small (domestic) CBs that can be easily replaced. They are equipped with exchangeable overcurrent trip relays. Moreover, in order to adapt a CB to the requirements of the circuit it controls, and to avoid oversized cables, the trip relays are generally adjustable. The trip-current setting I_r or I_{rth} (both designations are in common use) is the current above that the CB will trip. It also represents the maximum current the CB can carry without tripping.

The value of I_r must be greater than the maximum load current, but less than the maximum current permitted in the circuit. The thermal trip relays are generally adjustable from 0.7 to 1.0 times I_n, but when electronic devices are used for this duty, the adjustment range is greater; typically 0.4–1 times I_n. For example, a CB equipped with a 320A overcurrent trip relay, set at 0.9, will have a trip-current setting of $I_r = 320 \times 0.9 = 288$A and the adjustment range is from 224 (0.7×320) to 320A.

Short-Circuit Relay Trip-Current Setting (I_m) Short-circuit tripping relays (instantaneous or slightly time delayed) are intended to trip the CB rapidly on the occurrence of high values of fault current. Their tripping threshold I_m is either fixed by standards for domestic type CBs, for example, IEEE and IEC, or indicated by manufacturer for industrial-type CBs according to related standards. Industrial CBs are provided with a wide variety of tripping devices that allow a user to adapt the protective performance of the CB to the particular requirements of a load. The performance of the CBs with thermal-magnetic and electronic protective schemes [84] is shown in Figure 6.26.

Rated Short-Circuit Breaking Capacity (I_{cu} or I_{cn}) The short-circuit current-breaking rating of a CB is the highest value of current that the CB is capable of breaking without being damaged. The value of current quoted in the standards is the rms value of the AC component of the fault current, that is, the DC transient component is assumed to be zero for calculating the standardized value. This rated value (I_{cu}) for industrial CBs and (I_{cn}) for domestic-type CBs is normally given in kiloampere rms.

6.6.5.3 Selection Criteria The choice of a CB is made in terms of the following [85]:

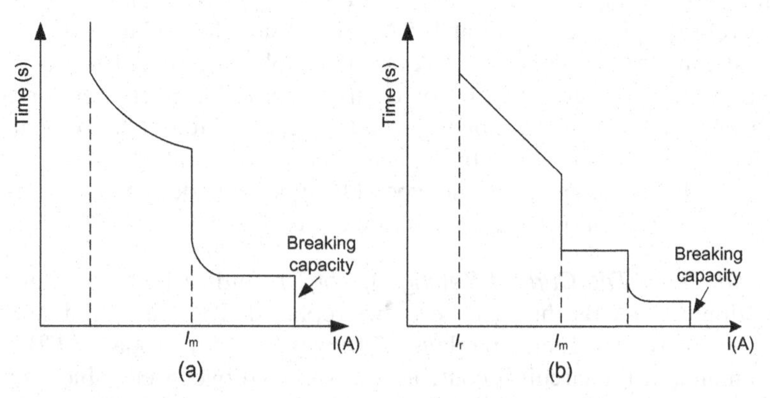

Figure 6.26 Performance of LV circuit breakers (short-circuit instantaneous relay trip-current setting vs. time). (a) Circuit breaker thermal-magnetic protective scheme; (b) circuit breaker electronic protective scheme.

- electric characteristics of the installation for which the CB is destined;
- its eventual environment: ambient temperature, in a kiosk or switchboard enclosure, climatic conditions, and so on. The rated current is defined for operation at a given ambient temperature, in general, 30°C for domestic-type CB and 40°C for industrial-type CB. The performance of these CBs depends basically on the technology of their tripping units;
- short-circuit current breaking and making requirements;
- operational specifications: discriminative tripping, requirements for remote control and indication and related auxiliary contacts, auxiliary tripping coils, connection into a local network, and so on;
- installation regulations, in particular, protection of persons;
- load characteristics, such as motors, fluorescent lighting, and transformers.

6.7 PROTECTION CLASSES

Protection classes for LV and MV electric equipment are identified by a code made up of the two letters IP (International Protection) followed by two digits. The first of two code digits identifies the degrees of protection against contact and ingress of foreign bodies. The second digit identifies the degrees of protection against ingress of water. The preferred protection classes for LV and MV distribution equipment and switchgear are summarized in Table 6.4.

6.8 SPECIFICATIONS AND IMPLEMENTATION OF EARTHING

Objectives and methods of distribution system earthing have been explained in Chapter 3. The implementation of equipment grounding with a goal of protecting life and property in the event of short-circuit faults and transients due to lightning and switching operations is presented in this section. The main elements of grounding system including three areas, MV, LV, and building services, are shown in Figure 6.27 as an example. It is seen that a number of earthing conductors are connected to the main earthing conductor directly as in building services area or indirectly through sub-main earthing conductors as in LV/MV areas. The sub-main earthing conductor, which joins together parts of several pieces of devices in an installation, is connected within this installation to the main earthing conductor. The main earthing conductor is connected to ground by using earth electrodes [86, 87].

The earth electrode is a conductor that is embedded in and electrically connected to the ground, or a conductor embedded in concrete, which is in contact with the earth over a large area. It has different types classified basically according to the form and cross section such as strip, bar, stranded wire, and tube electrodes. Also, different materials are used for electrode manufacturing such as (1) copper, which is suitable for earth electrodes in power installations with high fault currents because its electric conductivity is much higher than

TABLE 6.4 Protection Classes Commonly Used for LV and MV Equipment and Switchgear (IEC 529)

First Digit		Second Digit (Protection Against Contact and Bodies)					
Protection Against Contact and Ingress of Foreign Bodies		No Protection	Drip-Proof	Devices Tilted 15°	Spray-proof	Splash-proof	Hose-proof
		0	1	2	3	4	5
No protection	0	IP00					
Hand-proof (solid bodies >50 mm)	1	IP10					
Finger-proof (solid bodies >12 mm)	2	IP20					
Protection against solid bodies >2.5 mm, excluding tools and objects	3	IP30	IP31	IP32			
Protection against solid bodies >1 mm, excluding tools and wires	4	IP40		IP42	IP43		
Protection against harmful dust deposits (full protection against contact)	5	IP50			IP53	IP54	IP55
Protection against ingress of dust (full protection against contact)	6						IP65

Notes: Drip-proof is the protection against vertically falling water. Devices titled 15° means protection against water up to 15° to vertical. Spray-proof is the protection against water up to 60° to vertical. Splash-proof and hose-proof are the protection against water from all directions.

that of steel, and (2) galvanized steel, which is durable in almost all kinds of soil. It is suitable for embedding in concrete. Recommended minimum dimensions of earth electrodes and earth conductors made up of copper or steel are given in Reference 88. Depending on electrodes form and dimension, different arrangements of earth electrodes (radial, ring, meshed, or combination of these) can be used to satisfy a specific grounding resistance.

6.9 SAFETY AND SECURITY OF INSTALLATIONS

Safety installations, particularly in buildings that receive public or in which people are employed (offices, shops, factories, etc.), must be provided with the means for ensuring the safe evacuation of personnel in addition to

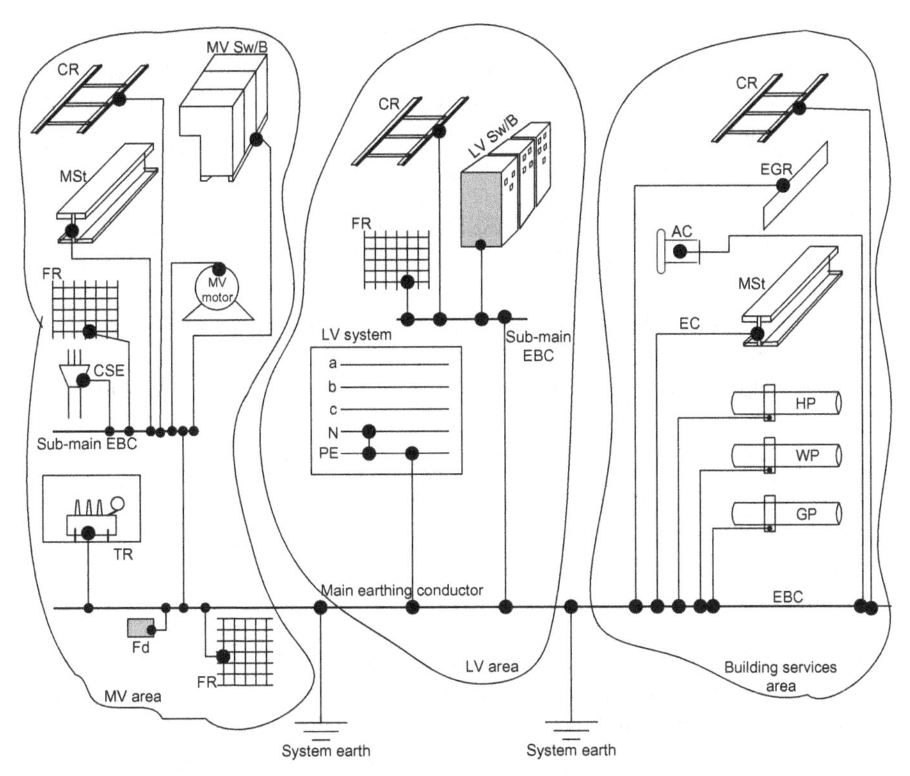

Figure 6.27 Grounding system with equipotential bonding between MV, LV indoor switchgear, and building services. CR = cable racks; LV Sw/B = LV switchboard; MV Sw/B = MV switchboard; GP = gas pipes; MSt = metal structure; FR = floor reinforcement; EBC = equipotential bonding conductors; TR = transformer room; CSE = cable sealing end; AC = air conditioners; HP = heating pipes; WP = water pipes; EGR = elevator guide rail; Fd = foundations; EC = earthing conductor.

- security and safety lighting,
- alarms and warning systems,
- automatic fire detection,
- fire-extinguishing systems,
- smoke evacuation, and
- air compressors for the pressure-operated fire-extinguishing system.

Regulations covering safety installations contain a number of conditions to be respected concerning their electric power sources:

- Duration time of an interruption: according to the case, the following options are imposed:
  - no break,
  - a break of less than 1 s, and
  - a break of less than 15 s.

- Autonomy demanded for the reserve-power source: in general, it corresponds to the time necessary to complete all safety operations for persons, for example, the time to evacuate an establishment receiving the public is 1 h minimum. In large apartment blocks, the autonomy of the source must be 36 h or more.

For switchgear installations, in addition to the aspects written above, some other factors in relation to safety must be accounted:

- interlocking mechanism (electric or mechanical) between earthing switches and CBs;
- interlocking mechanism between different power sources feeding the same bus bar, unless they are synchronized;
- the panel door is not allowed to open when the power is on;
- the panel door is not allowed to close if the CB is not correctly connected; and
- the dimensions of switchgear panels are in compliance with international standards.

Security of electric installations is satisfied by providing standby reserve-power supplies. Among many applications in which an interruption of power supply cannot be tolerated, the following may be cited:

- information technology (IT) systems;
- critical industrial processes, which necessitates the continuity of operation;
- telecommunications;
- surgical operating theaters;
- scientific research centers;
- ticketing, plane reservations, cash registers, and so on; and
- military business.

It may be noted that where several emergency-services standby sources exist, they can also be used as reserve-power sources. This is on condition that any one of them is available and capable of starting and supplying all safety and emergency circuits. Consequently, the failure of one of them does not affect the normal functioning of the others.

In order to satisfy the requirements of economical exploitation, the features below are imperative:

- Supply interruption is not tolerated in IT systems and industrial continuous-process operations.
- Period on conserving data in IT systems is 10 min.
- Autonomy is desirable for reserve-power supplies installations.

6.10 ASSESSMENT OF SWITCHGEAR

In assessing the merit of any particular switchgear design, the factors below must be considered:

- *Service Continuity:* The ability to provide a continuous and reliable operation not only during normal conditions, but also during and after a fault clearance or after routine maintenance is of the highest importance.
- *Operation Facilities:* It is often necessary to allow for independent control through different sections of the substation, either to isolate certain loads that have undesirable characteristics or to maintain the fault level of a system within a predetermined value.
- *Simplicity:* This aids safety and reliability by reducing the chances of faulty operation and making interlocking requirements simpler.
- *Extensibility:* The necessity for adding additional circuits is not always foreseen but almost invariably required. This must be possible at a minimum cost with a minimum interruption to the supply.

The above factors can in general be balanced one against the other but occasionally some salient feature of the electric system or of the geography of the substation may override these factors.

Example 6.5:

This example is just to illustrate the main idea of a suggested single-line diagram for an industrial distribution system providing the necessary security and continuity of supply for the critical loads.

An industrial plant has six load centers; five of them are operated at LV and one load center includes motors operated at MV. The utility source is at MV. As shown in Figure 6.28, the main MV switchboard is installed nearby two of the load centers. It includes

- two incoming feeders from two different sources; each one is connected to one section of the switchboard bus bar where the two sections are connected to each other through a bus coupler;
- two measuring sets, one for each incoming feeder;
- two outgoing feeders, each one feeds a specific load center through a substation transformer MV/LV;
- two outgoing feeders forming a loop to feed the other load centers; and
- one outgoing radial feeder to feed the motor control switchboard.

The critical loads (e.g., the industrial processes that cannot withstand source interruptions) are supplied by two substations fed from the sub-main indoor

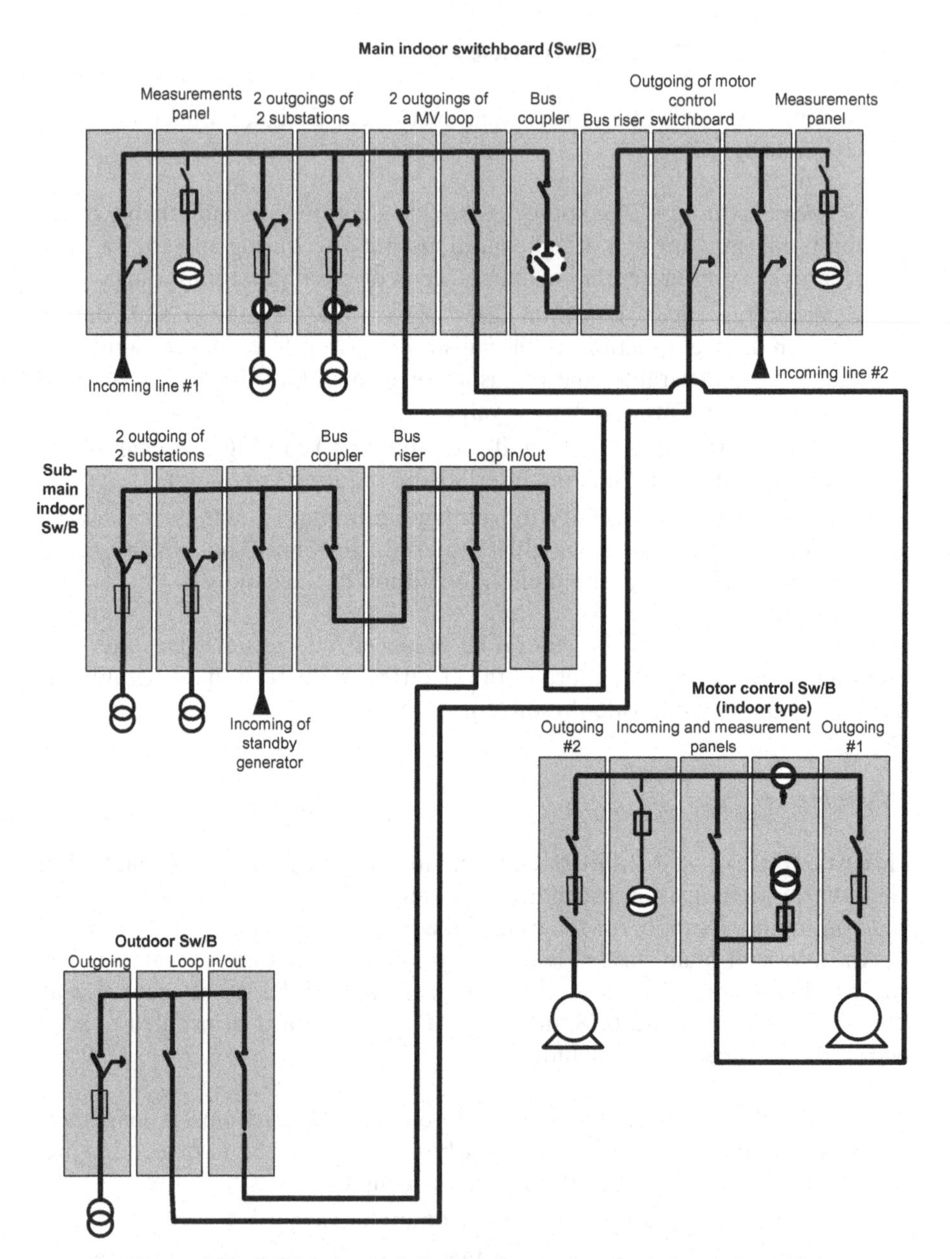

Figure 6.28 MV industrial distribution system (switchgear layout).

MV switchboard. This switchboard is connected to three sources: two sources from the two sides of the loop and the third is a standby generator to satisfy the required security.

According to the nature of loads and the geography of place, an outdoor MV switchboard connected to the loop is installed to supply services and auxiliaries. The MV motors are connected to the motor control switchboard, which is fed radially from the main MV switchboard.

6.11 STEPS FOR INSTALLING SWITCHGEAR

To install switchgear, a number of steps are followed. These steps can be summarized as below:

- Drawing the single-line diagram. It includes the following:
  - the power lines in which the power flows;
  - measuring instruments where their location on the diagram is illustrated together with the parameters to be measured;
  - protection devices; and
  - technical specifications of electric equipment encountered by the switchgear.
- Determination of available space to enable the manufacturer to decide whether the switchgear is one side I, or shape of Γ, or shape of Π.
- Providing the single-line diagram to the manufacturer to prepare the physical drawings of the switchgear.
- Examination of physical drawings by customer to be approved before starting of manufacturing.
- Building dimensioning and preparation of civil layout to indicate
  - location of switchgear,
  - the clearances, and
  - special requirements (trenches, ventilation, etc.).
- Accordingly, the civil engineering personnel prepare the civil design to start building construction.
- Switchgear installation.
- Testing.
- Commissioning and operation.

6.12 ARC FLASH HAZARDS

Arcing faults receive an increasing amount of attention as a particularly damaging and potentially dangerous type of fault. The arcing fault is a flashover of low electric current that flows through the air (high impedance). It releases a great deal of energy in the form of heat and pressure. The temperature can rise to 19,420°C (35,000°F [89]), and the intense heat from the arc can cause a sudden expansion of the air resulting in a blast with very strong air pressure that can vaporize the materials and damage the equipment. The arcing fault can persist longer and propagate. For example, a 3- to 4-in. arc in an LV system can become "stabilized" and persist for an extended period of time [90], and in enclosures the arc may propagate to the supply side of all devices in the same enclosure (Fig. 6.29) [91]. On the other hand, the enclosure magnifies blast that is forced to open its side, and the energy produced by the arc is transmitted toward the worker causing burn injuries or death. Some burn injuries that may happen due to electric arcs are shown in Figure 6.30 [92].

The arc flash hazard is defined as "a dangerous condition associated with possible release of energy caused by an electric arc" [93]. It is measured in terms of arc flash incident energy (AFIE), which is used to determine the appropriate level of personal protection equipment (PPE) and in terms of arc flash protection boundary.

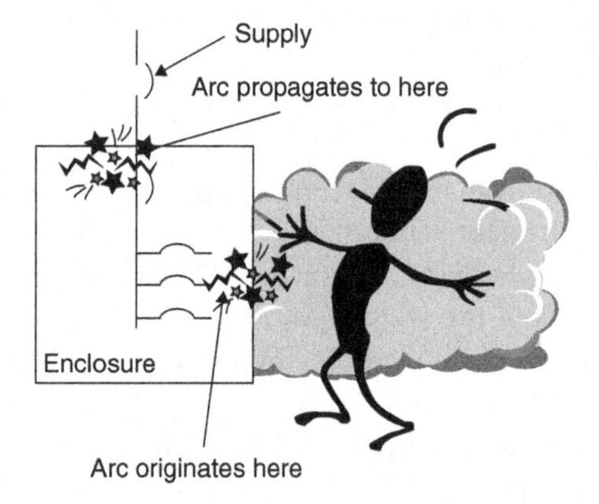

Figure 6.29 Arc propagation [91].

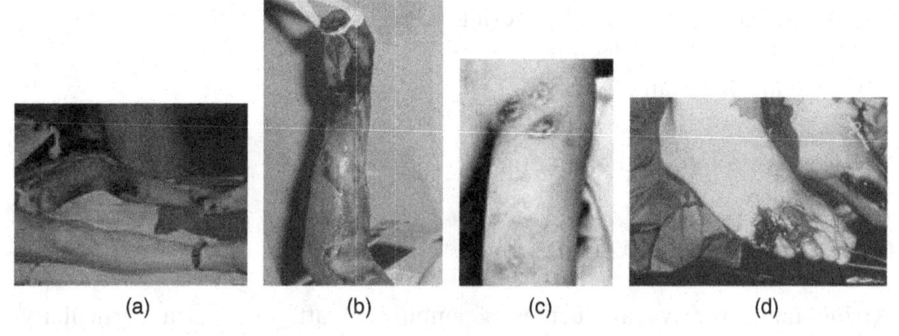

Figure 6.30 Burn injuries due to electric arcs [92]. (a) Leg burn; (b) hand burn; (c) shoulder burn; (d) foot burn.

6.12.1 Causes of Arcing Faults

Arcing faults are commonly caused by the following:

- foreign tools dropped into equipment that may cause short-circuit, produce sparks, and initiate arcs;
- entry of foreign bodies (e.g., rodent, snake, squirrel);
- misalignment of moving contacts and corrosion of equipment parts. This weakens the contact between conductor terminals and increases the contact resistance. So, heat is generated on the contacts and sparks may be produced leading to arcing faults with nearby exposed conductors;
- dirt contamination or dielectric breakdown. Dirt on insulating surfaces can provide a path for current, allowing it to flashover creating arc discharge across the surface. The flashover can also be created by condensation of vapor and water dripping on insulating surfaces;
- careless cover or device removal that gives an opportunity for accidental touching with live exposed parts producing arc faults;
- overvoltages across narrow gaps. If the air gap between conductors of phases or phase conductor and earth is narrow enough, the electric field intensity through the air gap may ionize the air during overvoltages producing arc faults; and
- improper maintenance procedures.

6.12.2 Arc Flash Consequences

Nonhuman Consequences:

- Process downtime due to accidents.
- Loss of product and lost revenue.
- Equipment damage.
- Occupational Safety and Health Administration (OSHA) citation and fines.

Human Consequences:

- Treatment of victims is very expensive. Victims may not retain their life quality.
- Loss of skilled manpower and loss of morale.
- Higher insurance costs.

6.12.3 Limits of Approach

For the safety of qualified persons who are working on energized equipment, the flash and shock boundaries are broken down as indicated in Figure 6.31 according to the definition stated in Reference 93.

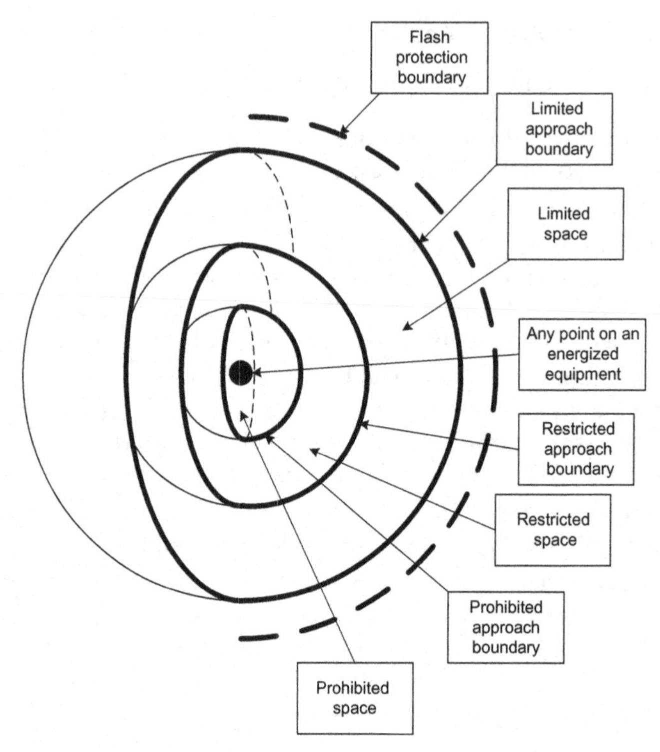

Figure 6.31 Arc flash approach limit regions.

- *Flash Protection Boundary:* It is an approach limit at a distance from the exposed live parts (arc source) within which the potential incident heat energy from an arcing fault on the surface of skin is 1.2 cal/cm^2 and causing a curable second-degree burn (Fig. 6.32). Appropriate flash-flame protection equipment must be utilized for persons entering the flash protection region.
- *Limited Approach Boundary:* It is an approach limit at a distance from the exposed live parts (arc source) within which a shock hazard exists. The person entering the limited approach boundary must be qualified to perform the job.
- *Restricted Approach Boundary:* It is an approach limit at a distance from the exposed live parts (arc source) within which the risk of shock is increased. The person entering this boundary must use the appropriate PPE and have an approved work plan from the authorized management.
- *Prohibited Approach Boundary:* It is an approach limit at a distance from the exposed live parts (arc source) within which the work is considered the same as making contact with the live part. The person entering this region must be well trained to work on live equipment.

Figure 6.32 Flash protection boundary [91] (a person at a distance D from arc source D_{fb} is the limit distance of the boundary).

Figure 6.33 Incident energy delivered by an arc [91].

6.12.4 PPE Hazard Risk Categories

Workers potentially exposed to arc flash hazard must be adequately protected. The severity of this hazard is measured by the amount of energy that an arc delivers to an exposed worker (Fig. 6.33). This "incident energy" is commonly measured by calorie per square centimeter or joule per square centimeter and its calculation provides a basis for selecting proper PPE (e.g., flash suits, arc hoods, and flame-resistant [FR] clothing).

Incident energy and flash protection boundary vary based on different parameters such as system voltage, arcing fault current, working distance from arc source, and fault duration. The hazard level depends on many system variables including equipment type, prospective bolted fault currents, and characteristics of upstream protective devices. Therefore, to properly select the levels

TABLE 6.5 Protective Clothing Characteristics (NFPA 70E 2009)

Hazard/Risk Category	Minimum Arc Rating of PPE (cal/cm^2)	Typical Protective Clothing Description
0	N/A	Untreated cotton, wool, rayon or silk, or blends of these materials with a fabric weight of at least 4.5 oz/yd^2
1	4	Arc rated FR shirt and FR pants or coverall
2	8	Cotton underwear plus FR shirt and FR pants
3	25	Arc rated FR shirt and pants or coverall, and arc flash suit selected so that the system arc rating meets the required minimum
4	40	Arc rated FR shirt and pants or coverall, and arc flash suit selected so that the system arc rating meets the required minimum

Note: Meltable fabrics and other similar synthetics are never permitted.
N/A = null.

of PPE to protect the workers, the potential arc flash hazard at a given system location should be analyzed.

Both too little and too much PPE are undesirable as too little PPE is insufficient to protect the workers and too much PPE may hinder movement and increase the risk level as well as increase the heat stress. The required PPE is determined by comparing the calculated incident energy to the ratings for specific combinations of PPE that are classified into five categories defined by National Fire Protection Association (NFPA) 70E 2009 as an example in Table 6.5.

For most protective clothing classes, other requirements for PPE are needed such as hard hat, safety glasses, hearing protection, leather gloves, work shoes, and face shields. The arc rating of each is determined according to NFPA 70E-2009 Std.

As seen in Table 6.5, there is no PPE class of rating larger than 40 cal/cm^2. NFPA 70E does not intend for work to be performed at locations with such high AFIE. Above these levels, arc blast may be as significant a concern as the arc flash where burn injuries are not the only concern but also hearing damage and internal injuries (collapsed lung, concussion).

6.12.5 Calculation Methods

Several methods for calculating the arc flash hazard have been developed. Three of these methods are briefly described below.

6.12.5.1 IEEE Standard 1584-2002 IEEE Standard 1584 [94] presents the most comprehensive set of empirical equations for calculating incident energy levels and flash protective boundaries. These equations are based on the available bolted fault current, voltage, clearing time, equipment type, earthing, and working distance. They cover systems at voltage levels ranging from 208 V to 15 kV and for available bolted fault currents ranging from 700 A to 106 kA, to sufficiently cover the majority of LV and MV installations.

This method can also determine the impact of certain current limiting LV fuses as well as certain types of LV breakers. The equations are rather complex if calculations are to be performed by hand, though the equations are easily implemented in a spreadsheet or in other computer software. The calculations are summarized in four steps as below:

1. *Arcing Current Estimation:* For LV applications <1 kV,

$$\log I_a = K + 0.662\log I_{bf} + 0.0966V + 0.000526G + 0.5588V(\log I_{bf}) - 0.00304G(\log I_{bf}) \quad (6.14)$$

For applications >1 kV,

$$\log I_a = 0.00402 + 0.983\log I_{bf}. \quad (6.15)$$

Converting from log,

$$I_a = 10^{\log Ia}, \quad (6.16)$$

where

I_a = the arcing fault current (kA),
K = −0.153 for open configurations
= −0.097 for box configurations,
I_{bf} = rms of the bolted fault current for three-phase symmetrical faults (kA),
V = the system voltage (kV),
G = the gap between conductors (mm) (Table 6.6).

2. *Normalized Incident Energy Estimation:* The normalized incident energy, based on 0.2-s arc duration and 610-mm distance from the arc, is given as

$$\log E_n = K_1 + K_2 + 1.081\log I_a + 0.0011G \quad (6.17)$$

and

$$E_n = 10^{\log En}, \quad (6.18)$$

where

E_n = incident energy normalized for time and distance (J/cm²),
K_1 = −0.792 for open configurations
= −0.555 for box configurations,

TABLE 6.6 Factors for Equipment and Voltage Classes [95]

System Voltage (kV)	Equipment Type	Typical Gap between Conductors (mm)	Distance × Factor
0.208–1	Open air	10–40	2.000
	Switchgear	32	1.473
	MCC and panels	25	1.641
	Cable	13	2.000
>1 to 5	Open air	102	2.000
	Switchgear	13–102	0.973
	Cable	13	2.000
>5 to 15	Open air	13–153	2.000
	Switchgear	153	0.973
	Cable	13	2.000

MCC = motor control center.

K_2 = 0 for unearthed and high-resistance earthed systems
= −0.113 for earthed systems, and
G = gap between conductors (mm) (Table 6.6).

3. *Incident Energy Estimation:* The incident energy at a normal surface at a given distance and arcing time is calculated in terms of the normalized incident energy by the relation

$$E = 4.184 C_f E_n \left(\frac{t}{0.2}\right)\left(\frac{610}{D}\right)^x, \tag{6.19}$$

where

E = incident energy (J/cm^2),
C_f = calculation factor = 1.0 for voltage > 1 kV
= 1.5 for voltage < 1 kV,
t = arcing time (s),
D = working distance from arc (mm), and
x = distance exponent as given in Table 6.6.

4. *Flash Protection Boundary:* The flash protection boundary is the distance from an arcing fault at which the incident energy is equal to 1.2 cal/cm^2, and a person without PPE may get a curable second-degree burn:

$$D_B = 610 \times \left[4.184 C_f E_n \left(\frac{t}{0.2}\right)\left(\frac{1}{E_B}\right)\right]^{\frac{1}{x}} \tag{6.20}$$

where

D_B = distance of the boundary from the arcing point (mm),
C_f = calculation factor = 1.0 for voltage > 1 kV
= 1.5 for voltage < 1 kV,
E_n = incident energy normalized,
E_B = incident energy at the boundary distance (J/cm^2); it can be set at 5.0 J/cm^2 (1.2 cal/cm^2) for bare skin,
t = arcing time (s),
I_{bf} = bolted fault current (kA), and
x = the distance exponent from Table 6.6.

6.12.5.2 NFPA 70E-2004 Equations for calculating arc current, incident energy, and flash protection boundary are given below [93]:

1. *Arc Current*
   For arc in box:

$$\begin{aligned} I_a &= 0.85 I_{bf} - 0.004 I_{bf}^2 && \text{for } V < 1\text{ kV} \\ &= 0.928 I_{bf} && \text{for } 1\text{ kV} < V < 5\text{ kV} \\ &= I_{bf} && \text{for } V > 5\text{ kV}, \end{aligned} \tag{6.21}$$

where

I_a = arc current (kA),
I_{bf} = bolted fault current (kA), and
V = system voltage.

2. *Incident Energy*
   For arc in open air, $V \leq 0.6$ kV, 16–50 kA short-circuit current:

$$E = 5271 D^{-1.9593} t[0.0016 I_{bf}^2 - 0.0076 I_{bf} + 0.8938]. \tag{6.22}$$

For arc in box, $V \leq 0.6$ kV, 16–50 kA short-circuit current:

$$E = 1038.7 D^{-1.4738} t[0.0093 I_{bf}^2 - 0.3453 I_{bf} + 5.9675]. \tag{6.23}$$

For arc in open air, $V > 0.6$ kV:

$$E = 793 D^{-2} V I_{bf} t, \tag{6.24}$$

where

E = incident energy (cal/cm^2),
I_{bf} = bolted fault current (kA), and
D = working distance from arc (inches).

3. *Flash Protection Boundary*
   When the arc current is 70.7% of the bolted fault current, the theoretical maximum arc power (MW) is half of the bolted three-phase fault MVA [96]. This is the base on which the equation of flash boundary calculation is deduced:

$$D_B = \sqrt{2.65 \times 1.732 \times V \times I_{bf} \times t}, \tag{6.25}$$

where

D_B = distance of the boundary from the arcing point (inches),
V = rated system voltage line to line (kV),
I_{bf} = bolted fault current (kA), and
t = arcing time (s).

In addition, NFPA 70E-2004 contains a method for selecting PPE that requires little or no calculation. Table 130.7(C) (9) (a) assigns "Hazard/Risk Category" values for typical work tasks that might be performed on common types of equipment, such as the insertion of starter buckets in a 600-V class motor control center (MCC). The Hazard/Risk Category values correspond to the five categories of PPE (Table 6.5) so that a worker may determine the level of clothing that is required by simply finding the appropriate work task in the table [97].

6.12.5.3 Computer Software It is more powerful to use software that provides an extensive array of capabilities to minimize the level of efforts required to obtain accurate arc flash analysis in compliance with both IEEE 1584 Std. and NFPA 70E. The software should have the following features that are essential for arc flash assessment:

- calculation of short-circuit current at each point in the system as well as the contributing currents from system branches;
- protection coordination where the time-grading setting of different protective devices should ensure proper selectivity. Results of arc flash analysis help to justify the setting of protective devices for both selectivity and arc flash protection;
- capability of arc flash analysis for different system configuration or system with changes in its parameters;
- capability of graphical display for the system single-line diagram associated with results of arc flash analysis, short-circuit currents, and protection units setting, and so on, at the different selected points in the system;
- capability of arc flash analysis at different operating conditions; and
- printing out the warning labels and required documents in compliance with OSHA and NFPA standards.

Therefore, the procedure of arc flash calculation can be summarized by the following steps:

- collection of system and installation data,
- determination of all system modes of operation,
- calculation of bolted fault currents,
- estimation of arc fault currents,
- defining protective device characteristics and arc duration,
- documentation of system voltages and equipment classes,
- selection of working distances,
- estimation of incident energy of all equipment, and
- determination of flash protection boundary for all equipment.

6.12.6 Selection of Calculation Method

It is necessary to understand each method before deciding, which one is best selected, that is, to ensure which method produces best results for a given situation. There is no single calculation method applicable to all situations. Therefore, for each situation, several principles may be applied to ensure that the best results are obtained [97]:

- The actual system conditions satisfy the method's range of applicability. Some of the calculation methods are based on empirical equations, that is, the equations are derived from test results and valid over the range of system conditions where testing was performed. So, the application of these equations cannot be extended to other conditions with a high degree of confidence.
- Applying device-specific equations is more accurate than applying general equations. For instance, general equations in IEEE 1584 Std. are based on testing over a wide range of system conditions but cannot accurately characterize the performance of each protective device in every possible situation. Consequently, to characterize current-limiting action of fuses or CBs, the use of general equations may be inadequate. Applying the specific equations of these devices if they are available is more accurate for characterizing such devices.
- Compliance with the latest updated version of standards and using newer methods for arc flash hazard analysis must be followed as the development and progress for achieving more accurate results are continuously taking place.
- Calculation of bolted fault current at a given location as well as the characteristics of the upstream protective device are the main variables to be considered when determining the level of arc flash hazard at that location.

- The level of arc flash hazard significantly depends on arcing fault current, fault clearing time, and system voltage level. Other parameters must be considered and determined before the incident energy and flash protection boundary are calculated. These parameters are working distance, gap between conductors, equipment configuration (e.g., in open air, or in box), and system earthing.
- Considering motor contribution to fault current is taken in the general equations in IEEE 1584 Std. but the simplified equations do not. As mentioned in Chapter 4, motors in distribution systems contribute to the bolted fault current on which the arcing fault current is based. So, when motor loads are present, the motor contribution adds to the arcing fault current as well. However, this portion of the arcing fault does not flow through the upstream protective device, and then does not affect the tripping time for devices with inverse-time characteristics. Incident energy levels and flash protection boundary may therefore be increased as the motor contribution increases the fault current without any corresponding reduction in fault duration.

6.12.7 Mitigation of Arc Flash Hazards

Different methods for arc flash hazard reduction such as reducing the arcing current, increasing the working distance, reducing the fault clearing time, and using arc flash detecting relays can be applied.

6.12.7.1 Arcing Current Reduction Use of current-limiting devices can help limit thc current available for a fault, which in turn reduces the corresponding incident energy for clearing times of short duration (one to three cycles). The fault current must be in the current limiting range for these devices. Test data are required to provide the coefficients of IEEE 1584 simplified equations that can be used for determining the incident energy. Fault currents below the current limiting range are treated in the same manner as noncurrent limiting devices.

6.12.7.2 Increasing the Working Distance As in Equation 6.22, the incident energy is inversely proportional to approximately the square of the working distance (in open air). Therefore, increasing the working distance will highly reduce the incident energy. Working distance can be increased by using remote racking devices, remote operating devices, and extension tools.

6.12.7.3 Reducing the Clearing Time Reducing arc flash hazards means reducing AFIE levels. It becomes an increasingly important consideration in designing electric distribution systems. However, selective coordination of overcurrent protective devices is equally important. The best solution is to provide superior reduction of AFIE without sacrificing selectivity [98].

In an MV relaying system, lowering device settings can be achieved by changing the curve shape or lowering the time dial settings within the available setting range that will keep selective operation. Changes of LV protection system are more limited because of device characteristics.

The coordinated settings of protective devices, in accordance with the study, can temporarily be reduced for only the time during which on-line work is performed but this may result in nonselective operation for downstream faults during the maintenance operation.

Bus differential protection can help reduce the clearing time as it detects bus faults and quickly clears the fault to minimize damage (two cycles or less plus breaker response). Its installation is more costly due to the number of Cts that must be installed.

6.12.7.4 Use of Arc Flash Detecting Relays The arc flash detecting relay operates on the principle that it sends a trip signal to upstream breaker(s) as little as 2.5 ms after arcing fault initiation based on sensing the enormous and nearly instantaneous increase in light intensity in the vicinity of the arc fault. Thus, the tip signal is sent nearly instantaneously regardless of the magnitude of fault current. Arc flash relaying compliments existing conventional relaying, operates independently, and does not need to be coordinated with existing relaying schemes.

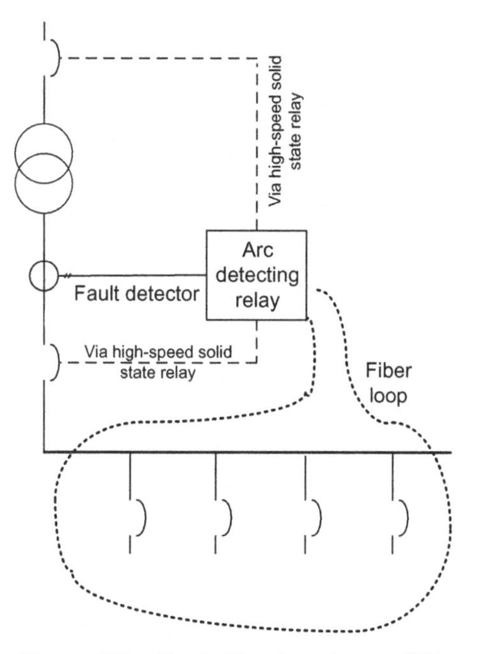

Figure 6.34 Single fiber loop layout [95].

The relay's sensitivity to light may be adjusted manually or automatically. Setting to automatic mode continually adjusts its threshold sensitivity to the relatively slow-changing background lighting levels that might result from opening a compartment door. Manual setting may be more appropriate where some normal low-level arcing might take place.

The optical fiber is used as arc flash sensor to detect the arc flash. It can be up to 60 m long and is routed through all compartments where an arc could potentially occur.

Manufacturers of optical arc flash system mostly recommend that the optical system is further supervised by single-phase fault detectors. This provides additional security at a cost of about 2 ms in operating time. The relay issues a trip signal when both the optical and the electric system indicates an arc flash fault [95]. As an example, Figure 6.34 shows the use of a single optical fiber sensor covering four separate feeders in a single loop. If an arc flash is detected and the fault detector threshold is exceeded in at least one phase, both high-side and low-side breakers are tripped via the high-speed solid-state tripping relays [95].

PART III

POWER QUALITY

CHAPTER 7

ELECTRIC POWER QUALITY

7.1 OVERVIEW

One of the main goals of the electric utility industry is to provide the consumers with a reliable and uninterrupted supply of electricity. Therefore, the planning of electric distribution systems is based on keeping the electricity supply in service all the time, that is, high level of continuity and availability. However, unscheduled outages caused by unpredicted failure events during the operation of electric distribution systems, leading to unavailability of electricity supply, can still be expected.

The unpredicted failure events can be classified into two types of failures as below:

Type I: System structure failure caused by short circuits, faulted cables, failed switches, failed protective gear, and so on. They lead to a complete interruption of supply. Thus, if the consumers have loads that cannot tolerate such failures, they must install backup generation and/or energy storage system (ESS) in order to alleviate the problem [99].

Type II: System function failures due to changes in power factor, magnitude and waveforms of voltage and current, and frequency of supply. Such system failure, that is, the supply does not meet a predetermined standard, is defined as power quality.

Electric Distribution Systems, First Edition. Abdelhay A. Sallam, Om P. Malik.

Published 2011 by John Wiley & Sons, Inc.

In practice, power electronics is being increasingly involved in industrial applications and in most daily life applications. This can be inferred from, for instance, the use of variable speed drives (VSDs), computer numeric controls (CNCs), programmable logic controls (PLCs), computers, and, in general, electronic devices. Such equipment is very sensitive to supply quality. For example, very short perturbations, measured in milliseconds, in the supply voltage may affect sensitive equipment, resulting in significant losses in performance and productivity.

Therefore, the utilities face problems not only in the distribution system reliability and power interruptions but also in the power quality. They need to provide the consumers with electricity at the highest possible quality that the consumers are demanding. A study of the nature and causes of power quality problems, performed by the Electrical Power Research Institute (EPRI), is reported in Reference 100.

ESS is one program that can be applied to mitigate the power quality problems. It should be studied from the economic point of view. The study reported in Reference 101 includes the cost of poor power quality and the ability of power exchange between the utilities and industrial organizations. Also, the technology used for designing the ESS should be identified from the technical point of view [102].

7.2 POWER QUALITY PROBLEMS

Electricity supply represents one of the most essential basic services for the support of industrial, commercial, and residential applications. From the point of view of the electricity consumers, it is required that this basic service be available at all times (i.e., high level of continuity) and also enable all the consumers' electric equipment to work safely and satisfactorily (i.e., a high level of power quality).

If electricity is referred to as a product, it will be completely different from any other product because of its intangible and transient nature. Electricity as a product exists for an instant at different points of delivery and at the same instant it comes into existence at different points to be used. The characteristics of this product are different at each point of delivery. In addition, its quality depends on the way of production and the way of feeding users' equipment. The manufacturers of equipment define power quality as the characteristics of power supply that are required to make end-user equipment work properly. These characteristics can be very different depending on the type of equipment and manufacturing process. Thus, an EPRI power quality workbook defines power quality as any problem manifested in voltage, current, or frequency deviations that result in failure or mal-operation of utility or end-user equipment [103]. Also, International Electromechanical Commission (IEC) standard (1000-2-2/4) defines power quality as the physical characteristics of the electric supply provided under normal operating conditions that do not disrupt or disturb the customers' processes.

The power quality problems can be classified into three classes [104]:

i) *Problems Generated by the Electric Utility Causing Problems at the Consumers' Premises:* These problems seem to be related to voltage regulation, location and sizing of capacitor banks, line design, transformer sizing, and so on. The electric utility is supposed to provide the consumers with an ideal voltage (pure sinusoidal waveform with constant magnitude and frequency). Some odd harmonic components may exist due to transformer magnetizing currents only. These harmonics must be within the limits prescribed in the standards.
ii) *Problems Generated by a Consumer Causing Problems to Other Consumers:* In practice, most power quality problems are generated by consumers' equipment such as VSDs, rectifiers, electric welders, arc furnaces, motors, and motor starters.
iii) *Problems Generated by a Consumer Causing Problems to His or Her Own Equipment:* Usually, these problems arise due to the presence of harmonics, heavy unbalanced loading, poorly connected equipment, inadequate wiring of low-voltage (LV) network, switching power supplies, and so on.

Transients, voltage sag or swell, voltage unbalance, and voltage distortions are the nature of the problems. These distortions result from a wide variety of events ranging from switching events within the end-user facility to faults hundreds of kilometers away on the utility transmission line [105]. The power quality variations due to these problems can be categorized as transient disturbances, fundamental frequency disturbances, and variations in steady state. Each category can be defined by certain characteristics as given in Table 7.1 [106]. Its causes are listed in Table 7.2.

The main power quality disturbances shown in Figure 7.1 are described by the following technical terms in accordance with the IEEE standards coordinating committee 22 (power quality) recommendations [107].

Transients: These pertain to or designate a phenomenon or quantity varying between two consecutive steady states during a time interval that is short compared with the timescale of interest. A transient can be a unidirectional impulse of either polarity, or a damped oscillatory wave with the first peak occurring in either polarity.

Sags: A decrease in root mean square (rms) voltage or current at power frequency for durations of 0.5 cycles to 1 min. A voltage sag of 10% means that the line voltage is reduced to 10% of the nominal value. Typical values are 0.1–0.9 pu.

Swells: A temporary increase in rms voltage or current of more than 10% of the nominal value at power system frequency, which lasts from 0.5 cycles to 1 min. Typical rms values are 1.1–1.8 pu.

TABLE 7.1 Categories of Power Quality Variations [101]

Major Category	Specific Category	Defining Characteristics
Transient disturbances	• Impulse transients	• Unidirectional Typically <200 μs
	• Oscillatory transients	• Decaying oscillations
	◦ Low frequency	<500 Hz
	◦ Medium frequency	500–2000 Hz
	◦ High frequency	>2000 Hz
Fundamental frequency disturbances	• Short-duration variations	• Duration 0.5–30 cycles
	◦ Sags	10%–90% nominal
	◦ Swells	105%–173% nominal
	• Long-duration variations	• >30 cycles
	◦ Undervoltages	
	◦ Overvoltages	
	• Interruptions	• Complete loss of voltage
	◦ Momentary	<2 s
	◦ Temporary	2 s to 2 min
	◦ Long term	>2 min
Variations in steady state	• Harmonic distortion	• Continuous distortion (Vor I) Components to fiftieth harmonics
	• Voltage flicker	• Intermittent variations in 60-Hz voltage magnitude; frequency component <25 Hz
	• Noise	• Continuous high-frequency component on voltage or current; frequency >3000 Hz

TABLE 7.2 Power Quality Variation Categories and Causes [101]

Category	Method of Characterization	Causes
Impulse transients	Magnitude, duration	Lightning, load switching
Oscillatory transients	Waveforms	Lightning, line/cable and capacitor switching, transformer and load switching
Sags/swells	Waveforms, rms versus time	Remote faults
Undervoltages/ overvoltages	rms versus time	Overloading of feeder/motor starting, load changes, compensation changes
Interruptions	Duration	Breaker operation/fault, clearing, maintenance
Harmonic distortion	Waveforms, harmonic spectrums	Nonlinear loads, system response characteristic
Voltage flicker	Magnitude, frequency of modulation	Intermittent loads, arcing loads, motor starting
Noise	Noise, coupling method, frequency	Power electronic switching, arcing, electromagnetic radiation

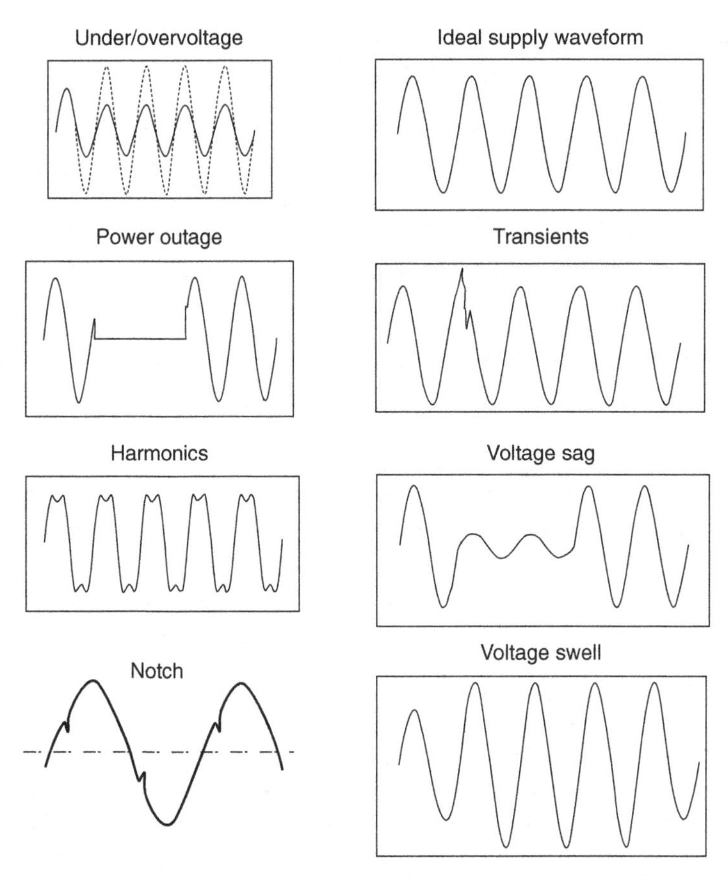

Figure 7.1 Graphical illustration of power quality events [101].

Undervoltage: Refers to a voltage having a value less than the nominal voltage for a period of time greater than 1 min. Typical values are 0.8–0.9 pu.

Overvoltage: When used to describe a specific type of long-duration variation, this refers to a voltage having a value greater than the nominal voltage for a period of time greater than 1 min. Typical values are 1.1–1.2 pu.

Interruptions: The complete loss of voltage (below 0.1 pu) on one or more phase conductors for a certain period of time. Momentary interruptions are defined as lasting between 0.5 cycles and 3 s, temporary interruptions have a time span between 3 and 60 s, and sustained interruptions last for a period longer than 60 s.

Harmonics: Sinusoidal voltages or currents having frequencies that are multiples of the fundamental power frequency. Distorted waveforms can be decomposed into a sum of the fundamental frequency wave and the harmonics caused by nonlinear characteristics of power system devices and loads.

Interharmonics: Voltages and currents having frequencies that are not integer multiples of the fundamental power frequency. Interharmonics are mainly caused by static frequency converters, cycloconverters, induction motors, and arcing devices, and can have the effect of inducing visual flicker on display units. Power line carrier signals are also considered as interharmonics.

Voltage Flicker: Voltage fluctuations are systematic variations in the envelope or a series of random voltage changes with a magnitude that does not normally exceed the voltage ranges of 0.9–1.1 pu. Such voltage variations are often referred to as flicker. The term flicker is derived from the visible impact of voltage fluctuations on lamps. Among the most common causes of voltage flicker in transmission and distribution systems are arc furnaces.

Notch: A periodic transient reduction in the magnitude of the quasi-sinusoidal mains voltage. It lasts less than one half-cycle and usually less than a few milliseconds. Notching is caused mainly by power electronics devices that draw a heavy load current during a small portion of the sine wave. Frequency components associated with notching can, therefore, be very high, and measuring with harmonic analysis equipment may be difficult.

Noise: Unwanted electric signals that produce undesirable effects in the circuits of the control systems in which they occur. (The control systems include sensitive electronic equipment in total or in part.) Noise is a high frequency, low-current, low-energy waveform superimposed on the sine wave of the alternating current (AC) mains [108]. The frequency of noise can range from low kilohertz into megahertz region. This low-level interference is typically characterized by a voltage less than 50 V and an associated current of less than 1 A. Noise is not a component-damaging anomaly but can be very costly in the form of data errors, lost data, or downtime. Potential noise sources in electric distribution systems include motors, transformers, capacitors, generators, lighting systems, power conditioning equipment, and surge protective devices (SPDs).

7.2.1 Typical Power Quality Problems

Some examples of power quality problems are listed below to illustrate the confusion and misconception that typically occurs when consumers experience problems:

- Switching an air conditioner "on" may cause a sag in the voltage, which might dim the lights momentarily. However, plugging in a coffeepot to the same receptacle as a personal computer might cause voltage sag that could scramble data every time the coffeepot is turned "on" or "off."
- Industrial equipment with microprocessor-based controls and power electronic devices that are sensitive to disturbances are affected by poor power quality. Control systems can be affected by momentary voltage sag or small transient voltages, resulting in nuisance tripping of processes.

Furthermore, many of these sensitive loads are interconnected in extensive network and automated processes. This interconnected nature makes the whole system dependent on the most sensitive device when a disturbance occurs.

IEC standard 61000-4-11 defines the limits on magnitudes and durations of various voltage disturbances that are acceptable to a switch-mode power supplies load. Similarly, the Computer Business Equipment Manufacturers Association (CBEMA), presently known as Information Technology Industry Council (ITIC), has defined the operational design range of voltage for most information technology equipment (ITE) by the curve shown in Figure 7.2. The equipment is typically designed to withstand and operate normally during disturbances as long as the event is within the function region of the curve. It depicts the ability of the equipment to withstand large voltage variations (100–500% under/over nominal voltage) for short durations (given in microseconds).

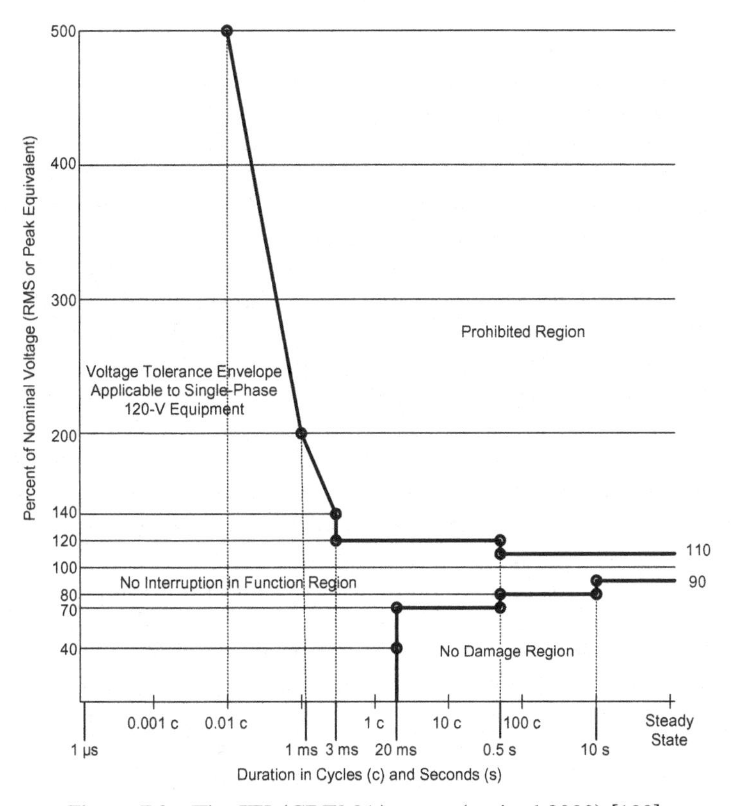

Figure 7.2 The ITI (CBEMA) curve (revised 2000) [109].

- The intervention of power electronics in industrial processes may generate harmonic currents that result in voltage distortion. For instance, VSDs can generate harmonics. These harmonics are subject to being magnified when exciting resonance occurs with LV capacitors causing equipment failure.
- A voltage unbalance was observed at a manufacturing plant. A three-phase autotransformer was used to boost the LV because the distribution transformer tap setting was low. One phase was used to supply power to the adjacent main office. The load of this office was large enough to cause unbalance at the plant. This unbalance resulted in longer motor starting times causing visible light flicker nuisance.
- A motel repeatedly experienced damage to window air conditioners of the individual rooms. At first, the utility was blamed because of not providing overvoltage protection. By studying the problem in depth, it was found that the electronic circuitry of the control board in the conditioner unit did not have any surge protection. It was replaced by the same type of device at each time of damage, so the problem was sustained.

7.2.2 Case Studies

It is essential to perform some investigation to diagnose the power quality problem and determine its causes. This necessitates carrying out some measurements and monitoring of supply parameters [110]. A study must be done to identify the measured supply parameters and the period of monitoring (short and long periods). Also, the utility experts and consultant engineers must determine the amount of data to be enough for the system analysis by using simple and accurate data acquisition systems.

As examples, three case studies have been done by the National Power Laboratories (NPL), the Canadian Electricity Authority (CEA), and EPRI.

The NPL monitored a sample of 130 sites (17% residential, 24% commercial, 31% industrial, 18% multistoried buildings, and 10% institutional) over 6 years (1990–1995) and measured the line-to-neutral voltage at the wall receptacle.

The CEA study is concerned with service territories of 22 Canadian utilities to monitor and measure the voltage at the service entrance panels. Samples were taken at 550 sites including residential, commercial, and light industrial consumers. It was monitored over a period of 4 years (1991–1994) [111].

The EPRI case study focused on the description of power quality levels on primary distribution systems in the United States not on the end users as in the NPL and CEA studies [112–114]. EPRI took samples of 277 sites monitored over 27 months. The measured parameters were the voltage and current of different feeders at different locations and with lengths ranging from 1 to 80 km. The feeders supply different types of loads: residential, commercial, and industrial. The results of the study compared with the ITI (CBEMA) curve, including the complete measurement database, show that there are several sag

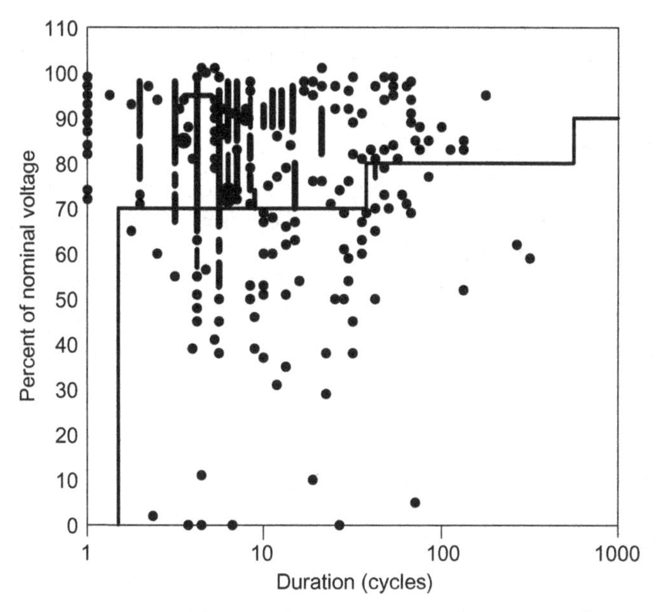

Figure 7.3 ITI (CBEMA) curve for voltage sag study [115].

events each year lasting for four, five, or six cycles below the ITI (CBEMA) curve (solid line in Fig. 7.3) [115].

7.3 COST OF POWER QUALITY

The cost analysis on power quality gives an excellent guideline for evaluating the effect of power quality on utilities and consumers' equipment. It helps to determine the best method used to mitigate power quality problems. Meanwhile, the cost/benefit study of power quality will provide a guidance to improve and increase the electricity production. Therefore, the cost analysis is applied to guarantee the safe, reliable, and economical power supply to consumers with good quality electricity, increasing economic profits of both utilities and consumers.

The costs associated with power quality problems arise from lost production as well as other related disruptions suffered by consumers such as equipment damage, startup costs, and labor wages. These costs are largely dependent on the consumers' activities that are impacted, the time of event occurrence, and its duration.

The difficulty of cost estimation is due to the need to gather a pool of data and information about utility and consumer's equipment, recorded parameters, monitoring periods, and so on. In addition, the cost varies from one consumer to another and from utility to utility.

The cost analysis is based mainly on the study of the relationship among power supply quality, quality cost (QC), and economic profit [116].

7.3.1 Power Supply Quality

As mentioned above, poor power quality may cause disruption of consumers' processes leading to a loss of revenue. Therefore, minimizing the process downtime caused by poor power quality is of great interest to the consumers. On the other hand, the consumer's process may affect the supply quality, and the utility is concerned with minimizing this effect. A partnership that brings together the utility, consumer, and equipment manufacturer is clearly needed to improve power quality.

7.3.2 QC

QC is defined as the sum of costs needed to improve power supply quality and reduce losses caused by poor power quality. These costs are cost of protective activities, cost of monitoring and inspection, cost of internal faults, and cost of external faults.

Cost of Protective Activities: It is the cost of all requirements to avoid supplying of poor power quality from the utilities. These requirements include planning for improving power quality, reliability management, data acquisition system, inspection of equipment, improving techniques, replacing/repairing the faulty equipment, and staff training.

Cost of Monitoring and Testing: It is the cost of detection and inspection required for monitoring the power quality during transmission, transformation, and distribution of electric power. It includes cost of equipment inspection, cost of measurements and meter's calibration, cost of installation, and cost of running tests.

Cost of Internal Faults: It is the cost of sum of losses caused by faulty equipment and the cost of manipulating these faults (e.g., fault analysis, equipment repair, decrease of electricity sold).

Cost of External Faults: External faults such as lightning may cause overvoltage in the lines feeding the consumers. This in turn may lead to damage to household electric appliances (televisions, recorders, refrigerators, computers, etc.) that should be paid by the utility.

QC can be quantified by calculating the following indices:

- Percentage of QC to production value (QCPV):

$$\text{QCPV} = \frac{\text{Total QC for power supply}}{\text{Total utility production value}} \times 100 \tag{7.1}$$

- Percentage of QC to income of electricity sold (QCIES):

$$\text{QCIES} = \frac{\text{Total QC for power supply}}{\text{Income of electricity sold}} \times 100 \tag{7.2}$$

- Percentage of QC to production cost (QCPC):

$$\text{QCPC} = \frac{\text{Total QC for power supply}}{\text{Total production cost}} \times 100 \tag{7.3}$$

- Percentage of fault cost to production cost (FCPC):

$$\text{FCPC} = \frac{\text{Cost of failures (internal + external)}}{\text{Total production cost}} \times 100 \tag{7.4}$$

- Percentage of failure cost to production value (FCPV):

$$\text{FCPV} = \frac{\text{Cost of failures (internal + external)}}{\text{Total production value}} \times 100 \tag{7.5}$$

- Percentage of failure cost to income of electricity sold (FCIES):

$$\text{FCIES} = \frac{\text{Cost of failures (internal + external)}}{\text{Income of electricity sold}} \times 100. \tag{7.6}$$

7.3.3 Economic Profit

The economic profit is the criterion used to evaluate economically the revenue of power quality enhancement to both the utility and the consumers. Three indices are used for each to evaluate the profit.

Economic Profit Indices for Utility:

- Quality profit to utility (QPU):

$$\text{QPU} = \text{total power quality revenue} - \text{utility QC} \tag{7.7}$$

- Percentage of utility's profit to QC (UPQC):

$$\text{UPQC} = \frac{\text{QPU}}{\text{Utility QC}} \times 100 \tag{7.8}$$

- Percentage of utility's income to QC (UIQC):

$$\begin{aligned}\text{UIQC} &= \frac{\text{Utility power quality revenue}}{\text{Utility QC}} \times 100 = \frac{\text{QPU} + \text{utility QC}}{\text{Utility QC}} \times 100 \\ &= (\text{UPQC} + 1) \times 100\end{aligned} \tag{7.9}$$

Economic Profit Indices for Consumers:

- Quality profit to consumers (QPC)

$$= \text{Quality revenue gained by consumer} - \text{consumer's QC} \tag{7.10}$$

where the quality revenue gained by consumer

= Revenue from decreasing running cost + revenue from increasing productivity

- Percentage of consumer's profit to QC (CPQC):

$$\text{CPQC} = \frac{\text{QPC}}{\text{Consumer's QC}} \times 100 \tag{7.11}$$

- Percentage of consumer's income to QC (CIQC):

$$\begin{aligned}\text{CIQC} &= \frac{\text{Quality revenue gained by consumer}}{\text{Consumer's QC}} \times 100 \\ &= \frac{\text{QPC} + \text{consumer's QC}}{\text{Consumer's QC}} \times 100 \\ &= (\text{CPQC} + 1) \times 100.\end{aligned} \tag{7.12}$$

The main basis of cost analysis can be summarized as shown in Figure 7.4.

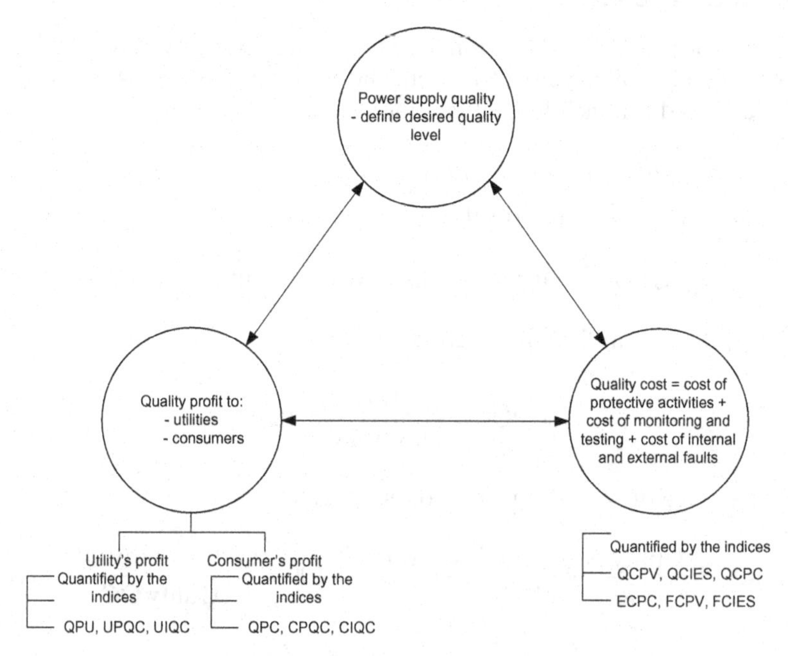

Figure 7.4 The main basis of cost analysis.

7.3.4 A Case Study

Studies for estimating the costs associated with power quality problems vary from case to case depending on many factors such as the nature of loads, distribution network structure, type of equipment used by the utility and consumers, and characteristic of events. Therefore, a case study to introduce a summary of results and what can be deducted from these results is presented here. This case study is an illustrative example and cannot be generalized to other studies.

A survey carried out by Duke Power [117] is shown in Table 7.3. A sample of 198 industrial and commercial consumers has been considered. The results are reported in terms of five types of reliability and power quality events. The cost of interruption is ultimately affected by the outage duration and type of power quality events. It is obvious from the results that the long-duration outages have the largest effect, where 90% of total production is influenced and the revenue changes dramatically.

The cost of power for industrial and commercial consumers for 1-h outage on a summer afternoon without advance notice is illustrated in Figure 7.5. The commercial and industrial consumers of Duke Power surveyed had interruption costs ranging from $0 to $10,000 and from $0 to $1 million, respectively. Also, in this figure, greater than 35% of all industrial and 8% of all commercial consumers surveyed experienced an interruption cost of greater than $10,000 on a hot summer day. The sample size for this survey consisted of 210 large industrial and commercial consumers and 1080 small/medium industrial and commercial consumers. It may be fair to assume that most of the 210 large consumers surveyed will experience a loss of greater than $10,000 per interruption lasting 1-h and will experience at least the average costs listed in Table 7.3.

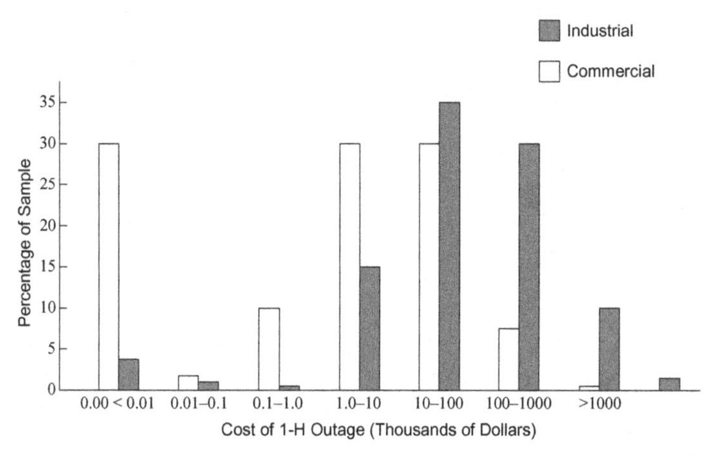

Figure 7.5 Difference in commercial and industrial customer interruption cost (Duke Power data) [117].

TABLE 7.3 Components of Outage Costs [117] **(Average of 198 Large Consumers in the Duke Power Service Territory)**

Cost Element	Outage without Notice 4 h	Outage without Notice 1 h	1-h Outage with Notice	Momentary Outage	Voltage Sag
Production impacts					
• Production time lost (hours)	6.67	2.96	2.26	0.70	0.36
• % of work stopped	91%	91%	91%	57%	37%
Production losses					
• Value of lost production ($)	81,932	32,816	28,746	7,407	3,914
• % of production recovered	36%	34%	34%	19%	16%
• Revenue change ($)	52,436	21,658	18,972	5,999	3,287
Loss due to damage					
• Damage to raw materials ($)	13,070	8,518	3,287	2,051	1,163
• Hazardous material cost ($)	323	269	145	136	90
• Equipment damage ($)	8,421	4,977	408	3,239	3,143
Cost to run backup and restart ($)					
• Cost to run backup generation	178	65	65	22	22
• Cost to restart equipment	1,241	1,241	171	29	29
• Other restart cost	401	368	280	149	74

TABLE 7.3 ***Continued***

	Outage without Notice				
Cost Element	4 h	1 h	1-h Outage with Notice	Momentary Outage	Voltage Sag
Savings ($)					
• Savings on raw materials	1,927	645	461	166	114
• Savings on fuel and electricity	317	103	85	12	9
• Value of scarp	2,337	874	450	228	140
Labor management during recovery					
Percentage of:					
• using overtime	33%	26%	25%	7%	6%
• using extra shifts	1%	1%	0%	1%	1%
• more working labor	3%	4%	4%	7%	4%
• rescheduling operations	4%	5%	5%	0%	0%
• others	1%	2%	2%	1%	0%
• not recovering	59%	62%	64%	84%	89%
Labor costs and savings ($)					
• Cost to make up production	4,854	1,709	1,373	254	60
• Cost to restart	665	570	426	192	114
• Labor savings	2,139	644	555	0	0
Average total costs ($)	74,800	39,500	23,100	11,000	7,700

7.4 SOLUTIONS OF POWER QUALITY PROBLEMS

First of all, the size of power quality problems must be defined by estimating the cost of disruptions caused to distribution system equipment (utilities and consumers' equipment) and the extent of its sensitivity to power quality. The equipment may be less sensitive to disturbances, allowing it to ride through the disturbances, that is, no need to apply a solution. Otherwise, a solution is needed, for example, by installing power quality devices that suppress or counteract the disturbances. A cost/benefit analysis of the different solutions is applied to enable the distribution system planners to make a decision of the most cost-effective solution.

The power quality devices are used to protect the electric equipment or to eliminate the source of disturbances or to mitigate the effect of disturbances. The devices that are commonly used for this application include the following:

- SPD
- shielding (SH)
- uninterruptible power supply (UPS)
- dynamic voltage restorers (DVRs)
- series capacitors (SCs)
- capacitor voltage transformers (CVTs)
- wiring and grounding (W&G)
- static var compensator (SVC)
- ESS
- backup generators (BCKGs)
- isolation transformers (ITRs)
- filters.

For each power quality event, the relevant choice of these devices is illustrated in Table 7.5.

The information in Table 7.4 can be reformed to illustrate the use of each device to protect the electric equipment against different power quality events (Table 7.5).

7.4.1 Examples of Power Quality Devices

7.4.1.1 SPDs The electric equipment in distribution systems may be exposed to internal or external surges. Internal surges are generated within a facility by the user's equipment. They result from switching processes, for example, switching inductive or capacitive loads, fuse, or breaker opening in an inductive circuit.

External surges are generated outside a facility and brought into the facility by utility wires. They result from fuse operation, power system switching, and lightning.

The SPDs protect the equipment against these surges by limiting the amount of undesired surge energy that reaches the equipment. The surge energy is diverted to a path rather than the equipment itself (neutral or earth).

TABLE 7.4 Relevant Solutions for Categories of Power Quality Events (Power Quality Assessment Procedures, EPRI CU-7529, December 1991)

Event Category	Method of Characterization	Causes	Power Quality Solution
Impulse transients	Magnitudes, duration	Lightning, load switching	SPD, filters, ITR
Oscillatory transients	Waveforms	Lightning, line/cable/capacitor/transformer/load switching	SPD, filters, ITR
Sags/swells	Waveforms, rms versus time	Remote faults	CVT, ESS, UPS
Undervoltages/overvoltages	rms versus time	Motor starting, load changes, compensation changes	DVR, CVT, ESS, UPS
Interruptions	Duration	Breaker operation, fault clearing, equipment failure, maintenance	BCKG, ESS, UPS
Harmonic distortion	Waveforms, harmonic spectrum	Nonlinear loads, system response, characteristic	Filters, ITR
Voltage flicker	Magnitude, frequency of modulation	Intermittent loads, arcing loads, motor starting	SVC, SC
Noisc	Coupling method, frequency	Power electronic switching, arcing, electromagnetic radiation	W&G, chocks, filters, SH

TABLE 7.5 Power Quality Events and Available Power Quality Devices

Power Quality Device	Power Quality Events							
	Transients		Sags/ Swells	UV/ OV	Interruptions	HD	Voltage Flicker	Noise
	Impulse	Osc.						
SPD	x	x						
UPS			x	x	x			
DVR			x	x			x	
SC				x			x	
CVT			x	x				
ITR	x	x				x		
ESS	x	x	x	x	x		x	x
BCKG					x			
SVC			x	x			x	
Filters	x	x				x		x

Osc. = oscillatory; UV/OV = undervoltage/overvoltage; HD = harmonic distortion.

The SPD is a nonlinear element acting as an isolating switch in the normal conditions where its resistance is very high. When the voltage increases and reaches a certain value called "clamping voltage," the SPD will change rapidly (in nanoseconds) from a very high resistance mode to a very low resistance mode. Then, the majority of surge energy is directed through SPD, and most of this energy is dissipated in its internal resistance (Fig. 7.6) as explained in Chapter 5.

7.4.1.2 BCKGs In large industries and for long-duration interruptions, backup generation is essential to supply at least the critical loads. It is common to use diesel-generator set with rating sufficient for feeding these critical loads such as emergency lighting system, electric lifts, industrial processes that cannot withstand long interruption, and hospitals.

Usually, the backup generator is used as a standby unit and connected to the distribution system of the industry at the main LV entrance (Fig. 7.7). An automatic transfer switch (ATS) can be used to automatically transfer the power source from the utility incoming feeder to the backup generator in

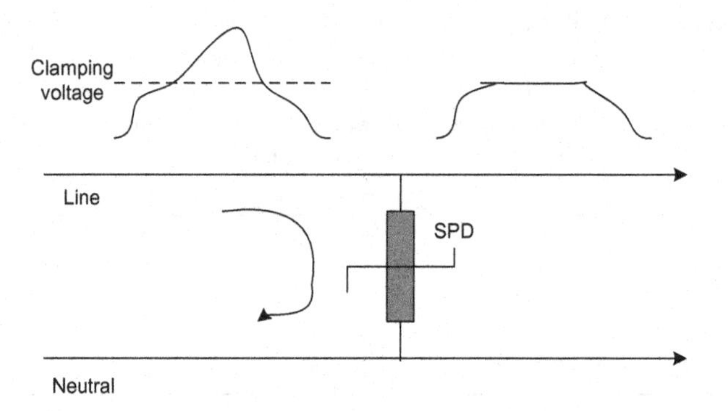

Figure 7.6 Function of SPD.

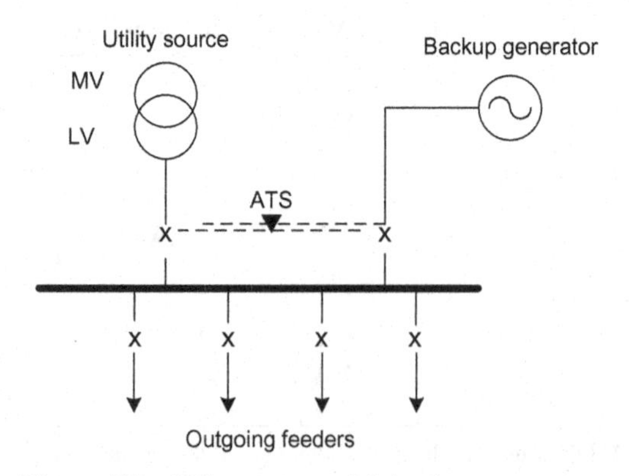

Figure 7.7 LV entrance with backup generator.

emergency cases. An electric interlock is provided between both sources to avoid parallel operation. Some of the outgoing feeders of uncritical loads may be disconnected to keep the power demand within generator rating that is mostly less than the rating of the utility source.

In many cases, the main LV bus bar is sectionalized into two sections with a bus coupler to increase the reliability. Each section is fed from a different utility source (Fig. 7.8). The outgoing feeders of critical loads are preferred to be connected to one of these two sections that is supplied by utility source in normal operation and backup generator in emergency operation. The circuit breakers A, B, C, and D must be controlled to satisfy the truth table given in Table 7.6. This table indicates on/off positions of generator circuit breaker D at different combinations of breakers A, B, and C positions. Position "on" is represented by "1" while "0" represents off position.

7.4.1.3 UPS The UPS is an alternative power source to supply power to the load during interruption or outage of the main power source (e.g., utility

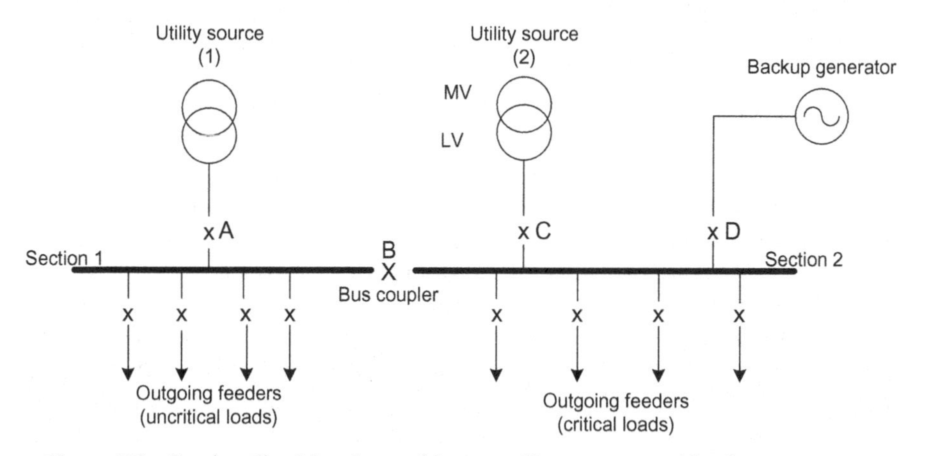

Figure 7.8 Sectionalized bus bars with two utility sources and backup generator.

TABLE 7.6 Circuit Breakers Positions "0" for Off and "1" for On

A	B	C	D
0	0	0	1
0	0	1	0
0	1	0	1
0	1	1	0
1	0	0	1
1	0	1	0
1	1	0	0
1	1	1	Not used

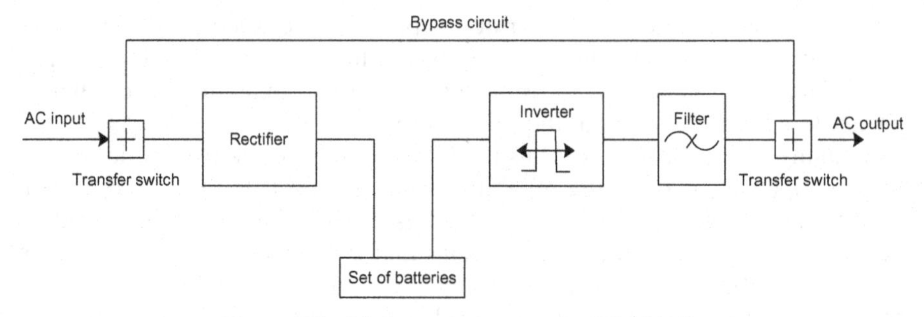

Figure 7.9 Main components of a UPS [119].

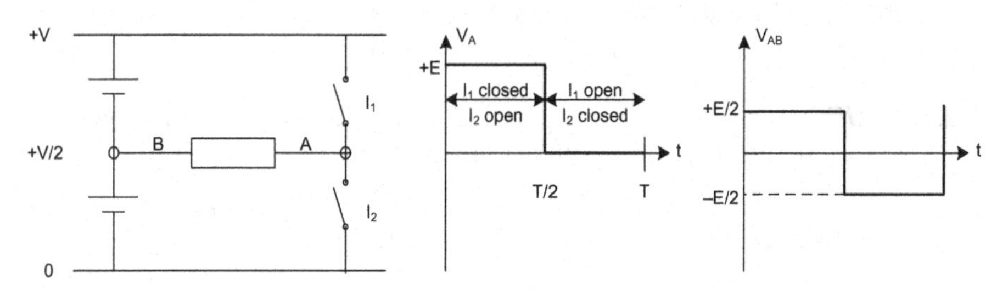

Figure 7.10 Principles of a half-bridge converter [118].

source). It includes a rectifier circuit to convert AC input power into direct current (DC) power. The DC power charges a set of batteries to store energy and an inverter to convert the DC stored energy back onto AC power for the load (Fig. 7.9). Six or twelve or twenty-four diode bridges can constitute the rectifier circuit depending on the desired level of wave distortion as seen in Chapter 10. From the point of view of frequency stability as well as voltage stability, the inverter that constitutes the UPS generator has performance superior to that of the mains. It is designed to generate sinusoidal voltage even when supplying nonlinear loads, that is, dealing with highly distorted currents. For instance, in a single-phase unit with a half-bridge converter, the square wave voltage appearing between A and B (Fig. 7.10) is filtered so as to obtain in the output of the unit a sinusoidal voltage wave [118, 119].

During normal operation, the utility supplies power to both the load directly bypassing the UPS unit and to the UPS to charge its batteries via the rectifier circuit. In an emergency operation, for instance an outage of utility power source, the UPS supplies power to the load fast enough (few milliseconds) to avoid any damage resulting from load interruption. This necessitates using an electronic transfer switch to change the power source to load.

UPSs are effective for microprocessor-based loads such as computer systems and PLCs where the loss of data is avoided. On the other hand, they have deficiencies where the transfer switch and rectifier are exposed to line disturbances in normal operating conditions. In emergencies, the operation time of UPS is limited by the capacity of batteries.

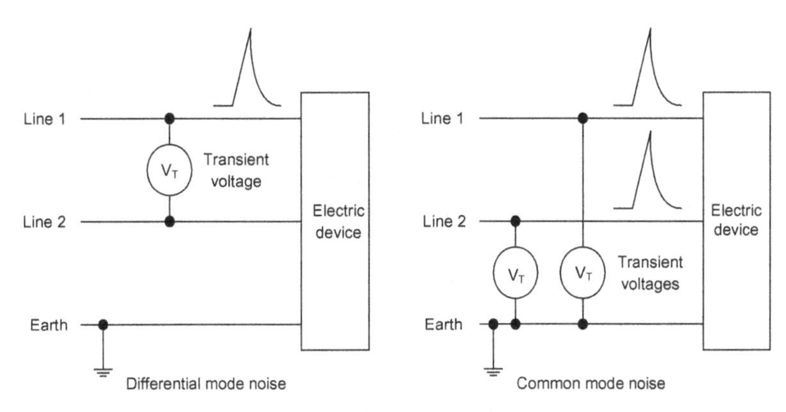

Figure 7.11 Differential and common mode noises.

The design of UPSs and, generally, ESS depends on the required mode of operation. Three modes of operation are considered: standby (off-line), on-line, and line interactive. The standby mode of operation means that the ESS operates only during the interruption time, while it operates full time in the case of on-line mode of operation. Line-interactive mode of operation includes both of these two modes.

7.4.1.4 ITRs They are generally composed of two separate windings with a magnetic shield between these windings to offer noise control. The noise can be transported to the electric device by electromagnetic coupling (EMC) in two basic ways: a differential mode noise and a common mode noise (as explained in Chapter 5, Section 5.8) (Fig. 7.11).

The ITR is connected between the power source and the electric device. Therefore, it carries the full load current, and thus must be suitably sized. The main benefit offered by ITRs is the isolation between two circuits, by converting electric energy to magnetic energy and back to electric energy, thus acting as a new power source.

7.4.1.5 ITR Operation Considering a high voltage, high-current transient is introduced into a power line by the direct and indirect (induced) effects of lightning activity or a switching surge. If these transients are differential mode, then the ITR will effectively pass these transients with little or no attenuation. This occurs because the ITR is designed to pass power frequencies in the differential mode, and the frequency make up of a lightning transient is such that most of the energy content is in the frequency components below a few tens of kilohertz, that is, within the pass band of most ITRs.

If, on the other hand, these transients are common mode, then a suitable shielded ITR will provide effective protection against such surges. This is because a common mode transient is split in two between a pair of power lines, and they proceed in the same direction. They flow into the transformer from

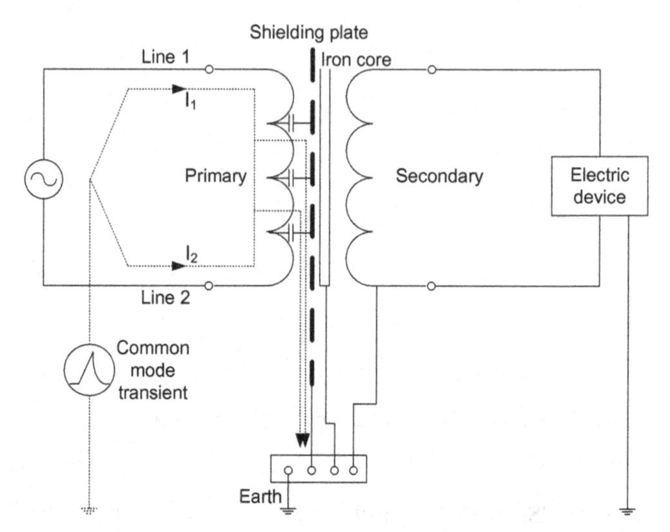

Figure 7.12 Common mode transient propagation [120].

its terminals and run in the primary coil, finally going out to the earth through the shield plate. At this moment, they run in the coil in opposite directions, canceling their inductive effects in the secondary side. Therefore, a common mode transient does not reach the secondary side (Fig. 7.12) [120].

It can thus be seen that the shielded transformer will provide effective protection against common mode surges provided the peak voltage does not exceed the insulation rating of the transformer. It also effectively provides no attenuation of differential mode surges.

7.4.1.6 Voltage Regulators (VRs) The function of VRs is to maintain the voltage at load within preset limits. During voltage sags, VRs increase the voltage to a desired level for sensitive loads, and, conversely, during overvoltages or swells, they decrease the voltage. Mostly, the usage of VRs is to mitigate the effect of sag events.

The common type used for regulating the voltage is a motor-driven variable ratio autotransformer. A motor is used to change the location of a slider on transformer winding providing a change of transformer ratio to increase or decrease the voltage levels (Fig. 7.13). The response time is slow, which may be inadequate for some loads and may not correct large short-term voltage variations.

7.5 SOLUTION CYCLE FOR POWER QUALITY PROBLEMS

Two key elements for achieving consumer's satisfaction must be considered. First, continuity of supply is one of the main aspects of distribution system planning provided by the utilities. Second, good power quality that is provided by a sharing action among utilities, consumers, and equipment manufacturers.

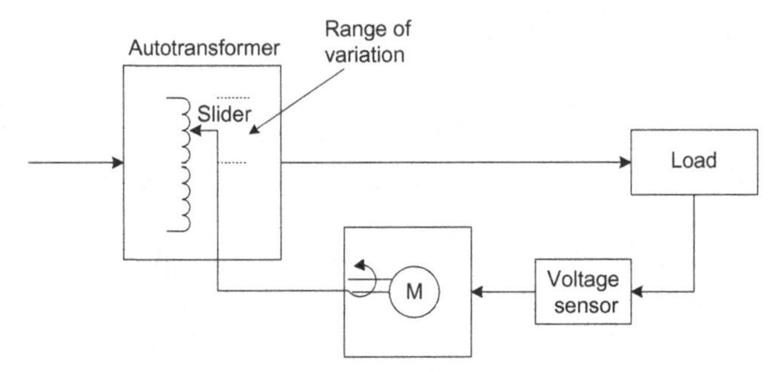

Figure 7.13 Motor-driven VR.

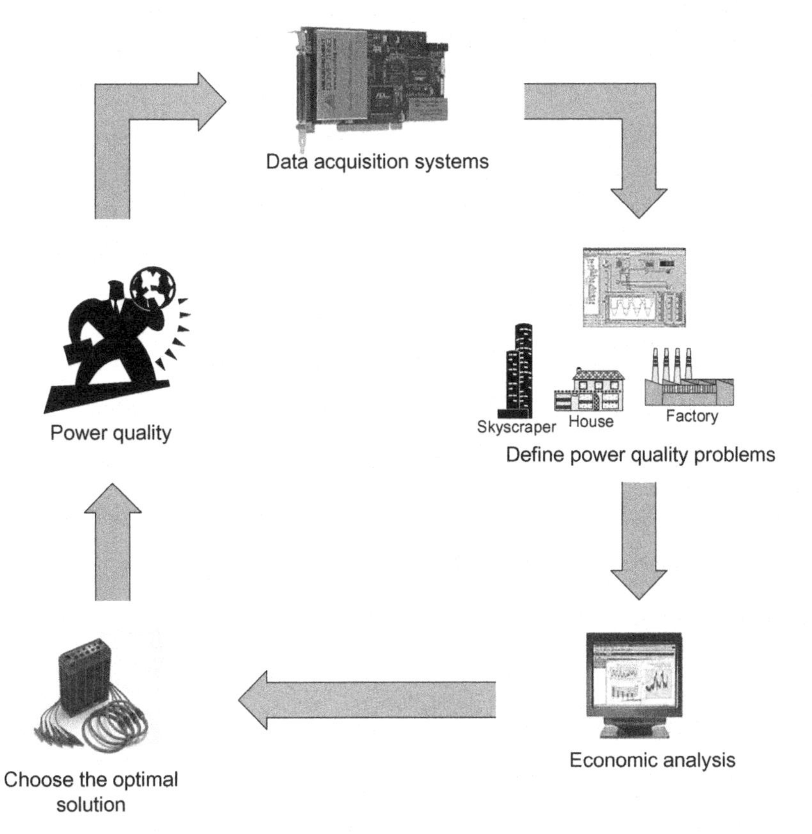

Figure 7.14 Main steps of power quality problems solution.

The main steps involved in solving power quality problems are shown in Figure 7.14 and described below.

Step 1: Monitoring and recording events; by using an accurate data acquisition system, the events can be recorded and collected for a specific period.

Step 2: Defining power quality problems; the events collected in step 1 are compared with system equipment specification to identify the events that exceed equipment performance limitations. It is beneficial to refer the equipment to its class, commercial, residential, and industrial, because it helps the economic analysis. The events are classified as mentioned before (sag, impulse, swell, etc.).

Step 3: Economic analysis; the events extracted in step 2 are analyzed technically and economically to accurately evaluate the size of the problem.

Step 4: Optimal solution; according to the results of step 3, the estimated cost caused by events is compared with the cost of alternative solutions. Then, the optimal solution can be decided.

Step 5: Solution implementation; the suggested solution is applied. The monitoring activities are maintained to confirm power quality.

CHAPTER 8

VOLTAGE VARIATIONS

As an extension to the previous chapter, the voltage quality events are briefly described in this chapter. Two types of voltage variations; voltage drop and voltage sag, are dealt with in more detail as well.

8.1 VOLTAGE QUALITY

The voltage at different load points in the network must be kept approximately constant. The allowable limits of the different types of voltage variations shown in the previous chapter, Figure 7.1, are given in Table 8.1 according to International Electromechanical Commission (IEC) or IEEE standards.

One of the main keys of power quality is the voltage. Many people translate the power quality as voltage quality. In the forthcoming sections, various types of voltage variations and how to deal with each type are introduced [121].

8.1.1 Voltage Drop

The equipment connected to a utility system is designed to operate at a specific voltage. It is difficult to supply power to each customer at a voltage exactly equal to what is written on the customer equipment nameplates. The main cause of this difficulty is that there is a voltage drop in each element of the power system: generation, transmission, and distribution, in addition to the

Electric Distribution Systems, First Edition. Abdelhay A. Sallam, Om P. Malik.

Published 2011 by John Wiley & Sons, Inc.

TABLE 8.1 The Limits of Voltage Variations

Type of Voltage Variation	Allowable Limit of Accepted Power Quality Level	Reference
Voltage drop	Up to 33 kV: $V_n \pm 10\%$ Above 33 kV: $V_n \pm 5\%$ for normal case $V_n \pm 10\%$ for emergency case	IEC 38/1983
Voltage unbalance	3%	IEEE 1159/1995
Voltage sag	10–90% of V_n from 3 s to 1 min 80–90% of V_n for duration >1 min	IEEE 1159/1995
Voltage swell	110–120% of V_n from 3 s to 1 min	IEEE 1159/1995
Overvoltage	110–120% of V_n for duration >1 min	
Voltage flicker	Duration: short period (up to 10 min), quality level 100%, MV 0.9, HV 0.8 Duration: long period (up to 2 h), quality level 80%, MV 0.7, HV 0.6	IEC 1000-3-7/1995
Frequency deviation	1% for nominal case 2% for emergency case	IEC 1000-2-4/1994

internal wiring of the customer's installation. The customer who has a large power demand or receives its power through large impedance is exposed to lowest voltage. This is because the voltage drop is proportional to the magnitude of demand current and the entire impedance between the source and the customer. Voltage profile along a feeder supplying residential loads is shown in Figure 8.1. It is seen that the first customer, nearest to the power source, has the least voltage drop, while the last and the farthest customer has the largest voltage drop regardless of the voltage drop resulting from the internal wiring of the customer installation.

To supply power to customers at a voltage of constant magnitude or within narrow limits, the power cost on utility side will be highly increased. On the other hand, if the supplied voltage is within broad limits to avoid the cost increase on utilities, the equipment must be designed to withstand a wide range of voltage variation. In this case, the equipment is expensive, that is, the cost on the customer's side is increased.

Therefore, a compromise between these two aspects is needed. Utilities and equipment manufacturers have cooperated in sharing the responsibility of fulfilling specific limits of operating voltage.

The American National Standards Institute (ANSI) has established a standard [123] that was formulated by both utilities and manufacturers, and its recommendations are followed by both. The standards for service voltages are illustrated in Figure 8.2. The utility must only meet the "service" requirements, and the customer's duty is to control the voltage drop in its circuit to operate the equipment at voltages within standard ranges. Two ranges are specified: range A which identifies the service voltage to be from 114 to 126 V and range

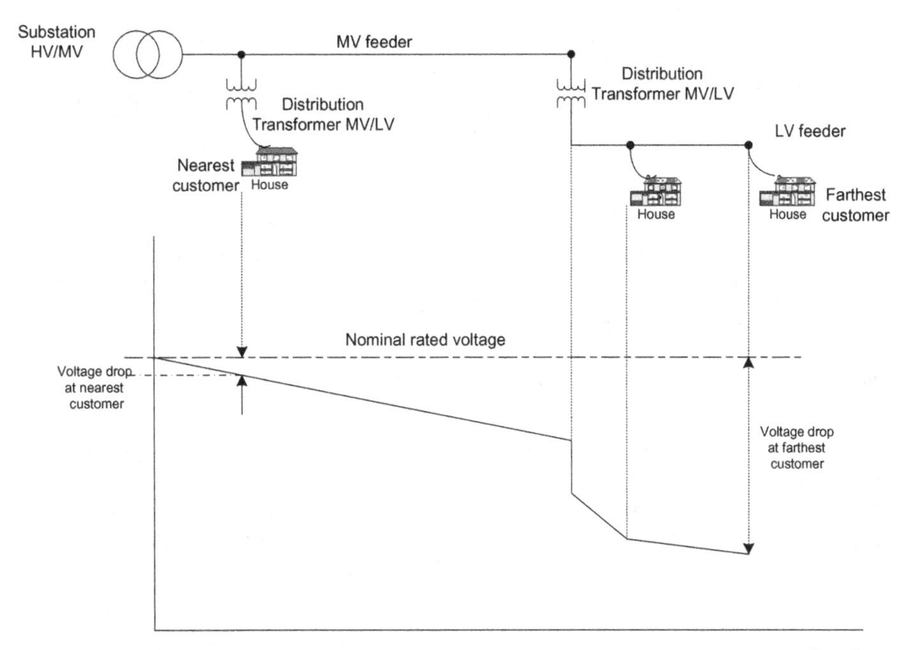

Figure 8.1 Voltage profile along a feeder supplying residential customers [122].

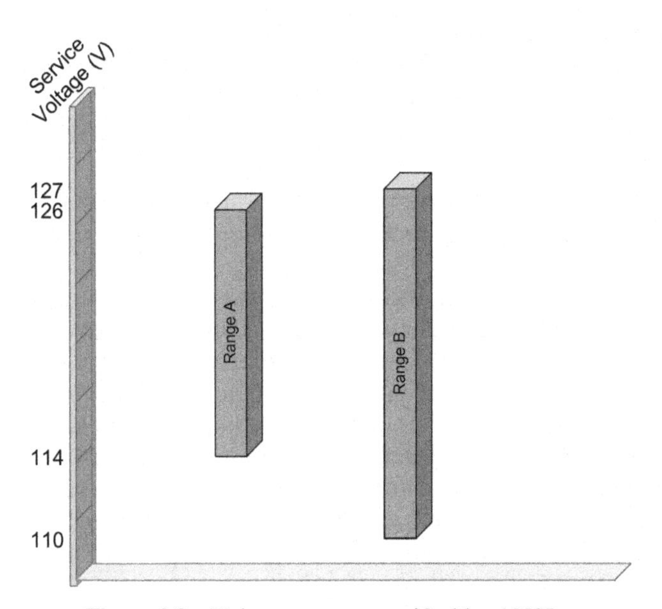

Figure 8.2 Voltage ranges specified by ANSI.

B allows the service voltage to be from 110 to 127V for limited duration in case of emergency operation and infrequent occurrence. Within a reasonable time, the voltage must be regulated to be within range A. Different methods of voltage drop reductions are explained in Section 8.2.

8.1.2 Voltage Sags

The term "sag" is used in the U.S. technical publications, while the term "dip" is usually preferred by the IEC [124]. The voltage sag is defined as a sudden reduction of alternating current (AC) voltage at a point in an electric power system followed by voltage recovery after a short period of time from a few cycles to a few seconds, 0.5 cycles to 1 minute's time, (IEC 61050-161). It is normally detected by the calculation of the voltage root mean square (rms), V_{rms}, over one cycle every half-cycle each period overlaps the prior period by one half-cycle.

As an example shown in Figure 8.3, there is sag if V_{rms} value falls below the sag threshold, which is a percentage of the reference value V_{ref} and set by the user according to the objective (typically ranges from 80% to 90% of V_{ref}, CENELEC EN 50160, IEEE 1159). The reference voltage V_{ref} is generally the nominal voltage for low-voltage (LV) power networks and the declared voltage for medium-voltage (MV) and high-voltage (HV) power networks.

The voltage sag is characterized by three parameters [125]:

1. The depth ΔV (or voltage sag magnitude): The depth of the voltage sag is defined as the difference between the reference voltage and the voltage sag value. It is usually expressed as a percentage of V_{ref}.
2. The duration Δt: It is the time interval during which the V_{rms} is lower than the sag threshold.
3. The amount of higher frequency components arises during sag voltage and the overshoot that occurs immediately after the sag (Fig. 8.4).

The depth ΔV and duration Δt are the most common parameters used to characterize the voltage sag. They are discussed in Section 8.3. In a three-phase system, the characteristics ΔV and Δt in general differ for each of the three phases. Therefore, the voltage sag must be detected and characterized separately on each phase. The value of ΔV is considered as the greatest depth on the three phases. On the other hand, it is commonly assumed that the sag starts when the voltage of the first disturbed phase falls below the sag threshold and ends when the voltages of all three phases are restored (equal to or above the sag threshold) [126]. A typical example of a three-phase disturbed system is shown in Figure 8.5.

8.1.2.1 Sources of Voltage Sag

Voltage sags are mainly caused by phenomena leading to high currents, which in turn cause a voltage drop across the network impedances with a magnitude that decreases in proportion to the

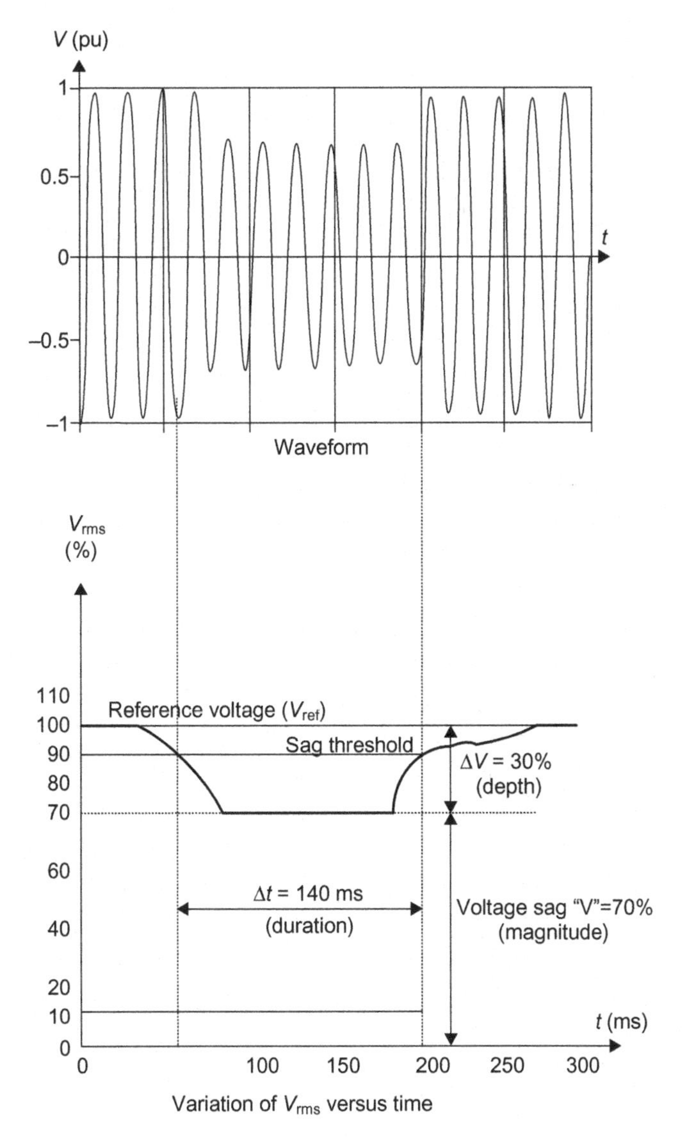

Figure 8.3 Voltage sag characteristic parameters [127].

electric distance of the observation point from the source of the disturbance. The common causes of sags are the following:

- Faults on the transmission (HV) or distribution (MV and LV) networks or on the installation itself. The duration of sag usually depends on the operating time of the protective devices (circuit breakers, fuses) to isolate the faulty section from the rest of the network.

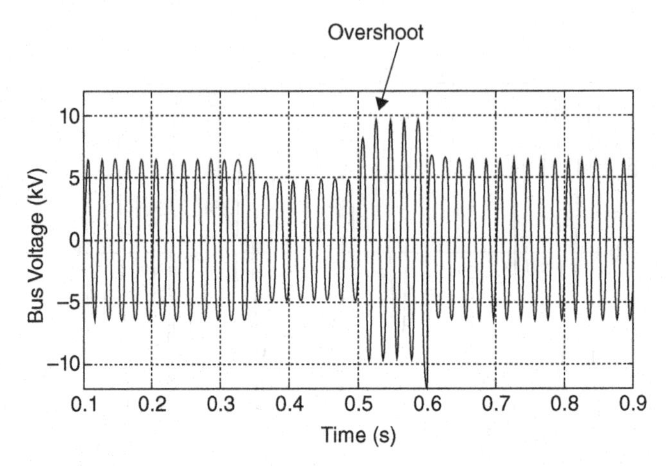

Figure 8.4 Waveform with high-frequency components and overshoots during and after the sag, respectively.

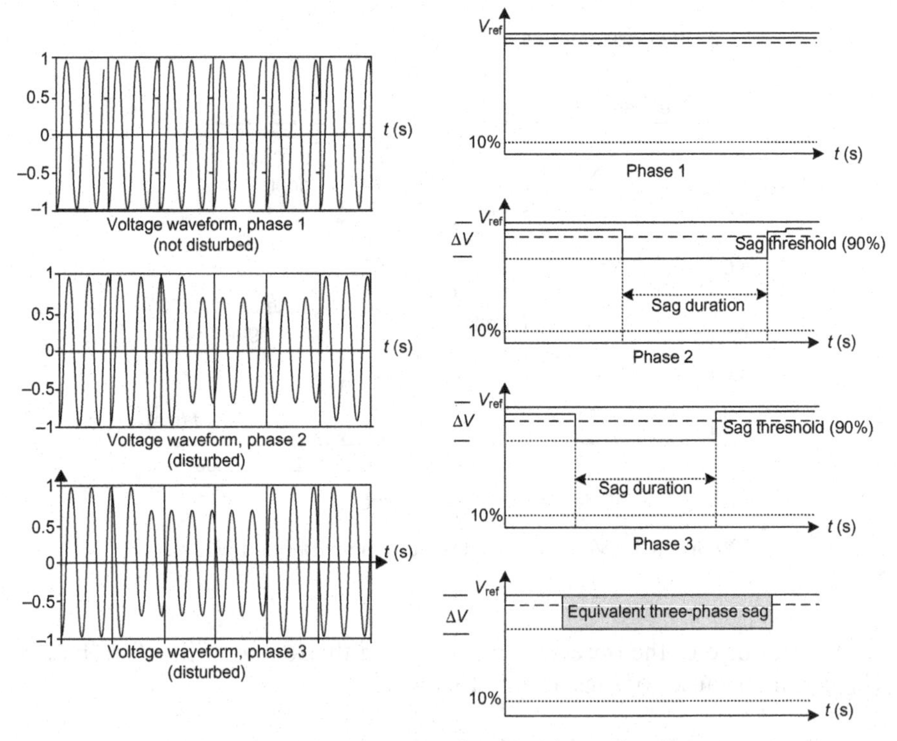

Figure 8.5 Voltage sag duration in a three-phase system.

- Switching of large loads compared with the short-circuit power. For instance, the starting of large motors inside an industrial facility can result in significant voltage sag. A motor can draw six times its normal running current, or more, while starting. This creates a large and sudden electric load that will likely cause a significant voltage drop tin the rest of the circuit on which it resides.

8.1.3 Flicker

Some electric equipment such as computers and electronic devices, are sensitive not only to voltage magnitude but also rapid changes in voltage, called "voltage flicker." The voltage regulation equipment augmented by utilities cannot compensate for instantaneous voltage fluctuations. The main reason of these fluctuations is the sudden application of low power factor loads, such as motors, welding machines, and arc furnaces. These loads do not demand constant current and generate variable voltage dips along the distribution feeder. This may cause disturbances noticed by other loads connected to the same feeder. For instance, the operation of arc furnace shown in Figure 8.6a may cause a flicker problem to other domestic loads. It is better to supply power by separate feeders as shown in Figure 8.6b.

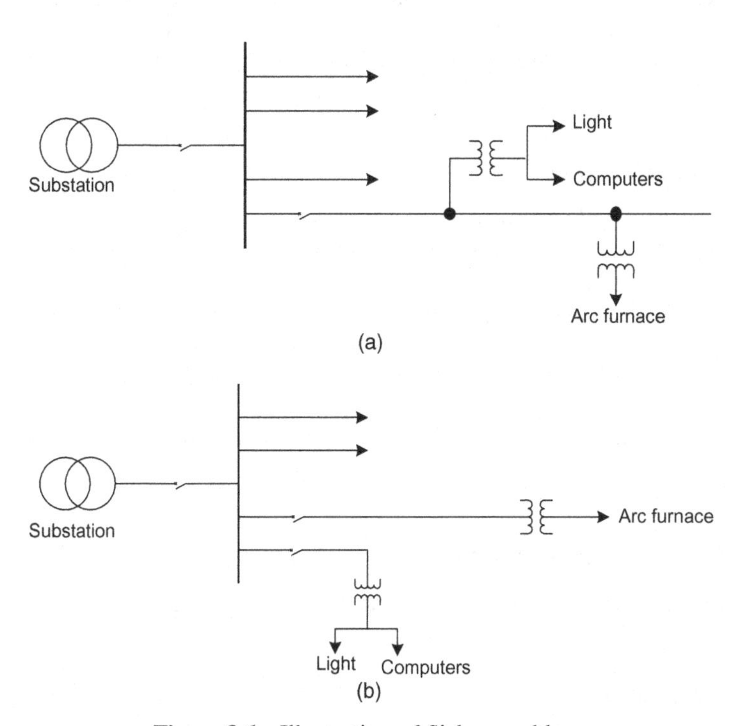

Figure 8.6 Illustration of flicker problem.

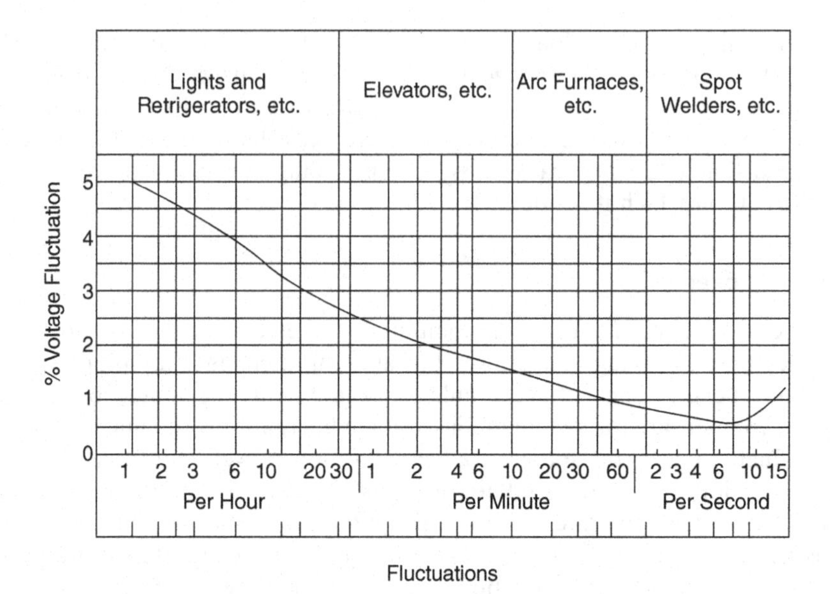

Figure 8.7 Borderline of irritation [122].

The borderline of irritation that indicates the maximum permissible voltage flicker before it becomes objectionable to the customer is shown in Figure 8.7. For example, the allowable number of flickers for elevators is three times per minute when the voltage variation is of order of 2%.

8.1.4 Voltage Swells

Voltage swell is an increase in the voltage of healthy phases when a ground fault occurs in one of the phases (Fig. 8.8). This voltage increase can be as much as approximately 30% for a four-wire, multigrounded system and over 70% for a three-wire system. Voltage swell is characterized by its magnitude and duration. Its magnitude depends on system grounding and is important for specifying the operation of surge arresters. The duration depends on system protection where it ranges from a few cycles to several minutes. Standards of maximum magnitude of swells for different grounding system designs are shown in Table 8.2 [123].

8.1.5 Transient Overvoltages

The transient overvoltages are classified into two types: impulse transients and oscillatory transients.

8.1.5.1 Impulse Transients They are mainly due to lightning strokes and are characterized by the rise and decay times. The lightning can strike any

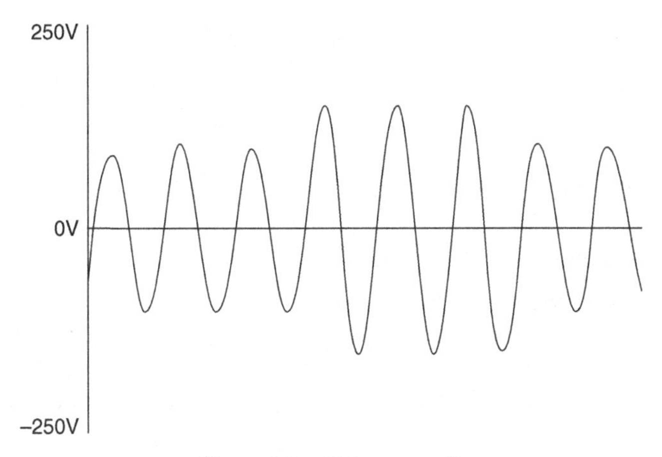

Figure 8.8 Voltage swell.

TABLE 8.2 Standards of Overvoltage Magnitude for Different Designs of Earthing System [123]

System	Overvoltage Magnitude
Unearthed	$1.82 \times E_{LE}$
Four-wire multi-earthed (spacer cables)	$1.5 \times E_{LE}$
Three or four-wire uni-earthed (open wire)	$1.4 \times E_{LE}$
Four-wire multi-earthed (open wire-gapped)	$1.25 \times E_{LE}$
Four-wire multi-earthed (open wire-MOV)	$1.35 \times E_{LE}$

Note: E_{LE} is the nominal line-to-earth voltage of the system.

place in the power system (unpredicted place) producing lightning surge currents that can flow from the power system into the loads causing damage and may cause flashovers of the utility lines.

The lightning arresters are used to protect the system equipment and loads against these surges. The arresters provide an easy pass to earth for the lightning surges. Usually, the earth is not a perfect conductor for impulses, so, two power quality problems as side effects with surge currents are found:

- The local earth potential is increased and will be higher than the potential of other earths. That is, there are two earth references at different potential. Therefore, any equipment connected between them is subjected to fail.
- The surges produce HVs in phase conductors as they pass through cable lines on the way to earth with lowest potential.

8.1.5.2 Oscillatory Transients The switching events on utility side or load side in the system are the most common reasons for these transients. For example, the switched capacitor banks used for improving the power factor and reducing the losses yield overvoltage oscillatory transients when switched. These transients propagate into the local power system and may be magnified on the load side if the natural frequencies of the systems are aligned.

8.2 METHODS OF VOLTAGE DROP REDUCTION

8.2.1 Application of Series Capacitors

8.2.1.1 Introduction Series compensation in distribution networks has been used for more than 50 years. However, series compensation at distribution voltage levels has never had its real breakthrough. One reason has been the relatively complicated and costly equipment for overvoltage protection and bypassing of line fault currents. Another reason has been the risk for different resonance phenomena. In recent years, however, a number of new factors are playing an increasingly important role in the interest for series compensation. Some of them are the following:

- transmission economy,
- quality of delivered electric power,
- environmental restrictions on building new lines, and
- cost for reactive power.

Other factors have at the same time made it possible to simplify the equipment. One example is the development of metal oxide varistor (MOV), which gives new possibilities for overvoltage protection. Another factor of importance is the general acceptance of series compensation at higher system voltages.

8.2.1.2 Basic Theories (Case No. 1) Assuming a simple radial feeder as shown in Figure 8.9 with the entire load concentrated at the end point, the following designations are used for system data:

- V_1 = voltage at feeding point,
- V_2 = voltage at load point,
- ΔV = voltage drop along the line,

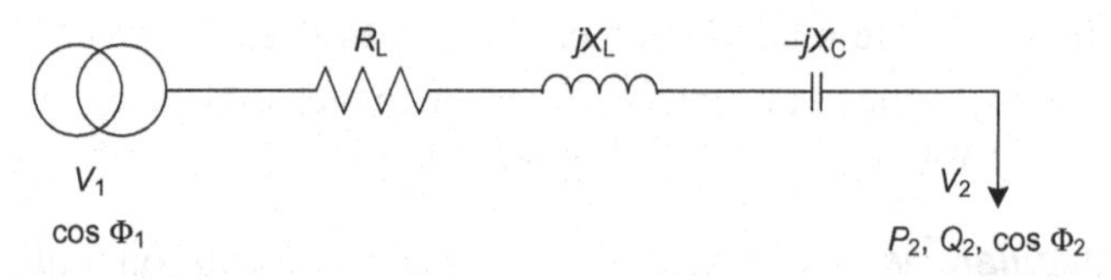

Figure 8.9 Single-phase diagram for a radial line with a series capacitor and a concentrated load at the line end.

- $\cos\Phi_1$ = power factor at feeding point,
- $\cos\Phi_2$ = power factor at load point,
- R = line resistance,
- X_L = line reactance,
- X_C = series capacitor reactance,
- P_2 = active power at load point,
- Q_2 = reactive power at load point,
- S_2 = apparent power at load point,
- I = line current.

A line of this type can be represented with a phasor diagram as shown in Figure 8.10a.

The voltage V_2 is less than V_1 and Φ_1 and Φ_2 are almost equal. It can be seen that the voltage drop, by which is meant the numerical difference between the two voltages, depends both on the resistance and the reactance of the line. If a series capacitor with a capacitive reactance of X_C is now installed, the phasor diagram shown in Figure 8.10b is obtained.

This diagram is drawn on the assumption that the voltage at the feeding point V_1 is maintained at a constant value and that Φ_2 and I are unchanged. When comparing Figure 8.10a,b, it can be seen that the series capacitor has leveled out the differences between the voltages V_1 and V_2, that is, reduced the voltage drop and at the same time, improved the power factor $\cos\Phi_1$ at the feeding end of the line.

It can be seen that if the phase angle Φ_2 is sufficiently large, it is theoretically possible to make the reactance X_C of the series capacitors so large that the voltages V_1 and V_2 will be equal (voltage drop = 0).

Should, on the other hand, the phase angle Φ_2 be close to zero, the series capacitor will only insignificantly reduce the voltage drop. Therefore, a

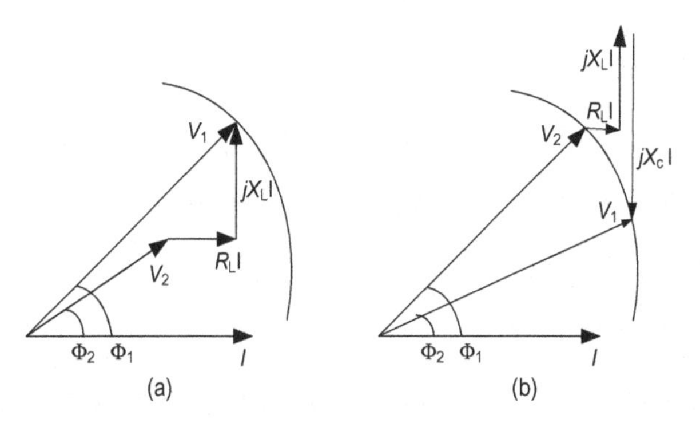

Figure 8.10 Phasor diagram for the line voltage for a simple radial line. (a) Without series capacitor. (b) With series capacitor.

necessary condition for a series capacitor to improve the voltage to a greater degree is that the load must be inductive. For a capacitive load, the series capacitor will reduce the load point voltage. For the sake of simplicity, it is assumed in the phasor diagram (Fig. 8.10b) that the load current I is of the same magnitude as prior to the insertion of the series capacitor. Generally, however, if the load remains constant, an increase in the voltage V_2 is accompanied by a reduction in the current I. The line losses will thus be diminished and the voltage drop will also be further reduced. The most common application for series capacitors in distribution networks is to improve the voltage profile and thereby increase the load-carrying capability of the line. As the voltage drop across the series capacitor and the line impedance is proportional to the line current, the voltage can be kept constant even in case of relatively rapid variations in the load. Another application is to strengthen the network for starting of motors. Improvement of the power factor $\cos \Phi_1$ means less required Q at the sending end. A consequence of the series capacitor installation is that the transmission losses will be significantly reduced. For all these applications, the capability of the series capacitor to compensate for the line inductance is being utilized.

8.2.1.3 *Reduced Voltage Fluctuations* In cases with flicker due to large variations of the load, a series capacitor will improve the voltage quality at the load end and along the line. It can be seen from the equation

$$\Delta V = [R \times P_2 + Q_2(X_L - X_C)]/V^2, \tag{8.1}$$

where

V = reference voltage,
ΔV = voltage drop along the line,
R = line resistance,
X_L = line reactance,
X_C = series capacitor reactance,
P_2 = active power at load end, and
Q_2 = reactive power at load end.

That is, if the resulting line reactance is small the influence from changes in reactive power consumption of the load will be minimal.

8.2.1.4 *Loss Reduction* The line losses in a distribution line are proportional to the square of the line current. By introducing a series capacitor in a line with inductive load, the voltage at the load end is increased. For constant load, the line current is reduced according to the equation

$$I = S_2/V_2, \tag{8.2}$$

where

S_2 = apparent power at load point and
V_2 = voltage at load point.

This affects the line losses as defined by

$$P_{loss} = R_1 \times I^2. \tag{8.3}$$

8.2.1.5 Illustrative Example A distribution line can be represented by its resistance R and its reactance X. The current causes a drop in voltage that is greater the further away the location is from the feeding point. By connecting series capacitors into the line, a corresponding capacitive voltage drop is obtained. Since the drop in voltage, both over the series capacitor and the line impedance, is proportional to the load current, voltage control is instantaneous and self-regulating. This means that the voltage can be kept constant even in the case of rapid variations in load. The voltage drop "ΔV" along the line can be calculated by the formula given in Equation 8.1.

Normally, the active voltage drop ($R \times$ P) is greater than the reactive voltage drop ($X \times Q$) in a distribution line. In case of overcompensation ($XC/X > 1$), the value ($X \times Q$) is negative, which means that the resistive voltage drop can also be compensated. Voltage profile for a series compensated line is shown in Figure 8.11.

Due to the series capacitor being bypassed in the case of overcurrent, it will be out of service when the short-circuit current flows into the line. Connecting series capacitors in a distribution line with inductive load reduces transmission losses. Since the voltage at the load point increases, the current is reduced in case of constantly transmitted power and the line losses are reduced as well.

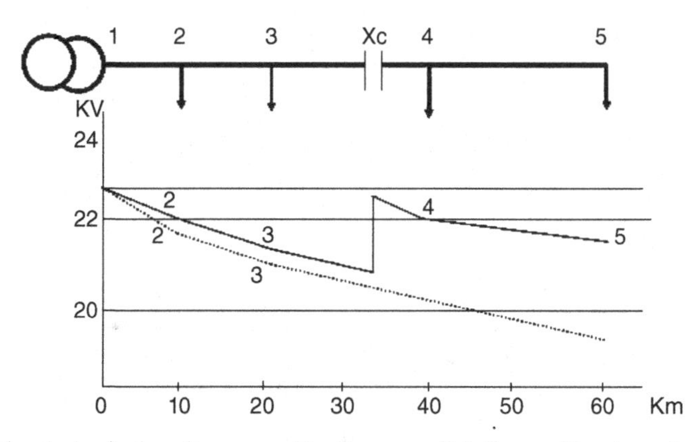

Figure 8.11 A typical voltage profile for a radial line without and with series capacitor.

8.2.1.6 Lateral Radial Feeder In the case of lateral radial feeder, shown in Figure 8.12a, each lateral can be considered as a load point on the main feeder with a lumped load equal to the summation of loads along this lateral and the system is reduced to be as shown in Figure 8.12b.

The location of series capacitor is determined at each load point to keep its voltage at a desired value "V_{sch}" (e.g., $V_{sch} = 98.5\%\ V_{ref}$ where V_{ref} is the rated voltage) or to be not less than the allowable limit ($V_{limt} = 95\%\ V_{ref}$). The calculated voltage at the load point P_{li} is considered as the sending voltage of the next section of the main feeder feeding the next load P_{li+1}. Each section is treated individually where the power flow differs from one section to another. For the laterals, each one is treated with the same method used for the main feeder. The calculated voltage at the compensated node of the lateral connection with the main feeder is considered as a source voltage of that lateral. The flowchart given in Figure 8.13 illustrates the main steps in determining the size of series capacitor and the corresponding voltage at each load point of a radial MV distribution feeder. Obviously, if the voltage at any load point is within the allowable limit, there is no need to insert a series capacitor.

Numerical Application A distribution system includes a radial overhead transmission line (OHTL) at 22 kV, 50 Hz, with a total length of 77 km. The power demand at six load points and the distances between them are as shown in Figure 8.14. The OHTL is aluminum with cross-sectional area (CSA) 240/40 mm^2. Its resistance and reactance per kilometer are 0.1329 and 0.32 Ω, respectively.

From data given above, it is found that $V_{ref} = 22$ kV, $V_{limt} = 20.9$ kV, and $V_{sch} = 21.68$ kV.

The calculation of voltages at load points without compensation is as below.

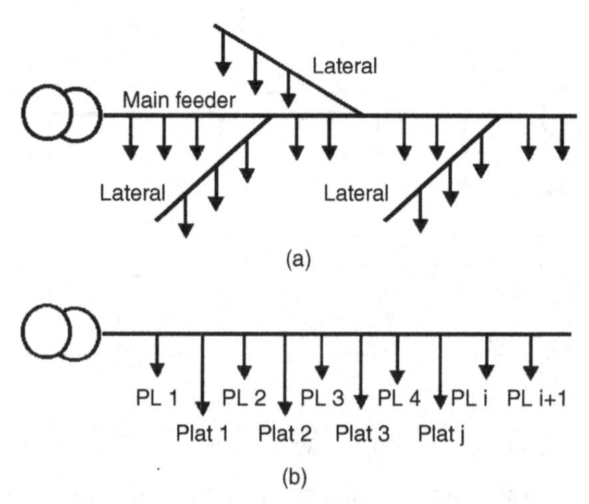

Figure 8.12 Studied system. (a) Lateral radial feeder. (b) Reduced lateral feeder.

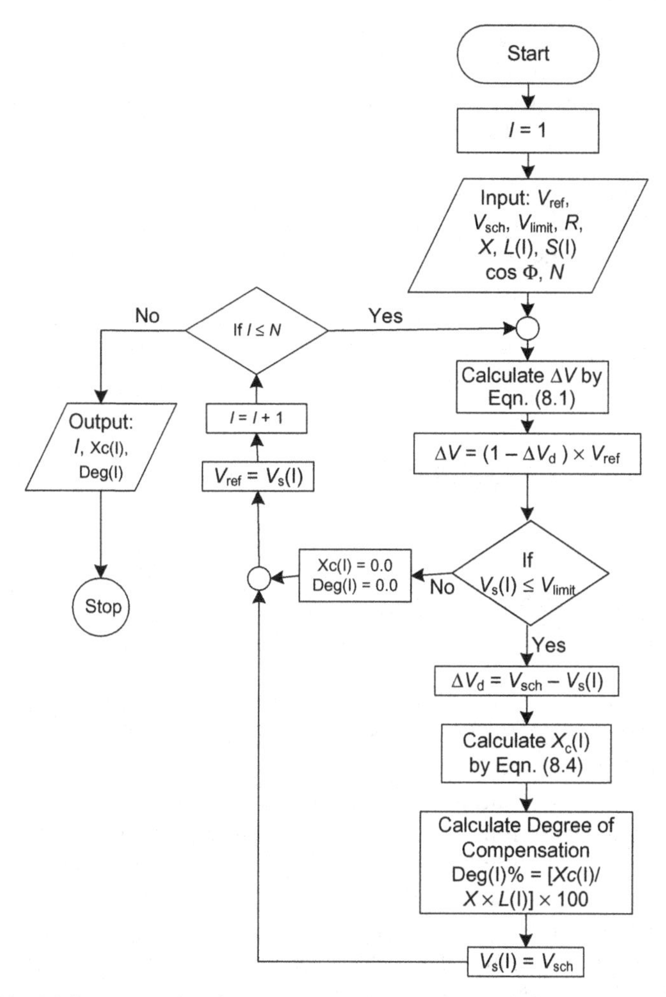

Figure 8.13 Main steps of series compensation calculation (N = no. of load points, $V_s(I)$ = source voltage at load point I).

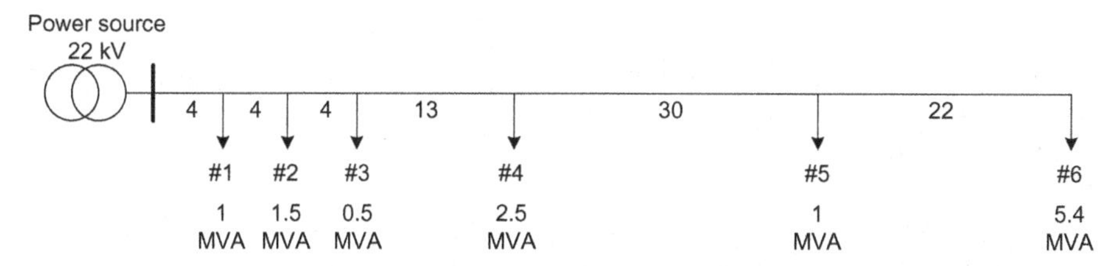

Figure 8.14 Line diagram of the case study.

At load point #1: It is the end of the first section of the feeder. The power flow into this section is $S(1) = 11.9\,\text{MVA}$ and the length $L(1) = 4\,\text{km}$. The source voltage at sending end $V_s = 22\,\text{kV}$.

By substituting into Equation 8.1, the voltage drop is

$$\Delta V_1 = [(11.9 \times 0.1329 \times 4 \times 0.8 + 11.9 \times 0.32 \times 4 \times 0.6)/22^2] \times 100 = 2.934\%.$$

Thus, the voltage at load point is given as $V_1 = (1 - 0.02934) \times 22 = 21.35\,\text{kV}$.

This voltage is considered as the sending end voltage of the next section (second section).

At load point #2: $S(2) = 10.9\,\text{MVA}$, $L(2) = 4\,\text{km}$, and $V_s = 21.35\,\text{kV}$.

Then,

$$\Delta V_2 = [(10.9 \times 0.1329 \times 4 \times 0.8 + 10.9 \times 0.32 \times 4 \times 0.6)/(21.35)^2 \times 100 = 2.854\%$$

and

$$\begin{aligned} V_2 &= 91 - 0.02854) \times 21.35 = 20.74 \text{ kV} \\ &= \text{sending end voltage of next section (third section section).} \end{aligned}$$

Repeating the same procedure, the voltages of the rest of load points are $V_3 = 20.19\,\text{kV}$, $V_4 = 18.49\,\text{kV}$, $V_5 = 15.4\,\text{kV}$, and $V_6 = 13.09\,\text{kV}$.

It is noted that the voltage at load point #1 is within the limits while the other load points is violating the limits and series compensation is needed.

The system calculations with compensation are as below:

The same steps are applied taking into account that the load points at which the voltage is less than the allowable limits will be compensated by inserting a series capacitor. The capacitor size can be calculated by applying Equation 8.1 with replacing the line reactance X_L by the net reactance $(X_L - X_C)$ and maintaining the voltage drop at a desired value.

Therefore,

$$\Delta V_{\text{desired}} = \frac{R \times P + (X_L - X_C)Q}{V^2} \times 100\%. \tag{8.4}$$

At load point #1: $V_1 = 21.35\,\text{kV}$, which is greater than V_{limit} and no need to compensate, that is, $X_c = 0.0\,\Omega$.

At load point #2: $V_2 = 20.74\,\text{kV}$, which is less than V_{limit} and must be raised up to $V_{\text{sch}} = 21.68\,\text{kV}$.

Accordingly, $\Delta V_{\text{desired}} = (21.35 - 21.68)/21.35^2 = -1.54\%$.

Substitution into Equation 4.5 gives $X_c = 3.06\,\Omega = 1.04 \times 10^3\,\mu\text{F}$.

Hence, the voltage at the load point #2 just before the capacitor connection point (BCCP) is 20.74 kV, while its value just after the capacitor connection point (ACCP) is 21.68 kV.

At load point #3: From Equation 8.1, it is found that

$$\Delta V = [(9.4 \times 0.1329 \times 4 \times 0.8 + 9.4 \times 0.6 \times 4 \times 0.32)/21.68^2] \times 100 = 2.387\%.$$

Then, $V_3 = 21.16\,\text{kV} > V_{\text{limit}}$ and there is no need to compensate, that is, $X_c = 0.0\,\Omega$.

At load point #4: $\Delta V = 7.709\%$ and $V_4 = 19.53\,\text{kV} < V_{\text{limit}}$. It must be raised up to 21.68 kV. The desired voltage drop is

$$\Delta V_{\text{desired}} = (21.16 - 21.68)/21.16 = -2.457\%.$$

By substituting into Equation 8.4,

$$X_c = 8.25\ \Omega = 0.39 \times 10^3\ \mu\text{F}.$$

Hence, $V_{\text{BCCP}} = 19.53\,\text{kV}$ and $V_{\text{ACCP}} = 21.68\,\text{kV}$.

Continuing the same procedures for load points #5 and #6 gives the following results.

At load point #5: $\Delta V = 12.188\%$, $V_5 = 19.04\,\text{kV} < V_{\text{limit}}$ and must be raised up to 21.68 kV. $\Delta V_{\text{desired}} = 0.0$ and $X_c = 14.92\,\Omega = 0.21 \times 10^3\,\mu\text{F}$. $V_{\text{BCCP}} = 19.04\,\text{kV}$ and $V_{\text{ACCP}} = 21.68\,\text{kV}$.

At load point #6: $\Delta V = 7.54\%$, $V_6 = 20.04\,\text{kV} < V_{\text{limit}}$ and must be raised up to 21.68 kV. $\Delta V_{\text{desired}} = 0.0$ and $X_c = 10.94\,\Omega = 0.29 \times 10^3\,\mu\text{F}$. $V_{\text{BCCP}} = 20.04\,\text{kV}$ and $V_{\text{ACCP}} = 21.68\,\text{kV}$.

Therefore, the line is compensated at load points #2, 4, 5, and 6 with capacitor size 3.06, 8.52, 14.92, and 10.94 Ω, respectively, as shown in Figure 8.15. A summary of results is given in Table 8.3 and the voltage profile along the line is shown in Figure 8.16.

8.2.2 Adding New Lines

By adding new lines with a goal of reducing the equivalent resistance and reactance of the feeder, the voltage drop is reduced. So, the voltage drop value depends on the number of parallel lines representing the feeding system.

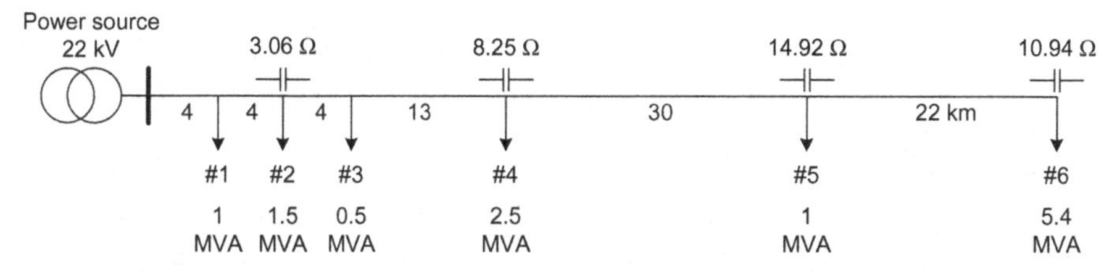

Figure 8.15 Series capacitor locations on compensated line.

TABLE 8.3 Summary of Results

Load Point No.	1	2	3	4	5	6
Voltage without compensation	21.35	20.74	20.19	18.49	15.4	13.09
Voltage with compensation						
BCCP	21.35	20.74	21.16	19.53	19.04	20.04
ACCP	21.35	21.68	21.16	21.68	21.68	21.68
Capacitor size						
(Ω)	0.0	3.06	0.0	8.25	14.92	10.94
(micro-Farad) × 10^3	0.0	1.04	0.0	0.39	0.21	0.29

BCCP = before capacitor connection point; ACCP = after capacitor connection point.

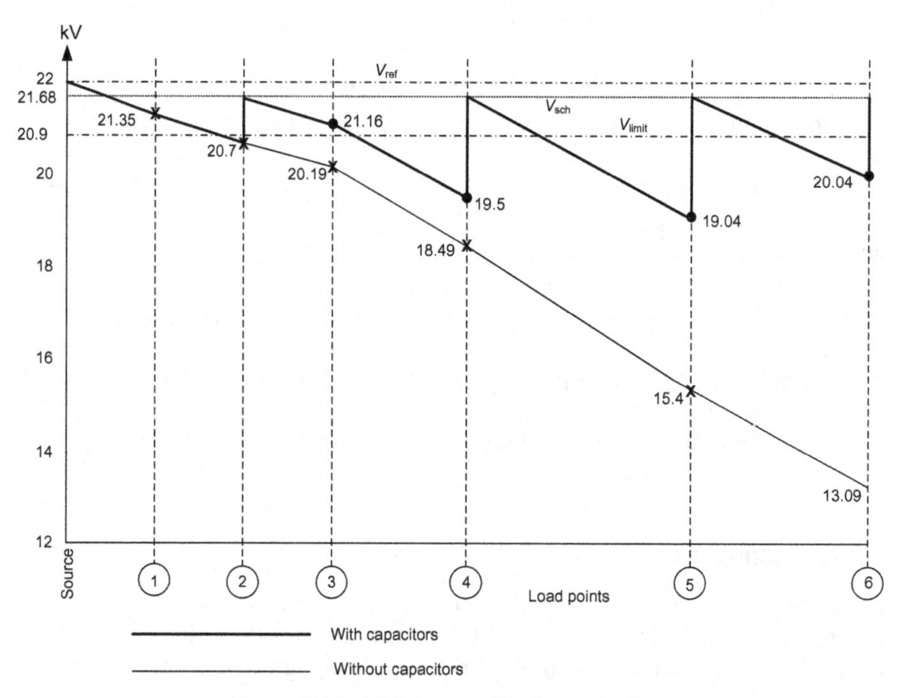

Figure 8.16 Voltage profile along the line.

Therefore, it is important to determine the number of parallel lines for each section of the feeder where the voltage drop at the end of this section is within the allowable limits.

The voltage drop is calculated in case of using one line of the feeder, but the voltage drop in this case was out of limit. Also, by using two lines of the feeder, the voltage drop was out of limit, and so on, by using three, four, five, and six lines of the feeder the voltage drop was out of allowable limits.

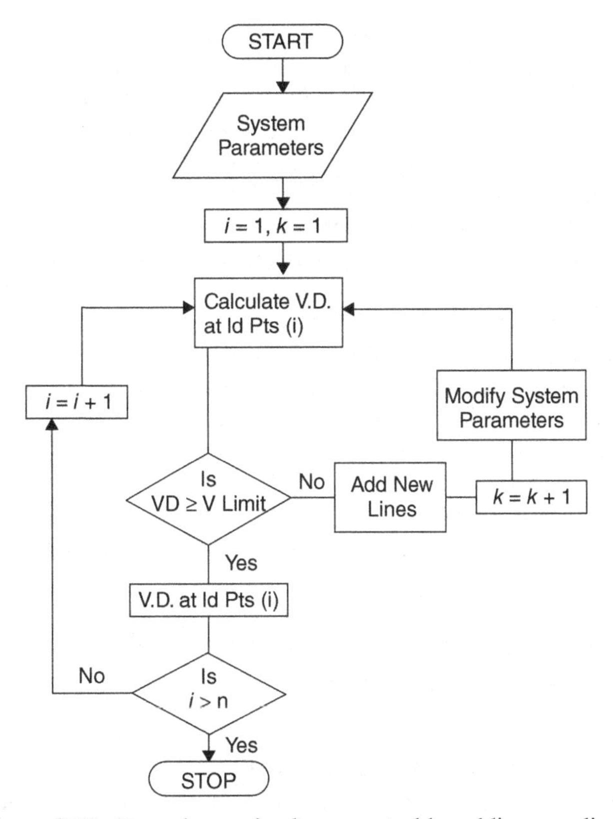

Figure 8.17 Procedures of voltage control by adding new lines.

A flowchart summarizing the steps of voltage drop calculations when adding new lines is shown in Figure 8.17. The calculations are terminated when getting the number of new lines at which the voltage drop at all nodes of the feeder is within the allowable limit.

8.2.3 Regulating the Voltage

The basic function of an electric utility to supply power to its customers appears to be rather simple. Voltage to be supplied is defined within a predetermined range. Unfortunately, the laws of physics enter in, with system losses and voltage sags that must be corrected and regulated in order to stay within the voltage range. There are choices to be made in how to regulate voltage; some of these choices can require significant capital expenditure.

For many utilities, demand for electricity continues to grow. New demand must be met with increased supply. Demand-side management and distribution automation look to utilize existing system resources more efficiently. Paybacks from lower system losses and deferred use of costly peak demand electricity justify purchase of new equipment.

Modular designs are favored, using products optimized for their function. Future benefits would likely be realized in ease of repair and improved availability when replacement is needed. Even in the real world of ongoing maintenance and occasional system or component failures, objectives of minimizing outage time and limiting equipment failures can be realized.

The use of simpler, basic, modular products increases the potential for benefits in product availability, performance optimization in product function, and reduced concentration on large engineering design development. Modular equipment offers the flexibility of modeling each feeder to supply service in an optimized range across most of the line or to the load center.

To meet today's as well as anticipated future needs, utilities will be focusing their efforts to better utilize equipment already in place and to carefully evaluate the inclusion of new equipment. Three factors to consider in new equipment purchase decision are the following:

- expected utilization level of the equipment,
- ease of use of the equipment and future maintenance schedules, and
- anticipated repair rate.

In the purchase of equipment to regulate substation voltage, the following product alternatives may be considered:

1. *Single-Phase Transformer with Load Tap Changers (LTCs) in One Package:* For short distribution feeders, it is entirely likely that voltage standards can be met using only the voltage control equipment in the substation, such as LTCs. The LTC can be either on-load or off-load. The tap can be automatically changed to regulate the voltage without interrupting the load when using on-load tap changing, while with off-load tap changing, the load must be disconnected and the taps changed manually. In both cases, the voltage is regulated within a range limited by the number of taps and the percentage of variation of each tap. It is a common situation where equipment maintenance or failure requires taking product offline, that is, de-energizing. With a transformer with LTC, if one phase of the tap changer fails or requires maintenance, the entire package is removed from service and all phases are affected.
2. *Single-Phase Regulators:* For long feeders or feeders with extremely heavy loads, it may be necessary to augment the substation equipment with voltage regulators located on the feeder. With single-phase regulators, if one phase goes offline, it can be replaced with another single-phase unit.

One spare regulator can roll through all three phases when maintenance is required; the same spare unit can function effectively on any of the system's three phases. Full maintenance can be completed without diminishing service anywhere on the system. Spare equipment may be available from emergency reserves, from another substation, or as new direct from the manufacturer or nearby distributor.

Operation of a single-phase regulator is essentially the same as for a three-phase device. The main difference is in the separation of the three phases into separate tanks. In separating the active tap-changing mechanisms, several benefits are seen. Each single-phase unit has an independent control to react individually to the phase voltage variances.

Where phase voltages are unbalanced, in single-phase loading on residential feeders, single-phase units can respond individually to these independent feeder models.

The single-phase step voltage regulators are tap-changing autotransformers (as explained in Chapter 7, Section 7.4.1). They regulate distribution line voltages from 10% raise (boost) to 10% lower (buck) in 32 steps of approximately 5/8% each. Voltage ratings are available for MV distribution systems, for example, from 2.4 kV (60-kV basic impulse insulation level [BIL]) to 34.5 kV (200-kV BIL) for 60- and 50-Hz systems. Internal potential winding taps and an external ratio correction transformer are provided on all ratings so that each regulator may be applied to more than one system voltage.

All voltage regulators are equipped with a bypass arrester connected across the series winding between the source and load bushings. This bypass arrester limits the voltage developed across the series winding during lightning strikes, switching surges, and line faults.

The disadvantage of voltage regulator is that it costs more and the voltage rise is restricted by the regulation percentage.

A combination of both LTCs and regulators can be used. Typical voltage profiles along a feeder supplying power delivered from a generator to a load through transformers with load tap changer and regulators are illustrated in Figure 8.18. These conditioners are necessary to be installed, particularly, for long feeders or short feeders supplying heavy loads. The profiles are plotted at heavy load and light load. In both cases, the system is adjusted to keep the voltage within standard limits.

The main steps of applying voltage regulators at load points along distribution feeders are given by the flowchart shown in Figure 8.19.

8.2.4 Applying Shunt Capacitors

The main use of shunt capacitors is in improving the power factor, which is the ratio of the active power to the apparent power as explained in the next chapter. It illustrates the relationship of the active power that is used to produce real work, and the reactive power required to enable inductive components such as transformers and motors to operate. As those components, with pronounced inductive characteristics, are common in industrial facilities and power networks, reactive power is usually inductive and current is lagging.

On the other hand, capacitive loads supply reactive power and the relevant current is leading. Most loads require the two types of power to operate. The

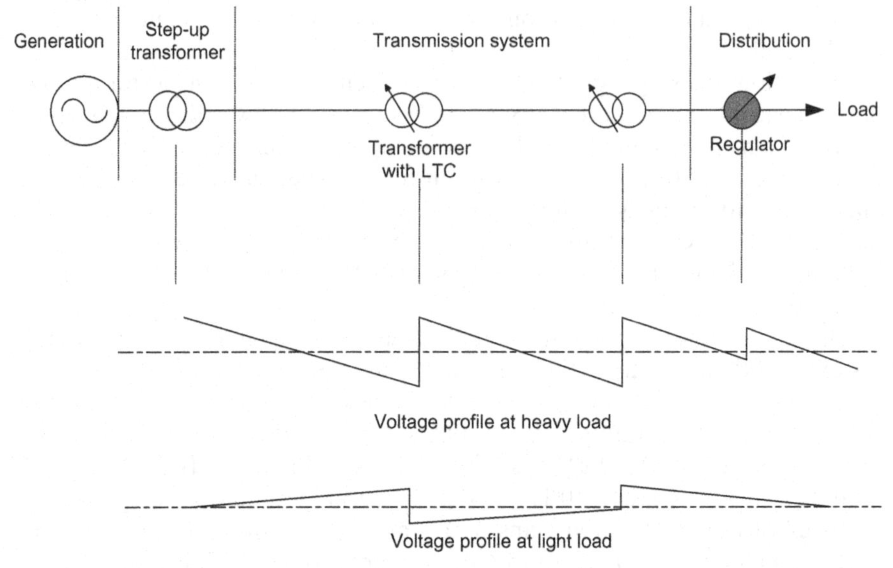

Figure 8.18 Typical voltage profile of a system at heavy and light loads [122].

reactive current flowing to a load will not affect the active power drawn by the load, but the supply circuit must carry the vector sum of the active and reactive current. The latter will contribute to the power dissipated in the supply and distribution network (losses), to the voltage drop, and to the network capacity requirements. A poor power factor will therefore result in lower energy efficiency and reduced power quality, in some cases affecting the product quality and the process productivity.

Many utilities employ shunt capacitor banks to help improve feeder voltage profile via power factor correction. The capacitors, decrease demand current, decrease system losses and reduce the voltage drop. Consequently, the voltage profile is improved. The capacitors are connected to the feeder at specific locations at which the voltage variation along the feeder does not violate tolerance of ±5% of nominal voltage (Fig. 8.20).

It has also been widely recognized that the application of shunt capacitors results in reduction of power and energy losses in the feeder. The problem of installing shunt capacitors on distribution primary feeders has been dealt with by many researchers. In applying their results to specific, physically based problems, however, most previous work suffers from the following:

a) Very restricted reactive-load distribution such as uniform-load distribution or a combination of concentrated and uniformly distributed loads have been extensively used in most analyses. Wire size of the feeder has been usually assumed to be uniform. It is apparent that solutions obtained under these assumptions may be far from what they are under real circumstances.

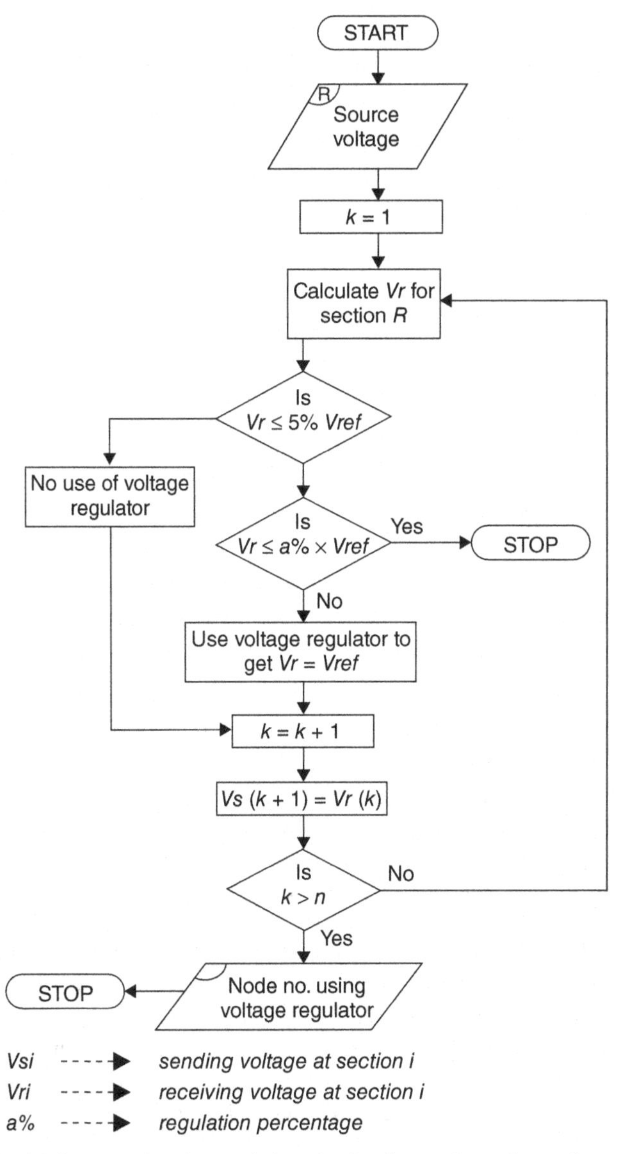

Figure 8.19 Main steps for determining the feeder nodes using voltage regulator.

b) Very often analyses were concerned with specified cases, the case of one capacitor bank, and two capacitor banks. Because of the lack of generality of those formulations, utility companies having different situations may not be able to apply those results to their systems. The primary objective is to remove certain unrealistic assumptions and to present very general and yet simple procedures for implementation so that they

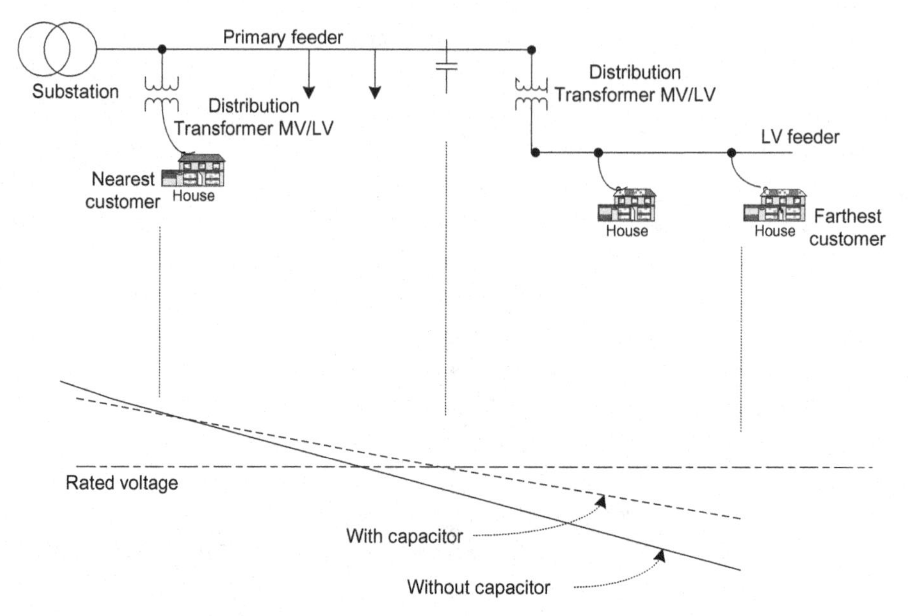

Figure 8.20 Voltage profile along a feeder with capacitors.

can be easily and readily modified, adapted, and extended depending on conditions of the system.

Shunt capacitors connected in parallel with lines are used extensively in distribution systems. Shunt capacitors supply the type of reactive power or current to counteract the out-of-phase component of current required by an inductive load. In a sense, shunt capacitors modify the characteristic of an inductive load by drawing a leading current that counteracts some or all of the lagging component of the inductive load current at the point of installation. Therefore, a shunt capacitor has the same effect as an overexcited generator, that is, synchronous condenser, or underexcited motor. Sometimes, at light load conditions, the capacitors may overcompensate and create an overvoltage lasting from a few seconds to several hours (Fig. 8.21).

By the application of shunt capacitor to a feeder, the magnitude of the source current can be reduced, the power factor can be improved, and consequently, the voltage drop between the sending end and the load is also reduced as shown in Figure 8.22. However, shunt capacitors do not affect current or power factor beyond their point of application.

Single-line diagram of a line and its voltage phasor diagram before the addition of the shunt capacitor are shown in Figure 8.22a,c and the same after the addition of capacitor are shown in Figure 8.22b,d. The voltage drop in feeders, or in short transmission lines, with lagging power factor can be approximated as

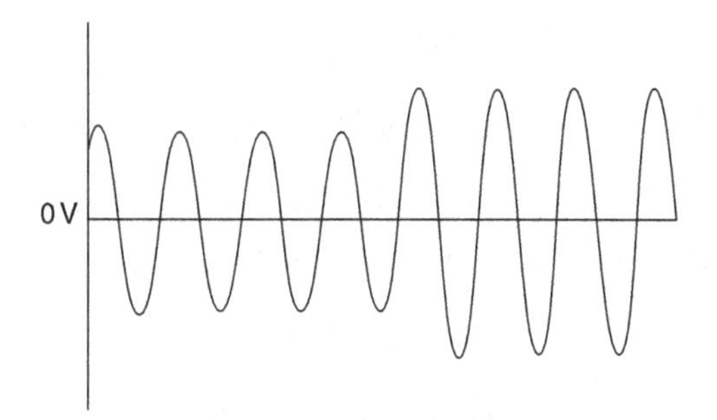

Figure 8.21 Overvoltages created by overcompensation.

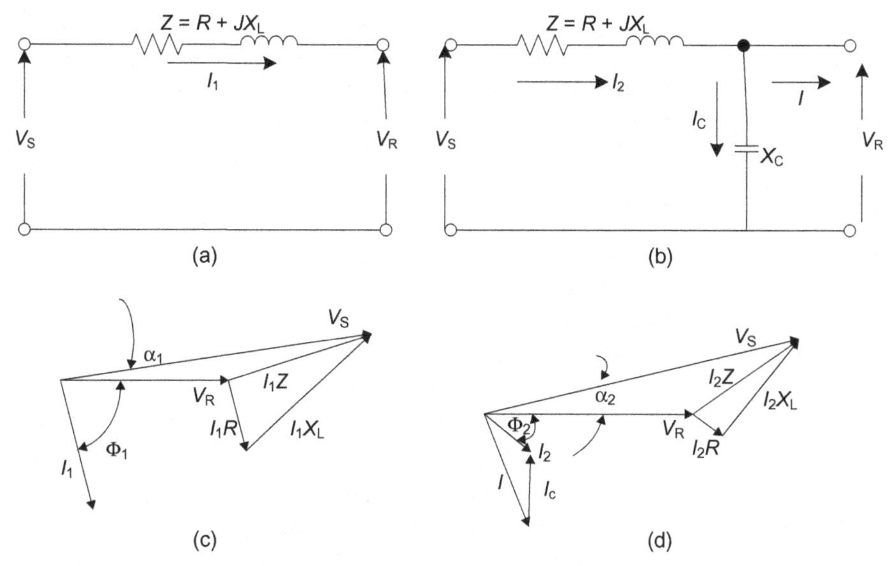

Figure 8.22 Voltage phasor diagrams for a feeder circuit of lagging power factor without (a,c) and with (b,d) shunt capacitors.

$$VD = I_R R + I_X X_L, \tag{8.5}$$

where

R = total resistance of feeder circuit (Ω),

X_L = total inductive reactance of feeder circuit (Ω),

I_R = real power (in-phase) component of current, and

I_X = reactive (or out-of-phase) component of current lagging the voltage by 90°.

When a capacitor is installed at the receiving end of the line, as shown in Figure 8.22b, the resultant voltage drop can be calculated approximately as

$$VD = I_R R + I_X X_L - I_C X_L, \tag{8.6}$$

where I_C is the reactive (or out-of-phase) component of current leading the voltage by 90°.

The difference between the voltage drops calculated by using Equations 8.5 and 8.6 is the voltage rise due to the installation of the capacitor and can be expressed as

TABLE 8.4 Comparison Between Alternative Methods for Voltage Drop Mitigation

Method of Voltage Improvement	Advantages	Disadvantages	Cost Comparison
Series capacitor	• Increases voltage • Reduces voltage variations • Instantaneous self-regulation • Improves the voltage profile • Reduces line losses • Increases the power factor	• Little effect in the case of high power factor, that is, resistive load	• Suitable for applying in radial lines
Shunt capacitor	• Increases voltage • Reduces line losses • Increases the power factor	• Little effect in the case of high power factor, that is, resistive load • No reduction in voltage variations • No self-regulating	• Somewhat cheaper, but it is not useful when the voltage drop is higher
New parallel lines	• Increase availability with parallel line • Increases voltage • Reduces voltage variations • Reduces line losses	• Long installation period • Difficulties of obtaining right of way • Reconstruction of the feeder and receiver stations	• Considerably more expensive
Voltage regulator	• Slow voltage regulation	• Higher losses • Does not improve the power factor • More maintenance	• More expensive
Transformer with LTC	• Slow voltage regulation	• Does not improve the power factor • More maintenance	• More expensive

$$V_R = I_C X_L. \tag{8.7}$$

A comparison between the different methods used to reduce the voltage drop at the load points of a radial distribution feeder is shown in Table 8.4. It shows the advantages, disadvantages, and the cost of each method.

Of course, for solving a specific problem, comparative study between these methods must be done to choose the best one technically and economically.

8.3 VOLTAGE SAG CALCULATIONS

As mentioned in Section 8.1.2, three main key elements necessary to be calculated to characterize the voltage sag are the magnitude, the duration, and the amount of higher frequency components due to the overshoots occurring immediately after the sag. The most common reason that causes the sag rather than any other reason is the short circuit. So, it is imperative to give some explanation about voltage sag calculations due to the occurrence of a short circuit.

8.3.1 Sampling Rate

The voltage sag due to a short circuit has a short duration time, about two cycles. Then its value is restored to the prefault conditions. One cycle is considered as a window to record the instantaneous value of the voltage. The wave shape of the voltage within the window is in the form of pulses of a certain number of recorded samples "N." The voltage sag, V_{sag}, is the value of the voltage during the sag while the sag itself means the amount of voltage decrease.

8.3.2 Magnitude of Voltage Sag

The magnitude of voltage sag is expressed by either rms value or peak value or fundamental voltage component.

Calculation by rms Value: The rms value is defined as

$$V_{rms} = \sqrt{\frac{1}{N}\sum_{i=1}^{N} v_i^2}, \tag{8.8}$$

where N is the number of samples per window (one cycle) and v_i is the recorded value at each time interval (for each sample).

In case of voltage sag, its magnitude at time interval "k" (i.e., the kth sample) is expressed by the rms of all preceding points within the window.

Therefore, if the number of samples = N per window, then, the rms value of the voltage sag at the kth sample is given as

$$V_{\text{rms}} = \sqrt{\frac{1}{N} \sum_{i=k-(N-1)}^{k} v_i^2}. \tag{8.9}$$

The same procedure is applied to any chosen window, for example, the window of samples may be taken as half-cycle.

If the calculation is made once a cycle, the rms value is given as

$$V_{\text{rms}}(kN) = \sqrt{\frac{1}{N} \sum_{i=kN-(N-1)}^{kN} v_i^2}, \tag{8.10}$$

where k is the cycle order.

Calculation by Peak Value: The magnitude of voltage sag is expressed by the peak value. At the kth sample, the calculated peak is the maximum value recorded over all the preceding samples in the chosen window (one cycle or half-cycle). This peak value is given by the following expression:

$$V_{\text{peak}} = \max |v(t-\tau)| \text{ and } 0 < \tau < T, \tag{8.11}$$

where T is the time length of the window.

Calculation by Fundamental Component: Applying Fourier theorem, the fundamental component of the voltage sag is expressed as

$$V_{\text{fund}}(t) = \frac{2}{T} \int_{t-T}^{t} v(\tau) e^{j\omega_o \tau} d\tau, \tag{8.12}$$

where $\omega_o = \dfrac{2\pi}{T}$ and T is one cycle of the fundamental frequency.

This relation is a complex relation, that is, it gives the voltage as a vector (magnitude and direction). Therefore, the advantages of this method of calculations are (1) the change of voltage phase angle is calculated during the sag period and (2) the harmonic spectrum can be drawn in addition to calculating the phase angle of each harmonic order as well.

8.3.3 Duration of Voltage Sag

As it is mentioned above, the voltage sag is caused mostly by the occurrence of a short circuit in the system. At points close to the fault, the sag is deeper and vice versa. This illustrates the effect of fault on the voltage sag as a magnitude. But from the point of view of protection engineers, it is important to know the duration of this sag because it affects the system performance.

Of course, when a short circuit occurs, the protective relays will operate leading to circuit breakers tripping to isolate the faulty part from the rest of the network. Therefore, the duration of voltage sag is mainly determined by the fault-clearing time but not exactly because:

1. The protection equipment has different speed of tripping at different voltage levels of the system. It is faster to clear the fault in transmission and subtransmission systems (60–150 ms) than that used in distribution system (MV: 0.5–2 s, LV: depending on protection performance) where the protection against overcurrent is the main form. It requires some time grading, which increases the fault clearing time. An exception is systems using vacuum reclosers or current-limiting fuses where they are able to clear the fault within one cycle or less [128].
2. When the fault is cleared, the voltage sag does not reach the prefault value immediately.

This means that the duration of sag may be longer than the fault-clearing time if the fault occurs at lower-voltage level.

On the other hand, when the sag source is other than a short circuit, the duration is governed by that source. For instance, some loads such as electric motors draw a large inrush current as the voltage recovers at the end of a disturbance. This results in extending the duration of voltage sag.

Also, the coincidence of the instant at which the voltage is recorded and calculated with the instant at which the sag starts cannot be emphasized. Consequently, the window duration and the method of voltage sag calculation (once a cycle, twice a cycle, etc.) together with the proposed threshold will affect the sag duration measurement.

8.3.4 Voltage Sag Phase-Angle Changes

As shown in Equation 8.12, the voltage sag is a complex quantity, that is, it is defined by a magnitude and a phase angle. This phase angle is not constant but it may vary depending on the type of fault, the fault location, and the network configuration. For each combination of these three factors, a specific network with specific values of its elements is configured. Accordingly, the voltage at any point in the network (magnitude and phase angle) can be calculated.

Most of the applications are not concerned with the phase-angle changes of the voltage sag. They concern only with the magnitude of voltage sag except the application of power electronic equipment, in particular, the equipment whose operation is based on the phase-angle information for firing instants, for example, power electronic converters.

8.3.5 Illustrative Example

The voltage divider model shown in Figure 8.23 is used to illustrate the properties of voltage sag and the parameters affecting its magnitude in radial systems.

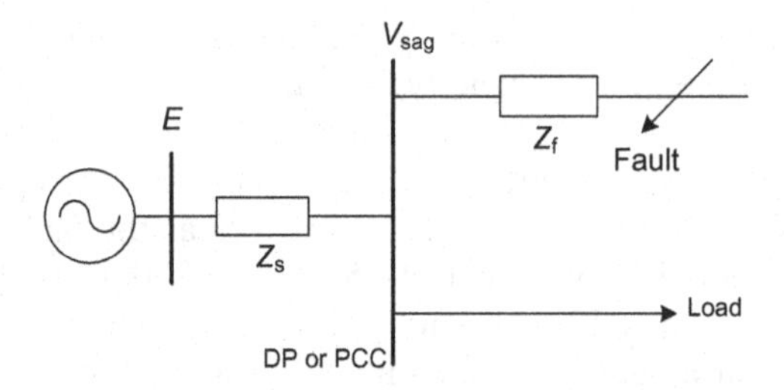

Figure 8.23 Voltage divider model.

Assume a source is feeding a point-of-common coupling (PCC) as a distribution point (DP), which in turn distributes the power to two equal loads through one outgoing feeder for each. Z_s is the source impedance at the DP and Z_f is the impedance between the DP and the fault. A temporary three-phase short circuit occurs at one of the two loads as shown in the single-line diagram (Fig. 8.23). The source voltage E is assumed to be 20 kV at a short-circuit level of 750 kVA and the values of impedances are

$$Z_s = 2\pi f \times 0.005\ \Omega$$

and

$$Z_f = 0.29 + j2\pi f \times 0.00251\ \Omega.$$

The voltage at DP is 6 kV under normal conditions.

The fault duration is 0.2 s (from 0.35 to 0.55 s)

The current before and during the fault is neglected, that is, there is no voltage drop between the load and the DP. Thus, the voltage sag magnitude at DP can be calculated by

$$V_{sag} = \frac{Z_f}{Z_s + Z_f} E. \tag{8.13}$$

An automatic circuit breaker opened the circuit at 0.5 s in the presence of the fault, and reclosed at 0.60 s while the fault had been removed at 0.55 s. The system was simulated and solved by using PSCAD software V.4.2. One cycle window with 128 recorded samples is used. The voltage waveform at the DP in the period from 0.1 to 0.9 s is shown in Figure 8.24. The general shape of the voltage waveforms indicates that before and after the fault, the voltage is at its steady-state value. When the circuit breaker opens the circuit, fault transient voltage appears due to the switching process until the reclosing of the circuit at fault removal, from 0.5 to 0.6 s. The rms values and peak values of voltage are shown in Figure 8.25.

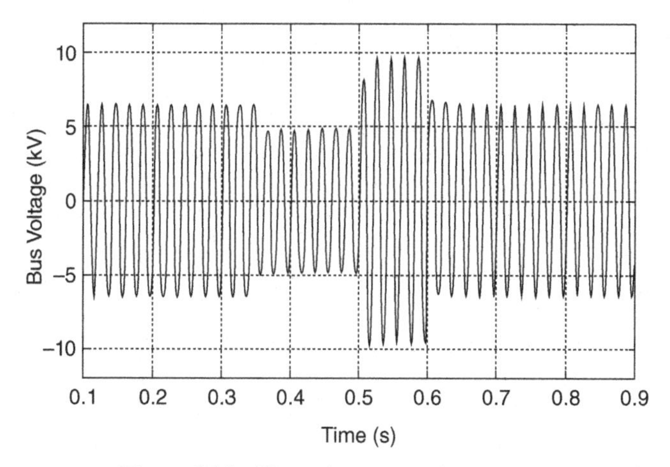

Figure 8.24 The voltage waveform at DP.

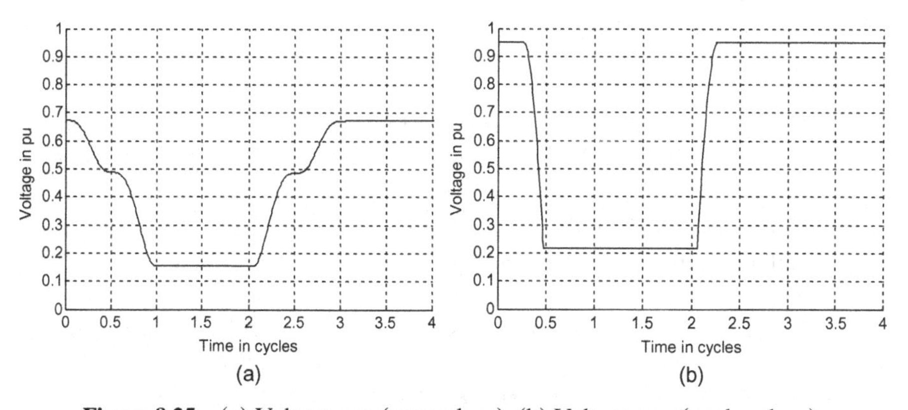

Figure 8.25 (a) Voltage sag (rms values). (b) Voltage sag (peak values).

In these studies, the parameters such as window length and system impedances (based on Eq. 8.13) have a significant effect on the voltage sag calculations as discussed below.

Window Length: The rms voltage values shown in Figure 8.25a are obtained by using a window of one-cycle length with 128 recorded samples. It is seen that the voltage does not immediately drop to a lower value but it takes one cycle for the transition. When reducing the window length to one-half cycle with 64 recorded samples, the voltage transition takes place in one-half cycle (Fig. 8.26a). Further reduction of window length is not useful since an oscillation in the results with a higher frequency (equal or twice the fundamental frequency) is produced as shown in Figure 8.26b.

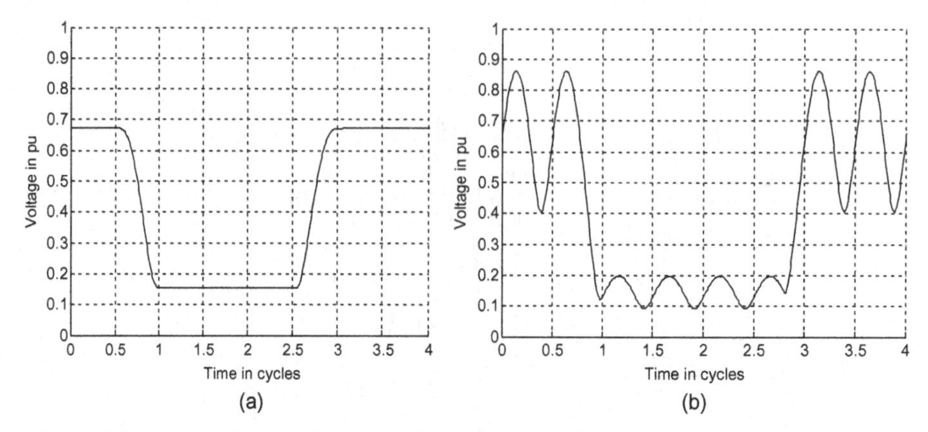

Figure 8.26 rms voltage values versus time in cycles. (a) At window length = one-half cycle. (b) At window length = one-quarter cycle.

System Impedances: Both Z_s and Z_f are affected by several parameters, which in turn affect the magnitude of voltage sag. These parameters are as follows:

1. *The short-circuit level at DP:* It is used to calculate the source impedance Z_s at the DP. Thus, with increasing the short-circuit level, Z_s is decreased and the sag becomes less severe, that is, the voltage sag is increased.
2. *The distance to the fault:* As the fault gets farther from the considered point, the value of Z_f increases and the voltage sag increases as well.
3. *The CSA of load feeders:* The impedance Z_f decreases with increasing the CSA of feeders (cables or OHTLs), and in turn the voltage sag, as defined in Section 8.3.1, is decreased, that is, the sag becomes more severe.
4. *Presence of step-down transformers between the DP and the fault:* The equipment at which the voltage sag is quantified usually operates at voltage lower than that at the DP where a transformer is connected. In this case, the voltage sag is higher and the sag is relatively small since the transformer has a large impedance to limit the fault level on the LV side.

Based on the divider model shown in Figure 8.23, the above analysis is concerned with three-phase faults and the impedances in this model are positive-sequence impedances. In practice, the faults may be single-phase or two-phase faults. In that case, the three-phases must be considered to characterize the sag in all three phases simultaneously. This can be done using symmetrical components as developed in References 125 and 129. The development includes a description of voltage sag characteristics that can be experienced at the end-user terminals when different faults occur at different locations. In addition, the presence of transformers between the DP and the equipment as mentioned above necessitates considering the effect of

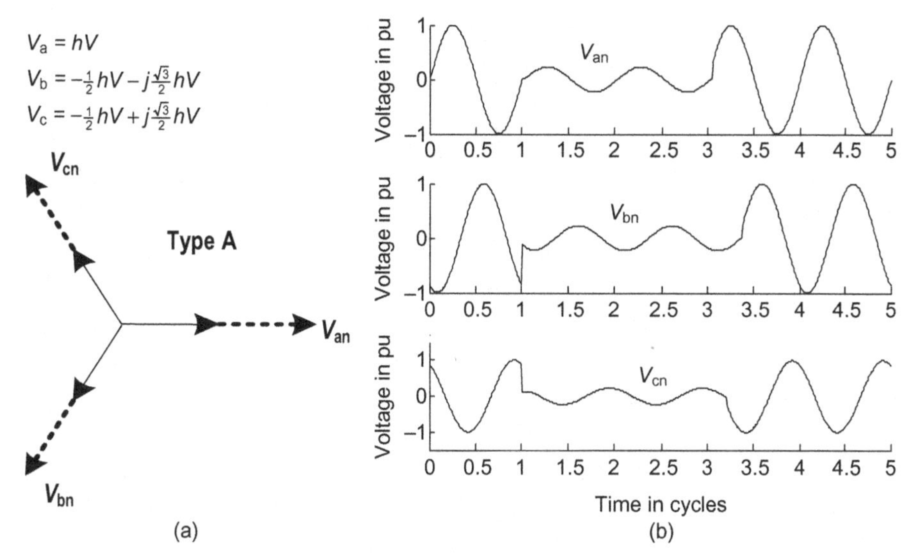

Figure 8.27 Sag type A. (a) Voltage phasor diagram; (b) voltage waveform. (------), Voltage before the fault; (——) voltage after the fault.

transformer and load connections. The sags are divided into seven types, A to G; type A refers to three-phase sags; types B, C, and D refer to single-phase and phase-to-phase faults; and three additional types E, F, and G correspond to two-phase-to-ground faults. This classification is useful for considering the voltage characteristics at end-user's location in case of fault occurrence. The sag types are illustrated by the phasor diagram, waveform, and phase voltage equations as below. The equations are given in terms of h (the sag magnitude, $0 \le h \le 1$) and V (phase voltage rms value). The waveforms are produced by applying each fault type to the divider model.

Sag type A is balanced sag due to a three-phase fault. The phasor diagram and voltage waveform are shown in Figure 8.27. The sag magnitudes and phase shift of the three-phase voltages are equal and given as

$$V_a = hV,$$
$$V_b = -\frac{1}{2}hV - j\frac{\sqrt{3}}{2}hV,$$
$$V_c = -\frac{1}{2}hV + j\frac{\sqrt{3}}{2}hV.$$

Sag type B is the sag due to a single-phase-to-ground fault and the equipment windings are wye connected. The phasor diagram and voltage waveform are shown in Figure 8.28. The phase voltage equations are

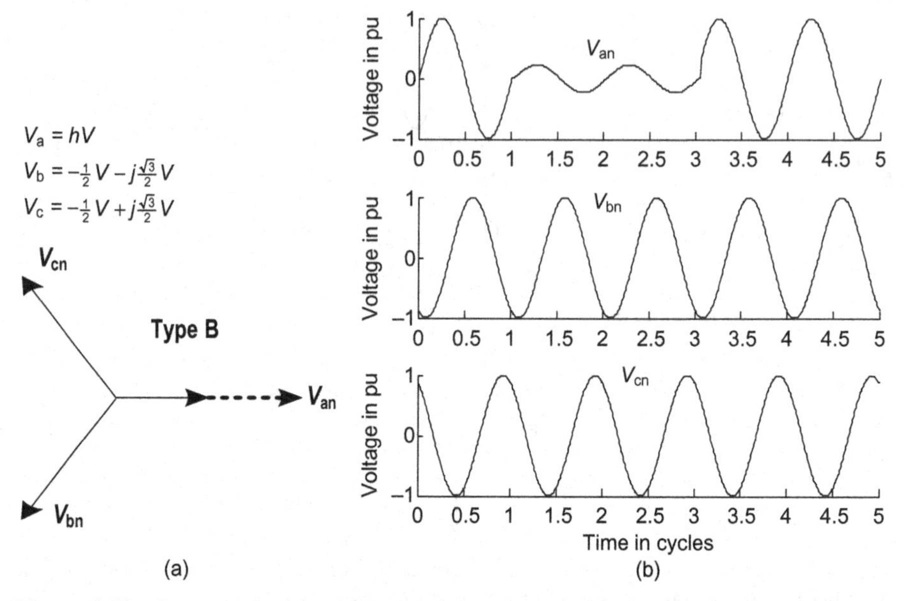

Figure 8.28 Sag type B. (a) Voltage phasor diagram; (b) voltage waveform. (------), Voltage before the fault; (——) voltage during the fault.

$$V_a = hV,$$
$$V_b = -\frac{1}{2}V - j\frac{\sqrt{3}}{2}V,$$
$$V_c = -\frac{1}{2}V + j\frac{\sqrt{3}}{2}V.$$

Sag type C is due to a single-phase fault and the equipment windings are delta connected, or due to a phase-to-phase fault behind a Δ-Y transformer. The phasor diagram and voltage waveform are shown in Figure 8.29. The phase voltage equations are

$$V_a = V,$$
$$V_b = -\frac{1}{2}V - j\frac{\sqrt{3}}{2}hV,$$
$$V_c = -\frac{1}{2}V + j\frac{\sqrt{3}}{2}hV.$$

Sag type D is due to a phase-to-phase fault, or due to a single-phase fault behind a Δ-y transformer, or a phase-to-phase fault behind two Δ-Y transform-

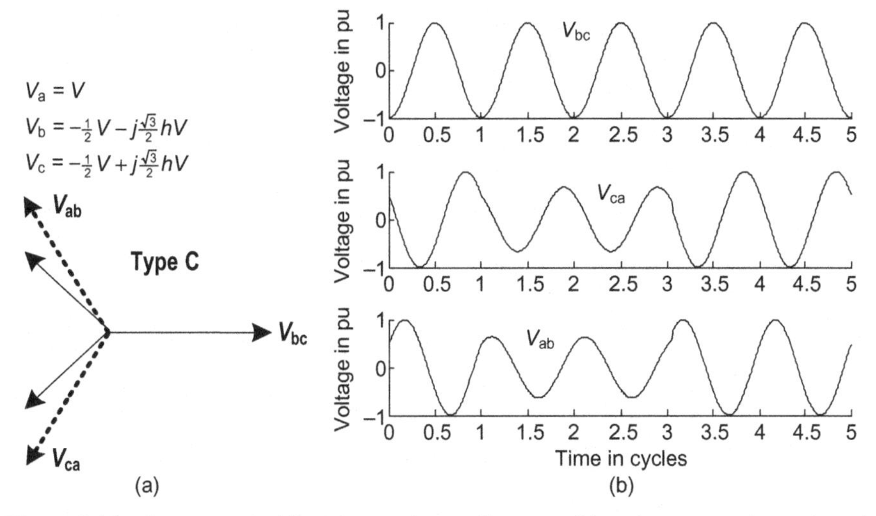

Figure 8.29 Sag type C. (a) Voltage phasor diagram; (b) voltage waveform. (------), Voltage before the fault; (——) voltage during the fault.

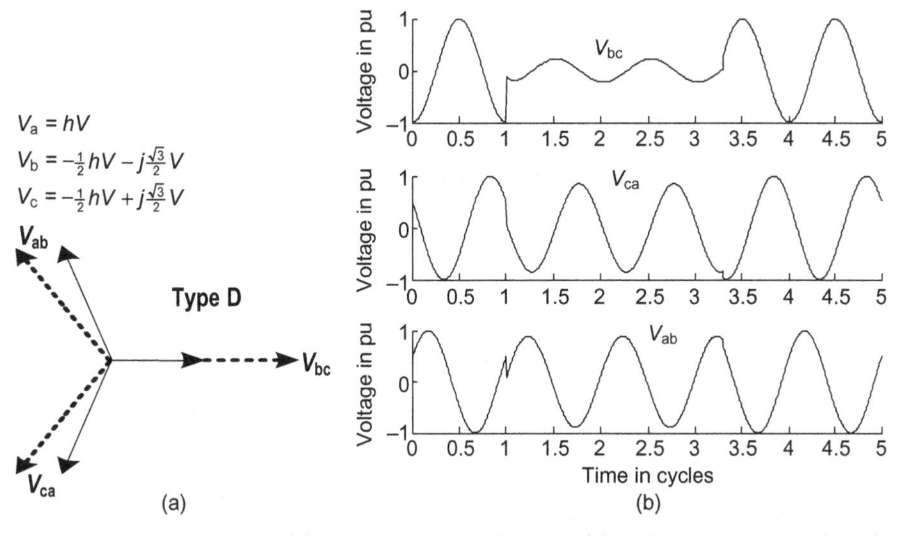

Figure 8.30 Sag type D. (a) Voltage phasor diagram; (b) voltage waveform. (------), Voltage before the fault; (——) voltage during the fault.

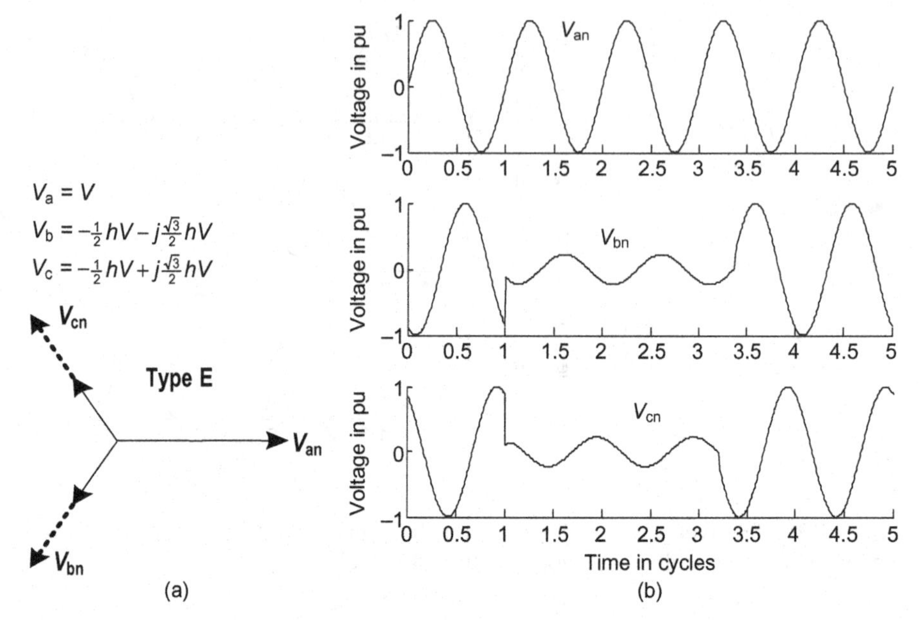

Figure 8.31 Sag type E. (a) Voltage phasor diagram; (b) voltage waveform. (------), Voltage before the fault; (——) voltage during the fault.

ers. The phasor diagram and voltage waveform are shown in Figure 8.30. The phase voltage equations are

$$V_a = hV,$$
$$V_b = -\frac{1}{2}hV - j\frac{\sqrt{3}}{2}V,$$
$$V_c = -\frac{1}{2}hV + j\frac{\sqrt{3}}{2}V.$$

Sag type E is due to a two-phase-to-ground fault and the equipment windings are wye connected. The phasor diagram and voltage waveform are shown in Figure 8.31. The phase voltage equations are

$$V_a = V,$$
$$V_b = -\frac{1}{2}hV - j\frac{\sqrt{3}}{2}hV,$$
$$V_c = -\frac{1}{2}hV + j\frac{\sqrt{3}}{2}hV.$$

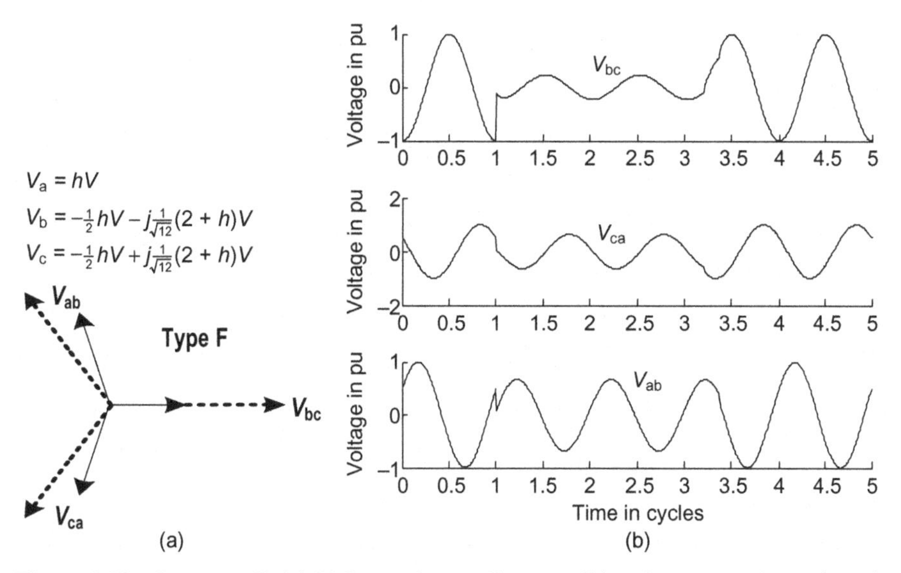

Figure 8.32 Sag type F. (a) Voltage phasor diagram; (b) voltage waveform. (------), Voltage before the fault; (——) voltage during the fault.

Sag type F is due to a two-phase-to-ground fault and the windings of the equipment are delta connected. The phasor diagram and voltage waveform are shown in Figure 8.32. The phase voltage equations are

$$V_a = hV,$$
$$V_b = -\frac{1}{2}hV - j\frac{1}{\sqrt{12}}(2+h)V,$$
$$V_c = -\frac{1}{2}hV + j\frac{1}{\sqrt{12}}(2+h)V.$$

Sag type G is due to a two-phase-to-ground fault and the windings of the equipment are wye connected. The phasor diagram and voltage waveform are shown in Figure 8.33. The phase voltage equations are

$$V_a = \frac{1}{3}(2+h)V,$$
$$V_b = -\frac{1}{6}(2+h)V - j\frac{\sqrt{3}}{2}hV,$$
$$V_c = -\frac{1}{6}(2+h)V + j\frac{\sqrt{3}}{2}hV.$$

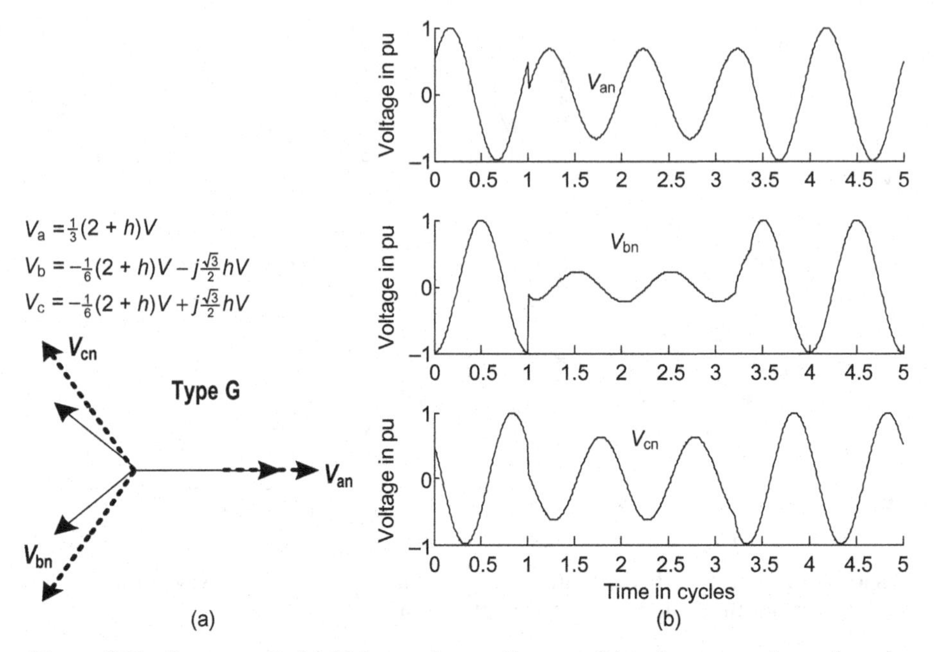

Figure 8.33 Sag type G. (a) Voltage phasor diagram; (b) voltage waveform. (------), Voltage before the fault; (——) voltage during the fault.

8.4 ESTIMATION OF DISTRIBUTION LOSSES

As the power demand increases, the voltage drop and losses are increased since the voltage drop is proportional to the demand current (as explained in Section 8.1.1) and losses are proportional to the square of demand current. Various types of distribution losses have been defined in Chapter 1, Section 1.4.5. The estimation of these losses is presented below.

It is important to estimate these losses as accurately as possible to achieve good planning and high system performance. Utilities need to develop loss estimation methods that are reliable enough to determine the specific actions for predetermined levels of losses. The nonavailability of comprehensive data of distribution system segments, such as substations, distribution transformers, MV feeders, and LV network, are some reasons for the difficulty in loss computation.

Total distribution losses D_L are the sum of technical losses D_{LT} and nontechnical losses D_{NLT}, that is,

$$D_L = D_{LT} + D_{NLT}. \tag{8.14}$$

D_L is calculated as the difference between purchased energy and sold energy. Therefore, once the technical losses are known, nontechnical losses are easy to compute as in Equation 8.14. Methods for technical variable losses are broadly classified as below [129, 130].

Top-Down Approaches: They are often referred to as benchmarking. The losses are computed by estimation from the zone substation to the consumers. These approaches are useful when the system data are insufficient. Losses are globally estimated by comparison with a similar system.

Bottom-Up Approaches: The computation can be done from the consumers to the zone substations. It necessitates the availability of a set of data, which includes consumers and feeders load curves, data of both MV and LV networks, data of transformers, and energy meters. Thus, the computation can be done with more accuracy by means of power flow simulations for different load levels using specialized software.

Hybrid Top-Down/Bottom-Up Approaches: These approaches can be used in case of providing detailed data of a part of the network and partial data of other parts.

8.4.1 A Top-Down Approach

Distribution technical losses can be estimated based on the calculation of loss factor (L_sF) [129]. Loss factor is the ratio of the average power loss (L_{av}) to the losses during peak load (L_{max}), that is,

$$L_sF = \frac{L_{av}}{L_{max}} = \frac{1}{L_{max}} \frac{\int_0^T L(t)dt}{T}, \tag{8.15}$$

where $\int_0^T L(t)dt$ is the instantaneous demand losses that represent the energy technical losses of the system (e) during a time period T. Thus, L_sF can be rewritten as

$$L_sF = \frac{e}{L_{max}T}. \tag{8.16}$$

As in Equation 8.16, the energy technical losses (e) can be determined if L_sF and L_{max} are known:

L_sF Estimation: It can be estimated by using either the load curve or a relation in terms of the load factor (LF) and a constant coefficient (k).

- If the load curve is available and applying the relation that gives the losses at time t, $L(t)$, in terms of the demand, $D(t)$ [99],

$$L(t) \cong C|D(t)|^2, \tag{8.17}$$

where C is a constant. The loss factor can be obtained by substituting Equation 8.17 into Equation 8.15

$$L_sF = \frac{1}{C|D_{\max}|^2}\frac{C\int_0^T |D(t)|^2\,dt}{T}. \tag{8.18}$$

The measurements are carried out incrementally every Δt (e.g., 15 min) and, therefore, Equation 8.18 can be written in a summation form rather than integration as

$$L_sF = \frac{\sum_{t=1}^{T} |D(t)|^2}{|D_{\max}|^2 T}, \tag{8.19}$$

where $D(t)$ and $D_{\max}$ (maximum demand) are measured values and can be determined from the load curve.

- In terms of LF and a constant coefficient k, the loss factor can be obtained after adopting a value of k by applying the relation

$$L_sF = k(LF) + (1-k)(LF)^2. \tag{8.20}$$

k is suggested to take a value of 0.3 [99].

The LF is the ratio of the average demand, D_{av}, to the maximum demand, $D_{\max}$, for a time period T,

$$LF = \frac{D_{av}}{D_{\max}} = \frac{1}{D_{\max}}\frac{\int_0^T D(t)dt}{T}, \tag{8.21}$$

where $\int_0^T D(t)dt$ is the instantaneous demand that represents the energy supplied to the system (E) during a period of time T. Thus, LF can be written as

$$LF = \frac{E}{D_{\max}T} \tag{8.22}$$

where both E and $D_{\max}$ can be obtained by direct measurements at the zone substations.

Determination of Losses During Peak Load ($L_{\max}$): One of the ways of determining $L_{\max}$ is to allocate the loads for each transformer by calculating the allocation factor (AF) [131]. AF is the ratio between the maximum demand measured at the substation, D_{max}, and the total capacity of all transformers, kVA_{total}:

$$\mathrm{AF} = \frac{D_{\max}}{kVA_{\text{total}}}. \tag{8.23}$$

Thus, the allocated load for each transformer (Ld_{TR}) can be obtained by

$$Ld_{\text{TR}} = \mathrm{AF}(kVA_{\text{TR}}), \tag{8.24}$$

where kVA_{TR} is the individual kVA rating of each transformer.

Consequently, the maximum loss calculation for each segment in the distribution system can be performed as below:

- Maximum losses for MV networks, $L_{\max,\text{MV}}$, can be obtained by applying one of the power flow techniques where the loads are allocated to each transformer in the network. Then, the energy losses in the MV network, e_{MV}, can be calculated for a period of time T by

$$e_{\text{MV}} = L_s F(L_{\max,\text{MV}})T. \tag{8.25}$$

- Maximum losses for transformers can be provided by the sum of core losses (L_{core}) and winding losses (L_{wdg}). L_{core} depends on the voltage applied to the transformer coils, while L_{wdg} depends on the load supplied by the transformer. Thus, the maximum losses for the transformers ($L_{\max,\text{TR}}$) and the energy losses (e_{TR}) can be calculated [131]:

$$L_{\max,\text{TR}} = L_{\text{core}} + (\mathrm{AF})^2 L_{\text{wdg}}, \tag{8.26}$$

$$e_{\text{TR}} = [L_{\text{core}} + L_s F(\mathrm{AF})^2 L_{\text{wdg}}]T. \tag{8.27}$$

- Maximum losses for LV networks, $L_{\max,\text{LV}}$: In LV networks, there are numerous number of circuits that may include meters, capacitor banks, and voltage regulators. To compute $L_{\max,\text{LV}}$, much comprehensive data and computational efforts are required. Accordingly, the energy losses, e_{LV}, will be

$$e_{\text{LV}} = L_s F(L_{\max,\text{LV}})T. \tag{8.28}$$

Therefore, the total energy loss, e_{total} is given as

$$e_{\text{total}} = e_{\text{MV}} + e_{\text{TR}} + e_{\text{LV}}. \tag{8.29}$$

Within the three categories of technical losses estimation, several techniques based on load flow calculations have been applied to calculate the different types of losses [132–135], to model the distribution feeders [136, 137], and to deduce approximated loss formulas [138, 139]. Artificial intelligent techniques such as fuzzy C-number and fuzzy regression algorithms have been applied to compute loss formulas as well [140].

CHAPTER 9

POWER FACTOR IMPROVEMENT

The application of shunt capacitors with the goal of reducing the voltage drop has been presented in Chapter 8. Another application of shunt capacitors as compensators to improve the power factor (PF) is given in this chapter. The basic material pertaining to this application is introduced in the forthcoming sections.

9.1 BACKGROUND

For a specific load, the PF is mainly affected by the reactive power. All alternating current (AC) electric machines (motors, transformers) have two types of power: active and reactive power.

The active power, P kW, of the load is entirely transformed into mechanical power (work) and heat (losses). The reactive power, Q kvar, is used to magnetize the circuits of the electric machines (transformers and motors). The apparent power, kVA, is the vectorial sum of these two powers and is the power supplied by the network. It is the apparent (or available) power S (kVA) of the loads and is the vectorial sum of P (kW) and Q (kvar).

Active and Reactive Current Components: A current component is associated with both the active and reactive powers. Current corresponding to the active power, I_a, is in phase with the network voltage and is transformed into mechanical power or heat. The current corresponding to the reactive power, I_r (or

Electric Distribution Systems, First Edition. Abdelhay A. Sallam, Om P. Malik.

Published 2011 by John Wiley & Sons, Inc.

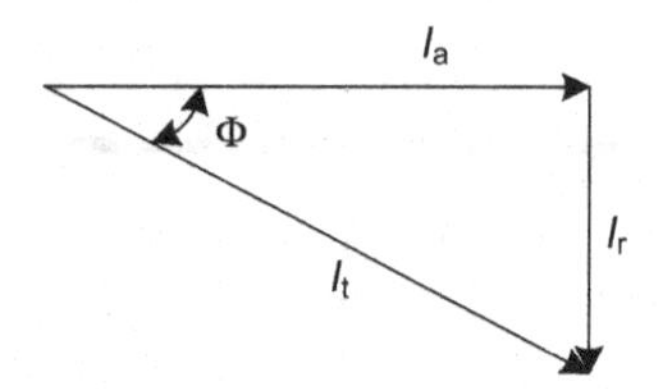

Figure 9.1 Current components.

magnetizing current), has a 90° phase shift with the active current, either backward (inductive load) or forward (capacitive load) and is necessary for magnetic excitation. The total current, I_t, is the phasor sum of I_a and I_r, and is the resulting current flowing through the electric line from the source to the load. This current is shifted by an angle φ compared with the active current (and voltage). For networks through which only a power frequency current (50 or 60 Hz) flows, that is, with no harmonics, these currents are vectorially composed of root mean square (rms) values as shown in Figure 9.1:

I_t = total current flowing in the network,
I_a = active current,
I_r = reactive current,
$I_t = \sqrt{I_a^2 + I_r^2}$,
$I_a = I_t \cos\varphi$, and
$I_r = I_t \sin\varphi$.

The $\cos\varphi$ in this case is known as the network PF.

Active and Reactive Power Components: The current diagram in Figure 9.1 also applies to powers (just multiply each current by the voltage of the network), which can be defined as

- active power $P = VI\cos\varphi$ kW, really used by the loads;
- reactive power $Q = VI\sin\varphi$ kvar, required for magnetic circuit excitation; and
- apparent power $S = VI$ kVA, supplied by the network to the loads. It is the vectorial sum of the above two powers.

The result is the vectorial graph shown in Figure 9.2:

S = apparent power,
P = active power,
Q = reactive power,
$S = \sqrt{P^2 + Q^2}$
$= VI_t$,

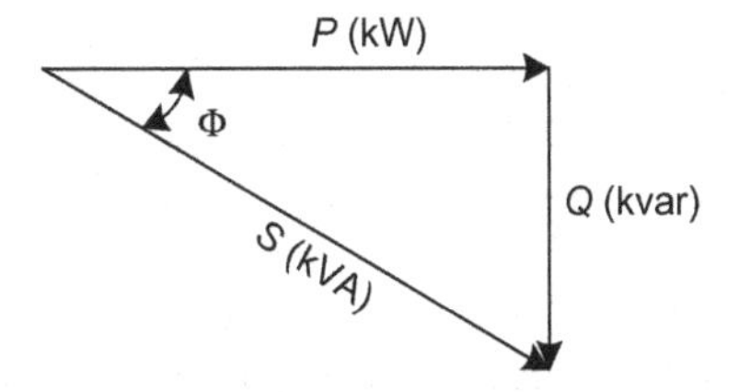

Figure 9.2 Power triangle.

$P = VI_t \cos\varphi = VI_a$,
$Q = VI_t \sin\varphi = VI_r$, and
$\cos\varphi = P/S = \text{PF}$.

General Definition: The PF is normally defined by the ratio

$$\text{PF} = P/S = \text{active power, kW/apparent power, kVA.}$$

It expresses the ratio between the active power, which can really be used for applications (mechanical, thermal) and the apparent power, VA, supplied by the network; that is, it measures the electric efficiency of the installation. It is thus an advantage for both electricity boards and users to have installations operating with the highest possible PF, that is, around the value of 1.

PF and cos φ: In the most common case when only a power frequency current (50 or 60 Hz) flows through the network, with no harmonics, this ratio equals $\cos\varphi$ (phase shift between current and voltage) of the installation:

$$\text{PF} = P/S = \cos\varphi.$$

This is only true for the fundamental. A general definition, which allows for the overall harmonic effect, is used when harmonics are present.

The PF is measured either using a $\cos\varphi$ meter, which gives an instantaneous measurement, or using a var-metric recorder to obtain the current, voltage, and average PF values for a set period of time (day, week, etc.).

PF and tan φ: The $\tan\varphi$ is often used instead of the $\cos\varphi$. In this case, the following ratio is calculated:

$$\tan\varphi = Q/P = \text{reactive power (kvar)/active power (kW).}$$

For a given period of time, this ratio is also a consumption one,

$$\tan\varphi = Q/P = W_r/W_a,$$

where

W_r = reactive power consumption (kvarh) and
W_a = active power consumption (kWh).

Reactive PF: It has a major technical impact on the choice of equipment and the operation of the networks. It, therefore, also has an economic impact.

For the same active power P, Figure 9.3 shows that the lower the $\cos\varphi$, the more apparent power, $S_2 > S_1$, has to be supplied, that is, the angle φ is large.

Yet again for the same active current I_a (at constant network voltage V) (Fig. 9.4), the lower the $\cos\varphi$ (angle φ is large), the more current, $I_2 > I_1$, has to be supplied.

Consequently, due to excessive available current, reactive power flow on distribution networks results in the following:

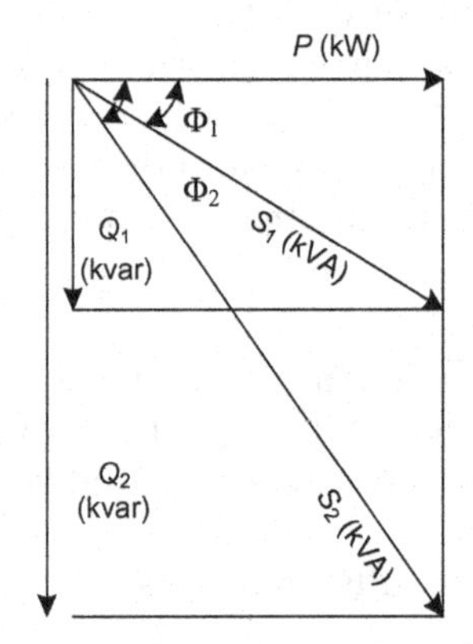

Figure 9.3 Apparent power at φ_1 and φ_2.

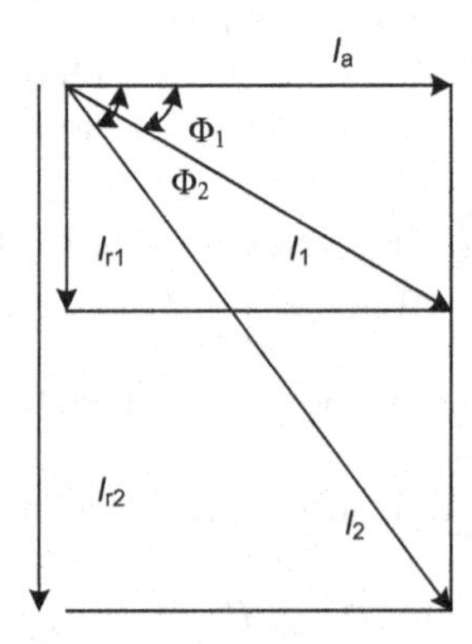

Figure 9.4 Apparent current at φ_1 and φ_2.

- transformer overloads,
- voltage drops at the end of the line,
- temperature rise in supply cables and thus in active power losses,
- oversizing of harmonic protection devices, and
- temperature rise in supply cables and thus in active power losses.

These four basic reasons mean that reactive power must be produced as close as possible to the motors and transformers to prevent it from being available on the network. Capacitors supplying inductive loads with reactive power are, therefore, used.

To encourage this and avoid oversizing of networks and risking voltage drops, utilities charge reactive power consumers heavily if they exceed certain thresholds.

Reactive Power and Network Components: An electric network is characterized by resistance (R), inductance (L), and capacitance (C):

- resistance R, which transforms electric power directly and irreversibly into heat;
- inductance L, which produces magnetic field; and
- capacitance C, which, most often, produces reactive power.

These elements are found in the network components with the following consequences on the reactive power.

Synchronous Machines: These machines are used as generators (of active power) when they transform mechanical energy into electric power; otherwise, they are used as motors. These machines can produce reactive power by varying the excitation.

In some cases, the machine does not supply any active power. It is then used as a synchronous condenser. Synchronous motors are rarely used as modulating devices.

Asynchronous Machines: They are different from the machines described above in that they draw from the network the reactive power they need. This power can be considerable: 25–35% of the active power at full load, much more at partial load.

Asynchronous motors, being the most common, are the main consumers of reactive power produced by industrial networks.

Lines and Cables: The L and C characteristics of overhead lines and cables are such that they consume reactive power at full load.

Transformers: Transformers consume reactive power that corresponds to approximately 5–10% of the apparent power, which flows through them.

Inductance: Inductances consume primarily reactive power such as current limitation inductances and stabilization inductances (for arc furnaces, fluorescent lamps).

TABLE 9.1 Power Factor for Common Devices

Devices	cos Φ	tan Φ
Standard asynchronous motor loaded at		
0%	0.17	5.8
25%	0.55	1.52
50%	0.73	0.94
75%	0.80	0.75
100%	0.81	0.62
Incandescent lamp	≈1	≈0
Noncompensated fluorescent lamps	≈0.5	≈1.73
Compensated fluorescent lamps (0.93)	0.93	0.39
Discharge lamps	0.4–0.6	2.29–1.33
Resistance furnaces	≈1	≈0
Induction furnaces with built-in compensation	≈0.85	≈0.62
Dielectric heating furnaces	≈0.85	≈0.62
Resistance welding machines	0.8–0.9	0.75–0.48
Arc welding single-phase static stations	≈0.5	1.73
Arc welding rotary sets	0.7–0.9	1.02–0.48
Arc welding transformers–rectifiers	0.7–0.8	1.02–0.75
Arc furnaces	0.8	0.75

Capacitors: By definition, capacitors generate reactive power at very high rate, which is why they are used for this application.

9.2 SHUNT COMPENSATION

Capacitors are used to produce the necessary reactive power locally and to prevent it from being available on the network. The result is an improvement in the PF of the installations; values of PF for a few common devices are given in Table 9.1. Implementation of this technique, known as compensation, is explained in the next section.

9.3 NEED FOR SHUNT COMPENSATION

Improving the PF of an installation (i.e., compensation) has many economic and technical advantages.

Economic Advantages: These advantages are linked to the impact of compensation on the installation and allow evaluation of compensation payback time. These advantages are the following:

- elimination of bills for excess consumption of reactive power;
- reduction of the subscribed power in kVA; and
- reduction of the active power consumption in kWh.

An example in the next section illustrates the economic advantage of compensation.

Technical Advantages: These advantages, resulting from the reduction in reactive current, are the following:

- A reduction in voltage drop: Loads with low PF supplied by overloaded medium-voltage (MV) distribution lines often cause voltage drops. These are prejudicial to correct operation of the loads even if the voltage at the supply end of the system is correct.
- The presence of a capacitor bank at line end will ensure reduction in voltage drop. Relative maintenance of voltage at line end is defined by the following formula:

$$\Delta V\% = \frac{X_{Lx}Q_C}{10^3 \times V^2}, \tag{9.1}$$

where

X_L = line inductance in Ω,
Q_C = reactive power of capacitor bank in kvar, and
V = rated voltage of capacitors in kV.

- A reduction of line losses at constant active power. Watt losses (due to resistance of conductors) are capitalized in the power consumption recorded by active power meters (kWh). They are proportional to the square of the current transmitted and fall as the PF increases. The graph in Figure 9.5 shows the percentage reduction of line losses at constant active power with improvement in PF. As an example, if before compensation $\cos\varphi_1 = 0.8$ and after compensation $\cos\varphi_2 = 0.9$, line losses are reduced by 26% (with constant active power).
- An increase of the active power available at the transformer secondary. The installation of means of compensation at the downstream terminals of an overloaded transformer allows the creation of a power reserve, which can be used in a possible extension of the plant. In this case, the transformer does not have to be changed and the corresponding investment can be postponed.
- An increase of the active power transmitted by lines with the same losses. An increase in activity often makes it necessary to transmit a larger active power in order to satisfy the power requirements of the loads. Installation of a bank of capacitors when the electric system is relatively weak ($\cos\varphi$ between 0.5 and 0.85) will permit transmission without modification to the existing power lines. The graph in Figure 9.6 shows the percentage increase in power transmitted with the same active losses as a function of improvement in PF. As an example, if before compensation, $\cos\varphi_1 = 0.7$

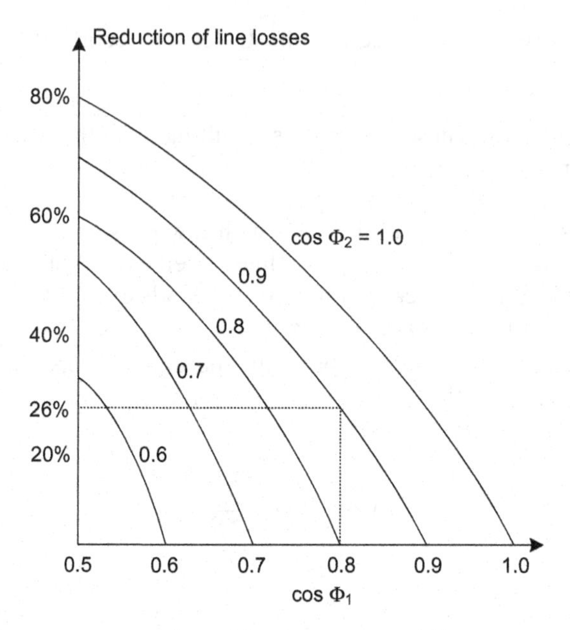

Figure 9.5 Percentage reduction of line losses versus PF at constant active power.

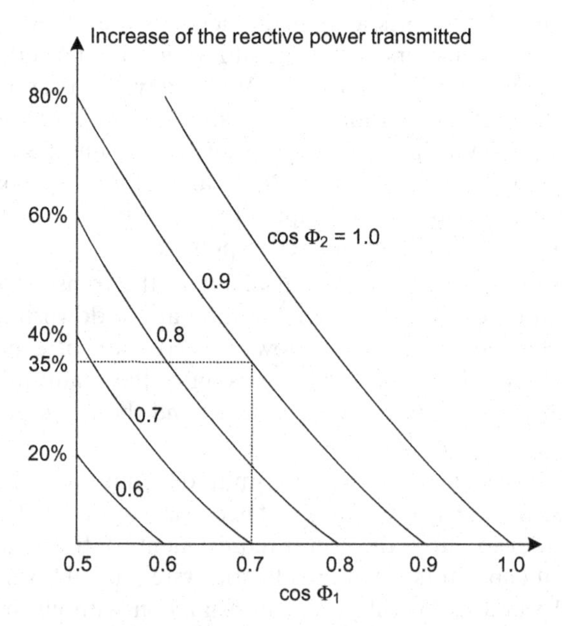

Figure 9.6 Percentage increase in power transmitted versus PF at constant active losses.

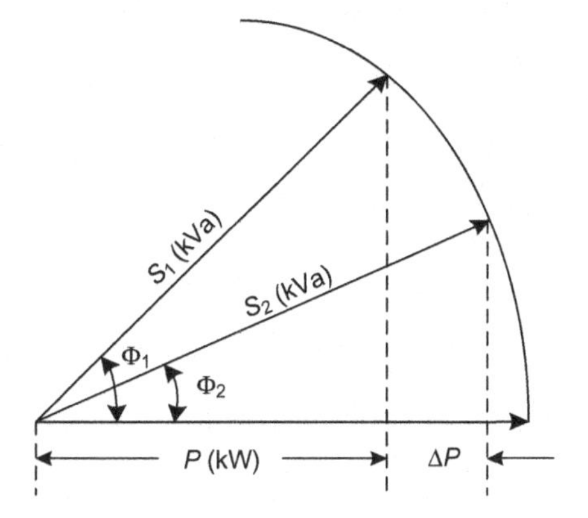

Figure 9.7 The increase of active power.

and after compensation $\cos\varphi_2 = 0.9$, there is a 35% increase in power transmitted (with constant active losses).

The increase of the active power transmitted by the lines with constant active losses can be calculated by using the vector diagram shown in Figure 9.7, where

$$\Delta P = S(\cos\varphi_2 - \cos\varphi_1) \tag{9.2}$$

and

$$Q_c = S(\sin\varphi_1 - \sin\varphi_2). \tag{9.3}$$

Economic Evaluation for Compensation: The economic advantage of compensation is evaluated by comparing the installation cost of the capacitor banks with the savings they produce.

The cost of the capacitor banks depends on several parameters:

- voltage level;
- installed power;
- breakdown into steps;
- control mode;
- quality of the protection;
- the capacitors can be installed either in low voltage (LV) or in MV;
- MV compensation is economical when the power to be installed is greater than 800 kvar;
- below this value, compensation will preferably be at LV.

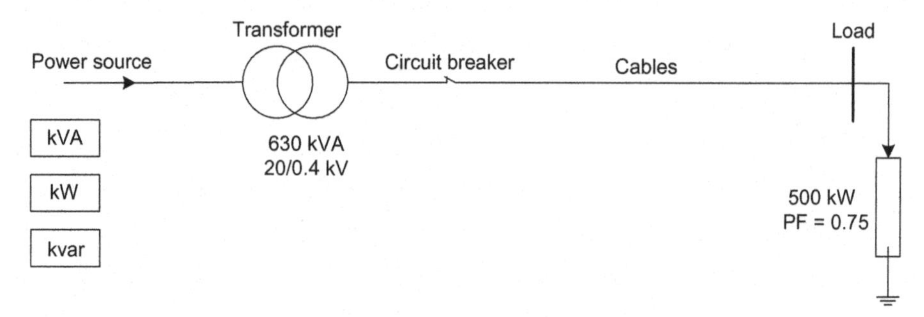

Figure 9.8 Supplying a load without compensation.

The following example shows an installation comprising a 20/0.4-kV transformer with a power of 630 kVA (rated apparent power). It illustrates how good PF means savings.

9.4 AN EXAMPLE

A 500-kW, PF 0.75 lag load is supplied from a power source at 20 kV through a transformer (630 kVA, 20/0.4 kV), a circuit breaker, and cables (Fig. 9.8).

Without PF Improvement: The apparent power "S" flowing into the transformer, neglecting the losses in the cable, is given as

$$S = \frac{P}{\cos\Phi} = \frac{500}{0.75} = 666 \text{ kVA.}$$

This means that the transformer is overloaded. The circuit breaker and cables must handle a current "I," which is calculated as

$$I = \frac{P}{\sqrt{3}V\cos\Phi} = \frac{500}{\sqrt{3}\times 0.4\times 0.75} = 962.28 \text{ A.}$$

Losses in the cables are proportional to the square of the current and equal to I^2R, that is, proportional to $(962.28)^2$.

The demand reactive power required for the load $Q_L = P\tan\Phi_1 = 500 \times 0.88 = 440$ kvar.

The consumer's payment is for the consumed active and reactive power.

With PF Improvement: Shunt capacitor banks are connected at load end (Fig. 9.9) to improve the PF to 0.93 lagging at the customer's terminals.

The apparent power "S" flowing into transformer now is 500/0.93 = 537.634 kVA. The transformer is not overloaded, leaving a 16% power reserve.

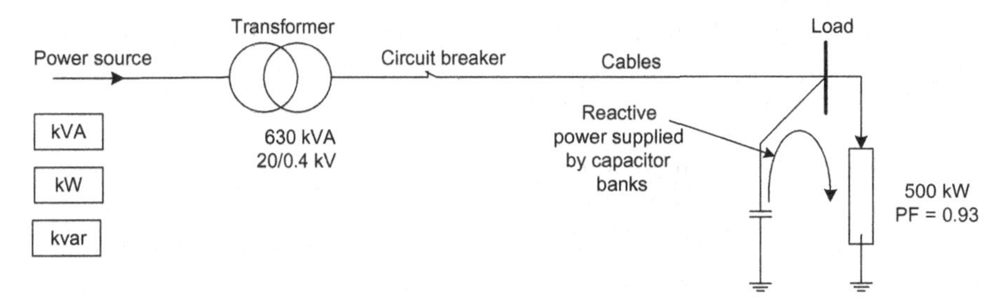

Figure 9.9 Supplying a load with compensation.

The circuit breaker and cables must handle a current that is reduced to 776.4A [$I = 500/(\sqrt{3} \times 0.4 \times 0.93)$]. The cable losses are also reduced because they are proportional to $(776.4)^2$.

The reactive power supplied by capacitor banks $Q_c = P(\tan\Phi_1 - \tan\Phi_2)$. Then, $Q_c = 500\ (0.88 - 0.395) = 242.5$ kvar. Consequently, the reactive power supplied by power source = 440 − 242.5 = 197.5 kvar. This means that the reactive power flow in the cables is reduced by about 55%.

The results of improving PF can be summarized as below:

- reducing the consumed reactive power and reducing the consumer's payment for the same consumed active power,
- avoiding utility penalties applied to consumers with low PF,
- avoiding overloading and oversizing of system equipment, and
- reducing line losses.

9.5 HOW TO DETERMINE COMPENSATION

Compensation of an installation is determined in four steps, namely,

- calculating reactive power
- choosing the compensation mode
  - global compensation for the entire installation
  - compensation by sector
  - individual load compensation
- choosing the compensation type
  - fixed by switching on/off a capacitor bank supplying a fixed amount of kvar
  - automatic by switching on/off "steps" by breaking up installed bank capacity and enabling adaptation in the installation's kvar needs
- allowing for harmonics.

This involves choosing the right equipment to avoid the harmful consequences that harmonics might have on the capacitors installed.

These steps are developed as below:

1. First step: calculating reactive power
   - *Calculation Principle:* The aim is to determine the reactive power Q_c (kvar) to be installed to increase the $\cos\varphi$ of the installation to reach a given target. This involves going from

     $\cos\varphi_1 \rightarrow \cos\varphi_2$ where $\cos\varphi_2 > \cos\varphi_1$;

     $\varphi_1 \rightarrow \varphi_2$ where $\varphi_2 < \varphi_1$;

     $\tan\varphi_1 \rightarrow \tan\varphi_2$ where $\tan\varphi_2 < \tan\varphi_1$;

     $Q_L \rightarrow Q_L - Q_C$ where Q_L is load reactive power; and

     $S_1 \rightarrow S_2$ where $S_2 < S_1$.

   This is illustrated in Figure 9.10.

   - *Calculation by Using Installation Data:* The power to be installed is calculated from $\cos\varphi_1$ or from $\tan\varphi_1$ noted for the installation. The aim is to move to a value of $\cos\varphi_2$ or $\tan\varphi_2$ thus improving the operation, as shown in Figure 9.10. Q_c can be calculated from the equation

$$Q_c = P_a(\tan\varphi_1 - \tan\varphi_2), \tag{9.4}$$

   which results from the figure, where

   Q_c = rating of the capacitor bank in kvar,
   P_a = active power of the load in kW,
   $\tan\varphi_1$ = tangent of the phase angle (V, I) of the installation, and
   $\tan\varphi_2$ = tangent of the phase angle (V, I) after installation of the capacitor bank.

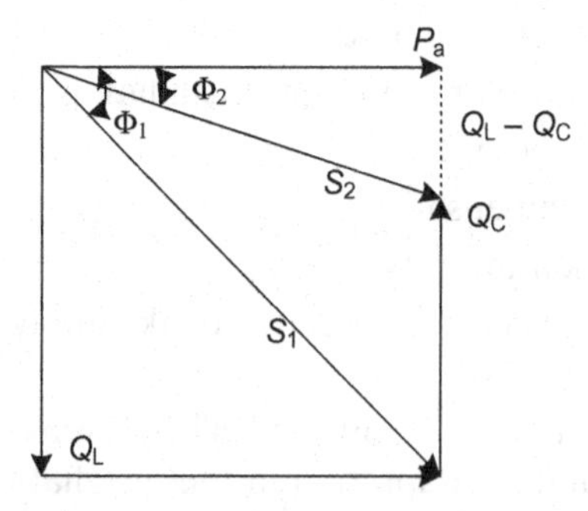

Figure 9.10 Determination of Q_C.

Q_c can also be calculated by using Table 9.2. The equation of Q_c can be rewritten as

$$Q_c = kP_a, \tag{9.5}$$

where k is a factor determined very easily from Table 9.2 if one knows $\tan\varphi_1$ or $\cos\varphi_1$ of the existing installation and $\tan\varphi_2$ or $\cos\varphi_2$, which is to be obtained. For instance, if the installation PF ($\cos\varphi_1$) is 0.71 and required to be 0.94 ($\cos\varphi_2$), then the factor k is equal to 0.63, which in turn means that the reactive power capacity required for compensation is 63% of the active power demand.

2. Second step: choosing the compensation mode
   - *Where to Install Capacitors?* Installation of capacitors on an electric network is referred to as the "compensation mode." It is determined by
     - the objective (eliminating penalties, relieving stress on cables and transformers, and improving network voltage),
     - the electric power distribution mode,
     - the load system,
     - the predictable effect of capacitors on network characteristics, and
     - the cost of the installation.

Reactive power compensation can be

- global
- by sector
- by individual load.

It is more economical to install capacitor banks in MV and high voltage (HV) for powers greater than approximately 800 kvar. Analysis of networks in various countries proves, however, that there is no universal rule.

The compensation mode depends on the energy policy of the countries and electricity boards. In the United States, compensation is mainly at MV for tariff policy reasons, whereas in Germany, compensation is at LV.

In France, the utility installs fixed banks on 63- and 90-kV networks (bank not broken up) and switched step banks (bank broken up) in MV source substations on 11-, 22-, and 33-kV networks. The latter's power can reach 4.8 Mvar at 22 kV.

- *Global Compensation:* The capacitor bank is connected at the supply end of the installation to be compensated and provides compensation for the entire installation. It is ideal for stable and continuous loads. For example, as in Figure 9.11,

TABLE 9.2 Capacitor Bank Sizing

Without Compensation			The Factor k (Capacitor kvar Output per kW) at Different Desired tan Φ or cos Φ												
tan Φ	cos Φ	tan Φ	0.75	0.59	0.48	0.46	0.43	0.40	0.36	0.33	0.29	0.25	0.20	0.14	0.0
		cos Φ	0.80	0.86	0.90	0.91	0.92	0.93	0.94	0.95	0.96	0.97	0.98	0.99	1.0
2.29	0.40		1.55	1.69	1.80	1.83	1.86	1.89	1.92	1.95	1.99	2.03	2.08	2.14	2.28
2.22	0.41		1.47	1.62	1.74	1.77	1.80	1.83	1.84	1.90	1.93	1.97	2.02	2.08	2.22
2.16	0.42		1.41	1.56	1.68	1.71	1.74	1.77	1.80	1.84	1.87	1.91	1.96	2.02	2.16
2.10	0.43		1.36	1.50	1.62	1.65	1.68	1.71	1.74	1.78	1.82	1.86	1.90	1.96	2.11
2.04	0.44		1.29	1.44	1.56	1.58	1.61	1.65	1.68	1.71	1.75	1.79	1.84	1.90	2.04
1.98	0.45		1.23	1.38	1.50	1.53	1.56	1.59	1.63	1.66	1.70	1.74	1.78	1.85	2.00
1.93	0.46		1.18	1.33	1.45	1.47	1.50	1.53	1.57	1.60	1.64	1.68	1.72	1.79	1.93
1.88	0.47		1.13	1.28	1.40	1.42	1.45	1.48	1.52	1.53	1.59	1.63	1.78	1.76	1.88
1.83	0.48		1.08	1.23	1.34	1.37	1.40	1.43	1.46	1.50	1.53	1.57	1.62	1.68	1.83
1.78	0.49		1.03	1.18	1.30	1.33	1.35	1.39	1.42	1.45	1.49	1.53	1.58	1.64	1.78
1.73	0.50		0.98	1.23	1.25	1.28	1.30	1.34	1.37	1.40	1.44	1.48	1.53	1.59	1.73
1.69	0.51		0.94	1.09	1.20	1.23	1.26	1.29	1.32	1.26	1.39	1.43	1.48	1.54	1.69
1.64	0.52		0.98	1.04	1.16	1.19	1.21	1.25	1.28	1.31	1.35	1.39	1.44	1.50	1.64
1.60	0.53		0.85	1.00	1.12	1.14	1.17	1.20	1.24	1.27	1.31	1.35	1.40	1.46	1.60
1.56	0.54		0.81	0.96	1.07	1.10	1.13	1.16	1.20	1.23	1.27	1.31	1.36	1.42	1.56
1.52	0.55		0.77	0.92	1.03	1.06	1.09	1.12	1.16	1.19	1.23	1.27	1.32	1.38	1.52
1.48	0.56		0.73	0.88	0.70	1.02	1.05	1.08	1.12	1.15	1.19	1.23	1.28	1.34	1.48
1.44	0.57		0.69	0.84	0.96	0.99	1.01	1.05	1.08	1.11	1.15	1.19	1.24	1.30	1.44
1.40	0.58		0.66	0.80	0.92	0.95	0.98	1.01	1.04	1.08	1.11	1.15	1.20	1.26	1.40
1.37	0.59		0.62	0.77	0.88	0.91	0.94	0.97	1.01	1.04	1.08	1.12	1.16	1.23	1.37
1.33	0.60		0.58	0.73	0.85	0.88	0.90	0.94	0.97	1.01	1.04	1.08	1.13	1.19	1.33
1.30	0.61		0.55	0.70	0.81	0.84	0.87	0.90	0.94	0.97	1.01	1.05	1.10	1.16	1.30
1.27	0.62		0.52	0.67	0.78	0.81	0.84	0.87	0.90	0.94	0.97	1.01	1.06	1.12	1.26

1.23	0.63		0.48	0.63	0.75	0.78	0.80	0.84	0.87	0.90	0.94	0.98	1.03	1.09	1.23
1.20	0.64		0.45	0.60	0.72	0.74	0.77	0.81	0.84	0.87	0.91	0.95	1.00	1.06	1.20
1.17	0.65		0.42	0.57	0.68	0.71	0.74	0.77	0.81	0.84	0.88	0.92	0.97	1.01	1.17
1.14	0.66		0.39	0.54	0.65	0.68	0.71	0.74	0.77	0.81	0.85	0.89	0.93	0.70	1.14
1.11	0.67		0.36	0.51	0.62	0.65	0.68	0.71	0.74	0.78	0.82	0.86	0.90	0.97	1.11
1.08	0.68		0.33	0.48	0.59	0.62	0.65	0.68	0.72	0.75	0.79	0.83	0.88	0.94	1.08
1.05	0.69		0.30	0.45	0.56	0.59	0.62	0.65	0.69	0.72	0.76	0.80	0.84	0.91	1.05
1.02	0.70		0.27	0.42	0.54	0.56	0.59	0.62	0.66	0.69	0.73	0.80	0.81	0.88	1.02
0.99	0.71		0.24	0.39	0.51	0.54	0.56	0.60	0.63	0.66	0.70	0.74	0.78	0.85	0.99
0.96	0.72		0.21	0.36	0.48	0.51	0.53	0.57	0.60	0.63	0.67	0.71	0.75	0.82	0.96
0.94	0.73		0.19	0.34	0.45	0.48	0.51	0.54	0.57	0.61	0.64	0.68	0.73	0.79	0.94
0.91	0.74		0.16	0.31	0.42	0.45	0.48	0.51	0.55	0.58	0.62	0.66	0.70	0.77	0.91
0.88	0.75		0.13	0.28	0.40	0.43	0.45	0.49	0.52	0.55	0.59	0.63	0.67	0.74	0.88
0.86	0.76		0.10	0.25	0.37	0.40	0.43	0.46	0.49	0.53	0.56	0.60	0.65	0.71	0.85
0.83	0.77		0.08	0.23	0.34	0.37	0.40	0.43	0.47	0.50	0.54	0.58	0.62	0.96	0.83
0.80	0.78		0.05	0.20	0.32	0.35	0.37	0.41	0.44	0.47	0.51	0.55	0.59	0.66	0.80
0.78	0.79		0.03	0.18	0.29	0.32	0.35	0.38	0.41	0.45	0.48	0.52	0.57	0.63	0.78
0.75	0.80			0.15	0.27	0.29	0.32	0.35	0.39	0.42	0.46	0.50	0.54	0.61	0.75
0.72	0.81			0.12	0.24	0.27	0.29	0.33	0.36	0.40	0.43	0.47	0.51	0.58	0.72
0.70	0.82			0.10	0.21	0.24	0.27	0.30	0.33	0.37	0.41	0.45	0.49	0.56	0.70
0.67	0.83			0.07	0.19	0.22	0.24	0.28	0.31	0.34	0.38	0.42	0.46	0.53	0.67
0.65	0.84			0.05	0.16	0.19	0.22	0.25	0.28	0.32	0.35	0.39	0.44	0.50	0.64
0.62	0.85			0.02	0.14	0.16	0.19	0.22	0.26	0.29	0.33	0.37	0.42	0.48	0.62
0.59	0.86				0.11	0.14	0.17	0.20	0.23	0.26	0.30	0.34	0.39	0.45	0.59
0.57	0.87				0.08	0.11	0.14	0.17	0.20	0.24	0.27	0.32	0.36	0.42	0.57
0.54	0.88				0.05	0.08	0.11	0.14	0.17	0.21	0.25	0.29	0.33	0.39	0.54
0.51	0.89				0.03	0.06	0.09	0.12	0.15	0.18	0.23	0.26	0.31	0.37	0.51
0.48	0.90					0.03	0.06	0.09	0.12	0.15	0.19	0.23	0.28	0.34	0.48

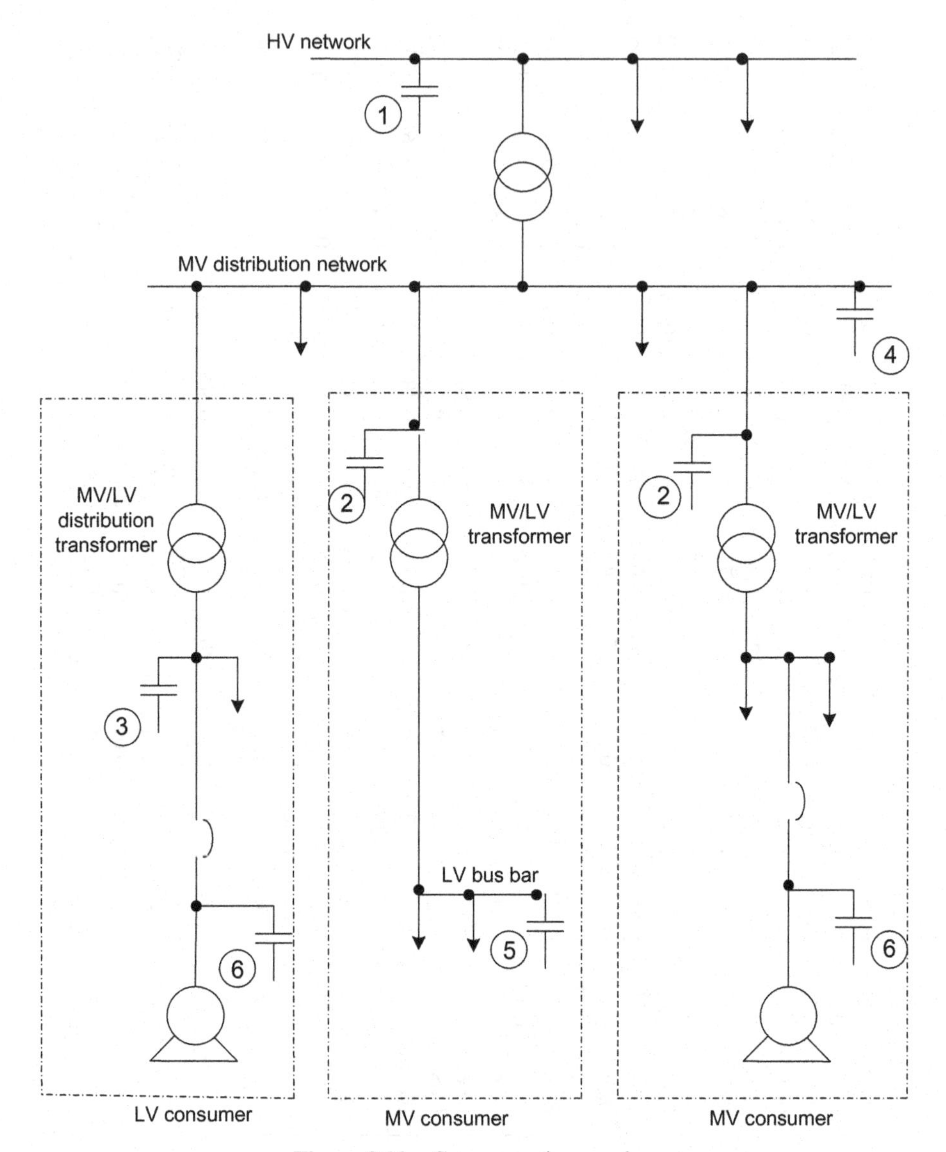

Figure 9.11 Compensation modes.

- HV bank on HV distribution network (1),
- MV bank for MV customer (2), and
- regulated or fixed bank for LV customer (3).

Advantages: This arrangement is more economical because all required kvar capacity is concentrated at one point. Furthermore, the load factor is taken into account in determining this capacity, which leads to optimized and thus smaller capacitor banks.

Disadvantages: The system no-load voltage is increased due to the overvoltage produced by the capacitor bank.

- *Compensation by Sector:* The capacitor bank is connected at the supply end of the installation sector to be compensated. It is ideal when the installation is extended and contains workshops with varying load systems, for example,
  - MV bank on MV network (4) and
  - LV bank per workshop for MV customer (5).

Advantages: This arrangement localizes capacitor banks near the reactive power consumers, which are divided into geographic zones. It acts on a larger portion of installation, in particular, on supply feeders. The load factor of each sector is taken into account to maintain a reasonable level of optimization of kvar capacity. Although it is more costly than global compensation, it is still a fairly economical scheme.

Disadvantages: The system may be exposed to a risk of overcompensation following large load variations within the sector. This risk can be avoided by means of automatic compensation with automatic capacitor bank.

- *Individual Compensation:* The capacitor bank is directly connected to the terminals of each inductive load (particularly motors). It is ideal when motor power is greater than the subscribed power. This type of compensation is an ideal solution, technically speaking, since the reactive power is produced where it is consumed in quantities to meet the demand, for example, individual bank per motor (6).

Advantages: This mode of compensation is technically ideal. The reactive power is produced at the place where it is consumed and in quantities corresponding exactly to the demand.

Disadvantages: It represents an expensive solution that completely ignores the possibilities offered by the load factor. So, this arrangement is reserved for large point consumers, for example, large motors.

3. Third step: choosing the compensation type
   - *The Types of MV Compensation:* The capacitor banks are connected onto the network. They can be either
     - fixed, that is, the bank is switched on/off, so there is a fixed kvar value (operation is of the "on/off" type) or
     - the breakdown into "steps" type with the possibility of (normally) automatic switching on/off of a certain number of steps. This provides automatic adjustment to needs.

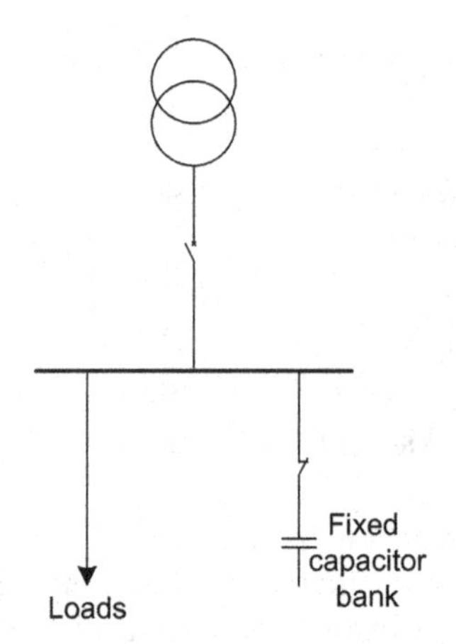

Figure 9.12 Fixed compensation.

- *Fixed Compensation:* The capacitors have a constant unit power and can be switched on and off:
  - manually by circuit breaker or switch,
  - semiautomatically by contactor, and
  - by control at inductive load terminals (motors or transformers).

  This type of compensation is used
  - when reactive power is low (<15% of upstream transformer power) and the load is relatively stable, and
  - on HV transmission networks for powers up to 100 Mvar (Fig. 9.12).
- *Automatic Compensation:* The compensation technique is commonly known as "switched steps." These capacitors are often used by very large industries (large installed capacity) and by electricity boards in substations. It regulates reactive power step by step. Each step is operated using a switch or a contactor (Fig. 9.13).

  The closing or release of the capacitor bank steps can be controlled by a var-metric relay. The relay regulates the different steps according to the measured cos Φ and reactive power requirements of the installation. Current data come from current and potential transformers, which must thus be placed upstream from the loads and capacitor banks.

4. Fourth step: allowing for harmonics
   Harmonic currents are present in HV/LV installations due to the presence of nonlinear loads (e.g., speed controllers, uninterruptible power supply [UPS], arc furnaces, lighting).

   Flow of harmonic currents in the network impedances creates harmonic voltages. The importance of harmonic disturbance on a network is assessed by the following:

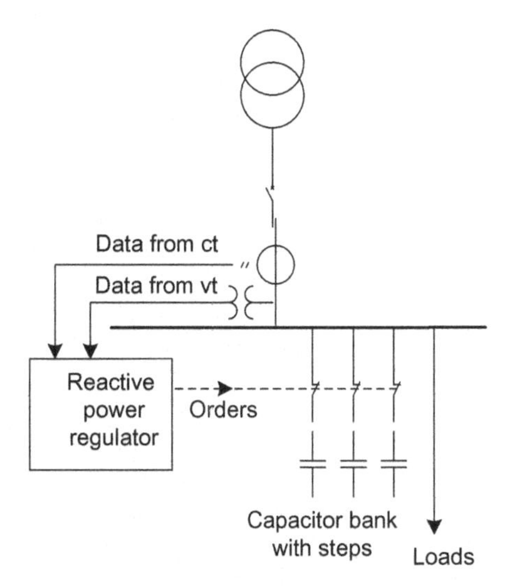

Figure 9.13 Automatic compensation.

- The individual rate (T) of the harmonic voltages, which measures the importance of each harmonic compared with the fundamental.
  For the h rank harmonic, this rate is

$$T(\%) = (V_{\mathrm{h}}/V_1) \times 100, \tag{9.6}$$

  where V_{h} is the rank harmonic voltage at the point considered and V_1 is the fundamental voltage.
- The global voltage distortion rate (D), which measures the thermal effect of all harmonics. This rate is

$$D(\%) = [\{\sqrt{\Sigma_2^{\mathrm{h}} V_{\mathrm{h}}^2}\}/V_1] \times 100. \tag{9.7}$$

The maximum value of h (maximum rank) is normally 19 or 25. In the same way, an individual rate and a current distortion are defined.

Risk of Resonance: When a network contains harmonics, the presence of a capacitor amplifies certain harmonics to some extent. Resonance appears, the frequency of which depends on the network impedance (or short-circuit power). More details about harmonics are presented in Chapters 10 and 11.

CHAPTER 10

HARMONICS IN ELECTRIC DISTRIBUTION SYSTEMS

10.1 WHAT ARE HARMONICS?

Any distorted nonsinusoidal periodic function can be analyzed by Fourier theorem to be represented as a sum of terms made up of

- a sinusoidal term at the fundamental frequency,
- sinusoidal terms (harmonics) having frequencies that are multiples of the fundamental frequency, and
- a direct current (DC) component when applicable.

The *n*th order harmonic (commonly referred to as the *n*th harmonic) in a signal is the sinusoidal component with a frequency that is *n* times the fundamental frequency.

The equation for the harmonic expansion of a periodic function is given as

$$Y(t) = Y_o + \sum_{n=1}^{\infty} Y_{\max,n}(n\omega t - \Phi_n), \tag{10.1}$$

Electric Distribution Systems, First Edition. Abdelhay A. Sallam, Om P. Malik.

Published 2011 by John Wiley & Sons, Inc.

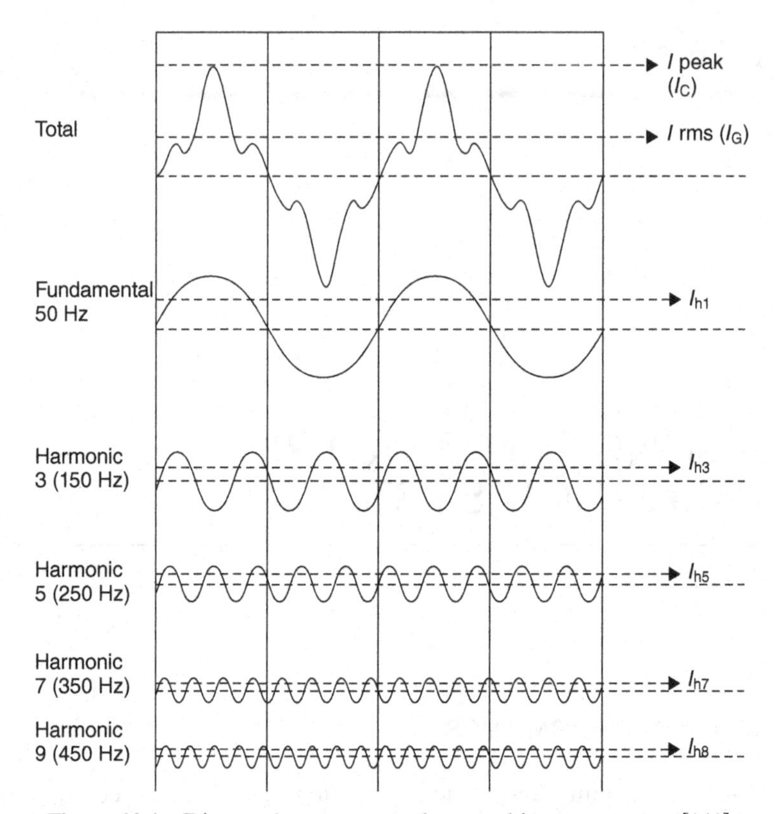

Figure 10.1 Distorted current waveform and its components [141].

where

Y_o = DC component,
$Y_{max,n}$ = maximum value (amplitude) of the nth harmonic,
ω = angular frequency = 2π times the fundamental frequency, and
Φ_n = initial displacement of the harmonic component at $t = 0$.

That is, a distorted periodic signal has the fundamental component at the fundamental frequency, say 50 Hz, plus second harmonic component at 100 Hz plus third harmonic component at 150 Hz plus, and so on. For instance, the current wave shown in Figure 10.1 is affected by harmonic distortion. It is the sum of fundamental component and superimposed harmonics [141].

Both electricity distribution networks and consumer equipment are affected by harmonic distortion of the voltage waveform. Harmonic distortion levels have increased rapidly in electric power systems in recent years due primarily to the increasingly widespread application of nonlinear semiconductor devices, which produce the majority of harmonic distortion [142].

In particular, switched-mode power supplies have over the last 20 years increasingly replaced transformer/rectifier power supplies in electronic equip-

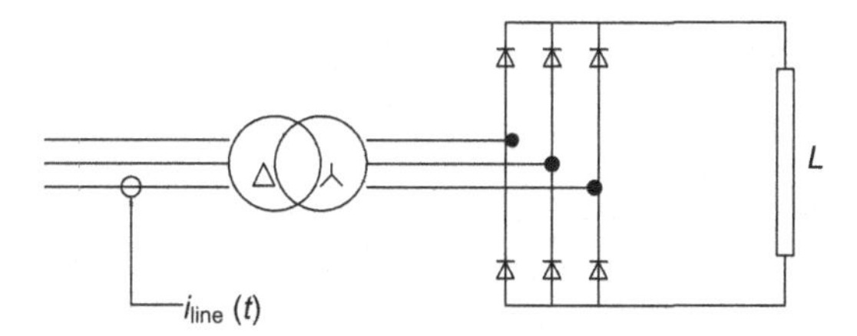

Figure 10.2 Example of a load drawing a nonsinusoidal current from the supply.

ment. These cause large fifth harmonic emissions, relative to the rated power consumption of the equipment, which are randomly oriented in-phase and so are strongly additive. Consequently, electricity supply systems are experiencing increasing background levels of harmonic level distortion at all voltage levels up to 400 kV. As harmonic voltage distortion increases, the risk of widespread problems arising from premature aging of equipment and overloading of neutral conductors grows [143, 144].

In response to that, many network operators are implementing monitoring systems to obtain a better overview of the present status of harmonic loads and of the related development.

Power system generators normally produce a clean sinusoidal waveform at their terminals. This sinusoidal waveform is regarded as the pure form of the alternating current (AC) voltage and any deviation from it is described as distortion. More and more types of loads absorb nonsinusoidal current from the power system. As an example, consider Figure 10.2 that shows a six-pulse thyristor bridge feeding a purely inductive load [145].

Imagine that the thyristor bridge is connected through a transformer to a clean sinusoidal supply system. The frequency of this sinusoidal voltage waveform is referred to as the fundamental frequency f_1. Upon connection of the load to the supply, a line current, denoted as $i_{\text{line}}(t)$, will flow. Figure 10.3 shows an approximation of the current waveform. From this figure, it may be seen that the current $i_{\text{line}}(t)$ deviates strongly from the sinusoidal current waveform, fundamental component $i_1(t)$, and thus "something" needs to be added to this sinusoidal component to obtain the current actually flowing.

In order to derive what this "something" is, consider Figure 10.4, which shows the resulting current waveform obtained by adding a component $i\_h_5(t)$ with a frequency equal to five times the fundamental frequency to $i_1(t)$.

It can be seen in Figure 10.4 that the resulting waveform $i_1(t) + i\_h_5(t)$ resembles closer to the line current waveform $i_{\text{line}}(t)$ than the waveform $i_1(t)$ alone. Continuing this approach, one can also add a current waveform $i\_h_7(t)$ to the already existing waveform $i_1(t) + i\_h_5(t)$.

In this, $i\_h_7(t)$ refers to a current component with a frequency equal to seven times the fundamental frequency. The waveforms $i_{\text{line}}(t)$ and $i\_h_7(t)$ and the waveform obtained by adding the current components $i_1(t)$, $i\_h_5(t)$ and $i\_h_7(t)$ are shown in Figure 10.5.

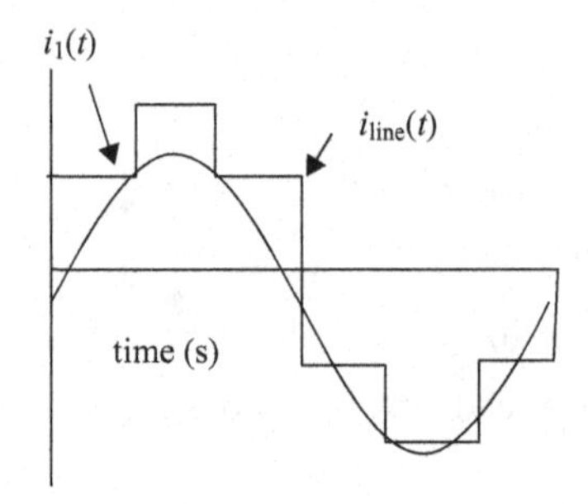

Figure 10.3 Line current $i_{line}(t)$ flowing in the supply branch of the six-pulse thyristor bridge [146].

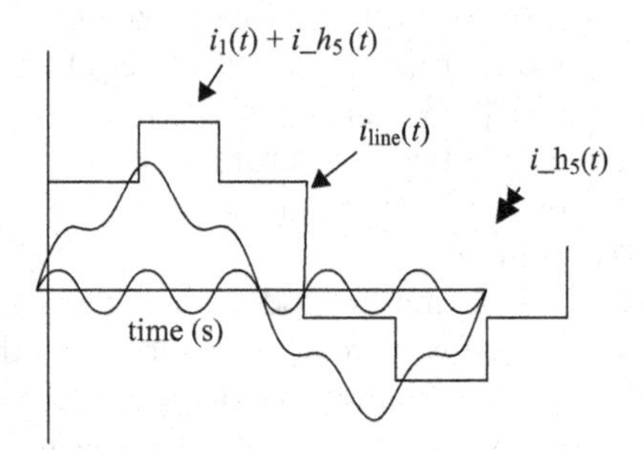

Figure 10.4 Comparison of current waveform $i_{line}(t)$ with the waveform $i_1(t) + i\_h_5(t)$ [146].

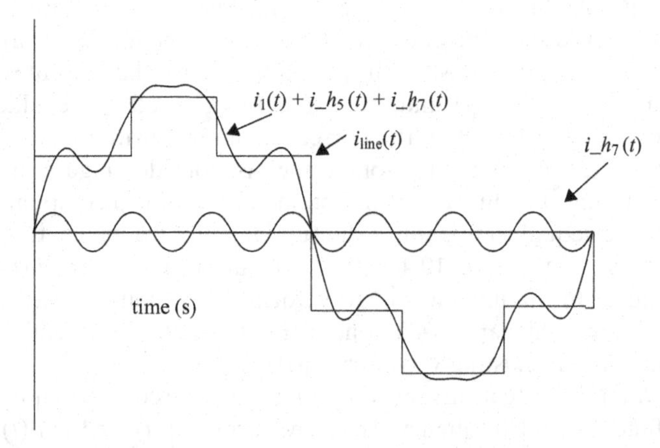

Figure 10.5 Comparison of the current waveform $i_{line}(t)$ with the waveform $(i_1(t) + i\_h_5(t) + i\_h_7(t))$ [146].

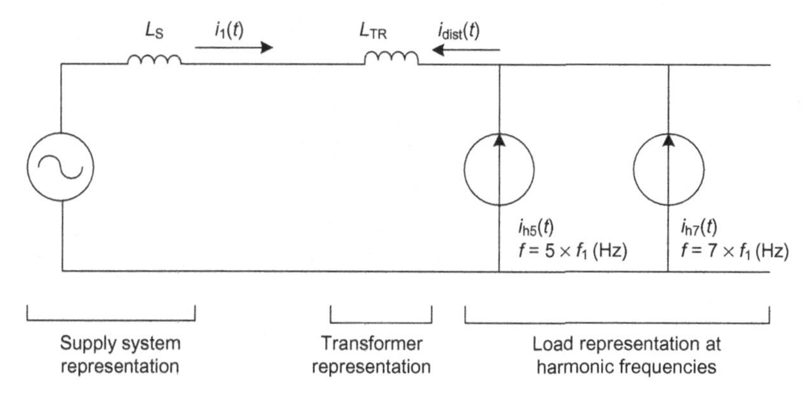

Figure 10.6 Schematic representation of the harmonic flow for the thyristor bridge load.

By adding the waveform $i\_h_7(t)$ to the waveform $i_1(t) + i\_h_5(t)$, the resulting waveform resembles even more to the line current $i_{line}(t)$ than it was the case for the addition of only $i_1(t) + i\_h_5(t)$. It may be shown that by adding still more current components $i\_h_i(t)$, each at a particular integer multiple of the fundamental frequency, the result can become as the measured line current $i_{line}(t)$.

The current components $i\_h_5(t)$, $i\_h_7(t)$, ... that have to be added to the fundamental current $i_{fund}(t)$ in order to compose the line current actually flowing, $i_{line}(t)$, are referred to as the integer frequency harmonic components of the current or simply "harmonics." They exist at frequencies, which are integer multiples of the fundamental frequency. Since it was assumed that the thyristor bridge is connected to a clean sinusoidal voltage at the fundamental frequency, it may be concluded that the load is the source of the harmonics. More specifically, the load is injecting current harmonics into the supply. This is schematically represented in Figure 10.6.

It should be noted that not all types of loads inject harmonics into the supply system. It depends on the types of load characteristics. While in the preceding figures the line current and its harmonic components have been represented by time domain waveforms, it is easier and more common to represent the harmonics by means of the current spectrum. This spectrum shows for each harmonic frequency the magnitude of the corresponding harmonic component present in the current analyzed. Possibly, the magnitude of the harmonic components is expressed as a percentage of the magnitude of the fundamental component. The horizontal axis shows generally the harmonic order, which is given by the ratio of the harmonic frequency over the fundamental frequency (Fig. 10.7).

Next to the integer frequency harmonic components discussed above, there also exists a class of harmonics that are not situated at integer multiples of the fundamental frequency. They are referred to as interharmonics. A typical example of a load producing interharmonics is a cycloconverter. This device

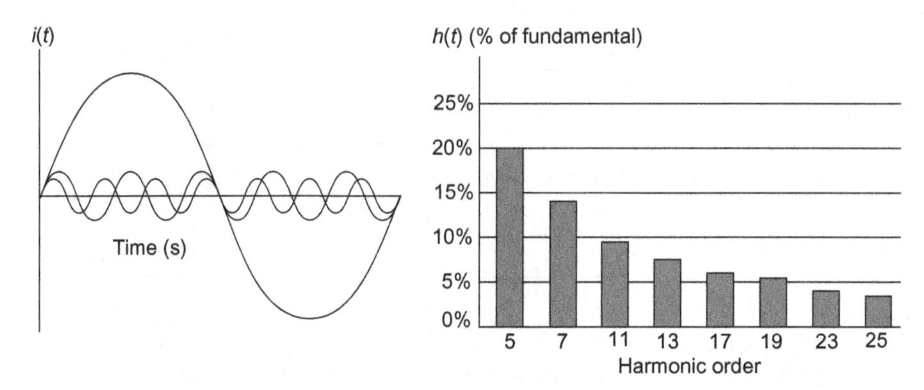

Figure 10.7 Representation of the current harmonics in the time domain (left) and in the frequency domain (spectrum) (right).

generates at its output a waveform with a variable frequency that can be used to drive, for example, a synchronous motor. The cycloconverter output frequency is situated in the low-frequency range, for example, up to 15 Hz. It should be noted that in general the magnitude of the interharmonic components present in the low-voltage networks is negligible. Therefore, they are not further discussed.

From the above explanation, harmonics are sinusoidal voltages or currents having frequencies that are multiples of the fundamental frequency, at which the supply system is designed to operate (e.g., 50 Hz).

An electric power supply system is inherently very passive and the generator outputs and voltages measured on the system would be nearly sinusoidal at the declared frequency if no load is connected to the system.

For a pure sinusoidal voltage wave, of frequency 50 Hz and 230 V amplitude, the spectrum is equal to zero for all frequencies except 50 Hz, for which the value is 230 V.

For a distorted voltage wave, the spectrum contains harmonic frequencies that are characteristic of the nature of distortion [147]. The fundamental is the component of the spectrum at which the network is designed to work. It is normally the first and greatest component of the spectrum. The term total harmonic distortion (THD) is used to describe the root mean square (rms) sum of the voltages of all harmonic frequencies that are present relative to the fundamental.

10.2 SOURCES OF HARMONICS

Historically, harmonics were mainly caused by magnetization nonlinearity, and recently, devices causing harmonics are present in all industrial, commercial, and residential installations as nonlinear loads.

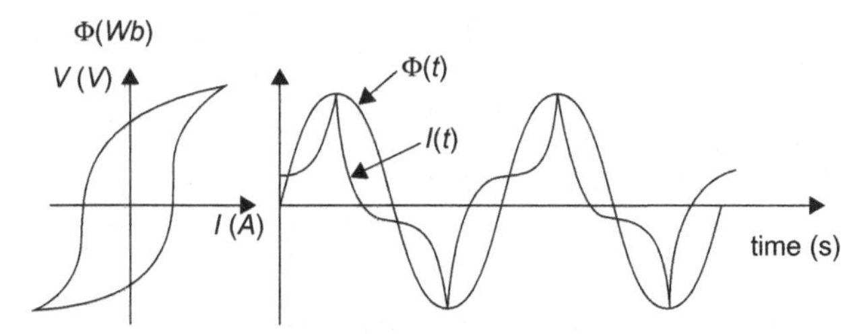

Figure 10.8 Illustration of the nonlinear magnetization characteristic of a transformer [146].

The load is said to be nonlinear when the current it draws does not have the same waveform as the supply voltage. Mostly, this load is represented by the device comprising power electronics circuits such as

- industrial equipment (welding machines, arc furnaces, induction furnaces, rectifiers),
- variable speed drives for asynchronous and DC motors,
- office equipment (PCs, photocopy machines, fax machines, etc.),
- household appliances (TVs, microwave ovens, fluorescent lighting, etc.),
- uninterruptible power supplies (UPSs), and
- saturation of equipment (essentially transformers) may cause nonlinear currents.

As an example, the magnetization characteristic of a transformer is shown in Figure 10.8.

When applying a purely sinusoidal voltage source (and thus a purely sinusoidal flux) to a transformer operating in its nonlinear region, the resulting magnetizing current is not sinusoidal. The resulting current waveform contains a variety of odd harmonics of which the third harmonic is the most dominant. It should be noted that the magnetizing current is in general a small percentage of the transformer's rated current and as such its effect becomes less pronounced as the transformer is loaded.

At present, power electronics-based equipment is the main source of the harmonic pollution in the low-voltage networks. Examples of such equipment include drives, UPSs, welding machines, PCs, and printers. In general, the semiconductor switches in such equipment conduct only during a fraction of the fundamental period. This is how this type of equipment can obtain their main properties regarding energy saving, dynamic performance, and flexibility of control.

However, as a result, a discontinuous current containing a considerable amount of distortion is drawn from the supply. Next, some typical load arrangements and the resulting harmonic distortion are given.

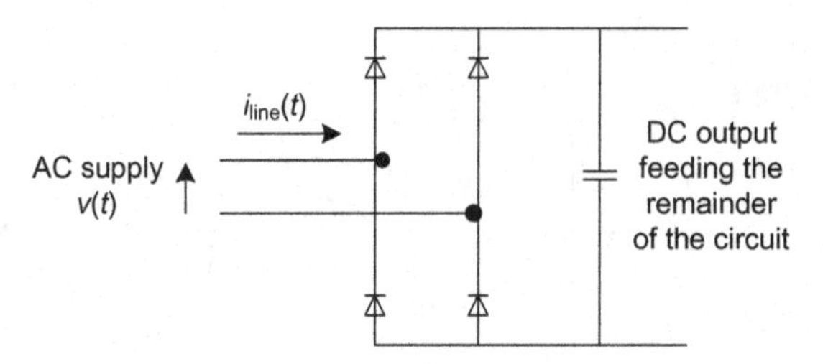

Figure 10.9 Single-phase rectifier with smoothing capacitor.

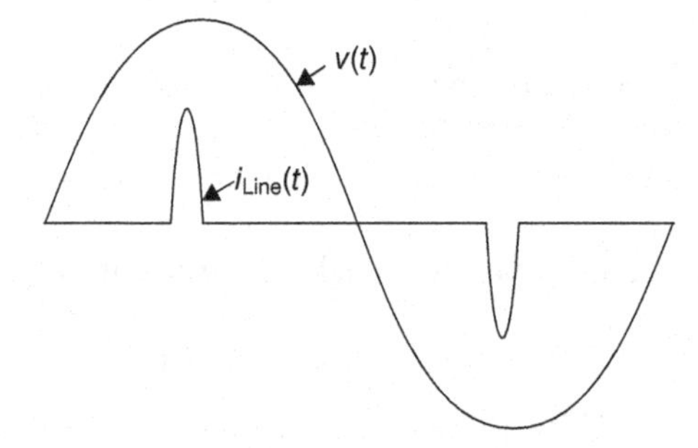

Figure 10.10 Supply voltage and current waveforms characterizing the single-phase rectifier bridge with smoothing capacitor [146].

- *Single-Phase Rectifier with Smoothing Capacitor (Fig. 10.9)*: This circuit topology is present in many supplies of single-phase devices such as computers, printers, and fluorescent lighting systems.

The current drawn from the supply is characterized by a sharp current rise and fall during only a fraction of the fundamental period. Typical voltage and current waveforms are presented in Figure 10.10.

The current waveform contains a considerable amount of odd harmonics, the magnitude of which may be higher than the fundamental current component. While the devices using this circuit topology generally have a small power rating, an increasing number of them are being used. This may result in an excessive amount of harmonic current flowing in the feeding transformer and the supply lines.

- *Six-Pulse Bridges:* Six-pulse bridges are commonly used in three-phase power electronics-based equipment such as drives (AC and DC) and UPSs. The switches used can either be controllable (e.g., isolated gate bipolar transistors [IGBTs] and thyristors) or uncontrollable (diodes). Depending on the equipment, the DC side of the bridge is connected to a smoothing capacitor, a smoothing inductor or both. The circuit topology

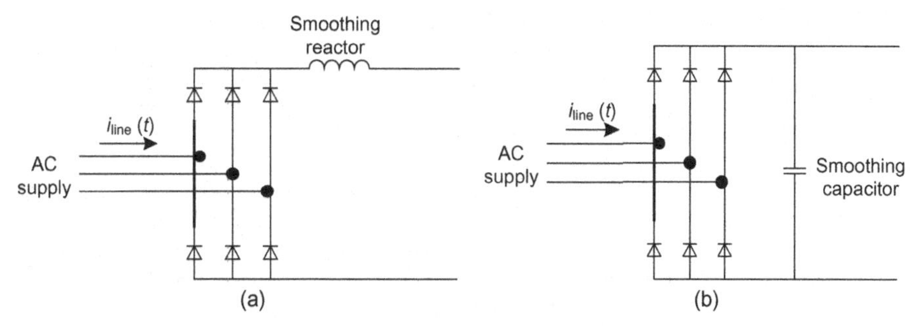

Figure 10.11 Circuit topology of a 3-ph diode with (a) a smoothing reactor and (b) a smoothing capacitor.

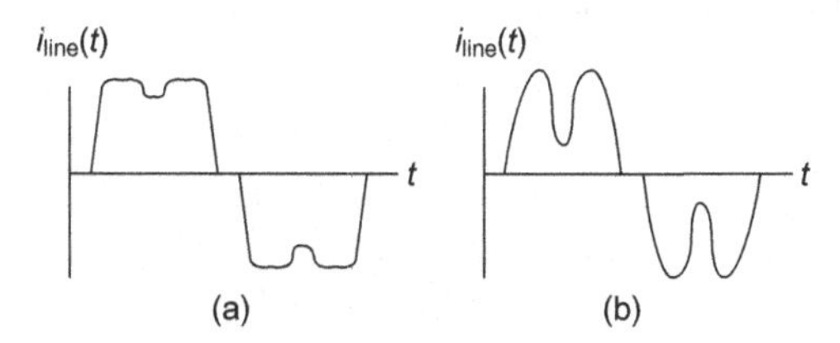

Figure 10.12 Line current waveform of a six-pulse diode bridge with (a) a smoothing reactor and (b) a smoothing capacitor.

for a diode bridge with a smoothing reactor is shown in Figure 10.11a and the topology for a diode bridge with a smoothing capacitor is shown in Figure 10.11b.

A typical waveform of the line current drawn by the circuit having a smoothing reactor is shown in Figure 10.12a and a typical waveform of the line current drawn by the circuit with a smoothing capacitor is shown in Figure 10.12b.

It may be shown that the current drawn by a six-pulse bridge contains harmonics of the order $h = (6j \pm 1)$ where j is an integer greater than or equal to 1. Thus, the line current contains harmonics of the order 5, 7, 11, 13, … . For a six-pulse diode bridge having a large smoothing reactor, the rms value of the harmonics is approximated by the expression

$$|Ih| = |I_1|/h, \tag{10.2}$$

where

$|I_h|$ = rms value of the harmonic with order h,
$|I_1|$ = rms value of the fundamental component, and
h = order of harmonic.

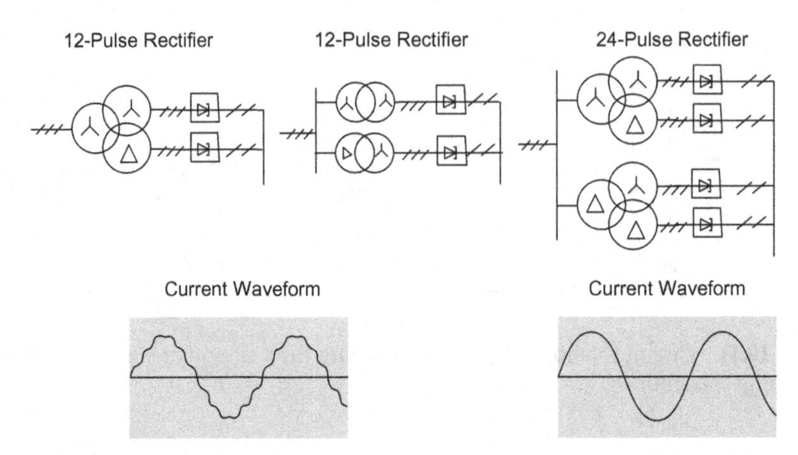

Figure 10.13 Line current waveform of 12- and 24-pulse drive [146].

For example, the fifth harmonic will have a magnitude equal to about 20% of the fundamental component. Six-pulse bridges that do not have a smoothing reactor or a rather small one will produce much higher harmonic currents than predicted by Equation 10.2. When the supply voltage is unbalanced, triple harmonics will also be present in the line current. Therefore, one could also choose a 12-pulse drive or 24-pulse drive rather than a six-pulse drive where the line current waveform is as shown in Figure 10.13.

- *Welding Machines:* A variety of welding machines exist. Many of them are of the single-phase type and are connected between two phases. They produce primarily odd harmonics, including the third.

From the preceding sections, it may be concluded that a lot of the equipment found in modern electric installations can be considered as a harmonic current source. The currents are injected into the supply system and give rise to voltage harmonics. This can be understood by considering the following electric installation and its equipment diagram for one harmonic frequency. In the following discussion, it is assumed that the supply voltage is initially not distorted.

The plant considered in Figure 10.14 has one drive that is connected through a transformer to the medium-voltage (MV) supply. In the equivalent scheme for harmonic evaluation, the supply is modeled by an impedance $Z_s = j\omega L_s$, where L_s is inversely proportional to the system fault level. The transformer is modeled by an impedance $Z_{TR} = j\omega L_{TR}$. In these expressions, j denotes the complex operator.

For simplicity, the resistive part of the supply system and the transformer are ignored. The drive is injecting a fifth harmonic current I_{h5}. Applying Ohm's law to the equivalent circuit allows for the determination of the resulting

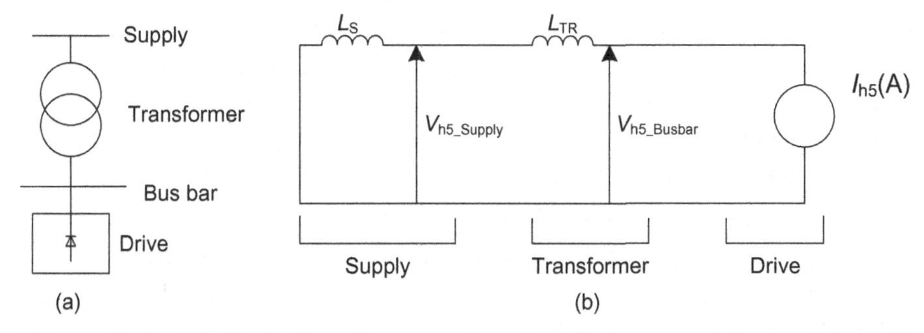

Figure 10.14 (a) Single-line diagram of an installation containing a drive. (b) Corresponding equivalent diagram for one harmonic frequency.

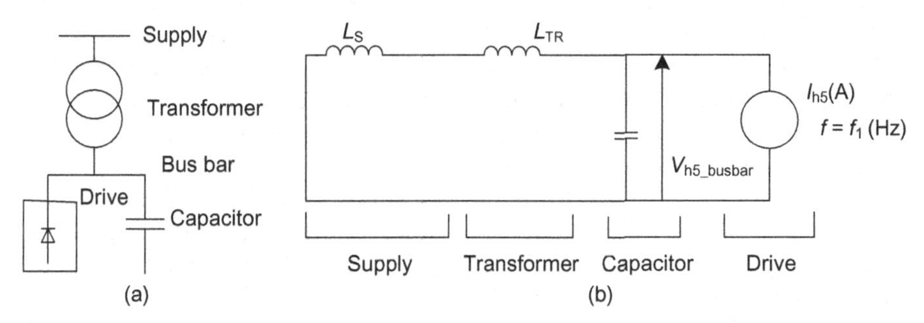

Figure 10.15 (a) Single-line diagram of an installation containing a drive and a capacitor bank. (b) Corresponding equivalent diagram for one harmonic frequency.

harmonic voltage at the bus bar V_{h5_busbar} and at the primary of the transformer V_{h5_supply}, that is,

$$V_{h5\_busbar} = (Z_{TR} + Z_s) \times I_{h5}, \tag{10.3}$$

$$V_{h5\_supply} = Z_s I_{h5}. \tag{10.4}$$

From Equations 10.3 and 10.4, it may be considered that the harmonic current injected by the drive gives rise to a distortion of the bus-bar voltage and the supply voltage. This may affect other loads that are connected to the same or neighboring plants that are connected to the same supply.

In many industrial plants, capacitor banks are often present for reactive power compensation. Imagine that this is the case in the plant discussed in Figure 10.14. The revised single-line diagram is shown in Figure 10.15a and the updated equivalent diagram is shown in Figure 10.15b.

In this case, the harmonic current is injected into the impedance that consists of a parallel circuit of reactors and a capacitor. The total parallel impedance is given by the expression

$$Z_{\text{parallel}} = [(Z_s + Z_{\text{TR}}) \cdot Z_c]/[Z_s + Z_{\text{TR}} + Z_C], \tag{10.5}$$

where

Z_{parallel} = parallel impedance of the reactors and the capacitor,
Z_s, Z_{TR} = supply and transformer impedance, respectively, and
Z_c = capacitor impedance, $Z_c = 1/j\omega C$.

In many cases, the supply impedance is negligible compared with the transformer impedance and can be omitted from Equation 10.5. In that case, the magnitude of the parallel impedance can be shown and given as

$$|Z_{\text{parallel}}(\omega)| = \omega \cdot L_{\text{TR}}/[1 - \omega^2 L_{\text{TR}} \cdot C]. \tag{10.6}$$

From this expression, it may be seen that there exists a frequency f_t (with $\omega_t = 2 \cdot \pi \cdot f_t$) for which the magnitude of the parallel impedance becomes theoretically infinite (denominator becomes zero). In practice, the magnitude at this frequency will be limited by the resistance of the circuit. An example of impedance versus harmonic order graph for a parallel impedance topology is shown in Figure 10.16.

In the example of Figure 10.16, the resonance frequency is situated near the seventh harmonic. Noting that the resulting voltage distortion at a particular harmonic frequency is the product of the injected current and the impedance at that frequency, it may be concluded that the voltage distortion will be very high when the parallel resonance is situated in a frequency range at which a large amount of harmonic currents are injected. This is for instance the case for the application discussed (the drive injects a considerable amount of fifth and seventh harmonic current and the resonance frequency is situated around these frequencies). Since the voltage supply becomes heavily distorted,

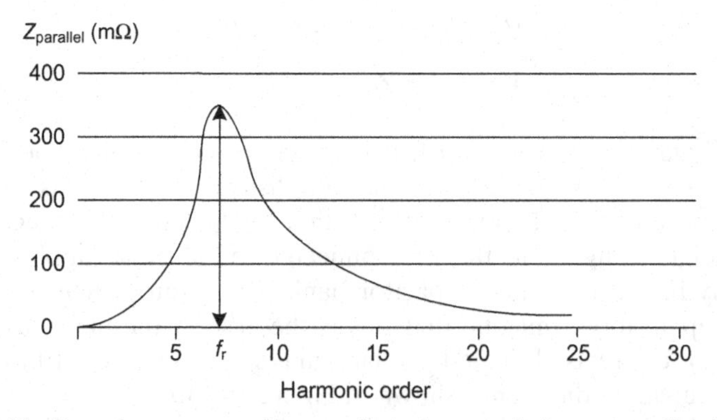

Figure 10.16 Impedance versus harmonic order graph for a parallel impedance topology.

it may affect the proper operation of the equipment being connected to the same supply (e.g., overvoltage on the capacitors and the drives in the plant considered and possibly problems in neighboring plants). In order to predict possible problems when harmonic generating equipment is installed in plants at which capacitors are present, it is useful to know the harmonic order at which resonance exists. A good approximation of this harmonic order is given as

$$h_t = \frac{f_r}{f_1} = \sqrt{\frac{S_{scT}}{Q_c}}, \tag{10.7}$$

where

h_t = harmonic order at which resonance exists,

S_{scT} = short-circuit power of the feeding transformer ($=S_T/U_{cc}$, where, S_T is the transformer power rating, U_{cc} is the transformer short-circuit voltage),

f_r = harmonic frequency and f_1 = fundamental frequency, and

Q_c = the connected capacitive power.

In the preceding discussion, it was demonstrated that the harmonic currents generated by equipment in one plant introduce voltage distortion on the supply to the plant. This distortion will increase if a parallel resonance exists within the plant. Consider now another plant that is connected to the same distorted supply. Suppose that this plant does not have any loads that produce harmonic currents but it does have capacitor banks for reactive power compensation. The single-line diagram of the plant considered (omitting the linear loads) is shown in Figure 10.17a,b shows the equivalent diagram for harmonic studies, including the harmonic voltages present on the supply.

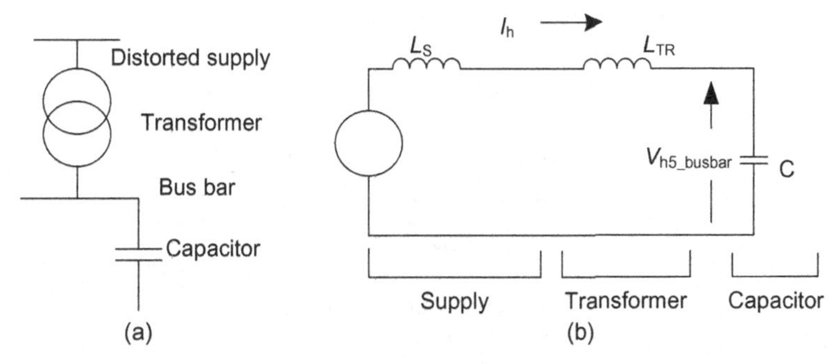

Figure 10.17 (a) Single-line diagram of an installation with a capacitor bank connected to distorted supply. (b) Equivalent diagram for harmonic studies.

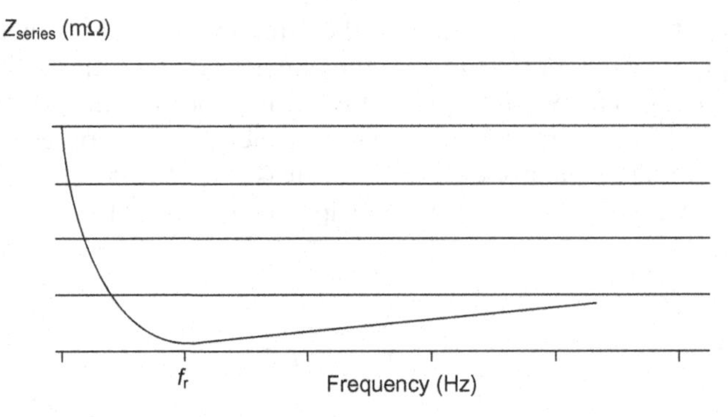

Figure 10.18 Impedance versus frequency for a series impedance topology.

It can be seen in Figure 10.17b that the harmonic voltages present on the supply give rise to harmonic currents flowing into the plant. This current is given by Ohm's law:

$$I_h = V_h / Z_{\text{series}h}, \tag{10.8}$$

where

I_h = current at harmonic frequency h,
V_h = voltage at harmonic frequency h, and
$Z_{\text{series}h}$ = impedance at harmonic frequency h.

From Equation 10.8, it is clear that the harmonic current flowing into the capacitor becomes very high if the series impedance at the corresponding frequency is low. Expression for the impedance as a function of the frequency, ignoring the supply impedance and the resistance of the circuit, is given in Equation 10.9:

$$|Z_{\text{series}}(\omega)| = \omega L_{\text{TR}} - (1/\omega C). \tag{10.9}$$

This equation shows that a resonance frequency f_t exists when the series impedance becomes theoretically zero. In practice, its value will be equal to the resistance of the circuit that tends to be small. The resonance considered is referred to as a series resonance. An example of impedance versus frequency graph for a series impedance topology is shown in Figure 10.18.

From the previous discussion, it may be concluded that the presence of a series resonance in the frequency range at which harmonic voltage components are present is likely to give rise to excessive harmonic currents flowing into the bus-bar system and into the capacitors connected to this system. As a result, the capacitors and other elements of the installation may become overloaded and break down. The harmonic order at which a series resonance is introduced when installing a capacitor bank in a plant can be found in Equation 10.7.

10.3 DISTURBANCES CAUSED BY HARMONICS

From the above explanation, it is now possible to illustrate with more concentration and in a simple way the effect of harmonics present in the distribution systems.

The supply of power to nonlinear loads causes the flow of harmonic currents in the distribution system. Voltage harmonics are caused by the flow of harmonic currents through the impedances of supply circuits (e.g., transformers and distribution system as a whole) (Fig. 10.19).

The impedance of a conductor increases as a function of the frequency of the circuit flowing through it. For each h-order harmonic current, there is, therefore, impedance Z_h in the supply circuit. The h-order harmonic current creates via impedance Z_h a harmonic voltage V_h, where $V_h = Z_h \times I_h$, that is, a simple application of Ohm's law. The voltage at point B is therefore distorted and all devices supplied downstream of this point will receive a distorted voltage. This distortion increases in step with the level of the impedances in the distribution system, for a given harmonic current.

For further understanding of harmonic currents in distribution systems, it may be useful to imagine that the nonlinear loads reinject harmonic currents upstream into the distribution system, in the direction of the source. Flow of the fundamental 50-Hz current in an installation subjected to harmonic disturbances is shown in Figure 10.20a, whereas the h-order harmonic current is presented in Figure 10.20b.

Consider a nonlinear load connected to a distribution system. Supplying this nonlinear load causes the flow in the distribution system of current I_1 (Fig. 10.20a) to which is added each of the harmonic currents (Fig. 10.20b) corresponding to each harmonic of order h.

Using once again the model of nonlinear loads reinjecting harmonic currents into the distribution system, it is possible to graphically represent this phenomenon as in Figure 10.21. It can be seen in this figure that certain loads cause harmonic currents in the distribution system and other loads are disturbed by them. Consequently, in distribution systems, the flow of harmonics reduces power quality and causes a number of problems, technically and economically.

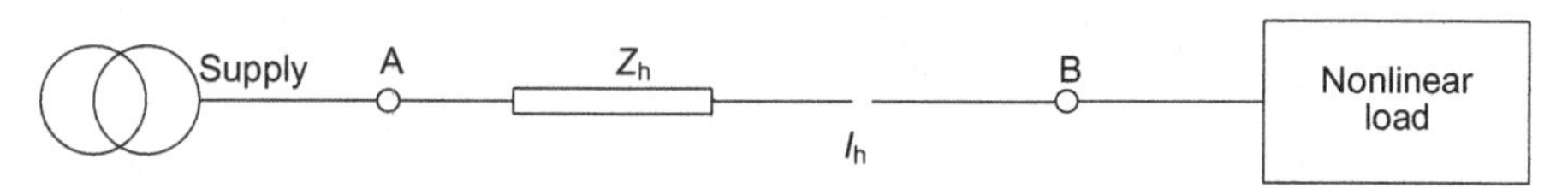

Figure 10.19 Single-line diagram showing the supply circuit impedance for h-order harmonic.

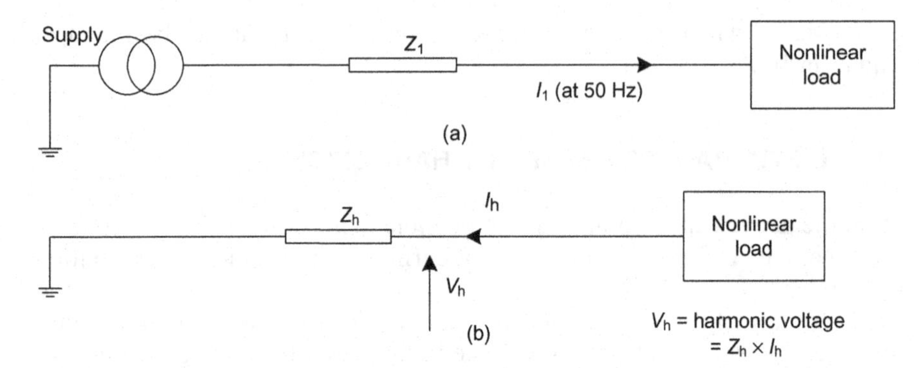

Figure 10.20 A distribution system connected to nonlinear load. (a) Flow of fundamental current. (b) Flow of harmonic currents.

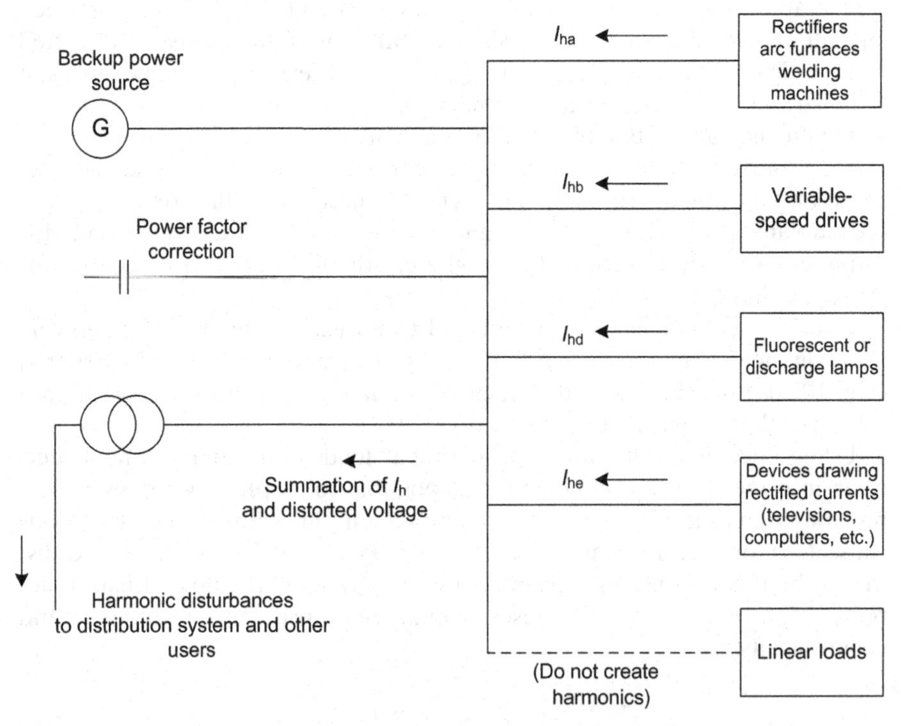

Figure 10.21 Flow of harmonic currents in a distribution system.

10.3.1 Technical Problems

Many problems are created by the harmonics. They have a technical impact as follows:

- overloads on the distribution system due to the increase in rms current;
- overload on the neutral conductor due to the summing of the third and its multiple-order harmonics created by single-phase loads;
- overloads, vibrations, and premature aging of generators, transformers, motors, and so on, and capacitors of power factor (PF) correction equipment;
- distortion of the supply voltage capable of disturbing sensitive loads; and
- disturbances on communication networks and telephone lines.

Further explanation for an increase in the rms value of the current and the neutral current as well is given below.

In the previous sections, it is outlined that many loads inject harmonic currents into the supply. These currents distort the supply voltage that may then give rise to harmonic currents at other locations, even when no harmonic generating equipment is present at these locations. It was also noted that resonance phenomenon, introduced by the interaction between the (inductive) supply system and capacitor banks, amplifies the harmonic distortion when it occurs around harmonic frequencies.

Harmonic pollution causes a number of problems. First effect is the increase of the rms value and the peak value of the distorted waveform. This is illustrated in Figure 10.22, which shows the increase of these values as more

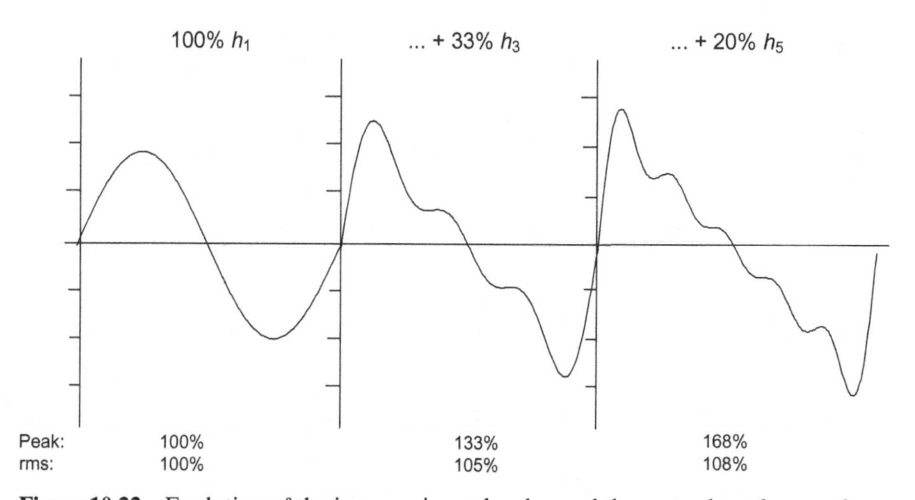

Figure 10.22 Evolution of the increase in peak value and the rms value of a waveform as more harmonic components are added [99].

harmonic components are added to an initially undistorted waveform. The rms value and the peak value of the undistorted waveform are defined as 100%. The peaks of the fundamental and the distortion components are assumed to be aligned. It may be seen that the distorted waveform that contains harmonics up to the fifth harmonic has a peak value that is 68% more than the value of the undistorted waveform and an rms value that is 8% higher.

The increase in rms value leads to increased heating of the electric equipment. Furthermore, circuit breakers may trip due to higher thermal or instantaneous levels. Also, fuses may blow and capacitors may be damaged. Energy meters may give false readings.

The winding and iron losses of motors increase, and they may experience perturbing torques on the shaft. Sensitive electronic equipment may be damaged. Equipment, which uses the supply voltage as a reference, may not be able to synchronize properly and either supply wrong firing pulses to switching elements or switch off. Interference with electronic communications equipment may occur.

In installations with a neutral, zero-phase sequence harmonics may give rise to excessive neutral currents. This is because they are in phase in the three phases of the power system and add up in the neutral. An example of this phenomenon is shown in Figure 10.23.

The excessive neutral current problem is often found at locations where many single loads (e.g., PCs, faxes, and dimmers) are used. The triple harmonics produced by these loads are of the zero-phase sequence type.

Overall, it may be concluded that an excessive amount of harmonics leads to a premature aging of the electric installation. This is an important motivation for taking action to mitigate harmonics.

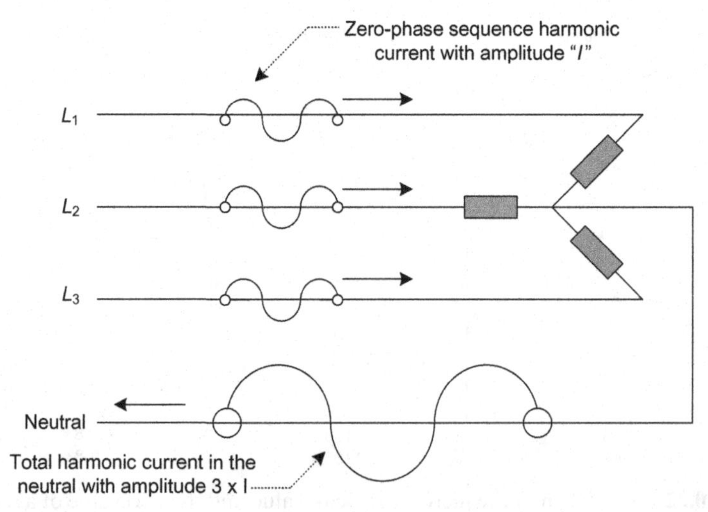

Figure 10.23 Illustration of the addition of zero sequence harmonics in the neutral.

10.3.2 Economical Problems

Harmonics have a significant economic impact, where

- premature aging of the equipment means that it must be replaced earlier, unless it was oversized to begin with;
- overloads on distribution system mean the level of subscribed power must be increased, with additional losses, unless the installation can be upgraded; and
- distortion of the current provokes nuisance tripping and shutdown of production equipment.

Of course, these extra costs in terms of equipment, energy, and productivity all contribute to reducing the competitiveness of the companies.

10.4 PRINCIPLES OF HARMONIC DISTORTION INDICATIONS AND MEASUREMENT

10.4.1 PF

It is defined as the ratio between the active power P and the apparent power S, that is,

$$\text{PF} = P/S \tag{10.10}$$

The PF is often confused with cos Φ, which may be defined as

$$\cos\Phi = P_1/S_1, \tag{10.11}$$

where

P_1 = active power of the fundamental and
S_1 = apparent power of the fundamental.

As described in Equation 10.11, $\cos\Phi$ applies only to the fundamental frequency. When harmonics are present, its value is different than that of the PF.

An initial indication that significant harmonic distortion exists is provided when the measured PF is not equal to $\cos\Phi$ (i.e., the PF is less than $\cos\Phi$).

10.4.2 rms Value

The rms value of a current or a voltage is calculated on the basis of the rms values of different harmonic orders,

$$I_{\text{rms}} = \sqrt{\sum_{h=1}^{\infty} I_h^2}$$

and

$$V_{\mathrm{rms}} = \sqrt{\sum_{h=1}^{\infty} V_h^2}, \tag{10.12}$$

where V_h is the voltage rms value of hth order harmonic and I_h is the current rms value of hth order harmonic.

10.4.3 Crest Factor

Crest factor is the ratio between the peak value of current or voltage and the corresponding rms value, that is,

$$K = I_{\mathrm{m}}/I_{\mathrm{rms}} \text{ or } K = V_{\mathrm{m}}/V_{\mathrm{rms}} \tag{10.13}$$

where

I_{m} = the peak value of current,
I_{rms} = the rms value of current,
V_{m} = the peak value of voltage, and
V_{rms} = the rms value of voltage.

For a sinusoidal signal, the crest factor is, therefore, equal to $\sqrt{2}$. For nonsinusoidal signals, the crest factor can be greater than or less than $\sqrt{2}$.

This factor is particularly useful in drawing attention to exceptional peak values with respect to the rms value.

A typical crest factor for the current drawn by nonlinear loads is much greater than $\sqrt{2}$. Its value can range from 1.5 to 2 or even up to 5 in critical situations. A very high crest factor indicates that high overcurrents occur from time to time. These overcurrents, detected by the protection devices, may cause nuisance tripping.

10.4.4 Power and Harmonics

The active power P of a signal distorted by harmonics is the sum of the active powers corresponding to the voltages and currents in the same frequency order. The expansion of the voltage and current into their harmonic components may be written as

$$P = \sum_{h=1}^{\infty} V_h I_h \cos\Phi_h, \tag{10.14}$$

where Φ_h is the displacement between the voltage and current of harmonic order h. (It is assumed that there is no DC component and for nondistorted signal, i.e., no harmonics, $P = V_1 I_1 \cos\Phi_1$ indicating the power of sinusoidal signal.)

Reactive power applies exclusively to the fundamental and is defined as

$$Q = V_1 I_1 \sin \Phi_1. \tag{10.15}$$

In *distortion power*, considering the apparent power,

$$S = V_{\text{rms}} I_{\text{rms}}.$$

In the presence of harmonics, it becomes

$$S^2 = \sum_{h=1}^{\infty} V_h^2 \cdot \sum_{h=1}^{\infty} I_h^2, \tag{10.16}$$

where V_h is the voltage rms value of hth order harmonic and I_h is the current rms value of hth order harmonic.

Therefore, in the presence of harmonics, the equation $S^2 = P^2 + Q^2$ is no longer valid. The distortion power D is defined as $S^2 = P^2 + Q^2 + D^2$, that is,

$$D = \sqrt{S^2 - P^2 - Q^2}. \tag{10.17}$$

10.5 FREQUENCY SPECTRUM AND HARMONIC CONTENT

10.5.1 Individual Harmonic Distortion

The amplitude of each harmonic order is an essential element for the analysis of harmonic distortion. Individual harmonic distortion is defined as the level of distortion, in percent, of order h with respect to the fundamental,

$$INDv_h = \frac{V_h}{V_1} \times 100\% \text{ and } INDi_h = \frac{I_h}{I_1} \times 100\%, \tag{10.18}$$

where $INDv_h$ is the individual harmonic distortion of hth order voltage harmonic, $INDi_h$ is the individual harmonic distortion of hth order current harmonic, and V_1 and I_1 are the fundamental rms values of voltage and current, respectively.

Plotting the amplitude of each harmonic order on a graph gives a graphical representation of the frequency spectrum; for example, Figure 10.24 shows the spectral analysis of a square-wave signal.

10.5.2 THD

For the current and voltage harmonics, THD_I and THD_V, respectively are given as

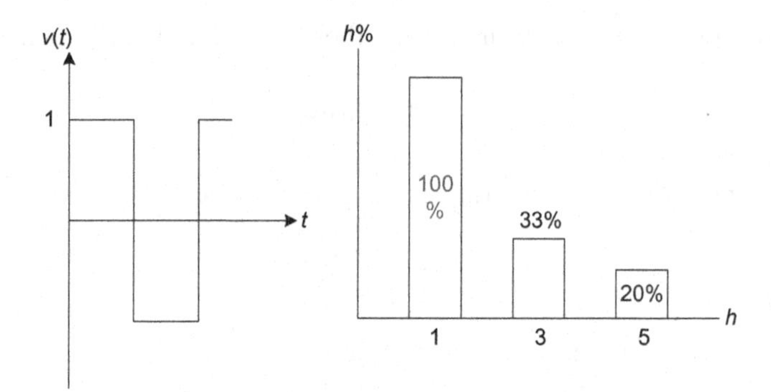

Figure 10.24 Frequency spectrum of a square-wave signal.

$$THD_{\mathrm{I}} = \frac{\sqrt{\sum_{h=2}^{\infty} I_h^2}}{I_1} \text{ and } THD_{\mathrm{V}} = \frac{\sqrt{\sum_{h=2}^{\infty} V_h^2}}{V_1}. \tag{10.19a}$$

THD_{I} is a measure of the amount by which a composite current waveform deviates from an ideal sine wave.

At light loads, that is, the fundamental component of load current is low; the THD value is high resulting in misleading of distortion levels. So, the THD may not be of significant concern in this case where the magnitude of harmonic current is low and its relative distortion to the fundamental frequency is high. To avoid this ambiguity, a total demand distortion factor (TDD) is used and is becoming accepted instead of THD factor. It is defined as

$$TDD_{\mathrm{I}} = \frac{\sqrt{\sum_{h=2}^{N} I_h^2}}{I_{\mathrm{R}}} \tag{10.19b}$$

where I_{R} is the rated rms load current.

The two factors THD and TDD are similar except the distortion in TDD is expressed as a percentage of the rated load current. On the other hand, it is practical and more realistic to express the distortion of current wave as a percentage of the designed value of rated load current or its maximum on which the power system design is based [148]. A typical example with measured parameters is given in Table 10.1 to illustrate that as the load decreases, TDD_{I} decreases while THD_{I} increases.

10.5.3 Relation Between PF and THD

Assuming the voltage is sinusoidal, that is, virtually sinusoidal, the following relations can be derived:

TABLE 10.1 ***TDD_I* Versus *THD_I***

Measured Parameters				
Total I, rms	Fund. I, rms	Harm I, rms	THD_I (%)	TDD_I (%)
936.68	936.00	35.57	3.8	3.8
836.70	836.00	34.28	4.1	3.7
767.68	767.00	32.21	4.2	3.4
592.63	592.00	27.23	4.6	2.9
424.53	424.00	21.20	5.0	2.3
246.58	246.00	16.97	6.9	1.8
111.80	111.00	13.32	12.0	1.4

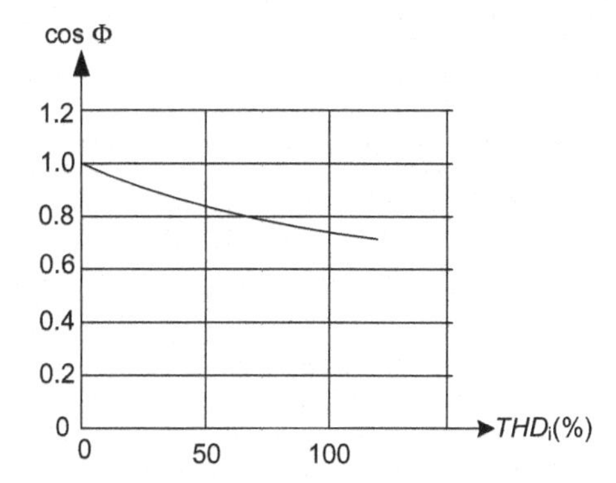

Figure 10.25 Variation of PF versus THD_I.

$$P \approx P_1 = V_1 I_1 \cos\Phi_1$$

and

$$\mathrm{PF} = \frac{P}{S} \approx \frac{V_1 I_1 \cos\Phi_1}{V_1 I_{\mathrm{rms}}}.$$

From the above equations,

$$\frac{I_1}{I_{\mathrm{rms}}} = \frac{1}{\sqrt{1+\mathrm{THD}_I^2}}$$

Hence,

$$\mathrm{PF} = \frac{1}{\sqrt{1+THD_I^2}} \tag{10.20}$$

Equation 10.20 is described graphically in Figure 10.25.

10.6 STANDARDS AND RECOMMENDATIONS

The harmonic pollution may not only affect in polluting the plant but may also disturb equipment in other plants. In order to limit this disturbance, maximum allowable distortion limits have been defined in standards and recommendations. Tables 10.2–10.5 show the values in the IEEE Std. #519 [148]. Also, the International Electromechanical Commission (IEC) has issued technical publications that outline assessment procedures to determine whether distorting loads may be connected to the supply system [149–151]. In many cases, the regulations impose a limit for the individual and THD of the voltage or current present at the point-of-common coupling (PCC) [87, 152–155]. The PCC is the location at which the plant is connected to the public power system (generally at the primary of the main transformer(s)). IEEE Standard #519-1992 allows

TABLE 10.2 Allowable Limits of Voltage and Current Distortion [148]

Maximum Voltage Distortion (IEEE Std. #519)			
	System Voltage		
Maximum Distortion (in %)	<69 kV	69–138 kV	>138 kV
Individual harmonic	3.0	1.5	1.0
Total harmonic	5.0	2.5	1.5

Note: For conditions lasting more than 1 h, shorter periods increase limit by 50%.

TABLE 10.3 Maximum Harmonic Current Limits (in %) for System Voltage > 161 kV (Odd Harmonics) [148]

I_{sc}/I_L	<11	$11 < H < 17$	$17 < H < 23$	$23 < H < 35$	$35 \leq H$	THD
<50	2.0	1.0	0.75	0.3	0.15	2.5
≥50	3.0	1.5	1.15	0.45	0.22	3.75

Note: Even harmonics are limited to 25% of the odd harmonic limits above.

TABLE 10.4 Maximum Harmonic Current Limits (in %) for System Voltage 69–161 kV (Odd Harmonics) [148]

I_{sc}/I_L	<11	$11 < H < 17$	$17 < H < 23$	$23 < H < 35$	$35 \leq H$	TDD
<20	2.0	1.0	0.75	0.3	0.15	2.5
20–50	3.5	1.75	1.25	0.5	0.25	4.0
50–100	5.0	2.25	2.0	0.75	0.35	6.0
100–1000	6.0	2.75	2.5	1.0	0.5	7.5
≥1000	7.5	3.5	3.0	1.25	0.7	10.0

Note: Even harmonics are limited to 25% of the odd harmonic limits above.

TABLE 10.5 Maximum Harmonic Current Limits (in %) for System Voltage < 69 kV (Odd Harmonics) [148]

I_{sc}/I_L	<11	$11 < H < 17$	$17 < H < 23$	$23 < H < 35$	$35 \leq H$	TDD
<20	4.0	2.0	1.5	0.6	0.3	5.0
20–50	7.0	3.5	2.5	1.0	0.5	8.0
50–100	10.0	4.5	4.0	1.5	0.7	12
100–1000	12	5.5	5.0	2.0	1.0	15
>1000	15	7.0	6.0	2.5	1.4	20

Note: Even harmonics are limited to 25% of the odd harmonic limits above.
I_{sc} = maximum short circuit current at point-of-common coupling; I_L = maximum demand load current (fundamental frequency) at point-of-common coupling; TDD = total demand distortion in percent of maximum demand.

TABLE 10.6 Values of Individual Harmonic Voltages as a Percent of Nominal Voltage at the Supply Terminals for Order up to 25 [156]

Odd harmonics	Not multiples of 3	Order (h)	5	7	11	13
		Relative voltage (%)	6	5	3.5	3
	Multiples of 3	Order (h)	3	9	15	21
		Relative voltage (%)	5	1.5	0.5	0.5
Even harmonics	Order (h)	Order (h)	2	4	6–24	
	Relative voltage (%)	Relative voltage (%)	2	1	0.5	

TABLE 10.7 THD_V Limits for Some Loads [154]

Type of Load	Computers	Telecommunications	TV Studios	General Distribution	Industrial Equipment
Max. THD_V	5%	5%	5%	8%	8%

Note: Point of connection to the public AC system: <5–8%.

the values of THD of the voltage or current at PCC to not exceed 5% for MV distribution systems, and not exceed 3% for critical systems:

- hospital equipment,
- air traffic control systems,
- mainframe computers, and
- data processing systems.

The European Standard EN50160 has defined the limits of individual odd and even harmonics of the voltage at the supply terminals as given in Table 10.6 [154]. Also, Table 10.7 gives the maximum values of THD_V for some loads, which have been taken from standards IEC 61000-2-2 and 61000-2-4.

CHAPTER 11

HARMONICS EFFECT MITIGATION

11.1 INTRODUCTION

The negative effects of harmonics on distribution systems have been illustrated in Chapter 10. It is necessary to look for solutions to eliminate these harmonics or at least mitigate their effects so that they fall within the accepted limits set by the standards (e.g., IEEE #519, 1992, International Electromechanical Commission [IEC] 61000-2-2, IEC 61000-2-4).

The solutions can be classified into three classes. The first class is concerned with the distribution system planning to determine how the installation can be modified. The second class of solutions deals with the use of special devices in the power supply system such as inductors and special transformers. If the first and second classes of solutions are not sufficient to attenuate the harmonics, the third class of solutions requiring the use of filters can be applied. More details about these solutions are given in the following sections.

11.2 FIRST CLASS OF SOLUTIONS

11.2.1 Supplying the Loads from Upstream

This solution can be implemented by connecting the disturbing load upstream in the system regardless of the economics. It is preferable to connect the disturbing load as far upstream as possible (Fig. 11.1).

Electric Distribution Systems, First Edition. Abdelhay A. Sallam, Om P. Malik.

Published 2011 by John Wiley & Sons, Inc.

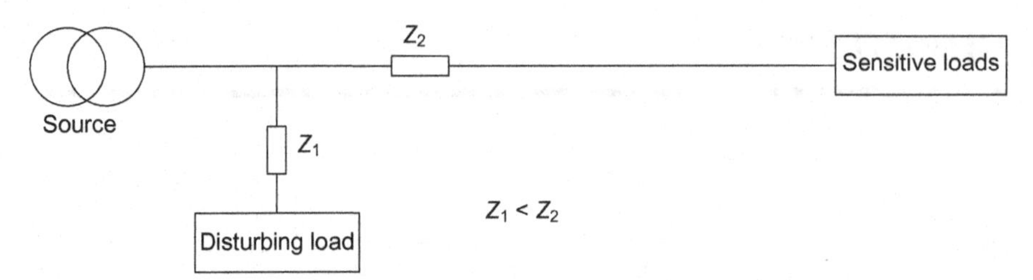

Figure 11.1 Disturbing load far from sensitive load.

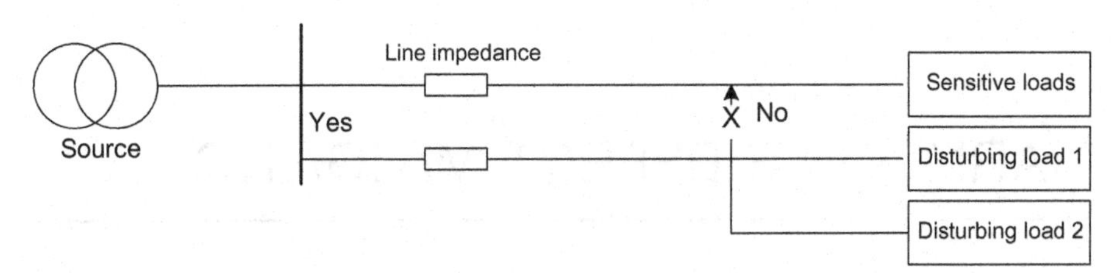

Figure 11.2 Grouping of disturbing loads.

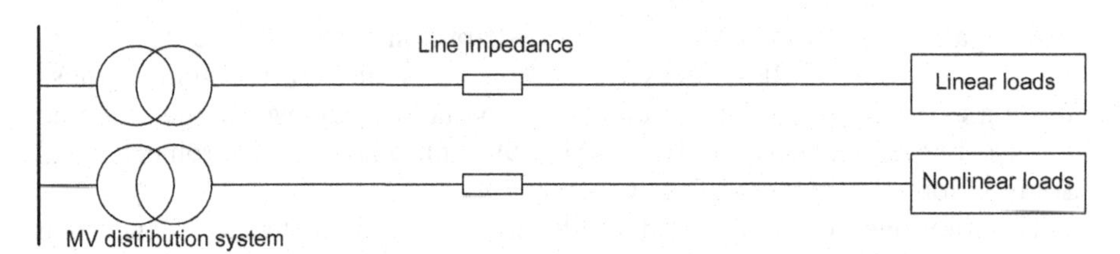

Figure 11.3 Supplying the loads from different sources.

11.2.2 Grouping the Disturbing Loads

Depending on the distribution system structure and the possibility of providing this solution, different types of loads should be supplied by different bus bars as seen in the single-line diagram of Figure 11.2. By grouping the disturbing loads, the possibilities of angular recomposition are increased as the vector sum of the harmonic currents is less than their algebraic sum. An effort should be made to avoid the flow of harmonic currents in the cables, because they cause voltage drops and temperature rise.

11.2.3 Supplying the Loads from Different Sources

In this method, additional improvement may be obtained by supplying different loads from separate transformers acting as different sources (Fig. 11.3).

11.3 SECOND CLASS OF SOLUTIONS

11.3.1 Use of Transformers with Special Connections

Special types of connections may be used in transformers to eliminate certain harmonic orders. The harmonic order elimination depends on the type of connection implemented as described below:

- Δ-Y-Δ connection eliminates the fifth- and seventh-order harmonics (Fig. 11.4),
- Δ-Y connection eliminates the third-order harmonic (the harmonic flows in each of the phases and loops back through the transformer neutral), and
- Δ-zigzag connection eliminates the harmonics looping back through the magnetic circuit.

11.3.2 Use of Inductors

In installations comprising variable-speed drives, the current can be smoothed by the installation of series inductors in the line. By increasing the impedance of the supply circuit, the harmonic current is attenuated. Use of harmonic inductors on capacitor banks is a means of increasing the impedance of the inductor and capacitor assembly for higher-order harmonics.

11.3.3 Arrangement of System Earthing

As it is explained in Chapter 3, TN systems are the systems having directly earthed neutral, and the exposed conductive parts of the loads are connected to the neutral conductor. One of the TN types is called TNC when the neutral and protection conductors are combined in a single conductor. In TNC systems, a single conductor, the protection neutral, ensures protection in the event of an earth fault and carries imbalance currents.

Under steady-state conditions, the harmonic currents flow through the protection neutral PEN. However, PEN has certain impedance, resulting in slight voltage differences (a few volts) between devices, which may lead to malfunction of the electronic equipment.

The TNC system must, therefore, be used only for the supply of power circuits on the installations and must never be used for the supply of sensitive

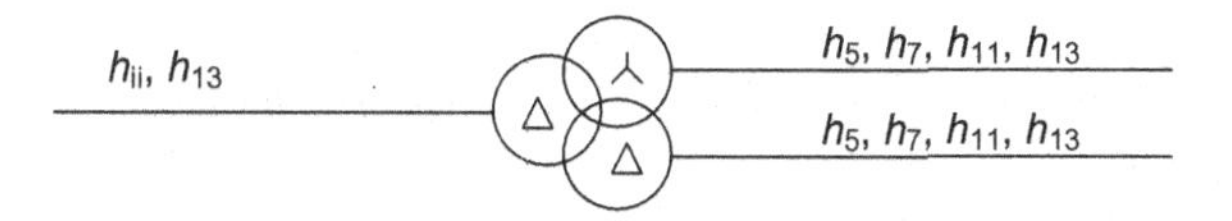

Figure 11.4 Δ-Y-Δ prevents h_5 and h_7 propagation.

loads. The neutral conductor (NC) and the protection conductor (PE) are completely separate, thus ensuring much more stable voltage on the distribution system.

11.3.4 Use of Six-Pulse Drive

One could also choose 12-pulse drives rather than six-pulse drives. These produce harmonics the order of which is given by the expression $n = (12 \cdot i \pm 1)$, where i is an integer greater than or equal to 1. Thus, the line current contains harmonics of the order 11, 13, 23 ... , each with a magnitude $|I_n| = |I_{fund}|/n$, which is an improvement over the spectrum generated by six-pulse equipment.

11.4 THIRD CLASS OF SOLUTIONS

When the harmonic levels are too high, a harmonic filter solution is needed. Traditionally, passive filters, active filters (AFs), and recently hybrid filters have been used.

11.4.1 Passive Filters

A passive filter consists of a series circuit of reactors and capacitors. Harmonic currents generated by, for example, a frequency converter are shunted by this circuit designed to have low impedance at a given frequency compared with the rest of the network.

The described function with a harmonics generator, impedance representing all other loads, a filter, and the medium voltage network are illustrated schematically in Figure 11.5. The equivalent circuit seen with the harmonics generator modeled as a harmonic current generator is shown in Figure 11.6. It includes the medium-voltage (MV) network, and the voltage distortion is represented by a harmonic voltage generator.

As the passive filters offer very low impedance at the resonance frequency, the corresponding harmonic current will flow in the circuit whatever its magnitude. Passive filters can get overloaded under which condition they will

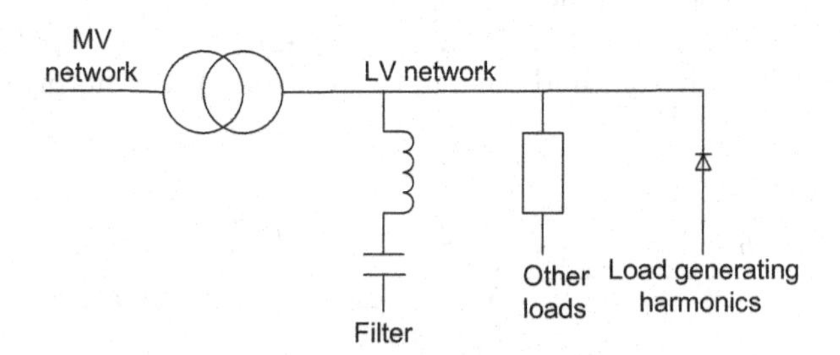

Figure 11.5 Passive filtering of harmonics.

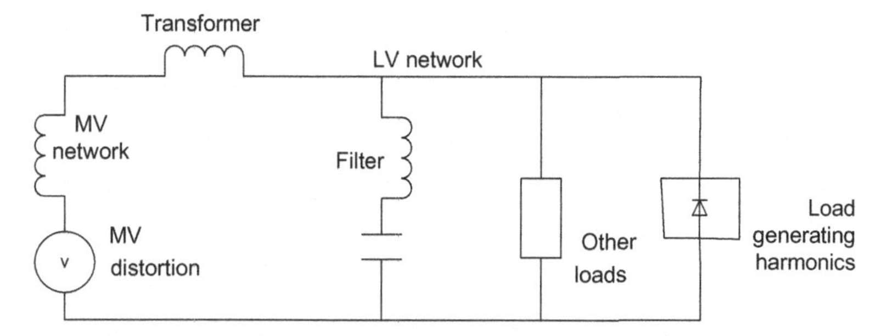

Figure 11.6 Equivalent circuit for passive harmonic filtering.

switch off or be damaged. The overload may be caused by the presence of unforeseen harmonics on the supply system or caused by structural modifications in the plant itself (such as the installation of a new drive).

Passive filter always provides a certain amount of reactive power. This is not desirable when the loads to be compensated are alternating current (AC) drives, which already have a good power factor (PF). In that case, the risk of overcompensation exists as a result of which the utility may impose a fine. The degree of filtering provided by the passive filter is given by its impedance in relation to all other impedances in the network. As a result, the filtration level of a passive filter cannot be controlled and its tuning frequency may change in time due to aging of the components or network modifications. The quality of the filtration will then deteriorate.

It is also important to note that a passive filter circuit may only filter one harmonic component. A separate filter circuit is required for each harmonic that needs to be filtered. In order to overcome the problems associated with traditional passive filters and in order to meet the continuing demand for good power quality, AFs have been developed for low-voltage (LV) applications.

11.4.2 AFs

AFs are systems employing power electronics. They are installed either in series or in parallel with the nonlinear load to provide the harmonic currents required by nonlinear load and thereby avoid distortion on the power system (Fig. 11.7).

The AFs inject, in opposite direction, the harmonics drawn by the load, such that the line current I_s remains sinusoidal. They are effective and recommended for the commercial installations comprising a set of devices generating harmonics with a total power rating less than 200 kVA (variable-speed drives, uninterruptible power supplies [UPSs], office equipment, etc.). Also, they are used for the situations where the current distortion must be reduced to avoid overloads.

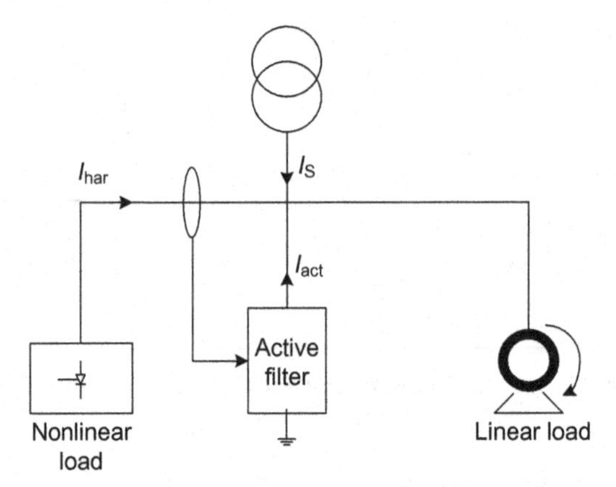

Figure 11.7 Active filter installation. I_s = source current; I_{act} = current injected by active filter; I_{har} = harmonic current generated by nonlinear load.

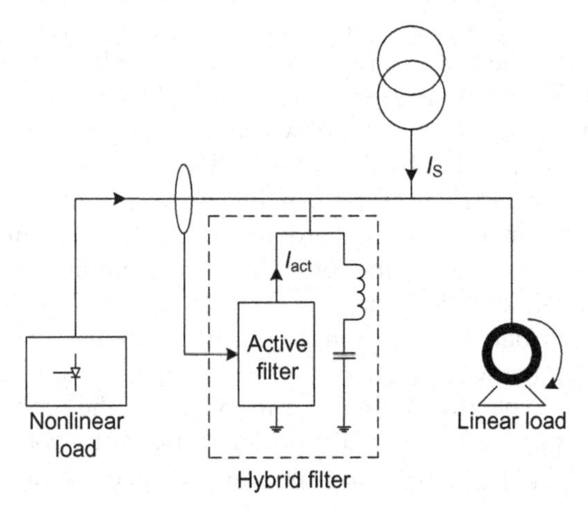

Figure 11.8 Hybrid filter installation.

11.4.3 Hybrid Filters

The two types of filters presented above can be combined in a single device, thus constituting a hybrid filter. This type of filtering solution combines the advantages of the existing systems and provides a high-performance solution covering a wide power range (Fig. 11.8).

The hybrid filters are adequate to be used for the following typical applications:

- industrial installations comprising a set of devices causing harmonics with a total power rating in the range of 200 kVA or more (variable-speed drives, UPSs, rectifiers, etc.);

- situations where voltage distortion must be reduced to avoid disturbing sensitive loads;
- situations where conformity with strict harmonic emission limits is required;
- installations where PF correction is required; and
- situations where current distortion must be reduced to avoid overloads.

11.5 SELECTION CRITERION

From the above explanation of filters and types of filters, it is possible to choose a suitable type of filter with the following criteria.

Passive filters offer both PF correction and large capacity for current filtering. Installations where the passive filters are installed must be sufficiently stable, that is, a low level of load fluctuations. If a high level of reactive power is supplied, it is advised to de-energize the passive filter when load levels are low.

Preliminary studies for a filter must take into account any capacitor banks and may lead to their elimination.

AFs compensate for harmonics over a wide range of frequencies. They can adapt to any load; however, their conditioning capacity is limited.

Hybrid filters combine the strong points of both passive filter and AFs.

11.6 CASE STUDIES

11.6.1 General

There are successive procedures that need to be followed to achieve an optimal solution of the harmonics problem. The cycle of solution is shown in Figure 11.9 and the procedures can be summarized as below:

1. Measure all electric parameters needed from the field of application such as voltage and current waveforms at different loading percentages to identify the degree of load nonlinearity, total harmonic distortion (THD), active and reactive power, and actual PF.
2. Simulate and model the distribution system in a valid form for studying and analyzing [157].
3. Analyze the system and write a report that explains the problems of PF and harmonics [158].
4. Specify and propose the solution.
5. Establish a procedure for procuring and commissioning the necessary equipment for the proposed solution.
6. Measure the electric parameters of the system to verify the effectiveness of the proposed solution.

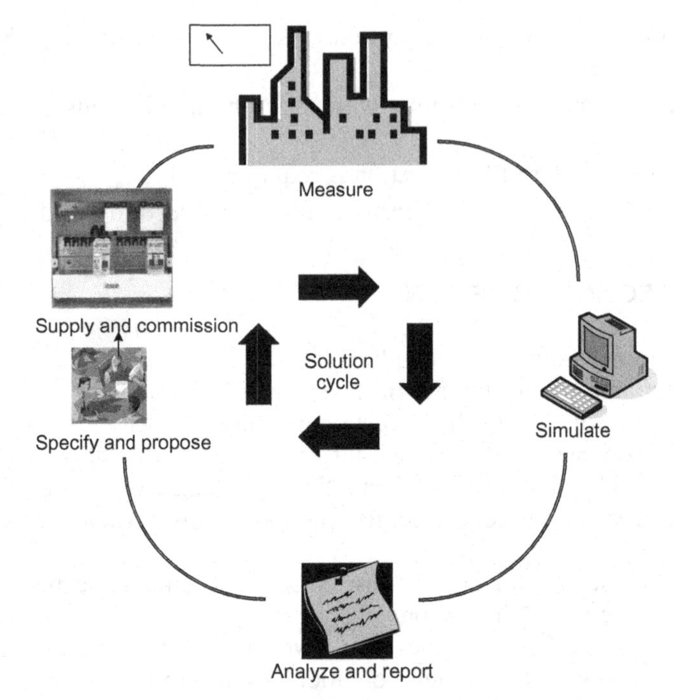

Figure 11.9 Main procedures of solution cycle.

11.6.2 Need for Shunt Capacitors

As explained in Chapter 9, the PF at load points of any distribution system must be improved to be not less than 0.9 as a common value of utilities. This improvement helps in the reduction of the power losses as the power loss is proportional to the square of current flow and the current is inversely proportional to the PF. The percentage of loss reduction resulting from PF improvement is given as

$$\text{\% of Loss reduction} = \left[1 - \left(\frac{\text{Actual power factor}}{\text{Desired power factor}}\right)^2\right] \times 100\%. \tag{11.1}$$

Thus, the heat produced by current flow through conductors is reduced particularly as the loss in the line conductors that can account for as much as 2–5% of the total load. The capacitors can reduce the losses by 1–2% of total load.

By adding the capacitors, the voltage will increase only by a few percentage points that is not a significant economic or system benefit. Usually, these capacitors, connected in shunt, are installed with the goal of PF improvement and not voltage rise. Also, they are rated to avoid PF overcorrection. Severe overcorrection can cause a voltage rise that can damage the insulation and the

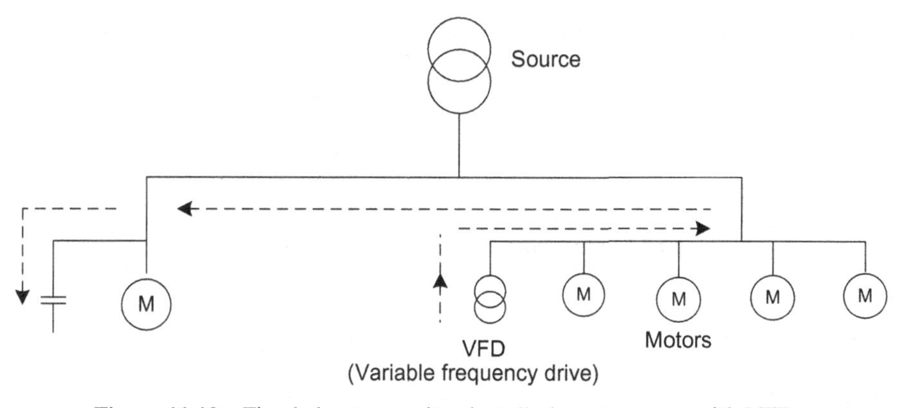

Figure 11.10 Fixed shunt capacitor installation at motors with VFDs.

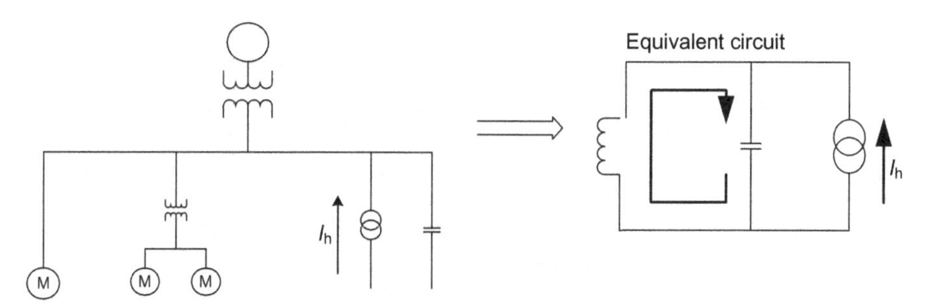

Figure 11.11 A network with shunt capacitor and its equivalent circuit. I_h = harmonic current.

equipment or result in utility surcharges. So, this aspect must be taken into consideration particularly when using large fixed capacitors at mains. Another aspect to consider, when installing fixed capacitors at motors with electronic drives, is the effect of motors located elsewhere. The harmonic currents they produce can flow back to the point of lowest impedance, the capacitor, and may cause premature failure (see Fig. 11.10).

Therefore, care must be taken to not apply capacitors in the presence of harmonics.

11.6.3 Effects of Harmonics on PF Capacitors

Capacitors have low impedance at high frequencies. This means that the capacitors absorb harmonic currents and consequently will heat up resulting in reduction of their life expectancy. Also, the voltage harmonics stress the capacitor dielectric and reduce the capacitor life expectancy. In addition, the parallel combination of capacitors with motor or transformer can cause resonance leading to magnified harmonic currents on the network (Fig. 11.11).

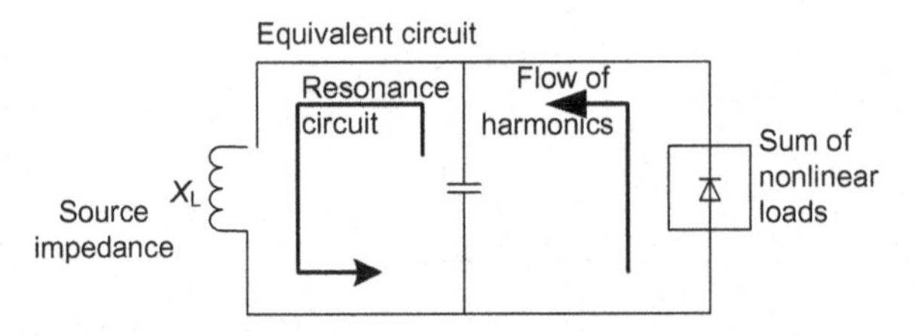

Figure 11.12 Equivalent circuit of harmonics current flow.

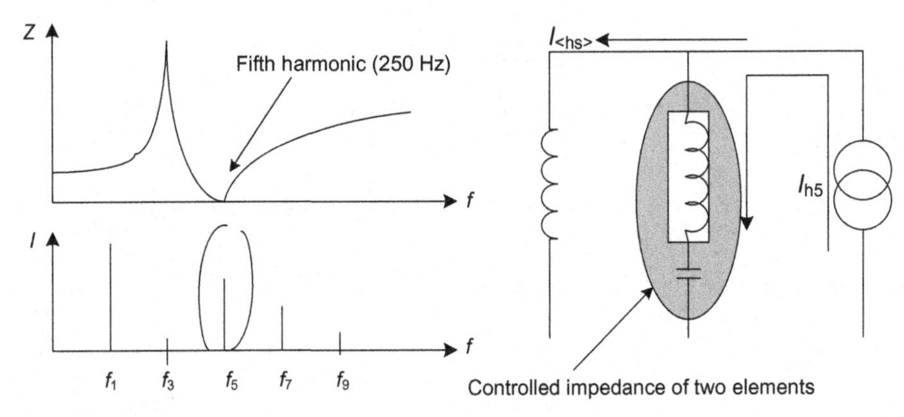

Figure 11.13 Network detuning.

A standard network and its equivalent circuit are shown in Figure 11.11. For example, consider a transformer rating of 1500 kVA and reactive power of 300 kvar compensated by 20% capacitive reactive power. The resonance frequency, f_r, can be calculated by applying Equation 10.7 as follows:

$$f_r = 50\sqrt{\frac{1500 \times 100}{300 \times 20}} = 250 \text{ Hz},$$

that is, the fifth harmonic.

The equivalent circuit of the fifth-order harmonic current is shown in Figure 11.12. An inductive element in series with the shunt capacitor can be added and the impedance of these two elements must be controlled to detune the network. Detuning the network is aimed at forcing the resonant point away from the naturally occurring harmonics. As seen in Figure 11.13, the circuit impedance Z has the minimum value at the fifth harmonic frequency (resonance frequency). Accordingly, the harmonic spectrum shows that the magnitude of fifth harmonic current I_{h5} is the largest one. The circuit must be detuned by controlling the impedance of two elements connected in parallel with the load to force the resonant point away from the fifth harmonic frequency.

11.6.4 PF Correction for a Pipe Welding Industry

An industrial plant manufacturing gas and oil pipes (Fig. 11.14) is taken as a real case study to illustrate the process of PF correction in the presence of harmonics.

This plant has six decentralized substations, 22/0.4 kV. Four substations have a transformer with the rating of 2000 kVA each and one substation with a rating of 1500 kVA for services. The sixth substation has a 2500-kVA transformer. The substations, connected together by an open-ring system, are supplied by the utility company through a two-out-of-three switchboard at 22 kV as shown in Figure 11.15. The 2500-kVA transformer feeds a machine for internal welding and some auxiliaries while one 2000-kVA transformer feeds a machine for external welding. The machines in the plant are controlled by programmable logic controllers (PLCs) and computer numeric control (CNC) in addition to the electronic motor drives. The supplier of internal and external welding machines asked for variable compensation with a maximum value of 2000 kvar to be connected with the 2500-kVA transformer at the low tension side to correct the PF and a variable compensator (from 0% to 40% of transformer rating) for each of the other substations. The problems met, particularly for substations feeding the welding machines, are the following:

- The conventional compensator consists of multistage capacitor banks. Practically, the maximum rating of each stage is 60 kvar. This means that a 2000-kvar compensation needed 33 electric steps. The switching time of each step is 30 s to be inserted into the system, that is, the switching time of 33 steps is 16.5 min to be connected to the system. On the other hand, the maximum pipe length is 12 m and the welding time of this pipe is 6 min

Figure 11.14 Pipe welding machine.

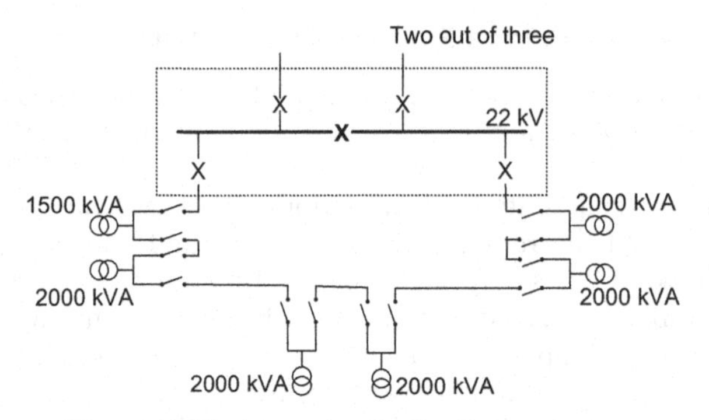

Figure 11.15 Plant electric distribution system.

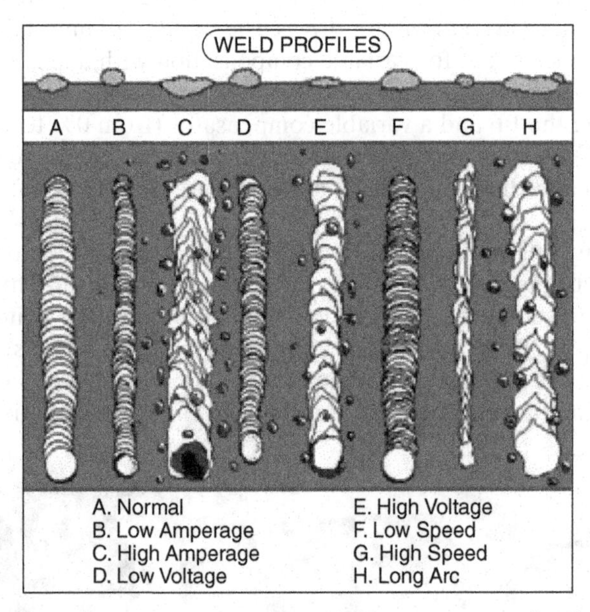

Figure 11.16 Weld profile at different parameters [159].

at maximum operating conditions (maximum diameter and maximum thickness).

So, the problem encountered is that the welding time is much less than the switching time of required compensation.

- The unsynchronized welding will cause unpredictable voltage and current fluctuations. The standard weld profile is greatly affected by the changes of supply voltage, the amperage, the machine speed, and the arc length as shown in Figure 11.16. It is necessary to keep the voltage constant during the machine operation to achieve the standard weld quality. Therefore, the importance of keeping the voltage constant is not only to enhance the distribution system performance but also to produce the accepted quality of plant products.

- Current harmonics and voltage spikes are produced during the welding operation as shown in Figure 11.17.

The solution to these problems requires harmonic cancellation, PF correction, var compensation, resonance elimination, simultaneous modes of operation, and 8-ms response time.

The solution provided by a supplier was a hybrid var compensator (HVC), which includes a passive filter and an AF called "AccuSine" as shown in Figure 11.18. The hybrid filter as mentioned in Section 11.4.3 offers the advantages of passive and active filtering solutions. It covers a wide range of power and performance as follows:

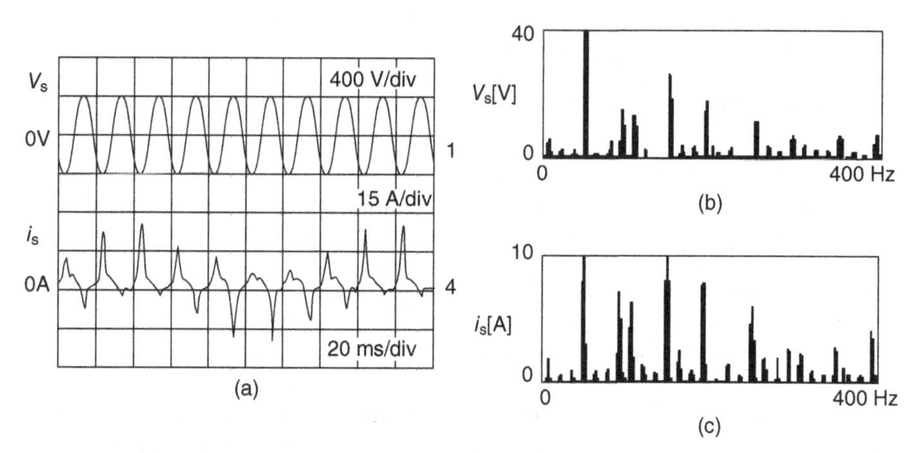

Figure 11.17 Voltage and current: waveforms and spectrum (real measurements recorded during system operation). (a) Mains voltage and current (zoom); (b) voltage spectrum; (c) current spectrum.

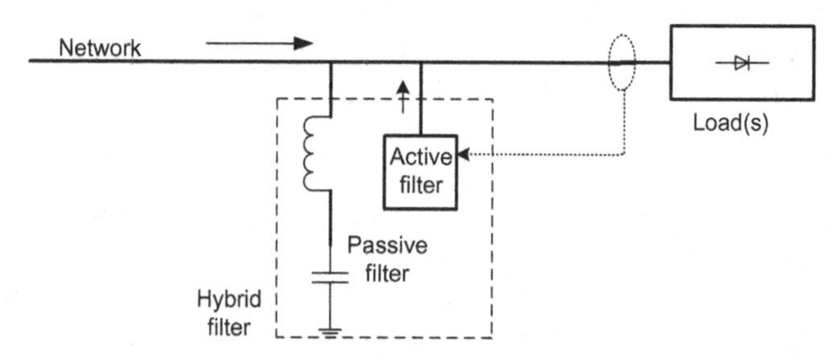

Figure 11.18 A hybrid filter connected to a network.

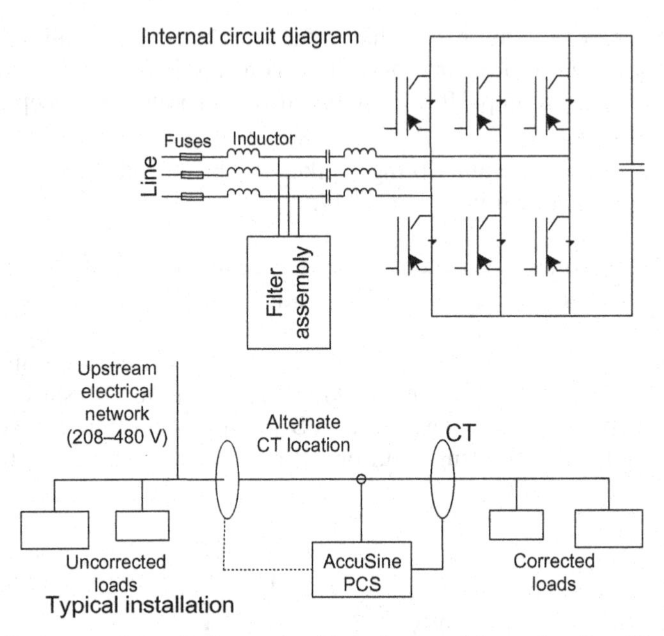

Figure 11.19 Internal circuit diagram and installation of the active filter "AccuSine"®.

- filtering on a wide frequency band (elimination of harmonics numbered 2–25),
- compensation of reactive power,
- large capacity for current filtering, and
- good technical–economic solution for network filtering.

The installation of "AccuSine" filter and its internal circuit diagram are shown in Figure 11.19. A picture of its construction is shown in Figure 11.20.

11.6.4.1 How the AF Works

1. *Harmonic Cancellation:* The filter cancels harmonics by dynamically injecting out of phase harmonic current, which yields a reduction of current wave distortion that, in turn, reduces voltage distortion. The filter is equipped with closed-loop control to allow for high accuracy and self-adaptive control.

It determines the harmonic compensation required by using current transformers to measure the harmonic current content in nonlinear load current waveform (Fig. 11.21a). The filter control logic removes the fundamental frequency component (50 or 60 Hz) from this waveform. The remaining waveform is then inverted and fires isolated gate bipolar transistors (IGBTs) to inject this waveform (Fig. 11.21b) onto the content as seen by the upstream

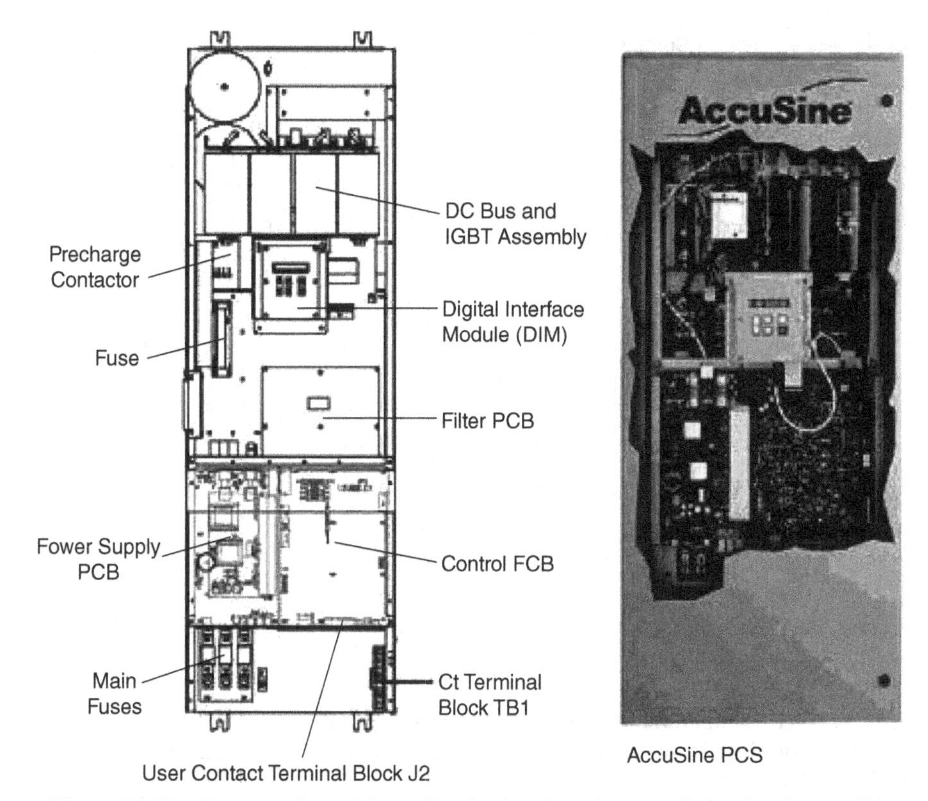

Figure 11.20 Construction of AccuSine® filter (reprint permission by Square-D).

electric system (Fig. 11.21c). Typical results indicating the effectiveness of filter use are given in Table 11.1.

2. *Resonance Elimination*: By dynamically removing harmonics from the network, no energy is present at the resonant frequency. The point of filter installation on the electric network determines where the harmonic elimination takes place. So, harmonic current cancellation on the network leads to the elimination of resonance conditions.
3. *PF Correction*: The filter is able to correct only the PF, or operate in a dual mode whereby current is injected to reduce harmonics, and any excess current capacity is used to improve the PF. PF correction is achieved by injecting current at the fundamental frequency (50/60 Hz). The filter is able to correct for either a leading or lagging PF. The result is a reduction in peak currents that frees system capacity and eliminates utility-imposed PF penalties. The load current waveform with poor PF and its correction after filter installation is shown in Figure 11.22a,b, respectively.
4. *Dynamic Var Compensation*: Large inductive inrush currents typically cause voltage sags that result in reduced productivity, poor process

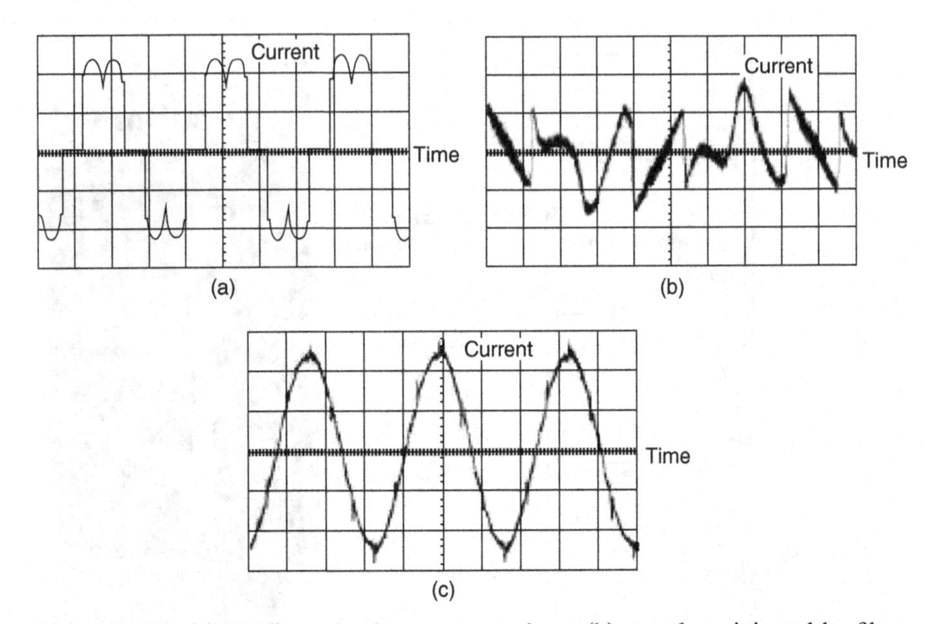

Figure 11.21 (a) Nonlinear load current waveform; (b) waveform injected by filter; (c) corrected current waveform.

TABLE 11.1 Amplitude of Current Components as a Percentage of Fundamental without/with Filter Installation

Order	Fund.	3	5	7	9	11
$I_{without}$%	100	0.038	31.66	11.48	0.435	7.068
I_{with}%	100	0.478	0.674	0.679	0.297	0.71

13	15	17	19	21	23
4.267	0.367	3.438	2.904	0.284	2.042
0.521	0.520	0.464	0.639	0.263	0.409

25	27	29	31	33	35
2.177	0.293	1.238	1.740	0.261	0.800
0.489	0.170	0.397	0.243	0.325	0.279

37	39	41	43	45	47
1.420	0.282	0.588	1.281	0.259	0.427
0.815	0.240	0.120	0.337	0.347	0.769

49	**THD (%)**
1.348	**35.280**
0.590	**2.670**

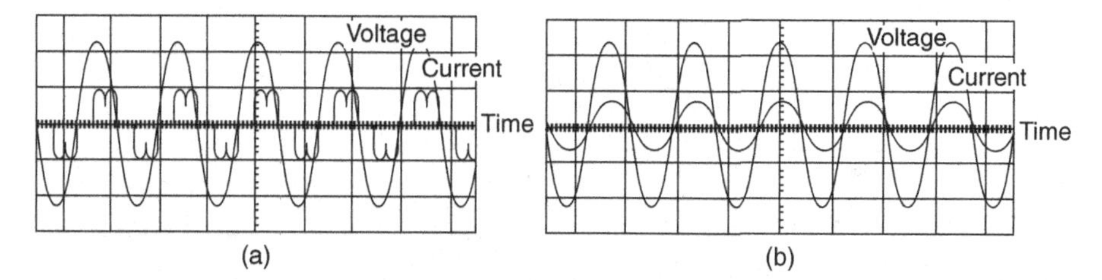

Figure 11.22 (a) Nonlinear load current waveform with poor power factor. (b) Corrected current waveform with imposed power factor and harmonic reduction after filter installation.

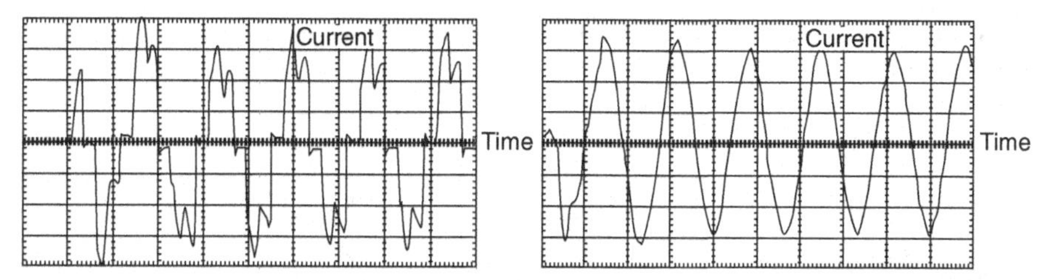

Figure 11.23 (a) Inrush current without filter installation. (b) Inrush current with filter installation (real measurements recorded before and after filter installation).

quality, and possible downtime due to undervoltage tripping of devices. The filter is able to inject peak current at two and half times its root mean square (rms) current rating for one cycle. For many applications, this level of compensation eliminates visible flicker and improves voltage regulation resulting in better productivity and quality. The inrush current waveforms without and with filter installation are shown in Figure 11.23a,b, respectively.

11.6.4.2 Application of Hybrid Var Compensator (HVC) System to Pipe Welding Industry The HVC system consists of four AccuSine (AF) units, with rating of 300A for each (1200A is equivalent to 830kvar) plus 1200kvar of fixed detuned capacitance. Once the 1200kvar is fully online, AccuSine will inject its 830kvar of lagging/leading vars as required with one cycle response to load changes. The values of compensation, fixed and variable, and the load changes with time are shown in Figure 11.24. As a result of HVC application, the voltage waveform is getting smoother when the THD of voltage is decreased from 3.1% to 1.1% and its rms value as measured is 386.9V (line to line). The PF is also improved from 0.75 lag to 0.98 lag.

11.6.5 Crane Applications—Suez Canal Container Terminal (SCCT)

The main electric substation (MES) at SCCT, Port Said—East, Egypt, is supplied by utility through 22-kV, 50-Hz, 17.5-km-long overhead transmission line

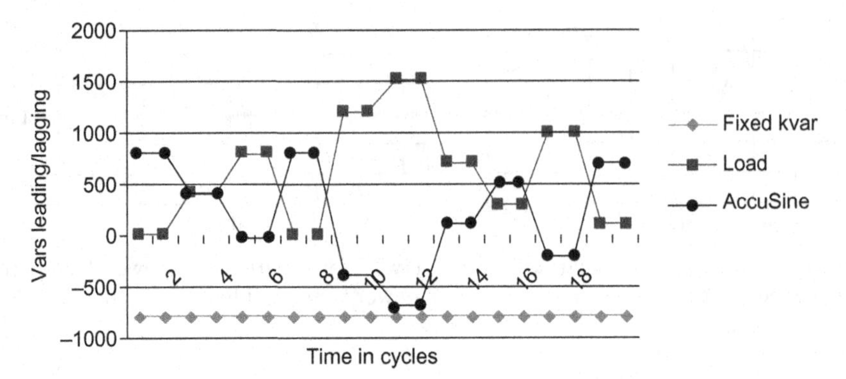

Figure 11.24 Compensation and load variation versus time in cycles (results of HVC installation).

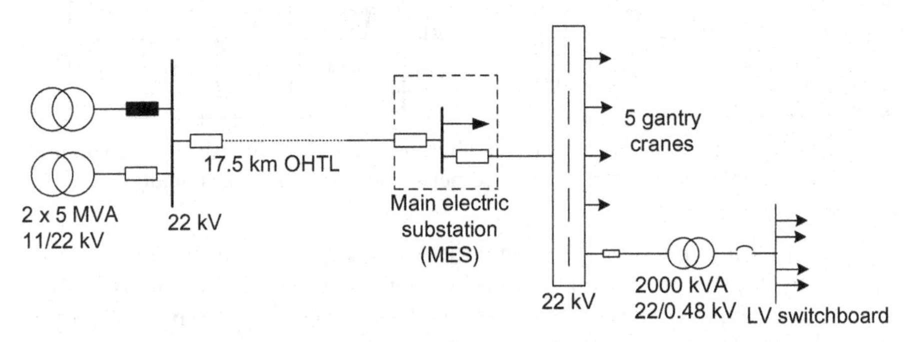

Figure 11.25 Schematic diagram of utility supply to SCCT.

(OHTL) as shown in Figure 11.25. The substation has a 200-kVA step-down transformer mainly to feed five cranes and some auxiliaries. The transformer has two secondary windings; the first supplies regenerative loads at 480V (1260-kW hoist, and DC supply to 360-kW trolley, 22-kW festoon, 496-kW gantry). The second secondary winding supplies all variable frequency drives (VFDs) and crane boom at 400V.

Each crane is equipped with PF correction and harmonic control system, which includes four steps (175kvar for each), fifth harmonic filter, stepped via contactor and provides about 140kVA of fifth harmonic filtering plus about 30kVA of seventh harmonic filtering per step.

11.6.5.1 System Problems The problems of the system described above can be categorized into two categories as below:

1. *Problems Related to System Performance:* The PF is highly variable as the load changes from no load to full load in cycles, not in seconds, and the on/off periods vary as well:

- The current has a large harmonic content with highly variable amplitude. Thus, the system does not have a steady-state condition where the variation occurs from cycle to cycle at variable on/off periods.
- The installed PF system is a conventional one in which
  - contactor switched banks are too slow, that is, the system responds in seconds and not in cycles;
  - the magnitude of the injected vars is either too small or too large, that is, the PF is not stable. It may result in a leading PF that raises the line voltage and may cause resonance;
  - it is difficult to attain the desired average PF; and
  - the system can filter only the fifth harmonic and, to some extent, the seventh harmonic as well. It does meet harmonic standards where it reduces the distortion by about 50%.

2. *Problems Related to Operational Issues:* The PF capacitors are not activated automatically. The operators can activate these capacitors manually by placing all of them online. The results of this method of operation are

- one crane raised the line voltage by about 2%,
- the PF is leading most of the time (0.375 leading), and
- THD% of voltage is close to 5% often with one crane in operation and exceeds 5% with two or more cranes in operation.

The measured and recorded parameters to diagnose exactly the system performance are shown in Figures 11.26–11.31. It can be concluded from these

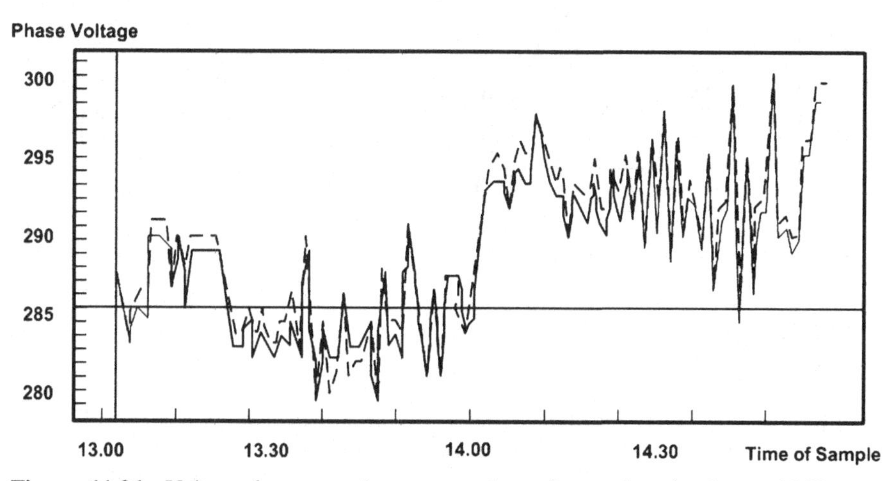

Figure 11.26 Voltage increase after connection of capacitor banks at 14:00 p.m. (maximum value: 301 V; minimum value: 279 V). (——), Voltage of phase #1; (- - - -), voltage of phase #2; (—— ——), voltage of phase #3.

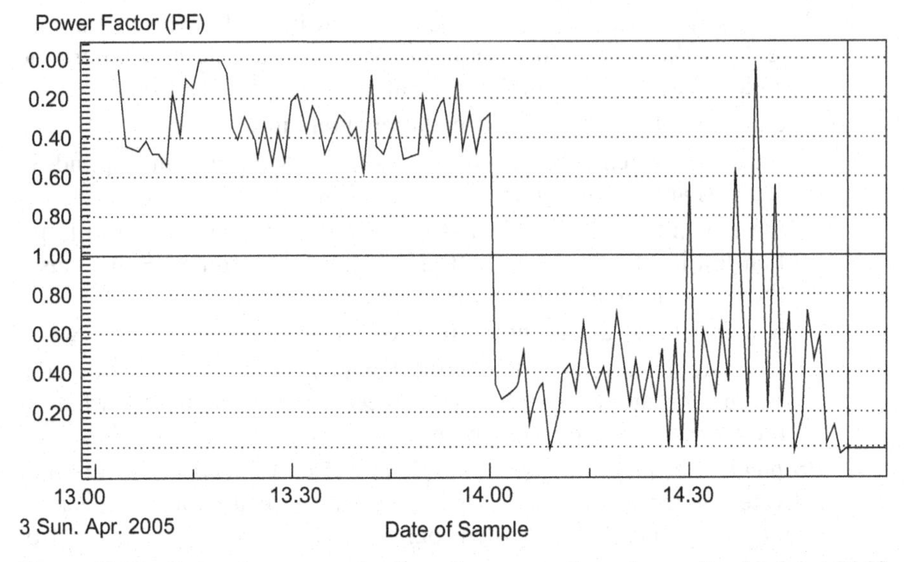

Figure 11.27 Power factor goes leading after connection of capacitor banks at 14:00 p.m. (maximum value: 0.00; minimum value: –0.01).

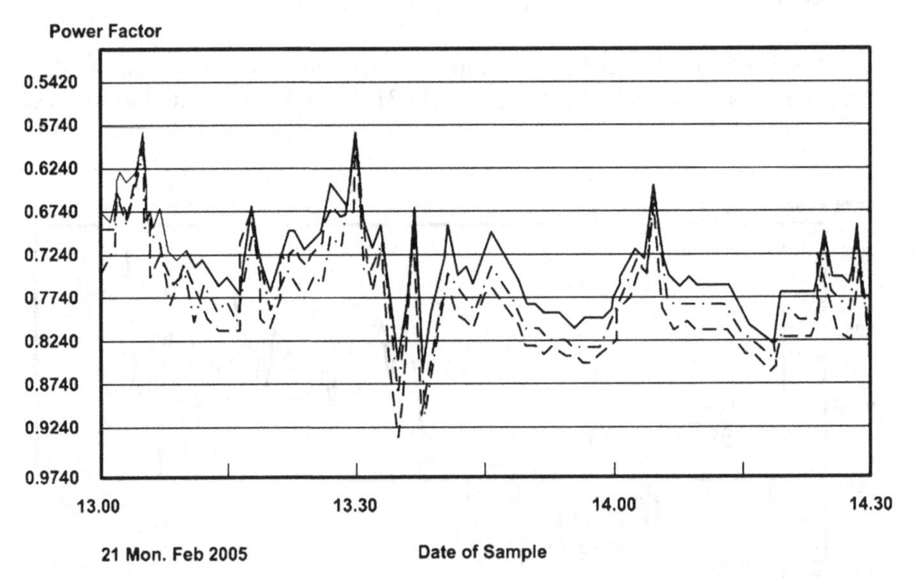

Figure 11.28 Power factor of each phase with no PF correction (average value: 0.78 lag). (——), PF: phase #1; (—— ——), PF: phase #2; (—— – –), PF: phase #3.

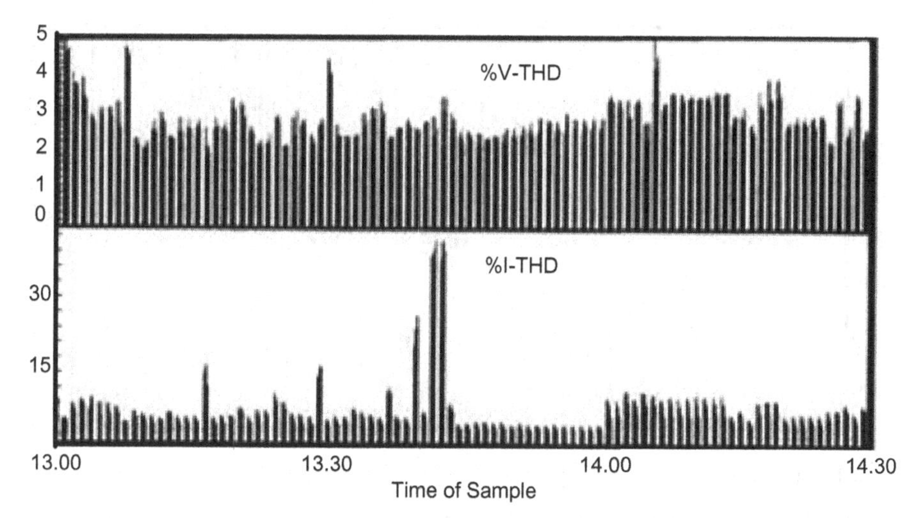

Figure 11.29 Total voltage and current harmonic distortion (%V-THD and %I-THD) for one crane with no PF correction. The maximum and minimum values of %V-THD are 5.7 and 0.0, respectively.

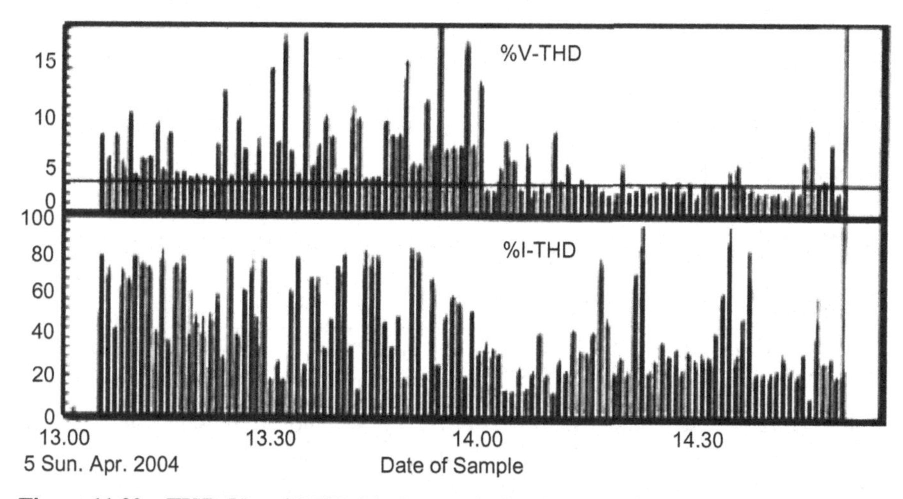

Figure 11.30 THD-V and THD-I before and after PF capacitors were added at 14:00 p.m. (maximum %V-THD: 19.5; minimum %V-THD: 1.9).

measurements that the PF is leading (the system is overcompensated). Consequently, the line voltage is raised by 10%. The resonance is more likely as the voltage rises and its frequency content is close to the seventh order. Also, as the voltage rises, adverse effects may occur on breakers such as pitting or welding of contacts leading to the probability of circuit breaker malfunction.

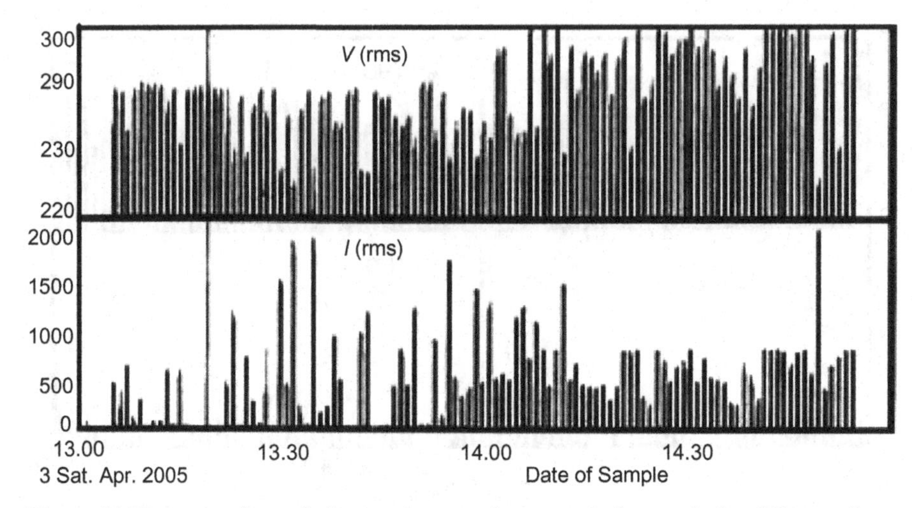

Figure 11.31 rms values of phase voltage and current, before and after PF capacitors were added at 14:00 p.m.

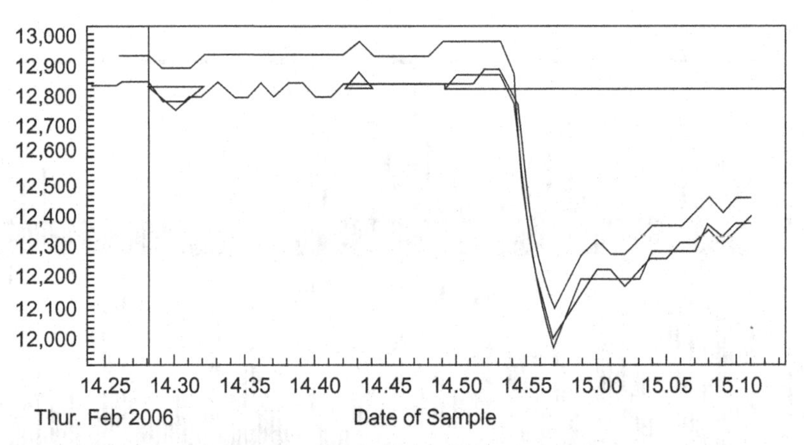

Figure 11.32 Phase voltage (at MV side) 14:55 switchover to AccuSine from PF capacitors (maximum value: 12,915 V; minimum value: 12,033 V).

11.6.5.2 Solution A resolution of the above problems was obtained by adding an AF "AccuSine" with a goal of voltage stabilization and reducing the THD_I and THD_V. Also, the PF must be maintained at 0.90 lagging at least and no resonance with utility substation capacitors.

The filter has been designed and applied. The results of application are shown in Figures 11.32 and 11.33. It is seen that there is no voltage rise, and THD_I is reduced to <5% and THD_V is reduced too to ≤3.5%. The measured values of PF give an average value of 0.88 lagging.

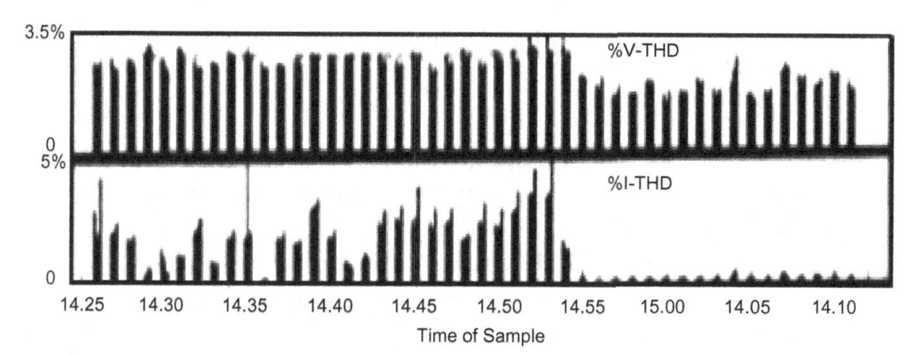

Figure 11.33 %V-THD and %I-THD with AccuSine activated at 14:55 p.m.

11.6.6 Principles to Specify AFs

It is important for the power engineers to specify accurately the main requirements of AFs ensuring a proper operation. So, the specifications are based on general principles written below and must comply with international standards applicable to this type of equipment.

11.6.6.1 Sites Concerned The specification applies to buildings or industrial sites where the circulation of harmonic currents may disturb electric installations. The sites are either new or existing sites. For new sites, the specification shall be used to apply a preventive solution to problems such as voltage distortion caused by use of electronic equipment (computers, lifts equipped with variable speed drives, battery chargers, fluorescent lamps, TV/radio transmitter–receivers, etc.), which are commonly used in any new installation. For existing sites, the specification shall be used to apply a corrective solution to problems encountered: violation of THD_V and THD_I limits, unexplained circuit breaker tripping, flicker of fluorescent lighting, cable overheating, and so on. Typical sites that are critically sensitive to such problems are high-rise commercial buildings, television and broadcast stations, and microprocessor- and real-time-based industrial processes.

11.6.6.2 Objectives and Distortion Limits AFs (also called active harmonic conditioners [ACHs]) shall be used to eliminate harmonic currents circulating within the electric installation. The value of THD_V should be limited to a given percentage of the fundamental voltage at 50 or 60 Hz in compliance with the standards (e.g., IEEE 519-1992, IEC 61000-2-2, IEC 61000-2-4, EN 50160-3).

11.6.6.3 System Description The AF shall maintain the source current in sinusoidal form by analyzing continuously the harmonic currents drawn by the load and injecting the proper current using IGBTs. It shall be compatible with any type of load and adapt automatically to any change in the load harmonic spectrum.

The AF mainly provides

- conditioning of harmonic currents within a certain range of orders, for example, from second to twenty-fifth order, in compliance with the standards;
- conditioning of phase displacement, cos Φ, for example, to be 0.94;
- capability of delivering its rated output harmonic current to the point-of-common coupling; and
- conditioning of third-order harmonic currents and multiples in the neutral conductor. This means the capability of delivering harmonic current three times the phase current.

11.6.6.4 Installation Modes At a given point in the installation, it is possible to connect several AFs in different configurations:

a) *Active Parallel Configuration:* This configuration is used to increase the conditioning capacity or dependability. As shown in Figure 11.34, three AFs are connected in parallel using a single set of current transformers (Cts).
b) *Cascade Configuration:* It is possible to connect several AFs in cascade to condition the harmonics at different levels of the distribution system (Fig. 11.35).
c) *Multifeeder Configuration:* The AF can be used to condition more than one feeder using a set of current transformers for each (Fig. 11.36).

11.6.6.5 Point of Connection Based on the power quality objectives, some of the installation modes in the former section can be connected at one or more of three possible points (Fig. 11.37):

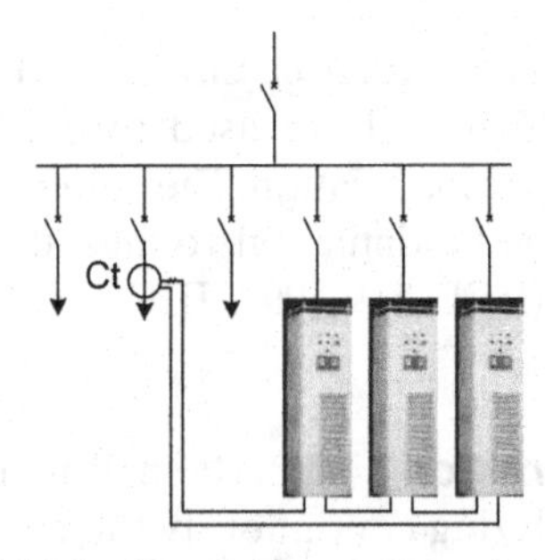

Figure 11.34 Parallel operation of three AFs.

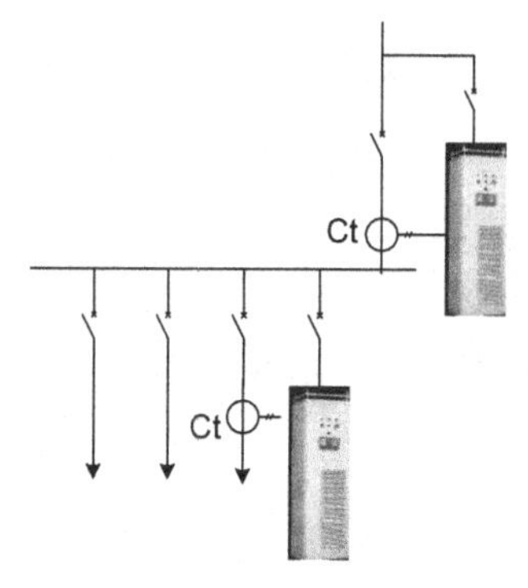

Figure 11.35 Two AFs in cascade configuration.

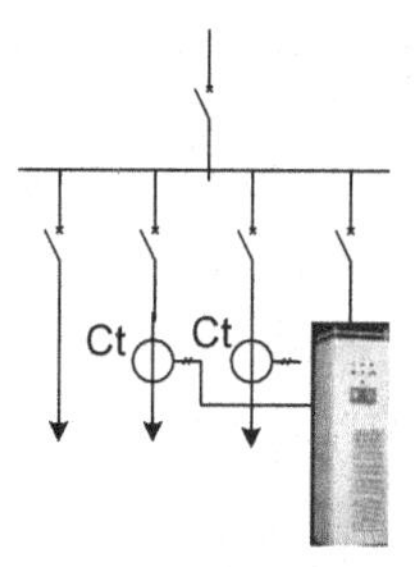

Figure 11.36 An AF for several feeders.

Point #1 (Global Conditioning): The AF is connected just downstream of sources, that is, at the main low-voltage switchboard (MLV Sw/B).

Point #2 (Zone Conditioning): The AF is connected at the secondary switchboard level to condition a set of loads.

Point #3 (Local Conditioning): The AF is connected directly to the terminals of each load.

11.6.6.6 Characteristics of AC Source The characteristics of AC electric installation should be specified so that the AF is suitable for connection to such installation. For example,

- single-phase or three-phase system,
- types of system earthing arrangements,
- nominal value of voltage, and
- frequency.

11.6.6.7 Protection A circuit breaker shall be installed close to the point of connection to the AC system to protect the connection cables. In addition, the AF shall be protected against thermal overload, internal short circuit, and

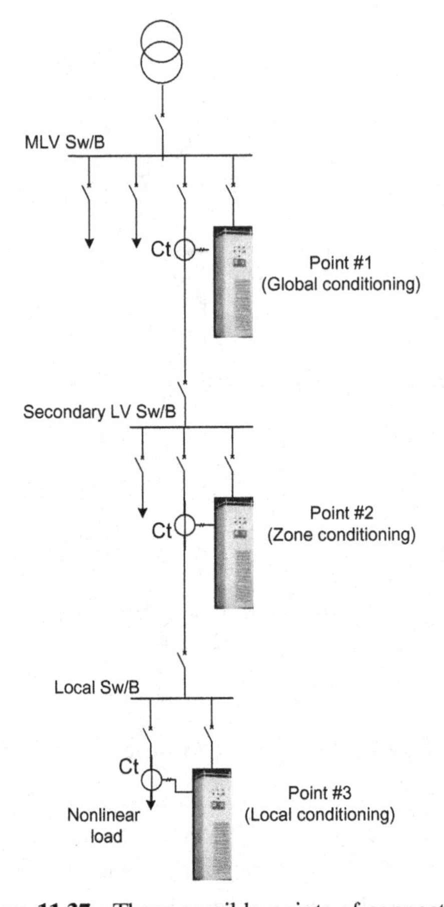

Figure 11.37 Three possible points of connection.

abnormal operation. The protection degree of the AF closure (e.g., IP30) shall be defined as well.

11.6.6.8 Environmental Conditions The AF shall be able to withstand the environmental conditions without damage or degradation of operating characteristics or life. These conditions can be specified by

- operating ambient temperature,
- relative humidity,
- altitude above sea level, and
- audible noise level.

It should be noted that all tests implied by the applied international standards shall be confirmed by certification from independent laboratories.

PART IV

MANAGEMENT AND MONITORING

CHAPTER 12

DEMAND-SIDE MANAGEMENT AND ENERGY EFFICIENCY

As discussed in Chapter 1, Section 1.5.2, demand-side management (DSM) is one of the non-network solutions used for meeting the demand and making more efficient use of existing distribution systems infrastructure without additional investments. With the development of advanced metering, two-way communications and the integration of behind the meter or house automation as part of smart grid, DSM is expanding to demand programs [160]. Therefore, more details about DSM programs and benefits as well as the energy efficiency are presented in this chapter and a brief description of the smart grid is provided in Chapter 13, Section 13.6.

12.1 OVERVIEW

Electric energy is an intermediate product. It is made from primary energy sources (gas, oil, coal) that are converted to electric energy that in turn is transported to the customer. The consumer purchases the electricity as an intermediate step toward some final nonelectric product. No one wakes up in the morning saying "I want to consume 10 kWh today." Instead, they want the products electricity can produce when applied through the actions of various appliances, for example, a cool home in summer, a warm one in winter, hot water on demand, vegetables kept fresh in the refrigerator, and watching TV. So, it is seen that most applications in life, whether industrial, commercial, or

Electric Distribution Systems, First Edition. Abdelhay A. Sallam, Om P. Malik.

Published 2011 by John Wiley & Sons, Inc.

residential, use electricity. This large-scale usage of electricity is due to various factors:

- easy to transmit over long distances with high efficiency;
- easy to produce at high efficiency;
- easy to distribute to consumers anywhere, any time, any amount; and
- wide variety of uses (heat, light, motors, etc.).

The consumption of electricity is growing continuously, and more generation capacity is needed to meet the increase in demand.

An illustrative example of the increase in the peak demand and energy supply in one country is shown in Table 12.1. The board of commissioners of public utilities of New Brunswick reported that electric energy required to meet the in-province load forecast will increase from 15,640 GWh in fiscal year (FY) 2003/04 to 17,683 GWh in FY 2014/15. During the same period, the maximum 1-h demand is forecast to increase from 3326 to 3604 MW [160]. Overall, from 2003/04 to 2014/15, the average annual growth in energy is

TABLE 12.1 Forecast Summary of the In-Province Customers of New Brunswick Power Distribution and Customer Service Corporation (Disco), Canada for the 10-Year Period 2005–2015 [160]

	Energy Supply			Peak Demand		
		Annual Increase			Annual Increase	
Fiscal Year	GWh	GWh	%	MW	MW	%
Actual						
2003/04	15,640			3,326		
Outlook 2004/05	15,710	70	0.4	3,719	147	4.4
Forecast						
2005/06	15,773	63	0.4	3,191	12	0.4
2006/07	15,881	108	0.7	3,200	9	0.3
2007/08	16,070	189	1.2	3,243	43	1.3
2008/09	16,069	1	0.0	3,233	10	0.3
2009/10	16,161	92	0.6	3,283	50	1.5
2010/11	16,439	278	1.7	3,342	59	1.8
2011/12	16,727	288	1.8	3,398	56	1.7
2012/13	17,029	302	1.8	3,462	64	1.9
2013/14	17,350	321	1.9	3,530	68	2.0
2014/15	17,683	333	1.9	3,604	74	2.1
Overall increase	2,043			278		
Average annual increase			1.1			0.7

expected to be 1.1% and the peak hourly demand will grow at an average annual rate of 0.7%.

On the other hand, economically, the cost of increasing generation capacity is always going up (cost of gas, cost of oil, cost of installation, etc.). Therefore, to achieve the optimal situation technically and economically, some obligations fall on both sides, the utilities and the consumers.

For utilities, each utility must utilize the generation, transmission, and distribution systems in an optimal manner and at the highest possible efficiency. This will help reduce the investment and, of course, the energy tariff as well.

On the consumer side, that is, the demand side, the electric energy demand must be reduced by managing the load and by applying "DSM" programs. Obviously, these programs differ from utility to utility and from country to country [161–164].

12.2 DSM

DSM is the planning, implementation, and evaluation of utility activities designed to encourage customers to modify their electricity consumption patterns with respect to both the timing and level of demand (kW) and energy (kWh). Thus, DSM mobilizes power sector resources on the customer's side of the electric services [165–167].

The utility DSM programs are categorized in two categories:

Category #1: Energy-Saving or Conservation Programs: These programs are activities designed to encourage the consumers to replace their less-efficient equipment by less-electricity consuming devices producing the same service (quality and quantity) and to change consumer behavior to reduce energy use.

Category #2: Load Management Programs: These programs are designed by utilities to reduce demand peak periods to make more economic use of existing resources through demand displacement over time. Load management programs are essential to utilities facing a substantial constraint in respect of producing power to meet the demand in real time.

The fundamental concepts of DSM for utility planners, engineers, and regulators concerned with power sector planning and regulation are presented in this chapter. These concepts are laid out in the order in which they would first be encountered in developing a systematic DSM program. Many of the steps in the DSM program development are iterative or serial, although many can also be carried out in parallel.

DSM relies on energy efficiency (i.e., reduction of kWh of energy consumption) and/or load management, that is, the displacement of demand to off-peak times to achieve the utility's demand curve objectives.

As indicated earlier, DSM is typically accomplished through utility sector programs as part of the planning and acquisition of electric resources. In this context, DSM is distinct from other activities that can achieve end-use energy efficiency. Such other activities may include the implementation of building codes and appliance efficiency standards by government bodies, the normal market response of consumers to energy price changes and the availability of new end-use technologies that use energy more efficiently.

12.3 NEEDS TO APPLY DSM

DSM programs assist the planners to achieve load reduction on a utility's system during periods of peak power consumption or allowing customers to reduce electricity use in response to price signals. These price signals are applied in different ways [168, 169]:

- *Time-of-Use Rates*: The price rate is raised to be two or four times higher than off-peak rates.
- *Real-Time Pricing*: The energy tariff is calculated on use time (each hour, each half day, daily, etc.). So, a technique to modify and recalculate the energy price online is required.
- *Direct Load Control*: The distribution system operator has the flexibility to shut down some consumer loads at the peak time according to preassigned contract.
- *Data Provided by Utilities*: Two types of data are requested from the utilities. The first type is data on each demand-side activity, while the second type is data on each supplier's system as a whole. A combination of the two types helps in comparing the effect of demand-side activities on the total system load. The main items of each type of data are the following:

  First Type:

  - Total saved power (kW) from annual peak for each demand-side activity. This item evaluates the saved power by lowering the peak demand for each activity annually. The sum of these values gives the total amount of not needed generating capacity. This indicates how much the demand-side activities are beneficial.
  - Total annual saved energy for each demand-side activity. The sum of these savings for all activities gives the total amount of annual saving of generated energy.
  - Identifying the percentage of consumers in a particular class to whom a specific demand-side activity is available.
  - The number of customers participating in demand-side activity at the time of annual peak demand.
  - Assessment of expenditures on DSM.

Second Type:

- The total amount of highest annual peak in kilowatt.
- The relative size of distribution system measured by the kilometer of distribution lines.
- Number of consumer classes.
- Contribution of total generation supplied from each producer. This helps utilities identify the qualified producers.

12.4 MEANS OF DSM PROGRAMS

The following six methods are the principle means by which modern DSM programs can influence customer's demand for electricity.

- *Peak Clipping*: The utility loads during peak demand periods are reduced. This can defer the need for additional generation capacity and result in a reduction in both peak demand and total energy consumption (Fig. 12.1a). Peak clipping can be implemented by direct control of consumers' appliances or end-use equipment.
- *Valley Filling*: In case of adding loads that previously operated on non-electric fuels (e.g., thermal energy storage), it is desired to supply these loads at off-peak load periods. It is preferable to apply valley filling when the long-run incremental cost is less than the average price of electricity. As shown in Figure 12.1b, the net effect is an increase in total energy consumption without increase in peak demand.
- *Load Shifting*: It involves moving load from on-peak to off-peak periods. It leads to a decrease in peak demand and no change in total energy consumption (Fig. 12.1c). Use of storage devices that shift the timing of some conventional electric appliances operation is one of the tools being used for load shifting. For instance, use of hot water storage enables the consumer to turn on the water heater at off-peak time. Also, applying time-of-use rates encourages the consumers to change the timing of end-use consumption from high-use, high-cost periods to low-use, low-cost periods.
- *Strategic Conservation*: By applying end-use efficiency techniques to reduce the end-use consumption, a decrease in both peak demand and total energy consumption can be achieved (Fig. 12.1d).
- *Strategic Load Growth (Load Building)*: In the event of new consumers or energy intensity increase, the overall sales are increased. The load is managed to keep the same load shape as shown in Figure 12.1e leading to an increase in both peak demand and total energy consumption.
- *Flexible Load Shape*: Instead of changing the load shape as in former methods, the loads as an option can be interrupted by utility when

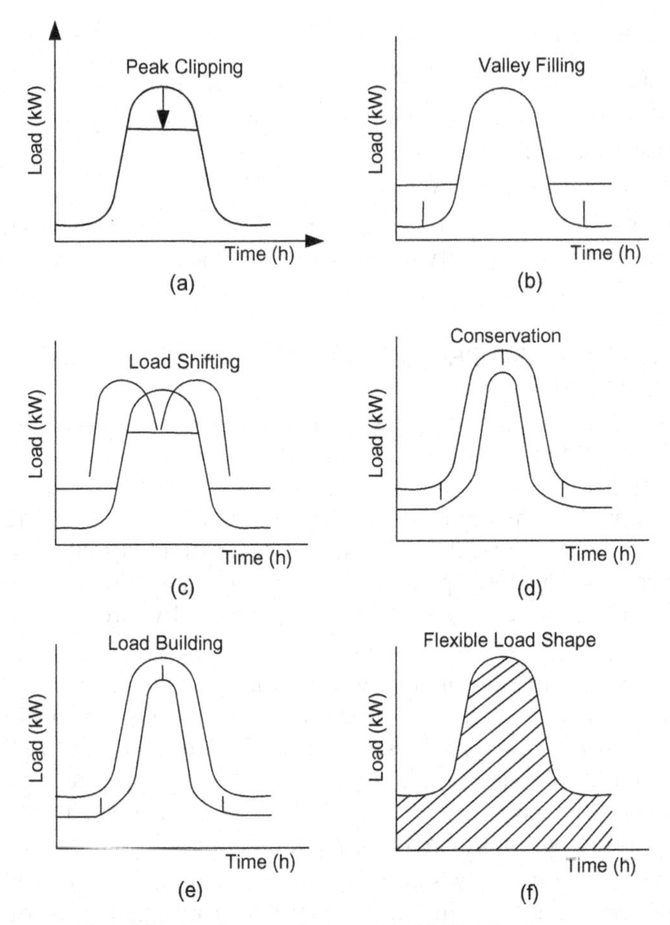

Figure 12.1 The load shape for standard DSM load management.

necessary aiming at reducing the peak demand but varying reliability or quality of service. This needs a preagreement from consumers [170].

The primary objective in each case is to manipulate the timing or level of customer demand in order to accomplish the desired load objective. For example, in the case of underutilized capacity, valley filling may be desirable. On the other hand, in countries with rapidly growing demand, peak clipping or strategic conservation can be used to defer costly new capacity additions, improve customer service, reduce undesirable environmental impacts, and maximize national economic benefits. It is important to note that while peak clipping, load shifting, strategic conservation, and flexible load shapes are considered as new resources that can help a utility meet the increasing demand of its customers, strategic load growth and valley filling are economic efficiency options for power systems with long-term expectations of surplus power.

At first, it may seem irrational for a utility to try to sell less of its product (electricity). However, load-reducing DSM may be an important strategy for providing electric services at least cost to customers and society because it is often less expensive to save energy than to produce it. In instances where regulations encourage market choices and socially optimal investments to coincide, utilities have come to understand that there are energy efficiency options on the customer's side of the meter that could be more cost-effective in meeting electric power needs than building expensive new power generation plants or transmission and distribution facilities.

From this experience came the recognition that a "negawatt" (a negative watt) of electricity saved through DSM is as good as a watt of generation capacity. For instance, if a utility manages to reduce electricity demand, it can postpone the construction of expensive new power plants or increase reliability. Alternatively, reducing total generation can eliminate the need to install costly environmental controls.

Capturing these benefits, however, requires utilities to review and rethink their traditional roles. Whereas utilities that rely solely on conventional supply-side resources such as power plants often view themselves as commodity producers, utilities that tap the potential of DSM must perceive themselves as service providers (i.e., they are in the business of meeting consumer needs rather than simply producing kilowatt-hours). Often, new government regulations or directives are needed to ensure that utilities will undertake DSM programs, because until recently, power sector regulation in most countries has been concerned only with the production of electricity and not the broader issue of the economically least-cost satisfaction of consumer demand for electric services.

Integrated resource planning, which is discussed in this chapter, is one approach governments often use to encourage the utilities to consider both supply-side and demand-side resources in their system planning.

12.5 INTERNATIONAL EXPERIENCE WITH DSM

Many countries and states have considerable experience with DSM in the power sector. Some of them have found that the power sector should be restructured, some should not, and others are in the process of restructuring.

The DSM programs are mostly coordinated by governments or agencies, and not by the utilities. In addition, these programs can be funded by taxes or other revenues, and not by the consumers, but in many countries, DSM is funded through electricity bills and utilities are involved in program delivery. International studies show that the benefits gained from DSM are high, in particular, for economy, supply security, and reliability. For instance, it is reported in Reference 171 that the total electricity consumption in the Southwest, the fastest growing region in the United States in terms of population and electricity use, could be cut 33% from the base case (business as

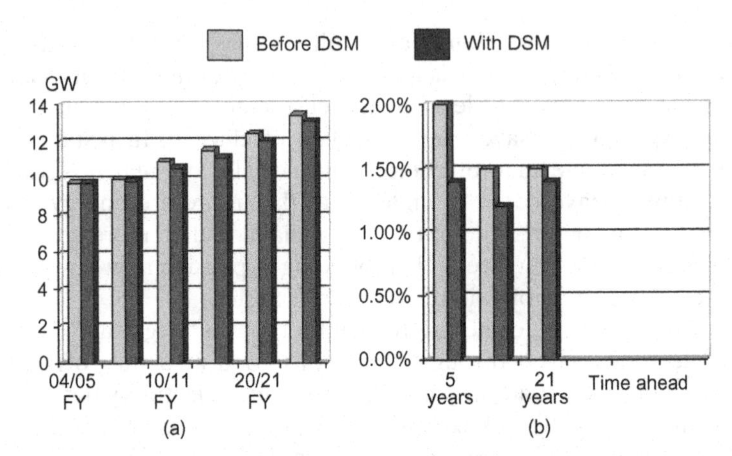

Figure 12.2 (a) Domestic system peak forecast before DSM. (b) Domestic system peak growth rate forecast with DSM [171].

usual) by 2020 with accelerated use of presently available energy efficiency technologies and other supportive policies. It could be achieved at an average cost of $0.02/kWh. In Europe, if energy efficiency programs were expanded to reach the entire Europe, electricity and gas consumption is expected to be reduced by 10% a year [171].

As reported by BC-Hydro [171], domestic system peak forecast before and with DSM program is shown in Figure 12.2a,b, respectively. It is seen that without DSM, the weather adjusted actual peak for FY 2004/05 is 9787 MW and expected to reach 13,417 MW in 2025/26 with growth rates of 2.0%, 1.5%, and 1.5% for 5, 11, and 21 years ahead, respectively. With DSM, the weather-adjusted peak for the same period is expected to grow from 9,787 to 13,074 MW with growth rates of 1.4%, 1.2%, and 1.4% at the same years ahead.

For meeting immediate needs to maintain system stability, the short-term measures for responding to demand are useful. These measures include price rationing and price reforms such as time-of-use pricing, real-time pricing, and interruptible pricing.

In contrast to conventional generation resources, DSM can lower stresses on the electric network at the point of consumption, improving the distribution and transmission systems reliability. Also, in contrast to high economic costs of rationing and long-term economic costs of adding supply, low-cost efficiency investments can lower the overall cost of electricity.

12.6 POTENTIAL FOR DSM APPLICATION

DSM has mainly two strategies, load management and end-use energy efficiency, for reducing peak demand and energy consumption. The potential for DSM application can be determined by evaluating the impact of these reductions on the cost. So, the main measure of DSM potential is the cost of

- peak demand savings and
- energy consumption savings.

12.6.1 Peak Demand Savings

Reduction in peak demand expresses the potential of load management strategy. Load management is of special interest for planning to help address the power shortage and to manage system resources efficiently over time. During power shortages, it helps to avoid some involuntary outages and heavy costs incurred during power supply disruptions. Therefore, the benefits of load management during power shortages are measured by the economic and environmental effects of particular strategies. Savings will include costs of avoided augmentation such as generation-related capacity, transmission, and distribution system expansion in addition to some operating cost savings. The amount of savings, of course, varies from system to system.

On the other hand, the cost of load management strategies also varies. For example, as reported in December 2005 [169], storage technologies such as air conditioners and heat storage electric boilers have been developed in East and South China that cost about $150 to shift 1 kW of load. Cement industries have enlarged the milling capacity to shift load from peak to off-peak periods. This strategy requires about $250 to shift 1 kW of load. This compares favorably to the cost of about $230/kW for peaking capacity. So, load management practices are especially well suited to reducing the high peak loads that last for relatively few hours.

12.6.2 Energy Consumption Savings

Both the electric utilities and users have a considerable potential for savings in the amount of energy consumption. The electric utilities should use higher-efficiency equipment (low-loss transformers, cables, and/or overhead lines with minimum losses by determining, accurately, their cross-sectional-area and length) and apply proper design of system structure aimed at reducing the overall system losses.

The end-use loads are generally classified into three classes: industrial, commercial, and industrial. To design or upgrade a building for any of these classes and, in particular, its electric installations, it is important to explore all of the elements that contribute to energy savings and that may or may not be selected. This depends on the installation; for example, for building containing an industrial process, the main area for savings lies in the production equipment.

However, there are some items that can normally be applied to the different load classes and may contribute to energy savings such as the following:

- *Heating, Ventilation, and Air-Conditioning (HVAC)*: HVAC systems are designed to maintain the inside temperature and ambient air at comfortable levels. In many buildings, HVAC is the first or second item in terms of energy costs. Some of the actions for energy savings involve heat loss minimization. This can be achieved by different tools such as accurate temperature setting; using efficient HVAC equipment (e.g., heating equipment as explained in Section 12.12.3); and proper insulation of the roof, doors, and windows.

- *Lighting*: It is an essential service in all buildings. It contributes to the safety and comfort of occupants and to the productivity of activities. It is often a very significant item of energy expenditure. The main approaches to saving energy with regard to lighting are described in Section 12.12.1.
- Reducing electric energy losses by improving the power factor (as explained in Chapter 9) and reducing the harmonic effects (further details are in Chapters 10 and 11).

12.7 THE DSM PLANNING PROCESS

To implement DSM programs, a comprehensive planning process is required like most other energy-efficiency projects. The success of this process is dependent on the presence of an institutional environment that is conducive to planning DSM programs because it has regulatory, financial, and technical dimensions.

The DSM planning process consists of several interrelated steps as shown by the flowchart depicted in Figure 12.3. Some steps are conducted iteratively

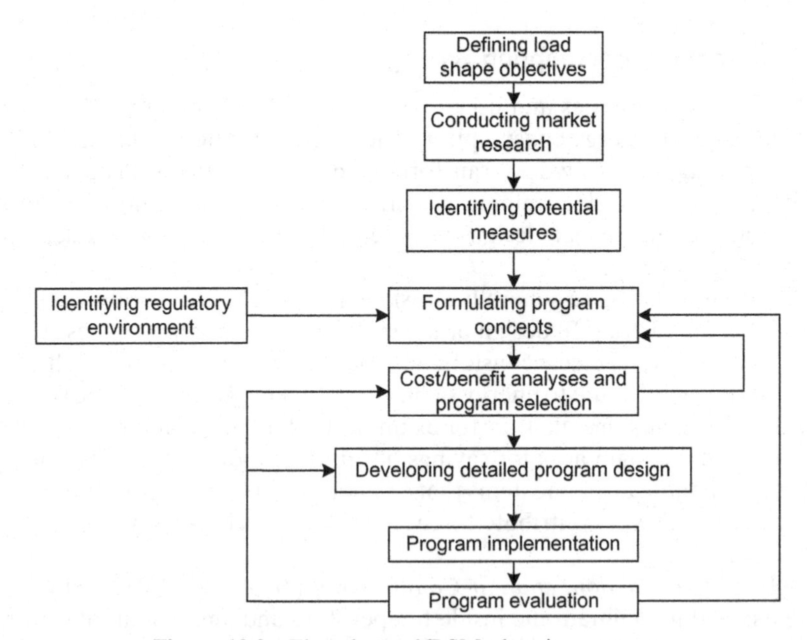

Figure 12.3 Flowchart of DSM planning process.

(e.g., the results of a pilot program may suggest changes to the program design, and the results of the program evaluation may guide the formulation of objectives for subsequent DSM programs). The basic framework for DSM planning comprises the main steps below:

- *Defining Appropriate Load Shape Objectives*: Based on the utility's requirements, load shape objectives are defined for the utility as well as for each target customer class or market segment. These objectives guide the program's design and facilitate its implementation and evaluation. Specific objectives are derived from broader objectives for the utility's financial performance and operational needs such as increasing system utilization or reducing dependence on critical fuels.
- *Conducting Market Research*: The goal of DSM programs is to enhance customer behavior by focusing on the technologies consumers use and by stimulating consumer participation. For instance, a sufficient number of residential customers must use high-efficiency air conditioners in order to make an air-conditioning DSM program feasible for utilities. If enough customers in fact use this technology, then DSM program targeting air conditioners can only be effective if certain conditions are met. For example, the program must take account of how and when the appliance is used; how and when a consumer decides to purchase or replace an air conditioner with a new, high-efficiency model; and whether consumers may agree to let the utility control the appliance, that is, load control, when the utility does not have enough generating capacity.

  In successful DSM programs, designers and marketing strategists must clearly identify a target population and take into account their values, actions, consumption patterns, and perceptions. Therefore, at this stage of the DSM process, program designers and implementers gather the pertinent data through a combination of customer surveys and focus groups, billing data analysis, and load research. These data are then used to establish a baseline that characterizes the most recent state of the electric utility and against which program impacts (i.e., the magnitude of energy savings) are estimated.
- *Identifying Potential Measures*: Based on the selected load shape objectives and the characteristics of each target market segment, various DSM measures are proposed and evaluated to determine their technical and economic potential. Technical potential refers to the impact of the measure if it were adopted wherever technically feasible. Economic potential refers to the impact of the measure if it were adopted wherever economically justified (including costs and benefits to the customer).

  Several examples of DSM measures for each of the six main load objectives are listed in Table 12.2.
- *Formulating Program Concepts*: Measures with promising technical and economic potential are grouped together and packaged into distinct DSM

TABLE 12.2 Example of DSM Measures [170]

DSM Load Objectives	Examples of Potential DSM Measures
Peak clipping	• Efficient air-conditioning • Water heater controls • Interruptible rates
Valley filling	• Electric vehicles • Targeted economic development • Security lighting
Load shifting	• Water heater control • Air conditioner control • Storage heating • Interlocks • Irrigation control • Time-of-use rates
Strategic conservation	• Building envelope efficiency • Efficient refrigeration • Efficient lighting • High-efficiency motors
Strategic growth	• New electric heating and cooling loads • Area development • Industrial electrification • Ecowatts[a]
Flexible load shape	• Interruptible loads • Dual-fuel heating • Standby generation • Priority service

[a] Ecowatts are additional electric loads that are usually realized through fuel switching from polluting fuels such as oil and benzene to "clean electricity." The rationale behind ecowatts is that

programs. Marketing strategies are then formulated for implementing the measures in the targeted customer segments by applying some common delivery mechanisms. Delivery mechanisms for DSM programs range from a centralized approach in which the utility is responsible for installing measures on the consumer's side of the meter, to a decentralized approach in which the utility only provides incentives for consumers, who are then responsible for identifying DSM opportunities and deciding whether or how they wish to take advantage of those incentives. Direct installation is the most centralized approach.

In its simplest form, direct installation entails utility staff visiting consumers and physically installing equipment that will modify consumer load shapes. Alternatively, a utility may hire a contractor to carry out this work.

- *Cost/Benefit Analyses and Developing Detailed Program Design*: Other activities in this step include making institutional and administrative arrangements for program implementation, cost-effectiveness analyses of the DSM measures to be offered through the program, designing systems for performance monitoring and participation/transaction tracking systems, and developing plans to evaluate the program's implementation and performance. Program designers must pay close attention to the following issues at this stage of program development:

- *Explicit Recognition of Uncertainty*: There is a great deal of technical, economic, and market uncertainty about the impacts of particular DSM programs. Program designers must thus explicitly recognize uncertainty and should take steps that reduce risks and increase the likelihood of program success. Four primary sources of uncertainty can be addressed in DSM planning and tested through pilot programs:
  1. Technical uncertainty arises from the lack of accurate and detailed information on the targeted applications, variability of the operational factors that characterize those applications, and the performance of the proposed DSM measures. For example, the benefits of a lighting efficiency program will depend on part of the power consumption of existing lamps, their hours of operation, the power consumption of the proposed replacements, and the technical feasibility of replacing the existing lamps with high-efficiency lamps (e.g., compact fluorescent lamps [CFLs] will not fit in all standard fixtures). It is difficult to develop accurate estimates of the average or typical values of these parameters without extensive fieldwork, much less capture their variability or distribution.
  2. Economic uncertainty is associated with program costs. For example, the estimated costs to install DSM measures and to administer and evaluate a program may be only general estimates; actual field experience must be acquired.
  3. Market uncertainty results from the range of potential consumer response to DSM programs. Implicit discount rates that characterize consumer decision-making criteria may vary widely among market segments, so that measures that may be attractive to some consumers are not attractive to others. Even if the economic attractiveness of measures could be accurately gauged during the program design process, knowing the technical impact of a DSM measure among all consumers for whom it would be economically attractive to participate is not enough to estimate program impacts accurately. For instance, it may take time for consumers to learn about, understand, or be convinced of the benefits of DSM program participation. Or, only limited resources may be available for program delivery, so that not all consumers who wish to participate can. Such factors can slow the market penetration or consumer acceptance of DSM programs.
  4. Uncertainty about the persistence of measures may be thought of as temporal dimension of uncertainty. The adoption of a measure does

not ensure its continued use. A consumer may remove a measure and replace it with the old equipment for any number of reasons. Alternatively, the lifetime of measures may not be as long as assumed prior to program implementation.

- *Attention to the Institutional Environment*: DSM program planning and implementation require the involvement and commitment of a broad range of groups such as regulatory institutions, utilities, trade allies (manufacturers, suppliers, and distributors of DSM measures), and end user (residential, commercial, industrial, and agricultural customers).

  The success of DSM programs will largely depend on the cooperation of these groups. Therefore, program designers must ensure that there are enough *incentives* for each of these groups to guarantee their participation in, and support of, the DSM program under consideration.

- *Program Implementation*: DSM is often implemented in a phased manner. If the technical and economic potential and likely acceptance of DSM measures were known with certainty, there would be no risk associated with DSM program implementation. Utilities would be informed of the program impacts and could compare the value of these impacts to the cost of the program before its full-scale implementation in order to determine the program's cost-effectiveness. However, as indicated earlier, there is uncertainty associated with DSM program design and implementation. Consequently, pilot programs are often used to obtain further information. In effect, they serve as additional market research and intelligence gathering vehicles that may develop more accurate program data and reduce the risks associated with full-scale program implementation.

  Pilot programs typically involve additional load research, customer surveys, and billing data analysis that facilitate the DSM program's evolution on a limited, but valuable, basis. They also provide an important demonstration function. Successful pilot programs can convince utilities, regulatory bodies, and consumers of the effectiveness and value of DSM programs under consideration.

  Based on the evaluation of pilot programs, it is possible to identify the best final design for a program to make it cost-effective and responsive to local conditions. As with pilot programs, full-scale programs include marketing, monitoring, and administration along with the actual delivery of various DSM measures.

- *Program Evaluation*: If DSM programs are to be utilized as true utility resources that defer conventional capacity expansion and reduce generation, program impacts must be quantified in terms of energy and demand savings. Evaluation is also critical to establish the optimal level of incentives for program implementation or participation. Impact evaluation determines the change in energy consumption patterns as a result of the program (i.e., kW and kWh saved). Process evaluation, on the other hand, examines the way in which programs are marketed and implemented. Market evaluation, which is sometimes distinguished from process evaluation, is used to assess why consumers choose to participate or not in a particular program, and leads to a reestimation of the program market potential and impacts, and program design and marketing techniques.

Program evaluation also provides an important feedback, or midcourse correction activity. It should include ongoing monitoring and program tracking to suggest adjustments as well as to verify program impacts.

12.8 EXPECTED BENEFITS OF MANAGING DEMAND

Reducing peak demand helps to reduce utility construction cost and investments as well as lower electric rates. So, the first benefit is that the average price of electric energy is lower [173].

Again, reducing peak demand together with energy consumption means that the existing distribution system is capable of supplying additional loads without additional investment. In addition, moving the demand to off-peak periods means saving money without saving energy.

The power systems that need peak load resource can benefit from peak clipping and load shifting DSM measures. On the other hand, the power systems that need additional base load resources can benefit from strategic conservation measures. These measures are mainly related to the end uses, that is, consumers' equipment. The equipment should be used in a way such that it saves energy at the same level of service utilizing the development of technology. This is expressed by equipment energy efficiency. Therefore, the combination of load-management programs "DSM" with end-use energy efficiency can maximize the effectiveness of both programs and lead to the greatest demand reduction.

12.9 ENERGY EFFICIENCY

Energy efficiency is defined as the amount of output per unit of energy input for a specific device. It depends on the physical performance of that device as end uses or energy services such as lighting, heating, cooling, and driving force. Also, for the utility side, the performance of its traditional functions such as generation, transmission, and distribution, must be at maximum efficiency.

The programs of energy efficiency can be applied to all energy sources with a goal of saving energy demanded by the consumers. These programs are also based on incentive polices delivered by the governments. So, the DSM initiatives belonging to utilities are complementary to energy-efficiency strategies initiated by government agencies.

12.10 SCENARIOS USED FOR ENERGY-EFFICIENCY APPLICATION

Four main scenarios can be used to apply the energy efficiency programs:

First: Educating the end users on opportunities available for improved efficiency to make them aware of the usefulness of these programs.

Second: Developing suppliers of end-use energy-efficient products and companies of energy service. This helps the availability of high-efficiency equipment in the market.

Third: A contract or agreement between the end users and the utilities to imply the commitment of carrying out the energy-efficiency programs.

Fourth: Providing financial incentives to end users to modify energy use or change end-use equipment.

12.11 ECONOMIC BENEFITS OF ENERGY EFFICIENCY

Energy efficiency as an investment is mostly cheaper, cleaner, safer, faster, more reliable, and more secure than the investment in new supply. In addition to reducing the need to construct new generation, transmission, and distribution facilities, improving efficiency also reduces maintenance and equipment replacement costs, as many efficient industrial technologies have longer lifetimes than their less efficient counterparts [174].

Relying on efficiency also avoids a number of costly risks associated with generation, such as lack of demand, cost overruns, interest rate risk, volatile fuel costs, technological obsolescence, catastrophic failure, and political and national security risks.

Efficiency can come online much faster than expanding energy supply, without any problems of surplus or shortage. Retrofitting motors and pumps, adding insulation to buildings, or even changing a lightbulb takes much less time than constructing a new power plant.

12.12 APPLICATION OF EFFICIENT TECHNOLOGY

This section concerns deeply with four groups of applications using electric energy sources that are responsible for a significant percentage of the total consumption. The four groups are mainly lighting, motors, heating, and pumps.

12.12.1 Lighting

As reported in Eurelectric, the energy consumption for lighting represents about 15% of the total electricity consumption. This amount of lighting is

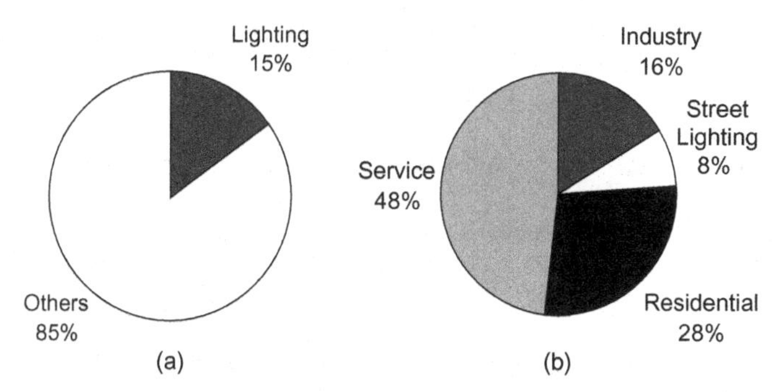

Figure 12.4 Energy distributions as percentage [174]. (a) Energy consumption distribution. (b) Light energy distribution.

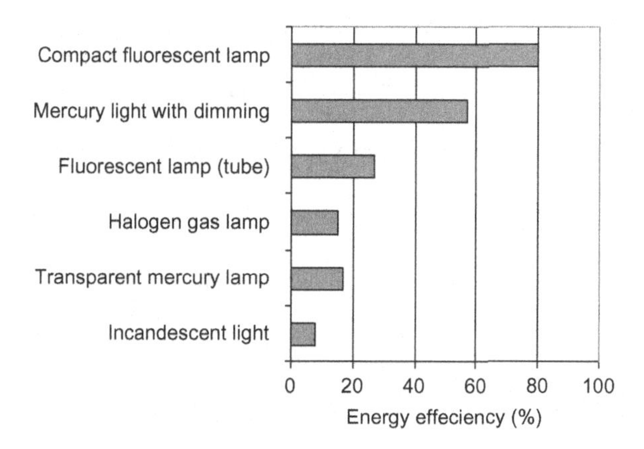

Figure 12.5 Efficiency of different light sources.

distributed over various sectors as shown in Figure 12.4. These percentages illustrate, as an example, the importance of lighting in saving money and energy by managing and using advanced technology [174, 175].

Different types of light sources are used commencing with the general incandescent lamps with lowest efficiency and lifetime to CFLs as one of the gas-discharge lamps with highest efficiency and longest lifetime. The efficiency of different light sources is indicated in Figure 12.5, and the field of applications for each source is given in Table 12.3. It is clear from Figure 12.5 that the development of lamps is continuing with the aim of increasing both energy efficiency and lifetime.

The indoor lighting is calculated by the following equation:

$$L_{\text{lght}} = \frac{N \times F \times U \times M}{A} \quad (\text{lx}), \tag{12.1}$$

TABLE 12.3 Field of Applications for Light Sources

Class of Lamps		Incandescent Lamp			Fluorescent Lamp			Mercury Lamp				Halide Lamp		Sodium Lamp		
		General Lamp	Reflector Lamp	Halogen Lamp	General Fluorescent Lamp	High Color-Rendering Properties	High Output Type	Transparent Mercury Lamp	Fluorescent Mercury Lamp	Reflector Mercury Lamp	Stabilizer Built-In Type	General Type	High Color-Rendering Type	High Pressure	Low Pressure	Xenon Lamp
Residence		✷	☆	✧	✷	☆	○	○	○	○	○	○	○	○	○	○
Office	General office	✧	✧	✧	✷	✧	☆	○	○	○	○	✧	✧	○	○	○
	High-ceiling office, lobby	☆	☆	☆	☆	✧	☆	○	☆	☆	✧	✷	☆	○	○	✧
	Single room, drawing room	☆	☆	✧	✷	☆	○	○	✧	○	○	✧	✧	○	○	○
Store	General stores	✷	✷	☆	✷	✷	☆	○	☆	✧	✧	✧	✧	○	○	○
	High-ceiling stores	☆	☆	☆	☆	☆	✷	○	☆	☆	☆	✷	☆	✧	○	✧
	Exhibit, showcase	✷	✷	✷	✷	✷	☆	○	✧	✧	☆	☆	☆	○	○	✧
Factory	Low-ceiling factory	✧	✧	☆	✷	☆	☆	○	✧	✧	✧	✧	✧	✧	○	○
	High-ceiling factory	✧	✧	☆	✧	✧	✷	○	✷	☆	☆	✷	☆	☆	○	✧
	Warehouse	☆	✧	☆	✷	✧	☆	✧	✷	☆	☆	☆	✧	☆	○	○

School	Classroom	✧	✧	✧	⬡	☆	✧	○	✧	○	○	✧	✧	○	○	○
Hospital	Operating room	☆	☆	✧	⬡	⬡	✧	○	○	○	○	○	○	○	○	○
Theater, hall	Spectator's seats	⬡	⬡	⬡	⬡	☆	✧	○	✧	✧	✧	☆	☆	○	○	✧
	Stage	⬡	⬡	⬡	⬡	☆	☆	○	✧	✧	✧	✧	✧	○	○	✧
Art museum, museum	General	⬡	⬡	☆	☆	⬡	✧	○	✧	✧	✧	☆	☆	○	○	✧
	Exhibits	⬡	⬡	☆	☆	⬡	✧	○	○	○	○	☆	☆	○	○	✧
Roads	Automobiles, exclusive roads	○	○	○	✧	○	○	✧	⬡	○	○	✧	○	☆	☆	✧
	Automobiles, exclusive tunnels	○	○	○	✧	○	○	✧	☆	○	○	✧	○	☆	⬡	✧
	Streets	✧	○	○	☆	○	○	☆	⬡	✧	✧	✧	✧	☆	✧	○
	Shopping streets	☆	○	☆	☆	✧	⬡	○	⬡	✧	✧	⬡	✧	☆	⬡	⬡
	Roads in resident area	☆	○	○	☆	○	○	✧	⬡	✧	○	✧	○	☆	○	○
Parking zone	Indoor	✧	✧	✧	⬡	○	☆	○	☆	✧	✧	✧	✧	☆	○	○
	Outdoor	✧	✧	✧	☆	○	○	✧	⬡	☆	✧	✧	✧	☆	✧	✧
Open space, park, garden		☆	✧	✧	☆	✧	○	✧	⬡	✧	✧	☆	✧	☆	⬡	✧
Floodlight lighting	Structure	☆	☆	☆	○	○	○	✧	⬡	⬡	☆	☆	☆	☆	✧	☆
	Advertisement	☆	⬡	⬡	☆	☆	☆	✧	⬡	⬡	✧	☆	☆	✧	○	☆
Sports	Indoor	☆	☆	⬡	☆	☆	☆	✧	⬡	☆	✧	⬡	☆	✧	○	✧
	Outdoor	☆	☆	☆	⬡	⬡	⬡	✧	⬡	☆	✧	⬡	☆	⬡	○	☆

⬡ = most suitable; ☆ = suitable; ✧ = not recommendable; ○ = not suitable.

Source: ECC.

where

L_{lght} = luminance (lx);
A = area of the room (m^2);
N = number of lamps;
F = luminous flux emitted from one lamp;
U (utilization factor)
= the ratio of luminous flux on working plane against that from the lamp. It depends on lamp position, light distribution, environment conditions, and so on; and
M (maintenance factor)
= predicting lowering rate of initial luminance with lapse of the working time. It depends on the maintenance policy.

The electric energy required for lighting can be calculated as follows:

$$E = \frac{N \times F}{\eta} \times t \quad \text{(Wh)}, \tag{12.2}$$

where

E = energy in Wh,
η = lamp efficiency, and
t = lighting time.

From Equations 12.1 and 12.2, the energy is given as

$$E = \frac{A \times L_{\text{lght}} \times t}{U \times M \times \eta} \quad \text{(Wh)}. \tag{12.3}$$

The calculated luminance from Equations 12.1 must be in correspondence with the standards given in Table 12.4 or substituting the standard value into Equations 12.1 to obtain the number of specific lamps. The standard illumination (lx) for different places is given in Table 12.4.

Power loss in the wires connected to the lamp is added to Equation 12.3 to obtain the actual power required. It is seen from this equation that the value of E (energy in Wh for lighting) can be reduced by reducing power loss in wires, lighting, and time of lighting and also by increasing distribution factor, maintenance factor, and lamp efficiency.

The desired change of these parameters can be achieved as follows:

- The reduction of distribution wire loss is verified by optimizing the design of wiring diagram and by specifying the junction's installation.
- The lighting luminance is kept proper by optimizing the lighting levels and considering the daylight.

TABLE 12.4 Illumination Standard for Different Places [174]

Illumination (lux)	Place	Operation
10–30	Outdoor (for passage and safety guard within compound)	
30–75	Indoor emergency staircases, warehouses, outdoor power equipment	Such as loading, unloading, load transfer
75–150	Entrance/exit, corridor, passage, warehouses involving operation, staircases, lavatories	Very rough visual operation
150–300	Electricity room and air-conditioning machine room	Rough visual operation such as packing and wrapping
300–750	Control room	Ordinary visual operation in general manufacturing processes, etc.
750–1500	Design and drawing rooms	Final visual operation in selection and inspection in textile mills, typesetting, and proofreading in printing factory, analysis, etc.
1500–3000	Instrument panel and control panel in control room, etc.	Exceedingly fine visual operation in manufacture of precision machines and electronic parts, printing factory, etc.

- Reducing the lighting time by switching off the unnecessary lamps and using automatic switches or timers for the outdoor lamps.
- Using high-efficiency lamps by choosing the best ones for the required application.
- Improving the utilization factor. It is directly proportional to room index and reflectivity of walls, floors, and ceilings as shown in Table 12.5. The room index "R_I" is calculated by the equation:

$$R_I = \frac{W \times L}{H(W + L)}, \tag{12.4}$$

where W and L are the width and depth of the room, respectively. H is the height (m) of the light source from the working plane.

The room index is maximum when $W = L$, that is, for square rooms.

- Improving the maintenance factor by cleaning the light source regularly and replacing the lamps on due time.

TABLE 12.5 Utilization Factor for Some Cases [174]

Ceiling Reflectivity (%)	80									50								
Wall Reflectivity (%)	60			30			10			60			30			10		
Floor Surface Reflectivity (%)	40	20	10	40	20	10	40	20	10	40	20	10	40	20	10	40	20	10
	Room Index Utilization Factor																	
0.6	0.45	0.42	0.40	0.31	0.30	0.30	0.26	0.25	0.25	0.41	0.39	0.38	0.30	0.29	0.29	0.25	0.25	0.25
0.8	0.56	0.51	0.49	0.41	0.39	0.38	0.35	0.34	0.23	0.51	0.48	0.47	0.39	0.38	0.37	0.34	0.33	0.33
1.00	0.63	0.57	0.55	0.47	0.45	0.44	0.41	0.40	0.39	0.57	0.53	0.52	0.45	0.44	0.43	0.40	0.39	0.38
1.75	0.71	0.63	0.60	0.55	0.52	0.50	0.48	0.46	0.45	0.64	0.59	0.57	0.52	0.50	0.49	0.46	0.45	0.44
1.50	0.76	0.68	0.64	0.61	0.56	0.54	0.54	0.51	0.50	0.68	0.63	0.61	0.57	0.54	0.53	0.52	0.50	0.49
2.00	0.85	0.75	0.70	0.71	0.65	0.62	0.64	0.59	0.57	0.76	0.70	0.67	0.66	0.62	0.60	0.60	0.58	0.56
2.50	0.91	0.79	0.74	0.78	0.70	0.66	0.71	0.65	0.62	0.80	0.73	0.70	0.71	0.67	0.65	0.66	0.63	0.61
3.00	0.95	0.82	0.76	0.83	0.74	0.70	0.77	0.69	0.66	0.84	0.76	0.73	0.76	0.70	0.68	0.71	0.67	0.65
4.00	1.01	0.86	0.80	0.91	0.79	0.75	0.85	0.76	0.71	0.88	0.80	0.77	0.78	0.75	0.72	0.78	0.72	0.70
5.00	1.09	0.88	0.82	0.96	0.83	0.77	0.91	0.79	0.78	0.91	0.82	0.79	0.88	0.78	0.78	0.82	0.76	0.73
10.00	1.13	0.93	0.86	1.08	0.90	0.84	1.05	0.89	0.82	0.97	0.87	0.83	0.94	0.85	0.81	0.92	0.84	0.80

12.12.2 Motors

Electric motors are used in many applications. They account for about 60% of the total electric energy consumption. In the service industry, the motors are used for air-conditioning, heating, escalators, and elevators. In the manufacturing industry, a very wide range of motor ratings are used, from few watts in servocontrol systems and information technology to several megawatts in pumps, compressors, cranes, and so on. Also, in household applications, the motors are used in some applications for a short time, for example, shaving machines, hair dryer, and vacuum cleaners, and in other applications continuously, for example, ventilators, refrigerators, and freezers. Efficiency is not a matter of major concern with the motors operating for a short time. In contrast, motors operating continuously should be efficient and managed properly.

The most common type of electric motor used in these applications is the induction motor. This type of motor is simple in construction, low cost, and reliable in addition to the availability of speed control over a wide range by power electronics.

To reduce the energy consumption of induction motors, three main factors are taken into consideration: (1) motor efficiency, (2) motor losses, and (3) power quality.

Variation of efficiency versus the load is shown in Figure 12.6b. It is seen that the efficiency reaches the highest level at loads of 60–100% full load. The motor losses are the iron, and friction and windage losses that are fixed and independent of load, and the copper (I^2R) and stray load losses that depend on the load as shown in Figure 12.6a.

It can be seen from Figure 12.6 that although the losses are higher at high loads, they do not have a significant effect on the efficiency.

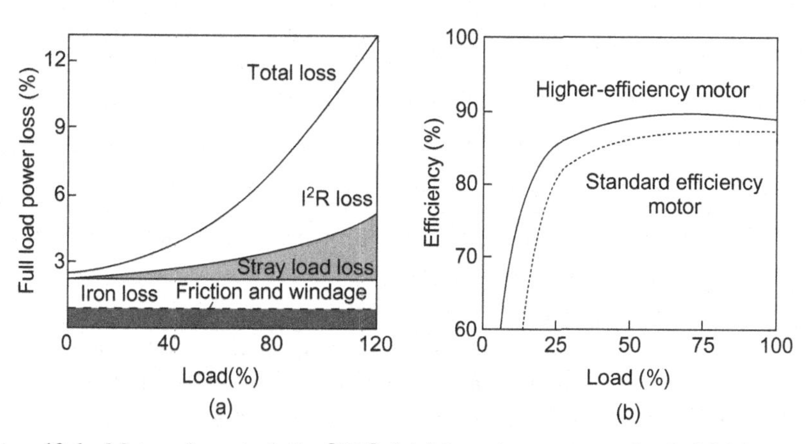

Figure 12.6 Motor characteristics [174]. (a) Motor losses versus load. (b) Motor efficiency versus load.

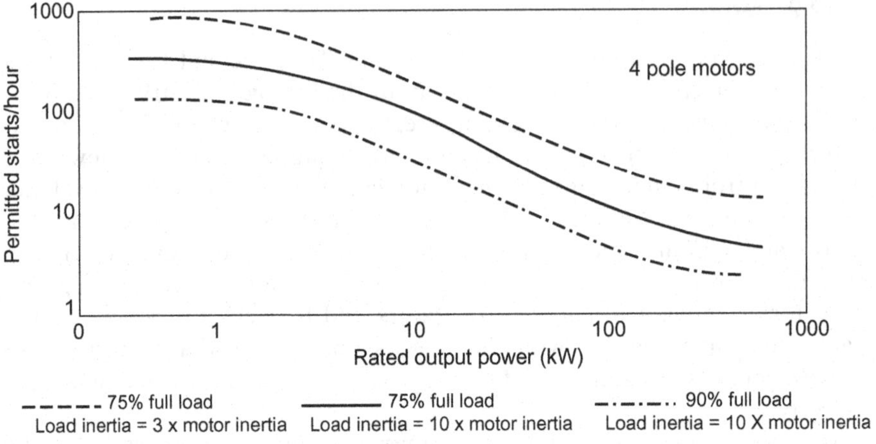

Figure 12.7 Allowable number of switching versus motor rated power [174].

The main elements concerning power quality and a negative effect on motor power consumption are the harmonic distortion and voltage fluctuation [176, 177].

Reducing Motor Energy Consumption: Different actions can be followed to reduce the energy consumed by the motors, such as the following:

- Shutting down the motor by switching off manually or automatically when it is not needed. On the other hand, the frequent switching increases wear on belt drives and bearings in addition to shortening the motor insulation life due to extra heating produced by the high starting current. Therefore, the number of switching on/off is limited depending on the motor rating. It is decreased as the rating increases, as shown in Figure 12.7.
- *Reducing Motor Load*: The approach that can be used to reduce the load on the motor depends on the nature and type of load. General basis to be followed for this purpose is illustrated below:
  - choosing a proper motor rating that is sufficient but does not exceed the load power;
  - in case of varying load, installing a number of motors in parallel with different ratings and selecting appropriate motor automatically;
  - avoiding use of low-quality spares;
  - maintaining a consistent load level;
  - ensuring the standards of load environment. The deviation of environmental operating conditions from the standards applied to motor design may cause an increase of power consumption;
  - using sensors on the load side to indicate any interruption and switch off the motor if needed;

- in some applications the motor as a driver is coupled with the load through a gearbox. The gearbox chosen must be of high efficiency and the lubrication and cleaning must be implemented regularly.

- *Reducing Motor Losses*: The reduction of motor losses can be achieved by applying the following measures of energy efficiency:
  - careful maintenance emphasizes the minimization of losses;
  - use of proper size of the motor to avoid higher losses when the motor operates at light load;
  - for light loads, the motor must be connected permanently in star connection to reduce the demand current; consequently, the losses are reduced. Another solution is by using energy optimizer.
  - high-efficiency motors are very effective for savings in all applications. The annual saving S is calculated by

$$S = \text{hrs} \times \text{kW} \times \%\ \text{F.L.} \times C \times (1/\eta_{\text{std}} - 1/\eta_{\text{hef}}),$$

where hrs is the number of operating hours, kW is the motor rating in kW, % F.L. is the average motor loading as a percentage of full load, C is the cost of 1 kWh, η_{std} is the efficiency of standard motor, and η_{hef} is the efficiency of the high-efficiency motor.

Replacing conventional motors that are still functional is not economically feasible. However, consistently replacing the present motors at the time of failure by high-efficiency motors should be considered.

Energy consumption per motor class and the average efficiency of conventional and high-efficiency motors in these four classes is shown in Table 12.6. These values show that high-powered motors are more efficient and that efficiency improvement by replacing motors is the highest for low-powered motors.

The annual energy cost of operating a motor depends on how often and how long a motor is used. Usually, this cost is higher than its purchase price. Therefore, the investment decision should always take the cost of energy consumption over the lifetime of the motor into account. The extra investment in a more expensive, high-efficiency motor often makes economic sense. Calculations show payback time of less than 3 years [178, 179].

TABLE 12.6 Motor Classes: Energy Use and Efficiency [178]

	Efficiency	
Power Range (kW)	Conventional Motor (%)	High-Efficiency Motor (%)
0.75–7.5	80	86
7.5–37	86	90
37–75	90	93
>75	95	96

- *Variable-Speed Drive*: In a variable speed drive, exactly the right amount of energy is delivered to the motor to obtain the required function properly. No energy is wasted, unlike traditional systems using mechanical mechanisms (e.g., breaking mechanisms), where excess energy is converted into useless heat.

  Given the significant energy (and cost) savings, together with other benefits of variable-speed drives, such as better control of industrial processes resulting in higher-quality products and services and lower global cost, they are being introduced on a very significant scale, both in new and in retrofitted installations.

12.12.3 Heating

Electric heating systems are the section of electrotechnologies describing high-power heating processes, which are sourced through electric energy.

In many cases, the electroheat technology is the only means for obtaining a certain product. For example, when objects consist of materials with different loss factors, interesting applications of microwave heating technology are possible, for example, pasteurizing pharmaceuticals and foodstuffs within their packages without burning the packaging materials.

In other cases, electroheat technology competes with concurrent technologies using fossil fuels, reducing industry's investment and operating costs and primary energy requirements.

Different methods of electric heating are applied depending on the required temperature for the desired application. These methods are arc heating, electromagnetic induction heating, high-frequency electric field heating, electron or ion beam heating, electromagnetic wave heating, and electromechanical power heating. For each case, the systems of electric-to-heat energy conversion are illustrated in Figure 12.8.

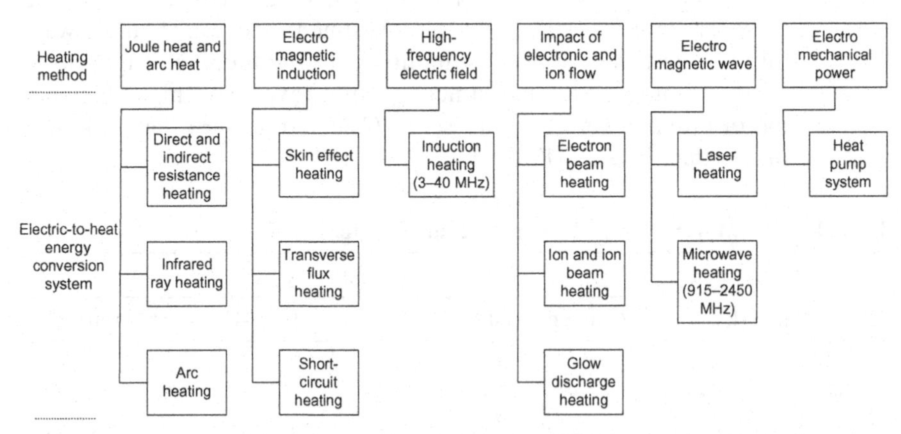

Figure 12.8 Energy conversion systems for different heating methods.

TABLE 12.7 Energy Consumption of a Gas Furnace and an Electric Furnace [178]

		η (%)	Useful Heating Power (kW)	Total Power (kW)	Consumption of Natural Gas (L/s)	CO_2 Emission Rate (g/s)
Gas furnace	A	40	100	250	6.7	12.4
	B	80	100	125	3.3	6.2
Electric furnace		95	100	105.3		11.7
						9
						1.6

TABLE 12.8 Comparison of a Hot-Air Oven and a Radio-Frequency Oven [178]

	Hot-Air Oven	Radio-Frequency Oven
Installed power	360 kW (gas + electric)	40 kW (electric)
Energy consumption	8640 kWh	120 kWh
Energy cost	150 €/day	9 €/day
Total investment cost	>900,000 €	525,000 €

As an illustrative example, a comparison is made between two gas furnaces (A and B with efficiency η 40% and 80%, respectively) and an electric furnace (efficiency 95%). They have a heating power of 100 kW. For the gas furnaces, the consumption of natural gas is expressed in liters per second. The rate of CO_2 emission is expressed in grams per second. The electric furnace is using the resistance heating as an electroheat technology. The comparison is shown in Table 12.7 [178].

It is noted that electric furnaces consume least power since they are most efficient. Three figures of CO_2 emission rate are given for the electric furnace depending on the way of producing electricity (e.g., using fossil fuel or others).

Another example of a comparison between two alternative techniques for gluing plastic components together is presented in Table 12.8 [178]. The first technique is hot-air heating; the second technique is radio-frequency heating. After application of the glue and bringing the parts together, the glue has to be heated. The figures in Table 12.8 prompt the following conclusion: The costs and energy use are lowest for the radio-frequency oven. Another advantage of the radio-frequency oven is it occupies five times less space than the hot-air oven.

To save energy in these applications, it is recommended to use electric heating for continuous, not intermittent, operation. This helps avoid waste of power due to cooling and reheating processes. Also, the heat insulation of the parts having significant heat loss is needed where temperature sensors can be used for measurements and control. In total, it can be said that electroheat processes are advantageous from the point of view of energy consumption and CO_2 emission as well [178].

12.12.4 Pumps

Pumps are one of the equipment consuming large amount of electric power. Therefore, the saving of energy consumed by pumps is an economical and technical effective factor; that is, it deserves attention. The efficiency of the pump itself is high, but as a pumping system (including pipes, valves, etc.) it is low. Centrifugal pumps are the most widespread type of pump. Simply, they involve an impeller that transfers mechanical energy, which is converted into potential energy represented by the pressure and kinetic energy represented by the flow rate, to the fluid (Fig. 12.9). The selection of pump size usually exceeds the application requirements. The pumps with excess head are selected to overcome any expected increase of head loss in the pipes. Also, pumps with excess capacity of flow rate are selected to be valid for future increase of supply and drainage quantities [180].

Many of these pumps are run continuously such as pump facilities for water supply and sewage. Their operation should be controlled according to the quantity and variation of water supply. In addition, pump facilities for industry in many cases are running inefficiently because of using valves to control flow rate. Therefore, it is expected to save considerable amount of electric power consumption if the pump operation is controlled adequately.

Pumps are specified basically by three elements: flow rate Q; total head H, which is the summation of static head H_{st} and dynamic head H_{dyn}; and rotating speed N (Fig. 12.10).

The pump is characterized by its specific speed N_s. The specific speed is constant for all impellers having the same shape and is calculated by the relation

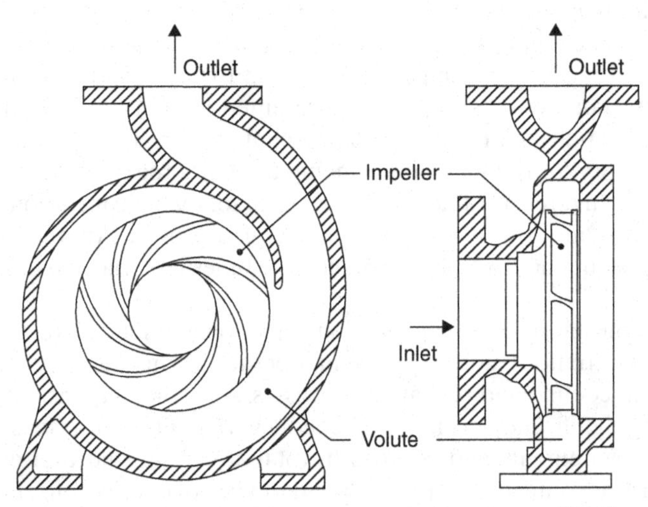

Figure 12.9 The main parts of a centrifugal pump [181].

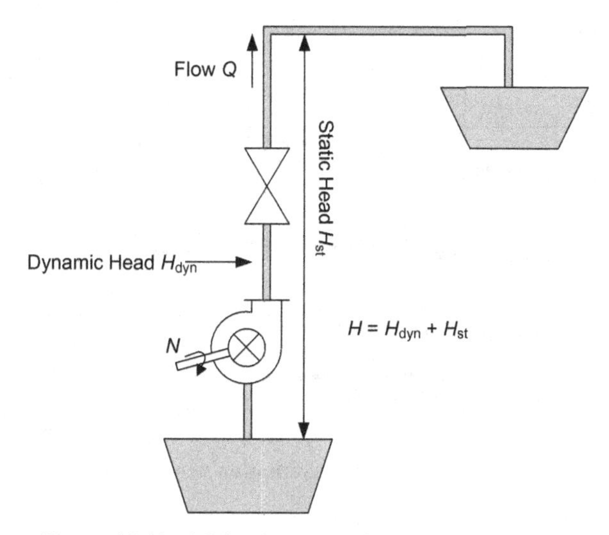

Figure 12.10 Main elements of pump specification.

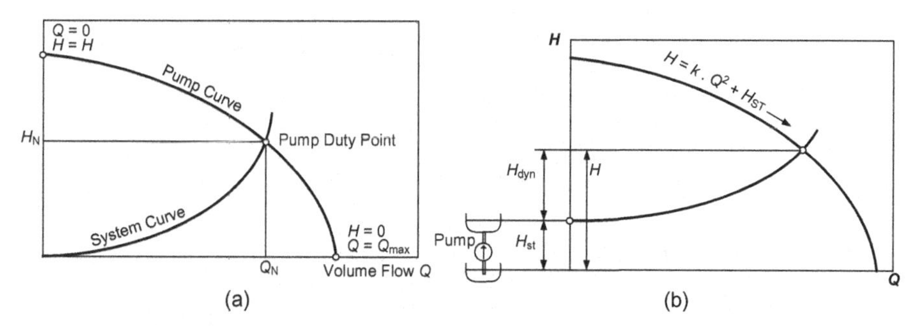

Figure 12.11 Pump characteristic (a) at $H_{st} = 0$ and (b) at $H_{st} > 0$.

$$N_s = N\frac{\sqrt{Q}}{H^{3/4}}K, \tag{12.5}$$

where N is the rotating speed (rpm), Q is the flow at point of maximum efficiency (m^3/min), H is the total head at point of maximum efficiency (m), and K is a factor = $\rho\ g/\eta$, ($\rho\ g$ is the specific weight and η is the overall pump efficiency).

The shaft power consumed by pump is given as

$$P = \frac{\rho \times Q \times H \times g}{\eta} = Q \times H \times K(\text{kW}). \tag{12.6}$$

12.12.4.1 Pump Characteristics The pump characteristic is expressed by the pump curve that shows the variation of flow rate Q versus total head H. The pump characteristics at $H_{st} = 0$ and $H_{st} > 0$ are shown in Figure 12.11 [182,

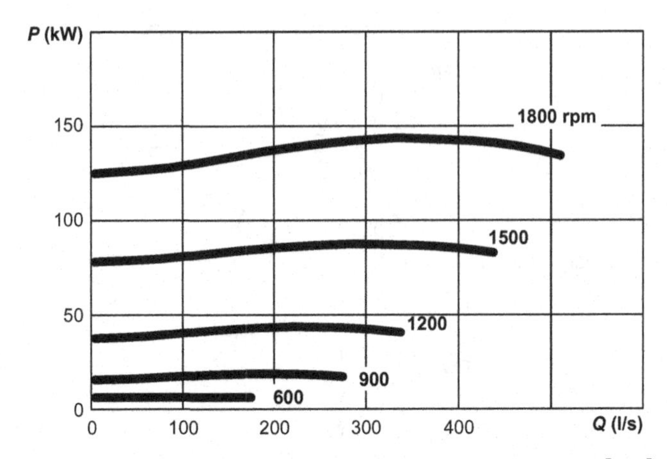

Figure 12.12 Pump power consumption versus flow rate [183].

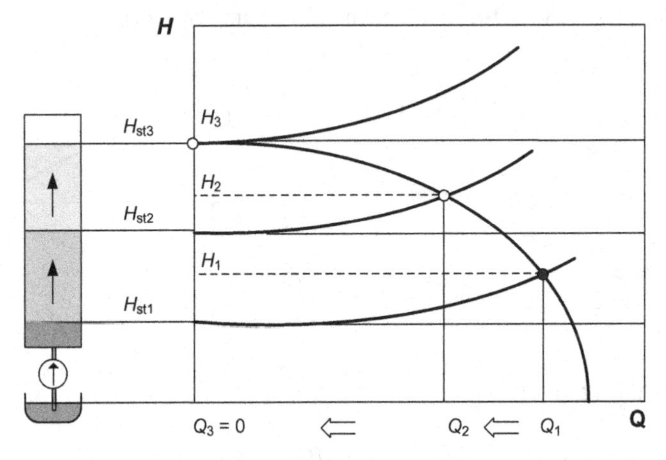

Figure 12.13 Pump operating point at different static heads.

183]. In this figure, the system curve represents the variation of system resistance against flow rate. The intersection of the pump curve and system curve gives the pump operating point (pump duty point).

The power consumption of a pump at rated speed changes with changing the flow rate as shown in Figure 12.12. From pump characteristic curves (pump and system curves), it can be seen that as the static head increases, the system curve is getting higher because the system resistance increases. Accordingly, the pump operating point moves in a direction that flow rate decreases. The decrease of flow rate from Q_1 to Q_2 to Q_3 when the static head increases from H_{st1} to H_{st2} to H_{st3} is shown in Figure 12.13.

The pumping system has three key elements: flow, static head, and frictional head. So, the possible opportunities for saving the electric power in pumps are

- reducing the flow and head requirements,
- selection of the most efficient pump type and size,
- maintaining the pumps and all system components to avoid efficiency loss, and
- efficient flow rate control.

12.12.4.2 Flow Rate Control It is common to find that the daily fluid demand from the pump is not constant. It may be due to the influence of various factors such as time, weather effects, unexpected events, and, generally, the change of operating conditions. Therefore, the control of flow rate according to the required signal flow pattern is an important aspect.

Flow rate control can be applied by using different methods such as throttling by a valve, by-passing by a valve, on–off pump control, variable-speed control of pump, and running of more than one pump in parallel or series. These methods are described below:

a) *Throttling by a Valve*: Throttling process increases the system resistance as shown in Figure 12.14 by the dotted line. Consequently, the pump operating point changes and flow rate becomes Q_2 instead of Q_1 ($Q_2 < Q_1$).
b) *By-Passing by a Valve*: As shown in Figure 12.15, when by-passing the pump by a valve, the head is decreased and the operating point moves from A to B to give flow rate Q_3. The system curve is getting higher and

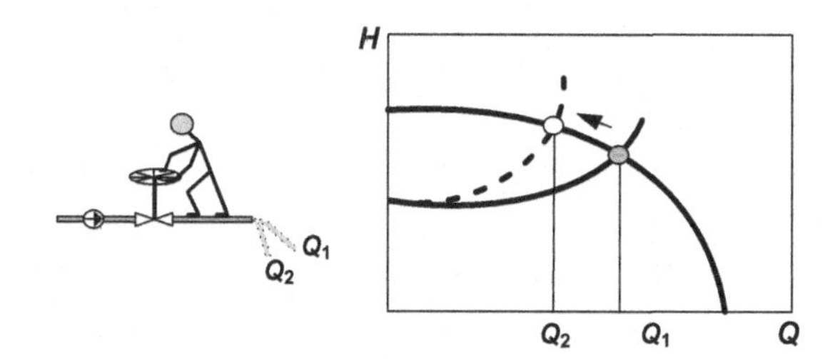

Figure 12.14 Pump characteristic when throttling the output by a valve.

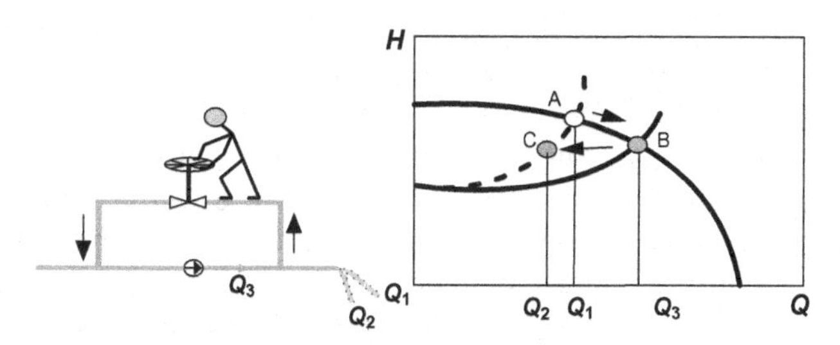

Figure 12.15 Pump characteristic when by-passing by a valve.

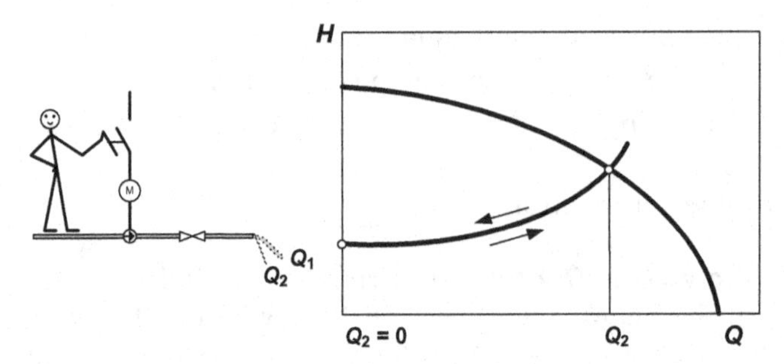

Figure 12.16 Pump characteristic with on–off control.

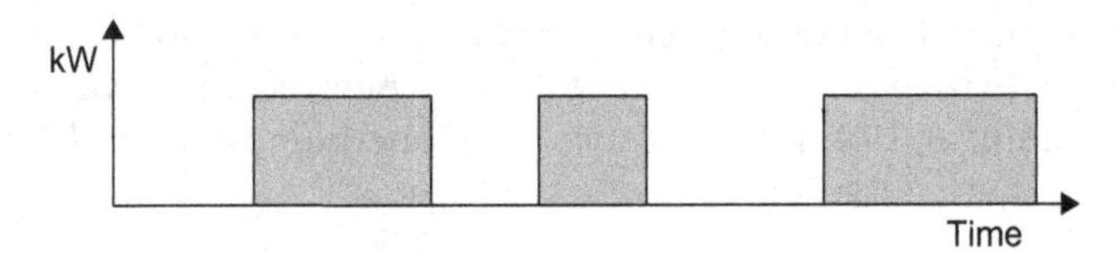

Figure 12.17 Pump power consumption.

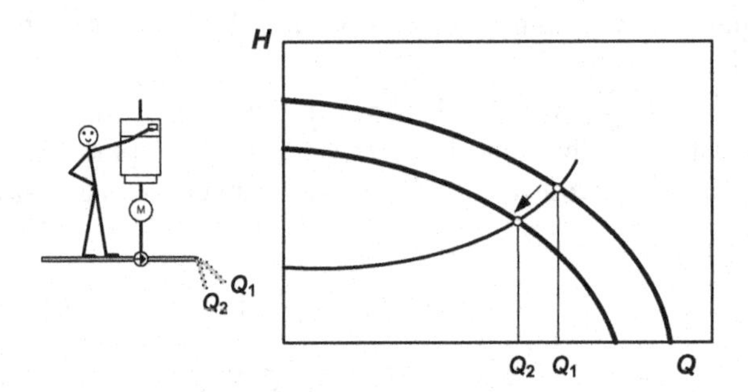

Figure 12.18 Pump characteristic with speed control.

the operating point moves to C (at the same head) giving a flow rate of Q_2.

c) *On–Off Control*: When the use of pump is clearly distinct between periods of need and no need, pumps may be stopped during unnecessary periods, that is, the pump is running by intermittent operation and its characteristic is shown in Figure 12.16. The power consumed by pump is only for operation periods (Fig. 12.17). This method can be applied when turning the pump on and off in long cycles to avoid the risk of fluid hammer effects and possible motor overheating.
d) *Variable-Speed Control*: This method is most effective and has several advantages such as smooth operation of pump at all ranges of flow rate (low and high) and large reductions in electric power cost. The change of pump speed changes the pump characteristic curve, which results in a change of flow rate (Fig. 12.18).

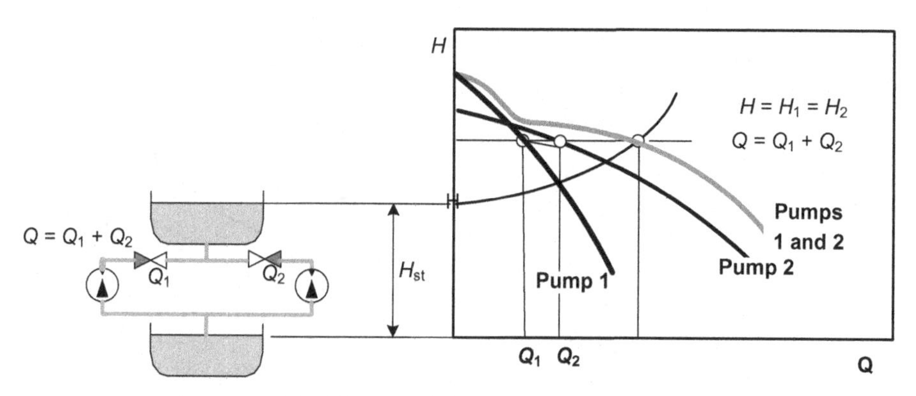

Figure 12.19 Characteristics of two pumps in parallel.

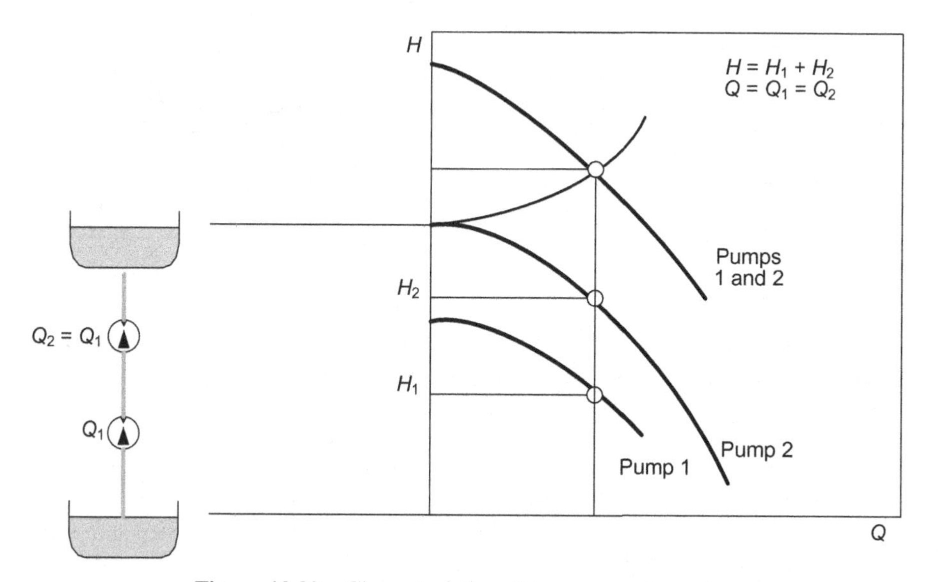

Figure 12.20 Characteristics of two pumps in series.

e) *Running More than One Pump*: A varied number of pumps are used according to flow rate fluctuation with a goal of reducing the total power consumption. The pumps are connected either in parallel (for flow rate control) or in series (for static head control) as shown by pump characteristics in Figures 12.19 and 12.20, respectively.

12.12.4.3 An Illustrative Example A water pump is running at the pump duty point of $Q = 10\,\text{m}^3/\text{m}$ and total head of $H = 10\,\text{m}$. The pump characteristic (head vs. flow rate) is shown in Figure 12.21. The flow rate is controlled to be changed to $7\,\text{m}^3/\text{m}$. Different methods of flow rate control are applied and the shaft power for each method can be calculated as below:

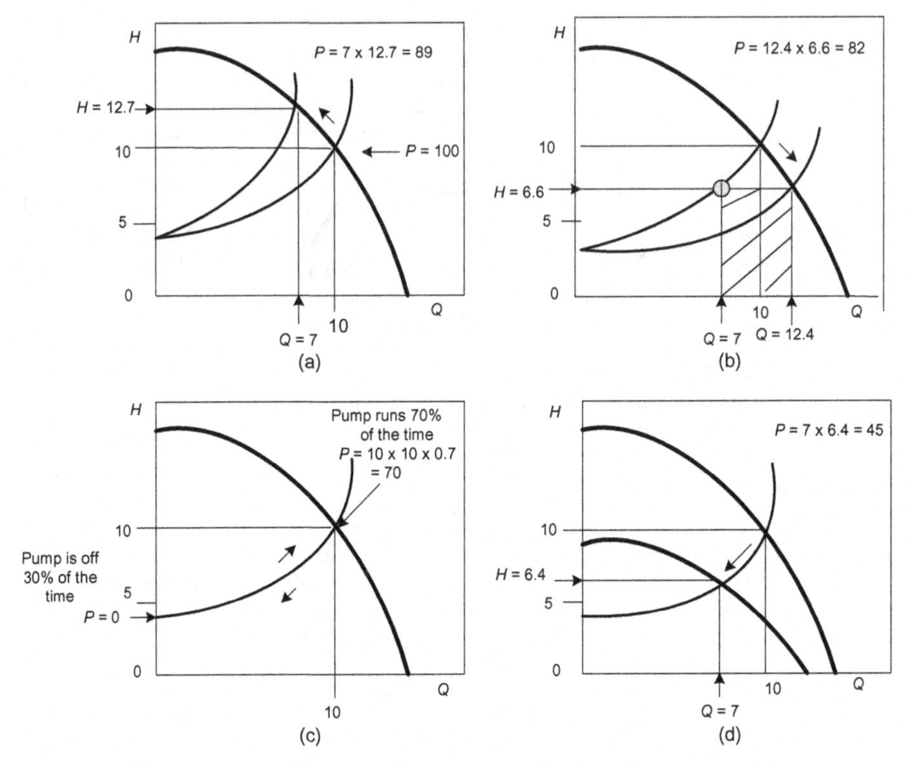

Figure 12.21 Pump characteristic at different methods of flow rate control. (a) Throttle control; (b) by-passing; (c) on–off control; (d) variable-speed control.

Assuming the factor K in Equations 12.6 equals unity, the shaft power P is given as $P = Q \times H$ (kW).

By throttle method, as shown in Figure 12.21a, $P = 7 \times 12.7 = 88.9 \approx 89$ kW.

Using by-passing method (Fig. 12.21b), $P = 12.4 \times 6.6 = 81.84 \approx 82$ kW.

Using on–off control (Fig. 12.21c), where the pump is off 30% of the time, $P = 0.7 \times 10 \times 10 = 70$ kW.

Using variable-speed control (Fig. 12.21d), $P = 7 \times 6.4 = 44.8 \approx 45$ kW.

It can be seen that variable-speed control is the most effective method.

CHAPTER 13

SCADA SYSTEMS AND SMART GRID VISION

13.1 INTRODUCTION

As discussed in Chapter 1 of this book, the distribution system at the medium-voltage (MV) or low-voltage (LV) levels is designed using different structures such as radial, double radial, open ring, and closed ring. Each one is selected on the basis of load priority and the desired reliability level. Accordingly, the system switching in case of emergency operating conditions is either manual or automatic, depending on the permitted duration of interruption periods and obviously the nature of loads. The distribution systems feeding the industrial and commercial loads must be automated in most cases.

Distribution system automation helps to raise the reliability level by decreasing the interruption duration, isolating the faulty parts, keeping the system equipment functioning, and automating the required switching processes during system operation. To what extent is it possible to decide the capability of automation applied to the distribution systems? It depends on the size of the network, the nature and amount of load, and the type of system equipment. When the distribution system grows and becomes more complicated, is the automation sufficient to achieve optimal performance and to operate satisfactorily and securely? This is the question to be answered by the following discussion.

In a commercial office building (e.g., 40 floors), the utility feeder comes into a main circuit breaker (CB) in the main switchboard located in the building

Electric Distribution Systems, First Edition. Abdelhay A. Sallam, Om P. Malik.

Published 2011 by John Wiley & Sons, Inc.

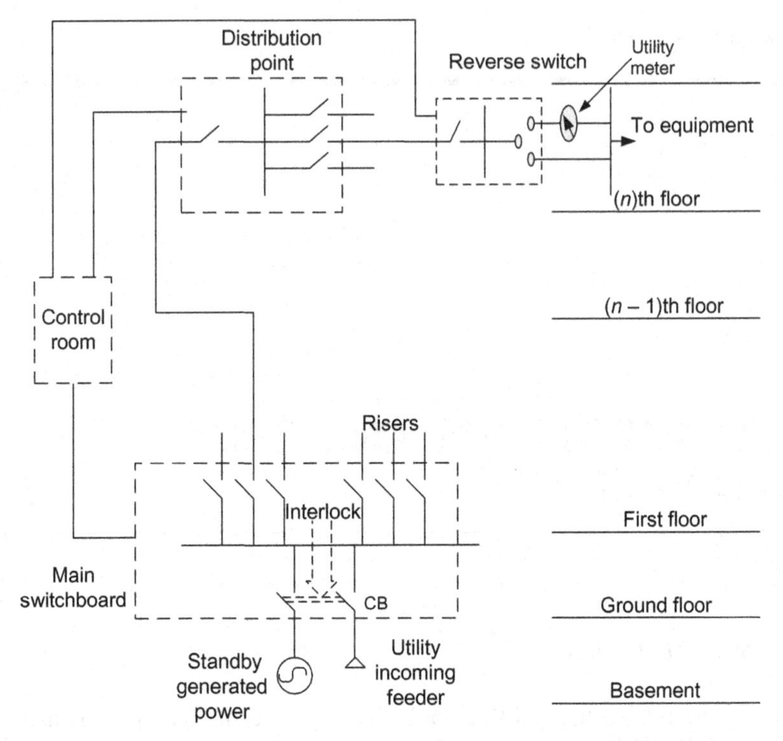

Figure 13.1 Schematic drawing of main lines of feeding a commercial building.

basement. The outgoing feeders from this switchboard go inside riser ducts to feed each floor through a distribution point (DP) (sub-switchboard). This DP distributes the power into the different apartments within each floor via utility meters of each apartment.

It is required to install standby generators as a backup to the utility supply to avoid the uncertainty in utility power reliability and damage in the event of power failure. In this situation, an interlock between the generators' switches (CBs) and the utility feeder CB is needed to avoid paralleling between them. In addition, it is necessary to install transfer switches on each floor to allow the standby generated power to flow into each apartment directly without passing through the utility meters (Fig. 13.1).

As a result of these requirements, a lot of equipment needs to be distributed throughout the building: interlock panel for each generator located in the control room near the main switchboard (utility switches), transfer-switches panel located at the DP of each floor, and the interconnection between the equipment and the control room. It is clear that the space required for the control room will be large, and the amount of wires needed between panels and remote points is also increased.

The operator monitors the system operation and the equipment functionality via the control room to know the position of each switch (open/closed/

emergency), to visually diagnose the system status during utility outage and to diagnose the fault. These monitoring functions, traditionally, need a very large number of indicating lamps and warning alarms. Imagine the difficulties and confusion that the operator is subjected to because of the large number of interconnecting wires and indicators. Also, do not forget the difficulties and time consumed in commissioning and testing the system. On the other hand, the load is increasing continuously and its nature may change resulting in probable redesign of the distribution system by adding new equipment or making changes in the wiring system that may be difficult to implement.

To mitigate these difficulties, each DP is equipped with a programmable logic controller (PLC) to perform the required control functions and at each location a user interface workstation is specified. The workstation has the capability of processing the data gathered from all metering points in the building through communication networks using communication cards. Data processing results in information about the system status, energy management system (EMS), economic dispatch, interconnection pricing of energy, and any other useful application the workstation is programmed to perform. The information about the system status can be sent to the PLC to execute the corresponding control commands while the other information is used for system planning, maintenance, and operation. With the workstation, the operator is able to display graphically on the screen the single-line diagram of the distribution system and any other visual display needed by utilizing user-friendly interaction through windows. This system with high data acquisition and reporting requirements (but with limited control) is defined as "SCADA."

The SCADA system can be easily adapted to any change in the application field as it can be integrated with other systems, for example, the protection system.

As explained in Chapter 5, consider a power system involving the protection relay as shown in Figure 13.2. In this diagram, the current and potential transformers send samples of phase currents and phase voltages to be used as inputs to the protection relay. In turn, if these inputs violate prespecified values, the relay sends a tripping signal to the CB to be opened. To guarantee a correct performance for this protection mechanism, it must be ensured that the protection relay is working properly. This can be verified by testing the relay using a generator as a simulator to generate the proper input signals and measuring the output (Fig. 13.3). The need for a simulator arises because the utility company cannot interrupt the main power lines with the goal of testing the relay. This test is applied from the protection point of view, but if it is needed to use this relay as a data source to the SCADA system, the situation becomes quite different. It needs to do another test to verify that the data generated by the protective functions is correctly communicated to the SCADA system. In this case, knowledge of the exact numerical representation of the data transmitted by the communication circuit is an important factor.

Disadvantages of the relay systems and the advantages of the SCADA systems can be summarized as below:

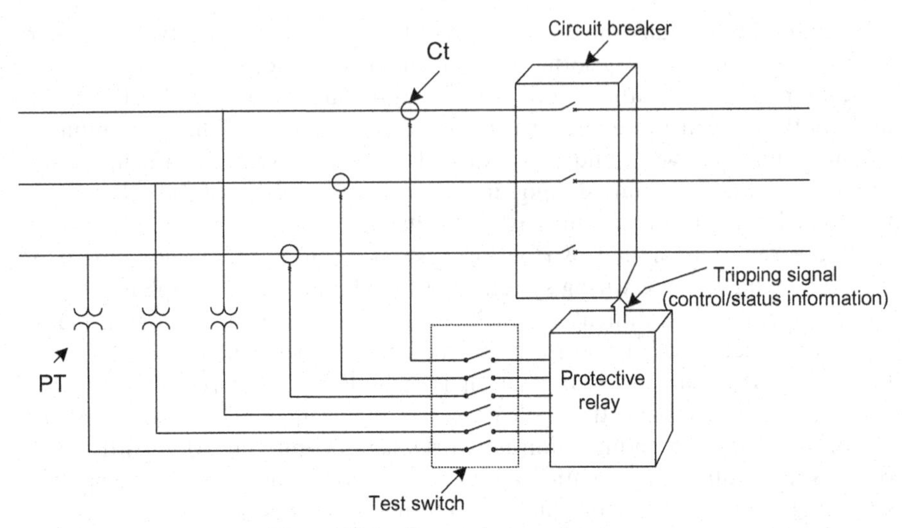

Figure 13.2 Protection mechanism.

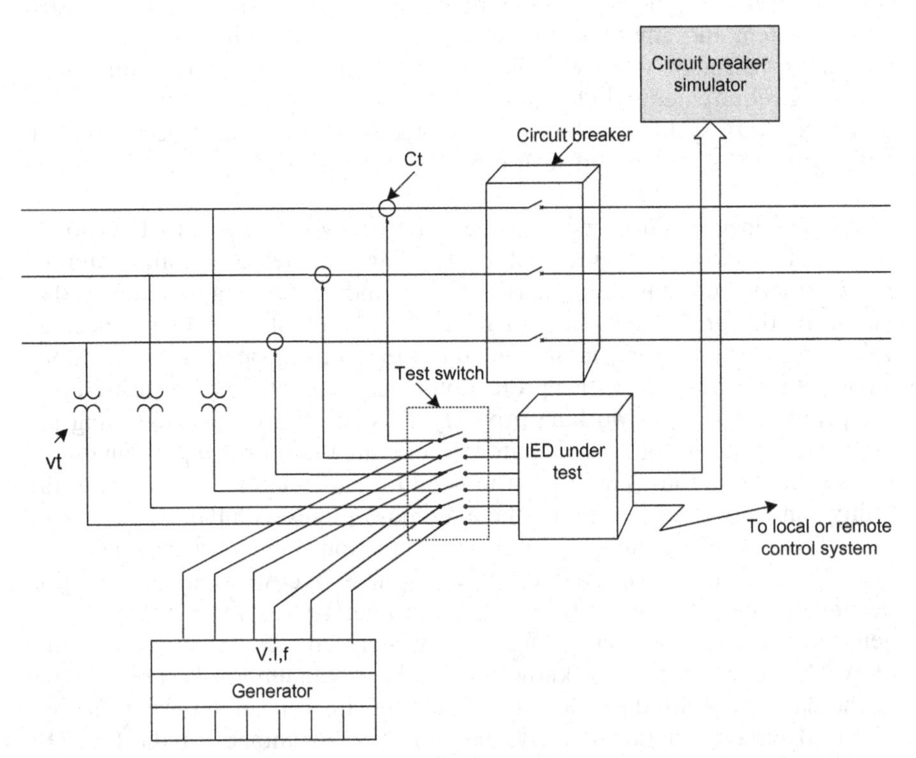

Figure 13.3 Testing of relay by using a simulator. V = voltage; I = current; f = frequency.

Disadvantages of the Relay Systems

- complicated control systems;
- expensive systems;
- systems need more space;
- control relays consume more power, generating more heat;
- relays are used only for on/off control;
- any change in the control program needs rewiring of the relays; and
- it is difficult to troubleshoot and diagnose the fault for complicated control systems.

Advantages of the SCADA Systems

- self-diagnostics and easily maintained;
- capability of arithmetic functions implementation;
- easy to program and reprogram;
- facility of communication with other controllers or a master host computer;
- the ability of PLCs to move past simple control to more complex schemes as proportional/integral/derivative (PID) control;
- industrial plant SCADA can be viewed as a distributed control system (DCS); and
- capability of graphical user interface (GUI) and visual display of system status.

More details about SCADA, definitions, components, communication, and others are given in the forthcoming sections.

13.2 DEFINITIONS

13.2.1 A SCADA System

SCADA system stands for supervisory control and data acquisition system. It refers to the combination of telemetry and data acquisition. It commences with measurement of the data by specific devices in the field of application and collected via intelligent electronic devices (IEDs), then transferring these data to a master station to implement the necessary processing and control algorithms. The results of processing are displayed on a number of operator monitoring screens, while the control actions are conveyed back to the field of application in real time (Fig. 13.4) [184].

13.2.2 Telemetry

Telemetry is the initial step in applying SCADA by defining the technique used for measuring the data (voltage, current, speed, etc.) from different

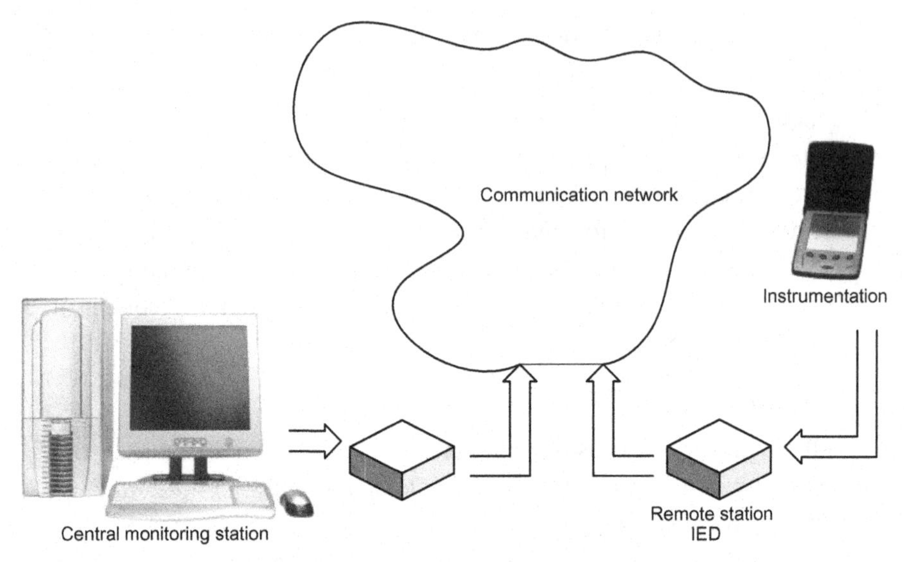

Figure 13.4 Elements of a SCADA system.

locations in the real-time process and transferring it to the IEDs such as remote terminal units (RTUs) or PLCs in another location through a communication circuit [185].

13.2.3 Data Acquisition

Data acquisition refers to the method used for accessing and collecting the data from the devices being controlled and monitored, and to be forwarded to a telemetry system ready for transfer to the various sites. The data may be analog or digital gathered by sensors such as ammeters, voltmeters, speedometer, and flowmeter. It can also be data to control equipment such as actuators, relays, valves, and motors.

13.3 SCADA COMPONENTS

The SCADA system consists of four components as follows.

13.3.1 Instrumentation (First Component)

This component refers to the devices used for monitoring certain parameters such as sensors and the devices used for controlling certain modules such as actuators. In general, these devices are connected to the equipment or machines being monitored or controlled by the SCADA system. Their main function is to convert the parameters from the physical form to electrical form as continuous (analog) or discrete (digital) signals readable by the remote station equipment (RTUs or PLCs).

13.3.2 Remote Stations (Second Component)

The measuring devices (first component) connected to the plant being monitored and controlled are interfaced to the remote station. Functions of remote stations are

- gathering data from the different devices in the plant being monitored and controlled;
- holding the data gathered in its memory and waiting for a request from the master station (master terminal unit [MTU]) to transmit the data; and
- receiving data and control signals from MTU and transmitting the control signals to the plant devices.

The remote station is either RTU or PLC. The RTU is used effectively in the event of difficult communications. Its inputs and outputs are shown in Figure 13.5. The RTU has digital/analog inputs and outputs with light emitting diode (LED) indication (selectable per channel), optically isolated for surge protection and also protected against short circuits. On the other hand, the PLC is usually expected to be already available in the plant processes, so it is of great worth to be used also in the SCADA systems. Both the RTU and PLC have been greatly improved recently.

13.3.3 Communication Networks (Third Component)

The geographically dispersed RTUs are connected to the MTU through a variety of communication channels, including radio links, leased lines, and fiber

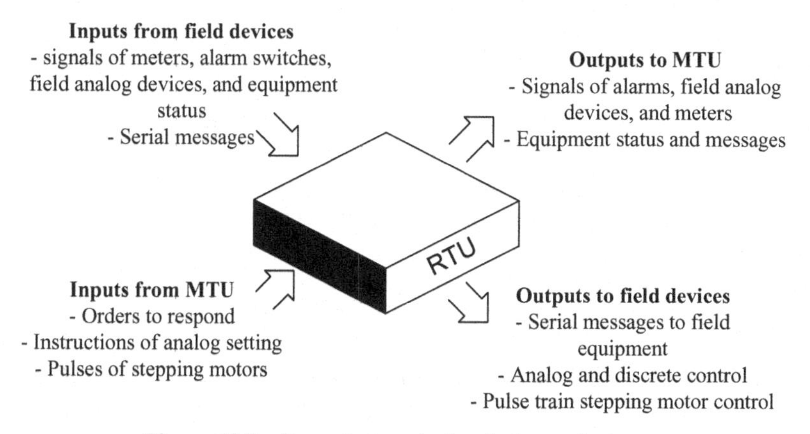

Figure 13.5 Remote terminal unit: inputs/outputs.

optics [186]. The design of both RTUs and MTU is largely affected by the availability limitation and high cost of communication channels.

The hardware and software design of both MTU and RTUs must guarantee that the information is transferred correctly from the RTUs to the MTU and vice versa, and not affected by the noise occurring randomly on the communication channel.

The configuration of communication system depends on

- number and location of RTUs;
- number of points at RTUs and required update rates; and
- available communication equipment, techniques, and facilities.

The SCADA communication techniques include modulation, multiplexing, message format, and information transfer.

Modulation: To convey information from one point (sending end [SE]) to another point (receiving end [RE]) through a communication channel, it must be modulated at the SE and demodulated at RE. This means that it is necessary to transmit a signal with a change from SE to RE where it is detected. This change can be carried out by either a change in signal amplitude or change of signal frequency or change of signal phase. The demodulator at RE detects the change in the signal and outputs the transmitted information.

If the information transmission is only in one direction, it is called "simplex channel," and if it is in the two directions but not simultaneously, it is called "half duplex." Finally, if the transmission is in the two directions and simultaneous, it is called "full duplex."

Multiplexing: In the SCADA system structure, it is needed to transmit information from different locations to one point, that is, multitransmitter to one receiver. Economically, the best method is to use a single communication channel for many pieces of information. This is known as "multiplexing" and can be carried out by one of two basic techniques: (1) frequency division multiplexing (FDM) and (2) time division multiplexing (TDM).

In the FDM technique, each piece of information is transmitted over a dedicated part of the available communication channel spectrum (Fig. 13.6). For the TDM technique, each piece of information is transmitted as part of a serial digital message over a separate span of time and demultiplexed by the receiver into the individual pieces of information (Fig. 13.7).

Message Formats: The transmission of information in both directions between the MTU and RTUs using TDM techniques requires the use of serial digital messages. All messages are divided into three basic parts:

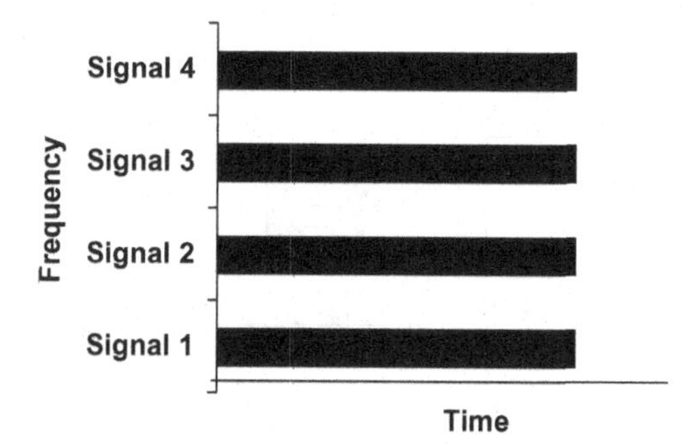

Figure 13.6 Frequency division multiplexing.

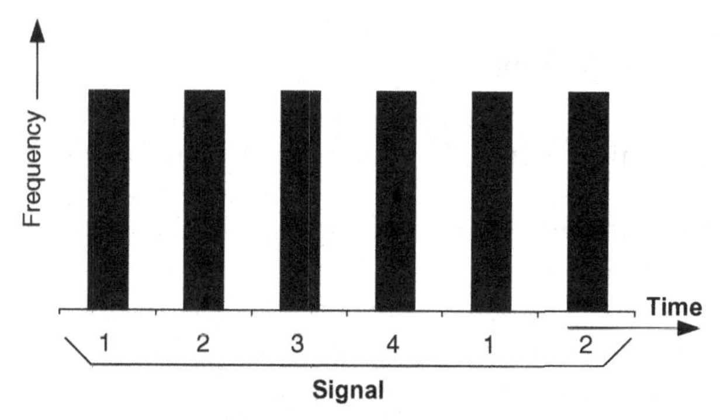

Figure 13.7 Time division multiplexing.

- message establishment, which provides the signals to synchronize the receiver and transmitter and the unique RTU address;
- information, which provides the data in a coded form to allow the receiver to decode the information and properly utilize it; and
- message termination, which provides the message security checks and a means of denoting the end of the message. Message security checks consist of logical operations on the data, which result in a predefined number of check bits transmitted with the message. At the receiver, the same operations are performed on the data and compared with the received check bits. The message is accepted if they are identical; otherwise, a retransmission of the original message is required.

Information Transfer: As it is clear in the preceding sections, the information in SCADA systems is transmitted from the RTUs to the MTU and also from the MTU to the RTUs. Therefore, it is not sufficient to just transmit the information correctly, but it must also be done securely. The techniques of checking the information errors must be applied.

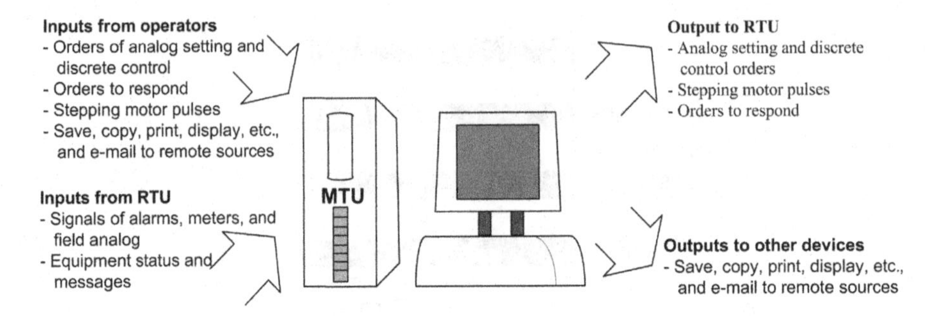

Figure 13.8 MTU inputs/outputs.

13.3.4 MTU (Fourth Component)

It is also called "central control station, or central station, or SCADA master." It is considered as the heart of the system where its main functions are

- making the communication, gathering data, storing information, sending information to other systems;
- processing the data gathered by remote stations to generate the necessary actions; and
- interfacing to the operators mainly via monitors and printers.

The inputs and outputs of the MTU are shown in Figure 13.8.

13.4 SCADA SYSTEMS ARCHITECTURES

Common features of the SCADA architecture and related hardware configurations together with the software design are described in this section.

13.4.1 Hardware

The SCADA system basically consists of a two-level hierarchy. The first level includes the IEDs, which are either RTUs or PLCs. As mentioned before, the IEDs gather the data from control devices (or sensors) in different locations and send it back to the SCADA master that represents the second level of hierarchy. The SCADA master processes the data and supplies the information again to the first level. The communication between IEDs and SCADA master is provided by using networks (local area network [LAN] or wide area network [WAN]) or field buses that are proprietary (e.g., Siemens H1) or nonproprietary (e.g., profibus). This configuration, known as "monolithic SCADA structure," is shown in Figure 13.9.

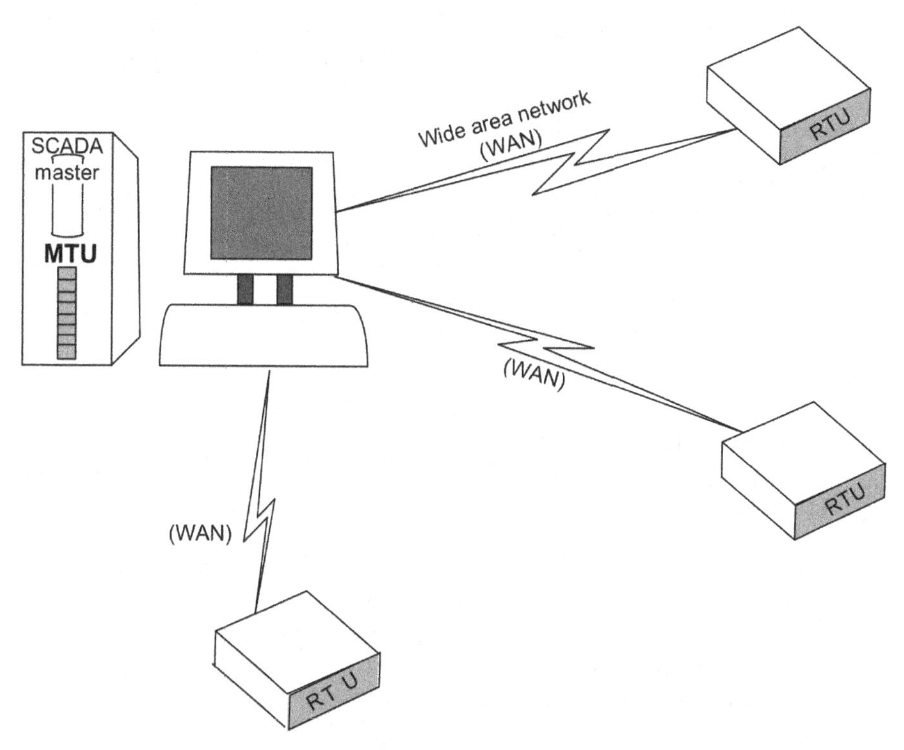

Figure 13.9 Monolithic SCADA architecture.

The interaction between the control devices (or sensors) in the field of application and the SCADA system has advantages and disadvantages as below:

Advantages

- Capability of recording and storing large amount of data.
- Huge number of IEDs over a wide area can be connected to the SCADA system.
- The operator can interact with the SCADA system through real data simulations.
- Availability of collecting different types of data from the RTUs.
- The data can be visualized remotely and displayed in a manner the operator desires.

Disadvantages

- The overall system becomes more complicated.
- The system needs more skill for design and programming.
- A number of wires are used for dealing with IEDs.

In the SCADA system itself, the communication between the first and second levels by field buses has advantages and disadvantages as below:

Advantages

- The wiring is reduced significantly.
- Awareness of the operator with what is going on at the IEDs level.
- More security for data entry and authorized persons.
- Installation and replacement of devices are easy.
- Less physical space is needed to install system devices.

Disadvantages

- System operation needs more trained and skilled persons.
- Any deficiency of communication network very much affects the system performance.

The SCADA system can be extended to a three-level control center hierarchy. The IEDs are grouped into a set of groups in the first level. Then, each group is communicated with a communication server (second level), which acts as a hub between the IEDs and the master station [187]. The master station (third level) can be a set of multi-operating stations connected together with the communication servers through a LAN. In this case, these operating stations can be distributed along the plant site as needed. In addition, each operating station can be specialized for certain processes. So, this configuration is more relevant to utility companies where the control of distribution systems by nature is a DCS. Occasionally, the DCS is similar to the SCADA system. It is usually used in factories and located within a more confined area. Also, it uses a high-speed communications medium such as LAN. A significant amount of closed-loop control is present on the system. The SCADA system covers larger geographic areas. It may rely on a variety of communication links such as radio and telephone. Closed-loop control is not a high priority in this system.

The configuration of a three-level SCADA hierarchy is known as "distributed SCADA architecture" and is shown in Figure 13.10.

When the communication medium between the SCADA master and the communication servers or RTUs is a WAN and satellite (Fig. 13.11), the SCADA configuration is known as "networked SCADA architecture." Also, the communication can be based on the Internet [188].

13.4.2 Software

The software is a multitask product based on a real-time database (RTDB) located in one or more servers. The servers are responsible for data acquisition and handling. The key features of the software are the following.

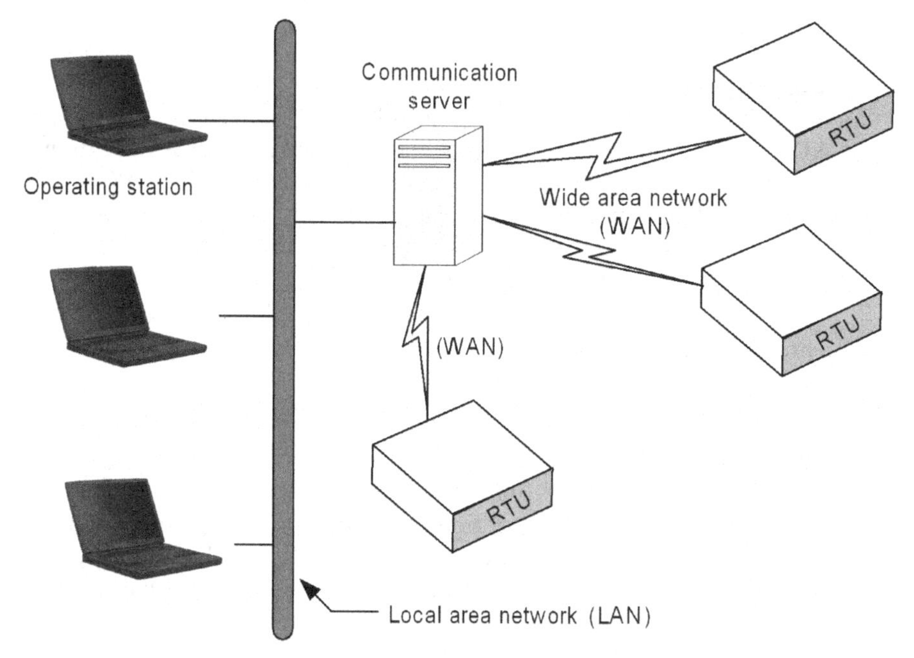

Figure 13.10 Distributed SCADA architecture.

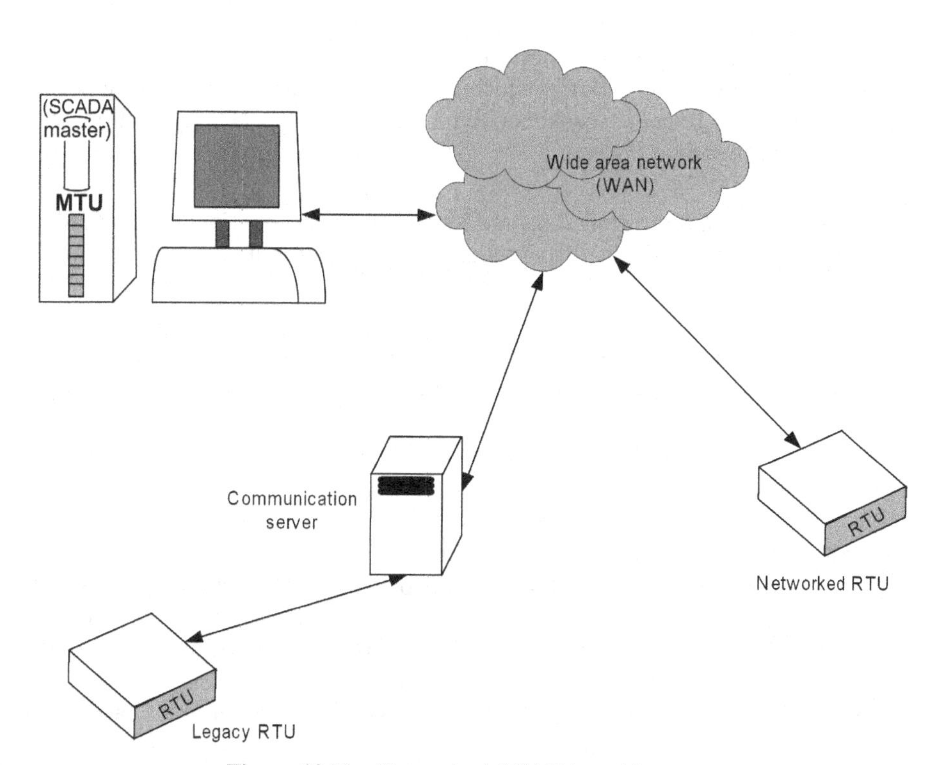

Figure 13.11 Networked SCADA architecture.

Interfacing (Application Interfaces/Openness): The provision of object linking and embedding for process control (OPC) client functionality for SCADA to access devices in an open and standard manner is developing. There still seems to be a lack of devices/controllers that provide OPC server software, but this is improving rapidly as most of the producers of controllers are actively involved in the development of this standard. OPC has been evaluated by the CERN-IT-CO group [189].

The products also provide

- an open database connectivity (ODBC) interface to the data in the archive/logs, but not to the configuration database;
- an easy way to use engineering tools;
- efficient application libraries of programs interface (APIs) supporting C, C++, and visual basic (VB) to access data in the RTDB, logs, and archive. The API often does not provide access to the product's internal features such as alarm handling, reporting, and trending;
- open protocols;
- links available to other commercial software and applications; and
- an open platform (e.g., Windows NT).

The PC products provide support for the Microsoft standards such as dynamic data exchange (DDE), which allows, for example, visualizing data dynamically in an Excel spreadsheet; dynamic link library (DLL); and object linking and embedding (OLE).

Database: The configuration data are stored in a database that is logically centralized but physically distributed and that is generally of a proprietary format.

For performance reasons, the RTDB resides in the memory of the servers and is also of proprietary format. The archive and logging format is also usually proprietary for performance reasons, but some products do support logging to a relational database management system (RDBMS) at a slower rate either directly or via an ODBC interface.

Scalability: Scalability is understood as the possibility to extend the SCADA-based control system by adding more process variables, more specialized servers (e.g., for alarm handling), or more clients. The products achieve scalability by having multiple data servers connected to multiple controllers. Each data server has its own configuration database and RTDB, and is responsible for the handling of a subset of the process variables (acquisition, alarm handling, archiving).

Redundancy: The products often have built-in software redundancy at a server level that is normally transparent to the user. Many of the products also provide more complete redundancy solutions, if required.

Access Control: Users are allocated to groups that have defined read/write access privileges to the process parameters in the system and often also to specific product functionality. The products support multiple screens that can contain combinations of synoptic diagrams and text. They also support the concept of a "generic" graphical object with links to process variables. These objects can be "dragged and dropped" from a library and included into a synoptic diagram.

Most of the SCADA products that were evaluated decompose the process in "atomic" parameters (e.g., a power supply current, its maximum value, and its on/off status) to which a tag name is associated. The tag names used to link graphical objects to devices can be edited as required. The products include a library of standard graphical symbols, many of which would, however, not be applicable to the type of applications encountered in the experimental physics community.

Standard windows editing facilities are provided: zooming, resizing, scrolling, and so on. Online configuration and customization of the man/machine interface (MMI) is possible for users with the appropriate privileges. Links can be created between display pages to navigate from one view to another.

Trending: The products all provide trending facilities and one can summarize the common capabilities as follows:

- the parameters to be trended in a specific chart can be predefined or defined online;
- a chart may contain more than eight trended parameters or pens and an unlimited number of charts can be displayed (restricted only by the readability);
- real-time and historical trending are possible, although generally not in the same chart;
- historical trending is possible for any archived parameter;
- zooming and scrolling functions are provided; and
- parameter values at the cursor position can be displayed.

The trending feature is either provided as a separate module or as a graphical object (ActiveX) that can then be embedded into a synoptic display.

Alarm Handling: Alarm handling is based on limit and status checking and performed in the data servers. More complicated expressions (using arithmetic or logical expressions) can be developed by creating derived parameters on which status or limit checking is then performed. The alarms are logically handled centrally; that is, the information only exists in one place and all users see the same status (e.g., the acknowledgment), and multiple alarm priority levels (in general many more than three such levels) are supported.

It is generally possible to group alarms and to handle these as an entity (typically filtering on group or acknowledgment of all alarms in a group). Furthermore, it is possible to suppress alarms either individually or as a complete group. The filtering of alarms seen on the alarm page or when viewing the alarm log is also possible at least on priority, time, and group. However, relationships between alarms cannot generally be defined in a straightforward manner. E-mails can be generated or predefined actions automatically executed in response to alarm conditions.

Logging/Archiving: The terms logging and archiving are often used to describe the same facility. However, logging can be thought of as medium-term storage of data on disk, whereas archiving is long-term storage of data either on disk or on another permanent storage medium. Logging is typically performed on a cyclic basis; that is, once a certain file size, time period, or number of points is reached, the data are overwritten. Logging of data can be performed at a set frequency, or only initiated if the value changes or when a specific predefined event occurs. Logged data can be transferred to an archive once the log is full. The logged data are time stamped and can be filtered when viewed by a user. The logging of user actions is in general performed together with either a user identifier (ID) or station ID. There is often also a video cassette recorder (VCR) facility to play back archived data.

Report Generation: One can produce reports using structure query language (SQL) type queries to the archive, RTDB, or logs. Although it is sometimes possible to embed Excel charts in the report, a "cut and paste" capability is in general not provided. Facilities exist to be able to automatically generate, print, and archive reports.

Automation: The majority of the products allow actions to be automatically triggered by events. A scripting language provided by the SCADA products allows these actions to be defined. In general, one can load a particular display, send an e-mail, run a user-defined application or script, and write to the RTDB.

The concept of recipes is supported, whereby a particular system configuration can be saved to a file and then reloaded at a later date.

Sequencing is also supported whereby, as the name indicates, it is possible to execute a more complex sequence of actions on one or more devices. Sequences may also react to external events. Some of the products do support an expert system, but none has the concept of a finite state machine (FSM).

GUI: It must satisfy the following chief characteristics [190]:

- open-system approach,
- seamless interface features,
- ease of user operation, and
- acceptability by the operators.

The main function of SCADA/GUI system is the visual presentation by which one can distinguish one system from the others. GUI forms the key core of any SCADA system. Its efficiency can easily determine the productivity of operations of any management system.

13.5 SCADA APPLICATIONS

13.5.1 Substation Automation

The distribution substation can be controlled and monitored locally or remotely from a central control room where the trend is toward what is called "unmanned substation." These days, because of the presence of powerful computers and intelligent devices, the substation operations are becoming completely automated by the use of such devices [191]. These devices (IEDs) are interconnected with the goal of collecting data and controlling the substation processes.

On the other hand, the SCADA system allows the substation operator to view and control the status of many aspects of the substation.

Therefore, the integration of SCADA system and substation automation system results in a real-time wide area monitoring and control of substation by controlling, for example, the power flow, power limits calculation, and substation operation. The level of integration depends on utility requirements [192].

Of course, the advanced control systems, system protection, and communication applications together with the SCADA system can significantly improve the capacity and reliability of the existing distribution networks. This means reducing the operation and construction costs, and optimizing the distribution system utilization.

The major trend of the architecture of both SCADA and automation systems is toward the distributed architecture because of its effectiveness and compatibility with the nature of distribution system, in addition to the following:

- increase of availability and modularity,
- no master controller is needed,
- peer-to-peer communication,
- good performance, and
- single-failure criterion principle.

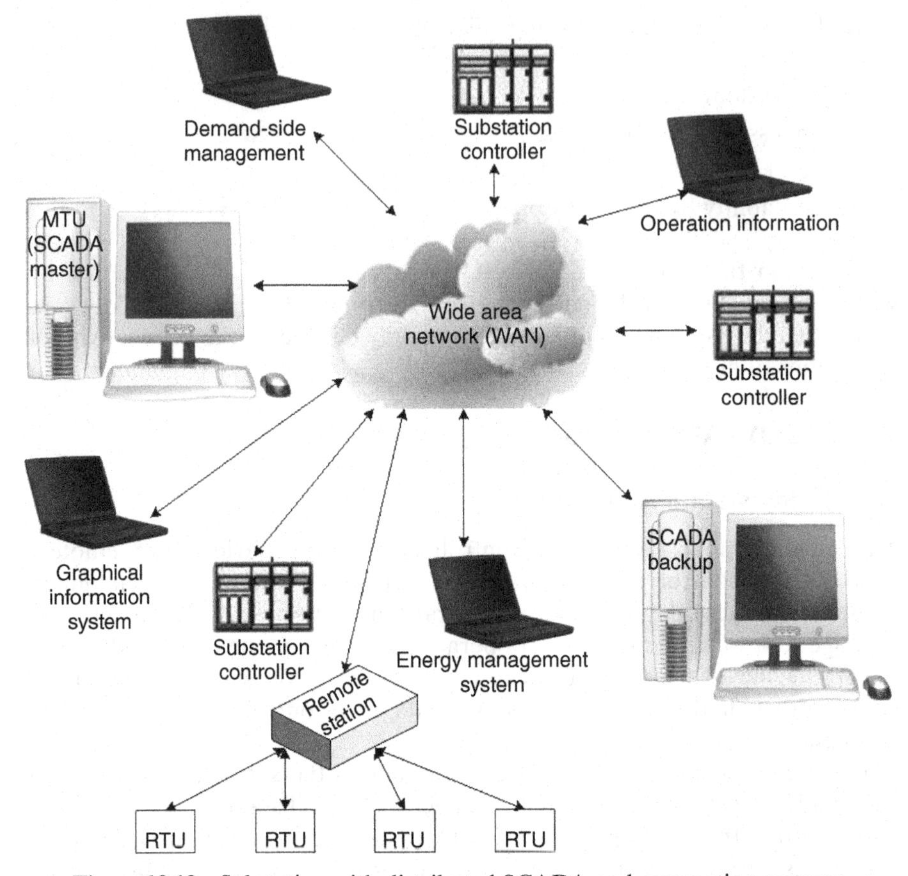

Figure 13.12 Substation with distributed SCADA and automation systems.

Schematic diagram of a distribution substation control and monitoring system is illustrated in Figure 13.12 [193]. This advanced system results in many gains against the conventional control system, such as

- saving in hardware,
- no separate local control board,
- no separate local sequential event recorder,
- no separate remote terminal unit "RTU,"
- no separate cross-connection cubicles between bay and station level equipment,
- no transducers,
- less signal cabling,
- easy to use,
- dynamic colors,

- window operating system,
- automatic pop-up of important information,
- information available when demanded,
- information for preventive maintenance,
- reports and listing, and
- support of applications with distributed architecture.

13.5.2 Commercial Office Buildings

The example of a commercial office building discussed at the beginning of this chapter and depicted in Figure 13.1 can be controlled and monitored by a SCADA system as follows:

- For each apartment at each floor, an IED represents the status of the reverse switch.
- An RTU collects data from a group of IEDs (e.g., one RTU per floor).
- In case of a large number of RTUs, they are divided into groups where each group communicates with a remote station that in turn communicates with an Ethernet bus (LAN).
- For the basement where the main switchboard is located, an IED describing the interlock status is connected to a dedicated RTU that communicates with the Ethernet bus.
- A central control station and a SCADA master station are connected with the LAN to control and monitor the system, respectively, as shown in Figure 13.13.

13.5.3 Power Factor Correction System

Power factor correction is implemented by inserting shunt capacitors into the system. It is supposed in this application as an example to use zone compensation, that is, the shunt capacitors are connected to the MV bus bars of the distribution switchboard responsible for distributing the power to the loads located in that zone as a DP.

Traditionally, the shunt capacitors are grouped into multistages where each stage encompasses a group of capacitors (capacitor bank) with a certain amount of kvar as a rating of this stage. Each stage may have a rating equal or not equal to the rating of other stages. It depends on the planner's study of the load and its nature of change. The capacitors at each stage are connected in delta and are connected to the DP bus bars through fuses and a three-phase switch. A system automatic controller receives samples of bus-bar currents and voltages via current transformers (Cts) and voltage transformers (vts), respectively, and the phase power factor as well. Accordingly, the controller sends a control signal to switch "on" or "off" the switch of each stage individually to

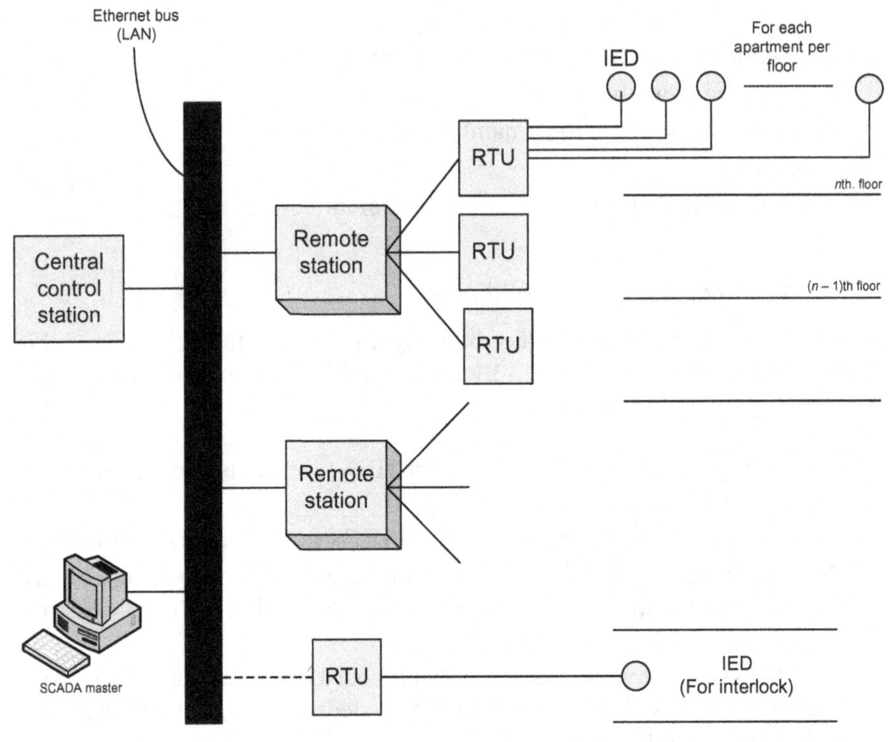

Figure 13.13 A SCADA system for a commercial office building.

achieve the preset power factor value. The system of power factor correction is protected against overcurrent, overvoltage, and over-temperature by using adequate relays. Any stage cannot be reenergized in less than a specific period (few minutes) to avoid the switching over and restriking voltages.

Applying the SCADA system to the power factor correction system, the overall system as shown in Figure 13.14 will be equipped with the following:

- Position switch: It allows each capacitor bank to be inserted into the system manually or automatically or to be disconnected (man/off/auto).
- Auto/SCADA switch is used to enable the system to operate automatically or with the SCADA system.
- The central control station (system controller): It controls the capacitor banks based on the voltage and power factor of the system. The control station is enabled solely when SCADA is disabled.
- Cts and vts: They are acting as sensors (IEDs) to send samples of system parameters (phase currents, phase voltages, line voltages, active power, reactive power, etc.) to both the protection and measuring systems.
- Protection system: It includes different types of relays to protect the system against abnormal operating conditions. A temperature sensor can

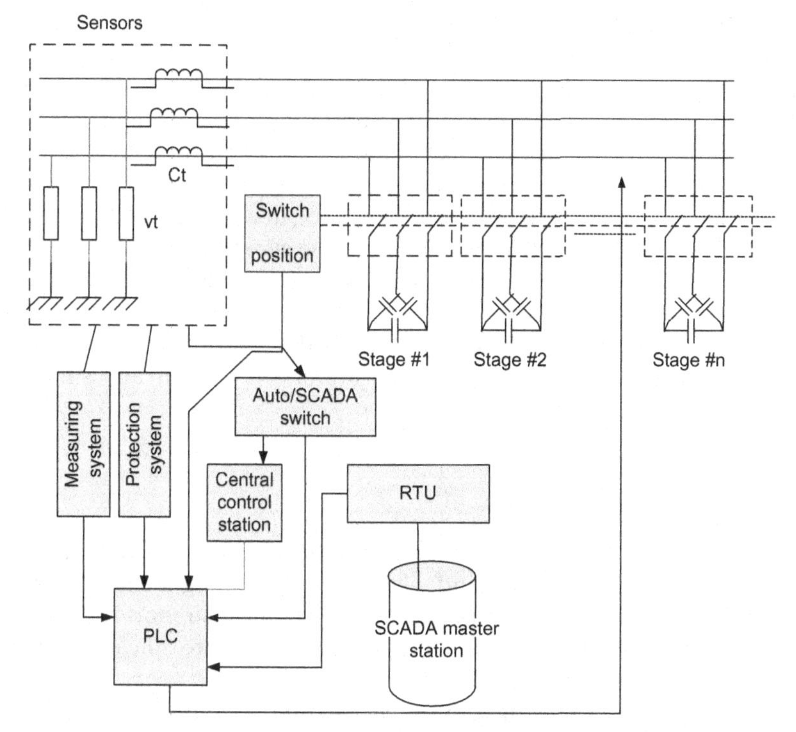

Figure 13.14 Application of SCADA system to power factor correction system.

be added when needed to protect the system against ambient temperature rise.

- Measuring system: It measures all parameters required according to the signals received from sensors and sends the measured values to a PLC.
- SCADA master station: It sends control signals to the PLC. In addition, it performs all functions of indication, alarm, reporting, and graphical display. The SCADA displays actual power factor, connected steps, pending operations, capacitor loss, load and reactive currents, and voltage total harmonic distortion (THD). It can also alarm for low power factor, excessive hunting, abnormal power factor, undervoltage, overcompensation, frequency not detected at start-up, overcurrent, overvoltage, overtemperature, high-voltage THD, capacitor overload, and capacitor low output.
- PLC: It controls the functional requirements of both system controller and SCADA deciding which capacitor bank stages are switched "on" and which ones are "off." This decision is based on the inputs from the measuring system, protection system, position switch, auto/SCADA switch, controller, and SCADA master stations.

13.6 SMART GRID VISION

Traditional SCADA systems are early smart grid technologies. However, the reach of SCADA is usually limited to substations, feeders, and few major distribution automation devices like remote-controlled disconnect switches. SCADA involves dedicated control center with communications to monitor and control equipment. At the distribution level, due to the number of points that are being monitored and controlled, and the capital costs, SCADA is typically not cost-effective at the substation level and rarely at the feeder level. The majority of distribution systems are not only monitoring and controlling but also satisfying different aspects on which the smart grid vision is based. The data managed by SCADA play an important role in any smart grid implementation.

13.6.1 Smart Grid Overview

Electric power grid, operating near capacity, needs critical improvements to support current and future demand while accommodating a changing mix of distributed energy resources. In addition, the grid must support the efforts to reduce carbon emissions and use the energy resources more efficiently through changes in energy use patterns.

Increasing customer demands, the need for additional facilities and security, and to keep costs under control will motivate utilities to operate differently and more intelligently. Nowadays, many utilities are more interested in implementing the smart grid in order to cope with changes and challenges.

The key aspects of the smart grid include the following:

- *Ability of the Grid to Self-Heal Following a Disturbance*: The present grid responds to prevent further damage through protection of assets following system faults. The proposed smart grid automatically detects and responds to actual problems focusing on prevention and minimizing consumer impact.
- *Power Quality*: Supplying power free from sags, swells, outages, and other power quality/reliability issues. Smart grid also has the ability to provide various levels of power quality at various prices.
- *Support for All Generation and Storage Options*: The present grid has a relatively small number of centralized generating plants. Numerous obstacles exist for interconnecting distributed and renewable energy resources (RES). Smart grid has the ability to interconnect very large numbers of those resources and storage devices to complement the large generating plants and satisfy "plug-and-play" convenience.
- *Better Asset Utilization and Efficient Operation*: The present grid has minimal integration of limited data with asset management processes and technologies. Smart grid has expanded sensing and measurement of grid

conditions to collect integrated data with asset management processes to effectively manage assets, costs, and implementing condition-based maintenance.

- *Increased monitoring* through low-cost sensors.
- *Customer Involvement*: In the present grid, consumers are uninformed and nonparticipative with the power system. In smart grid, consumers are informed, involved, and active through broad penetration of demand response.
- *Rapid Restoration*: The present grid is vulnerable to natural disasters and terrorism attacks, whereas the smart grid is resilient to attack and natural disasters with rapid restoration capabilities.

13.6.2 Smart Grid Concept

Smart grid is a concept not a computer system or some sort of hardware. It is about an intelligent electric delivery system that responds to the needs of and directly communicates with consumers. While there are many facets to the concept, smart grid is really about three things: distributed intelligence, digital communications, and decision software [194].

Distributed Intelligence: More efforts must be made to interconnecting distributed generation (DG), providing distribution system automation and optimization, and realizing customer involvement and interaction. So, it is necessary to have more intelligence and control for the entire power system, from generation to the meter on the customer's site. The level of control required to achieve this is much greater than in current distribution systems. It includes power flow assessment and voltage control and protection that require cost-competitive technologies, as well as new communication systems with more sensors and actuators than are presently available in the distribution system; for example, fixed and mobile devices can be used. Fixed devices are as follows:

- SCADA devices (as explained in the former sections) and distribution automation devices;
- automatic meter reading (AMR) devices to help determine the needs of all vital aspects, for example, daily workflow, workforce management, asset management, billing system, and self-register meter points [195];
- EMS; and
- energy technologies for monitoring and control, both for electric utilities and for consumers.

Mobile devices include

- geographic positioning system (GPS) devices,
- mobile computing devices (e.g., laptop PCs),

- mobile communications (e.g., cell phone), and
- voice and data dispatch radios.

Digital Communications: Two-way fast digital communications will be required throughout the smart grid to gather real-time data and for active grid management. This communication is applied between and among the utilities, the meter, utility's devices, and consumer's devices. Wide variety of digital communications can be used including

- wired and wireless telephone,
- voice and data dispatch radio,
- fiber optics,
- power line carrier,
- satellite, and
- Internet.

Decision and Control Software: The distribution system has numerous numbers of points (every power line and piece of equipment, every one of utility's consumers) that must be monitored and controlled. Therefore, immense amount of data must be organized, analyzed, and acted upon by using software that can be classified into two categories: decentralized and back office software:

i) Decentralized software: The collected data from each point in the distribution system is organized and analyzed individually using IEDs that can perform computations as well. These computations determine what data should be communicated where and what local control actions may be necessary.
ii) By deployment of IEDs and two-way digital communications, each of back office software solutions can be more powerful and effective. The following back office software are mostly used by utilities:
    - enterprise resource planning such as accounting and business systems, work and workforce management, performance and productivity management, and customer information systems (billing and payment, relationship management).
    - engineering and operations such as geographic information systems (GIS) [196], interactive voice response, and engineering analysis. Engineering analysis software may include
        - circuit modeling, reliability, and real-time distribution analysis;
        - outage management system;
        - demand-side management (DSM) system;
        - active distribution grid management: It includes three stages: (1) initial stage for the monitoring and control of DG and RES to facili-

tate greater connection activity, (2) intermediate stage to evolve a management regime that is capable of accommodating and defining significant amount of DG and RES, and (3) final stage at which the full active power is managed using real-time communication and remote control to meet the majority of the network services requirements [197].

13.6.3 Driving Factors

Power providers, researchers, lawmakers, and customers are all implicated in the move toward smart grids. According to the aspects and concept of smart grid, their action must be based on the driving factors below [197]:

- evolving polices and strategies encouraging low carbon generation, new and renewable energy sources, and more efficient use of heat energy;
- the need to understand and manage the technical challenges and opportunities for integrating new generation technologies into grids;
- the need for investment in end-of-life grid renewal in an innovative way to better position the networks for the next 50 years of operation;
- the need to handle grid congestions with market-based methods;
- the desire to deliver benefits to customers at the earliest opportunity;
- increasing both participation and awareness of customers in the energy field;
- the need to reduce uncertainty and risk to business-making investment decisions; and
- the progress in technology that allows improvements in operation and new services at reasonable costs.

PART V

DISTRIBUTED GENERATION

CHAPTER 14

DISTRIBUTED GENERATION

14.1 POWER SYSTEMS AND DISTRIBUTED GENERATION (DG)

The arrangement of power system (generation, transmission and subtransmission, and distribution) indicates that the power flow is a unidirectional flow from the location of generation to the distribution substations and terminates at the consumers' premises [198, 199]. Some consumers produce electricity using their own generation sources with the goal of feeding their loads or as backup sources to feed critical loads in case of emergency and utility outage (Fig. 14.1). These sources are defined as "distributed generation" (DG) in North American terms and "embedded generation" in European terms. Therefore, DG produces electricity at or near the place where it is used to meet all or a part of the customer's power needs. It ranges in size from less than 1 kW to tens or, in some cases, hundreds of kilowatts. DG technologies include three main types based on the type of primary energy sources: fossil fuel, renewable energy, and waste heat. Some common forms of DG are the following:

- Cogeneration systems or combined heat and power (CHP) systems. They produce both electric and thermal energy that can be used for industrial processes. CHP systems can be designed to use one or more types of fuel: fossil fuel, biomass, and waste heat. Such systems are fuel cells, microturbines, combustion engines, steam-cycle turbines, Stirling engines, and gasification digesters.

Electric Distribution Systems, First Edition. Abdelhay A. Sallam, Om P. Malik.

Published 2011 by John Wiley & Sons, Inc.

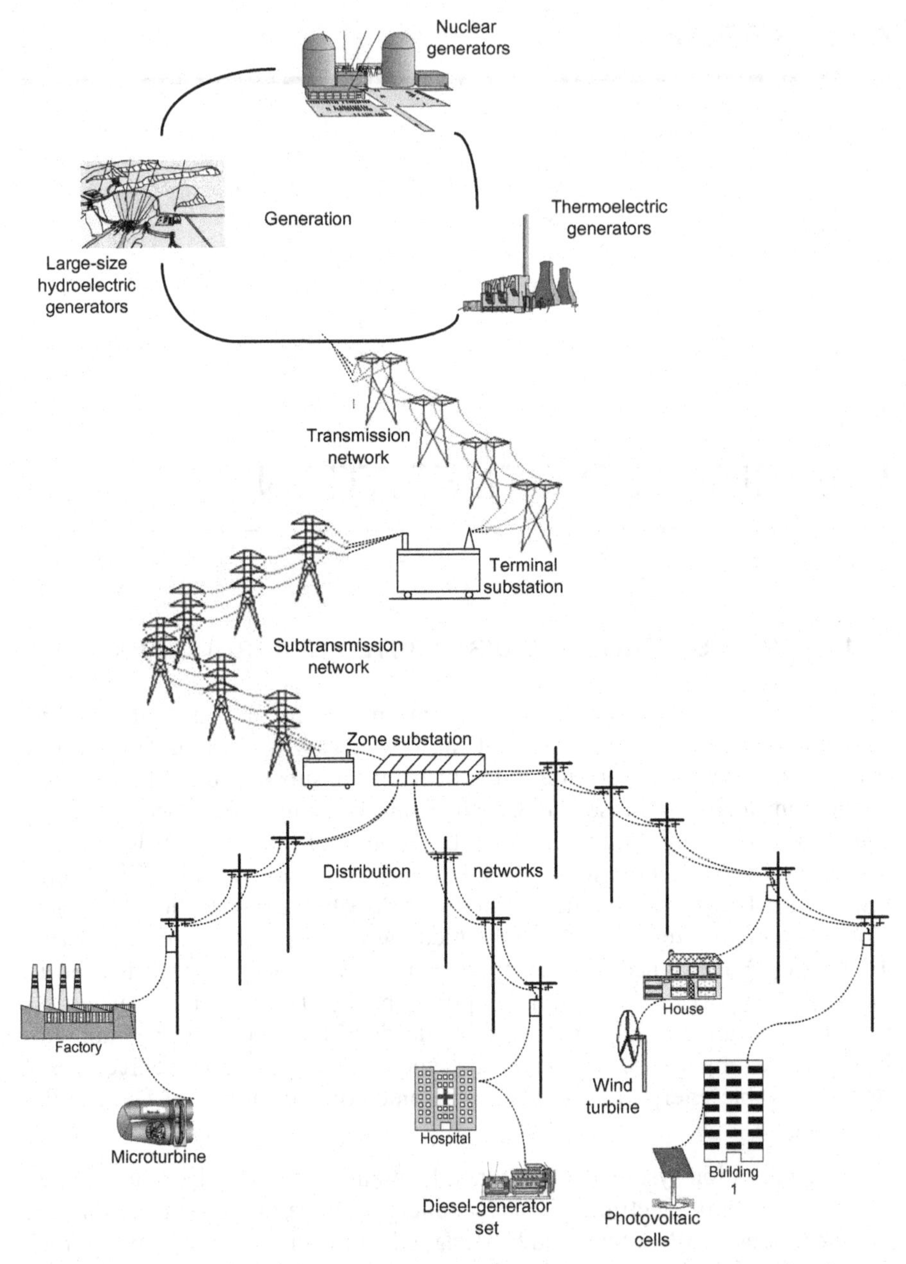

Figure 14.1 Power system arrangements with distributed generation [198].

- Renewable energy sources to generate electricity without using any of the former fuels, such as wind turbine, solar electric systems, and small hydroelectric generators.
- Reciprocating-engine generators and small combustion turbines that use diesel or natural gas.

DG systems have power capacity that usually exceeds the independent supplier's needs in addition to the available full capacity of backup units in normal operating conditions. The overall trend is concerned with efficient utilization of DG by its application for

- supplying electricity to small loads in remote locations where it may be more economic than establishing a new line to the load site;
- supplying heat energy and steam to hospitals and some industries from cogeneration systems;
- providing high power quality for electronic and sensitive equipment;
- backup power source during utility outages, in particular, for loads requiring uninterrupted power supply such as hospitals, banks, and data centers;
- peak-shaving programs where DG can be used during high-cost periods to supply consumers participating in the programs resulting in reduction of overall power cost;
- reduction of air emissions by using renewable energy sources;
- avoiding distribution system investments;
- providing excess capacity to utilities;
- dispatching DG to achieve most economical operation taking into account the priority of supplying independent producers; and
- reducing transmission and distribution (T&D) losses.

It is stated in Reference 200 that DG is expected to provide a more secure and reliable power system if it is connected to the distribution system and power is sold to the utilities. Therefore, it will be worth utilizing these distributed generating units for supporting utility distribution networks for both technical and economic reasons.

In connecting DG systems to the existing distribution network, problems, some technical and the others economical [201], may be faced. The technical problems are as follows:

- Some on-load tap changer transformers are not designed for reverse power flow.
- Increase of fault levels.
- Protection of distribution systems is not designed for reverse power flow.
- Nuisance tripping of some healthy parts in distribution systems.

- Existing networks are not designed for high voltage rise. So, voltage reduction schemes and network voltage control schemes are adversely affected by DG, especially if operating under voltage control, or if generator output changes rapidly.
- Metering equipment and communication system between meters and the data center should be modified.

Acceptable techniques for solving the technical problems are available in the literature. Economical or nontechnical problems may cause a significant barrier to connect DG with distribution system. This connection needs to be charged for (1) new assets and present connection changes, and (2) operation and replacement of system wide assets. Therefore, the problem is about determining which side is responsible for financial support, utilities, or independent energy producers.

An illustration of some technical problems to show how DG has an impact on the distribution system design is presented below [202].

Considering a four-wire distribution system, the neutral conductor is either multi-earthed (connected to earth several times), which is the characteristic for many medium-voltage (MV) grids in the United States and low-voltage (LV) grids in Europe, or earthed only at the source. A synchronous generator, as a DG unit, is coupled with the utility network through an interconnection transformer (Fig. 14.2). Several winding arrangements for interconnection transformers are possible (Δ-earthed Y, Δ-Δ, earthed Y-Δ, and earthed Y-earthed Y), each of which has its advantages and disadvantages, depending on the earthing technique of the utility. Focusing on the fault contribution of DG units when a line-to-earth fault occurs at one of the locations F_1, F_2, or F_3 is discussed below.

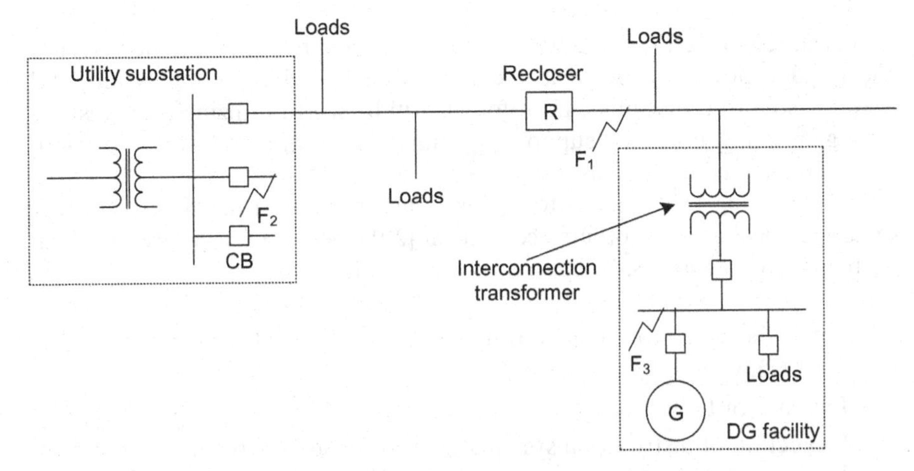

Figure 14.2 Utility and DG interconnected system.

In case the primary winding on the utility side of the interconnection transformer is not earthed, the DG unit does not supply any earth-fault current for a fault at F_1 or F_2. If the primary winding is an earthed Y, the DG unit can back-feed the distribution feeder, which might be a serious problem, certainly in four-wire multi-earthed distribution grids.

An earthed Y-Δ transformer creates additional earth current paths as is depicted in Figure 14.3. The fault currents are no longer flowing just in one path from the substation to the fault, but in several parallel paths even downstream of the fault. This can result in malfunctioning of protective relays and needlessly blowing line and transformer fuses. One common side effect is that the feeder breaker will trip for any line-to-earth fault on all feeders served from the same substation bus. The utility transformer might be overstressed repeatedly and eventually fail. In order to prevent the sympathetic tripping of the feeder breaker for line-to-earth faults on other feeders, directional overcurrent relays can be used [203]. The same considerations discussed above are also valid in case of a direct connection of DG units to the utility grid.

The fault currents can be calculated as described in Chapter 4, taking into account the distribution configuration, the type of fault, and the fault location (nearby or far from DG units). To reduce the fault currents, attention must be paid when choosing the type of earthing; for example, it is not always recommended to solidly earth the generator neutral. Solid earthing is, however, not possible for asynchronous generators because of third harmonic currents associated with this type of generator.

To understand how a DG unit can create an overvoltage during earth fault, consider an interconnection transformer with a Δ winding on the utility side

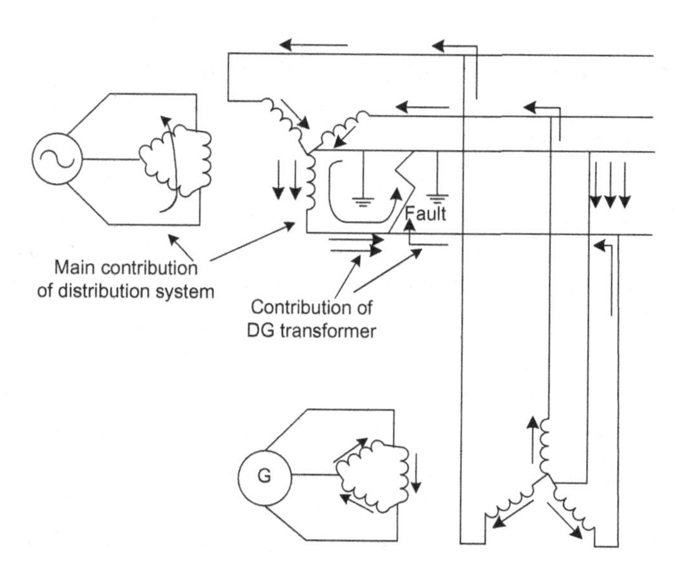

Figure 14.3 Line-to-earth fault contribution of a DG unit [202].

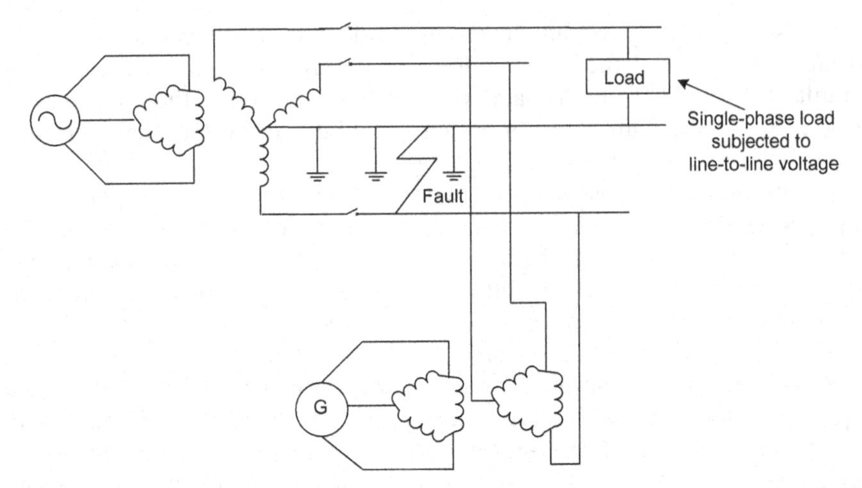

Figure 14.4 Loads subjected to line voltage during line-to-earth fault [202].

(Fig. 14.4). A permanent line-to-earth fault has occurred and the utility interrupting device has opened before the DG was disconnected.

This leaves an isolated system energized by the DG unit. There is no longer an earthed source on the utility side of the transformer. Now the potential of the neutral essentially equals that of the faulted phase. Any loads or voltage arresters connected between an unfaulted phase and neutral are subjected to a line-to-line voltage. This could damage the loads after just a few cycles.

In the forthcoming sections, some of the most common distributed power sources are discussed.

14.2 PERFORMANCE OF DISTRIBUTED GENERATORS

14.2.1 Microturbines

Small-scale microturbines can be used for DG, alone or in CHP systems. The generated electric power ranges from 30 to 400 kW to feed (local or nearby) loads or to be interconnected with utility grid. The output exhaust thermal energy can be used for heating purposes, which some loads need (industrial processes, hospitals, air-conditioning, etc.). Natural gas is mostly used as a primary fuel (more economic) for microturbines in spite of the validity of using a variety of fuels such as gasoline, diesel, kerosene, digester gas, and methane [204].

Compared with other fossil-fueled units and in case of interconnection with utility grid, microturbines have the following advantages:

- easy to operate in parallel with other units to supply large loads;
- produce stable and reliable power;

- flexible in connection to the utility grid;
- relatively, fast response to load variations since they do not need to build up pressure as in steam turbines or have high momentum as in reciprocating engines;
- use a variety of fuels;
- provide sufficient heat for water and space heating;
- low noise, vibrations, emissions, and maintenance; and
- low weight per horsepower.

On the other hand, microturbines have the following disadvantages:

- low efficiency (28–32%);
- high sensitivity to ambient air temperature, pressure, and humidity;
- need for highly skilled technicians for maintenance and repair;
- some designs need a mechanical device for initial run up to speed; and
- start too slowly for some applications.

The main elements of a microturbine are shown in Figure 14.5 to illustrate the basic idea of operation. A mixture of air and fuel gas compressed to a point

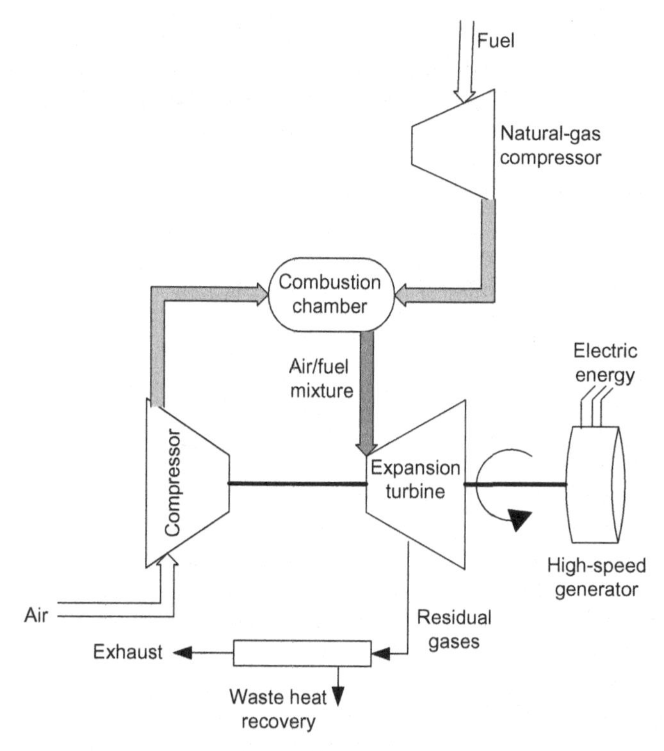

Figure 14.5 Microturbine systems.

of continuous combustion is passed into a combustion chamber. The fuel-to-air ratio is controlled and adjusted by the governor to produce complete combustion. As a result of combustion, the hot mix of air with burned fuel expands rapidly as it enters the expansion turbine. This turbine collects most energy out of the gas as it passes from one blade to another to drive both turbine shaft and auxiliary machines. A high-speed permanent magnet generator (90,000–120,000 rpm) is installed on the shaft to convert its kinetic energy (KE) into electric energy. When the process is completed, residual gases from combustion at high temperature are expelled. A part of the residual heat energy can be used for waste heat recovery and the rest is an exhaust. Some microturbines may need to have an auxiliary machine to act as a starter to bring the turbine up to the operational speed. The starting-up process can be done by operating the generator in a motoring mode, then acting in generation mode when the turbine reaches nominal speed. The generator voltage is rectified and delivered to an inverter connected to the grid (Fig. 14.6).

The microturbine must be equipped with a control system to provide the following functions:

- to maintain optimal temperature and speed characteristics when the generator is connected to the grid;
- to maintain constant voltage and frequency not affected by load variations in stand-alone generator operation mode;
- to control turbine speed to support direct current (DC)-bus voltage to ensure that the requirements of power production are satisfied at steady-state operation; and

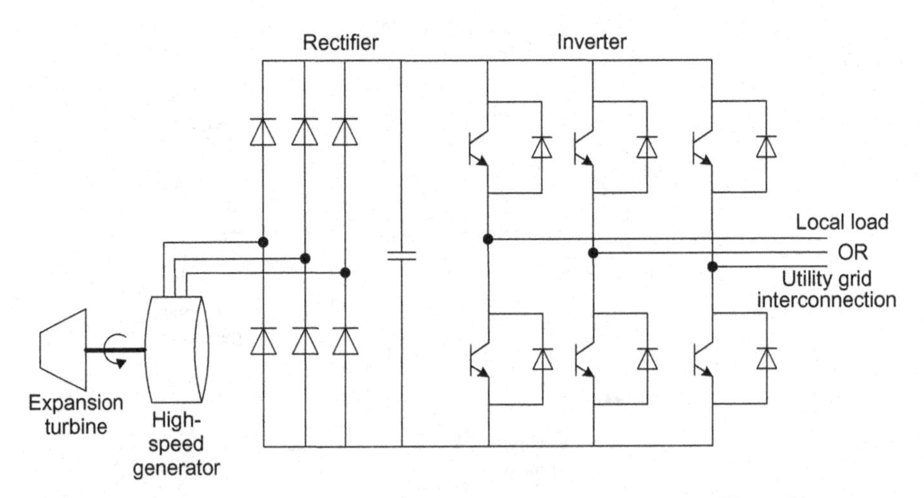

Figure 14.6 Generator interconnection to local loads or utility grid.

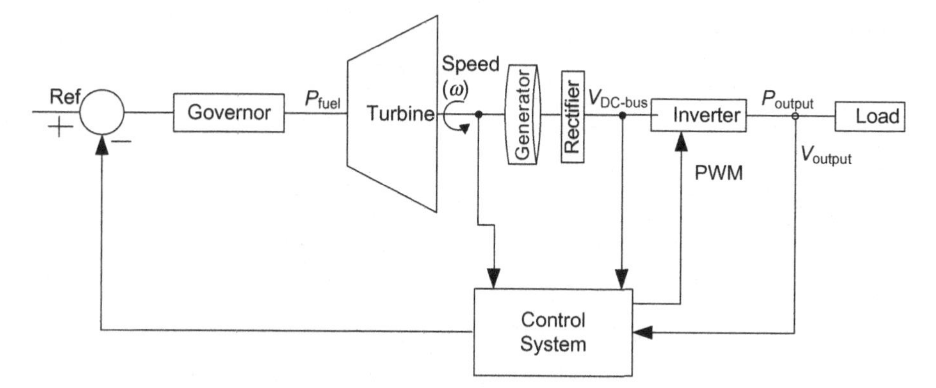

Figure 14.7 Block diagram of microturbine and control system.

- to control turbine speed in transient state when sudden load changes occur to keep the balance between energy delivered to and absorbed by the rotor.

Therefore, the control system receives signals (inputs) from turbine speed, DC bus voltage, and load power to send outputs to the governor to adjust fuel injection rate to the turbine to operate at a desired stable speed. In addition, the output of the inverter is controlled by pulse width modulation (PWM), such as sinusoidal PWM, to deliver clean output voltage (Fig. 14.7).

The microturbine speed control is not unlimited. The speed is restricted by three constraints: (1) minimum turbine speed is required for a minimum DC bus voltage, which is necessary to maintain a linear PWM; (2) microturbine shaft as a rotating mass carrying different weights at different points is subjected to vibrations and other mechanical constraints if the speed exceeds certain limit; and (3) by increasing turbine speed, the output power increases. This increase is limited by the available maximum fuel injection rate. The three constraints are illustrated in Figure 14.8 defining the operating zone of the microturbine (the hatched area).

14.2.2 Wind Turbines

The movement of air mass in the form of wind energy is the primary source of mechanical KE produced by the wind turbine to drive the generator installed on the turbine shaft. So, the wind farm (a group of wind power turbines, Fig. 14.9) is located in the area that has sufficient wind to produce the electric energy at the desired rate. This is a necessary and not sufficient condition for location selection. The selection criterion includes other aspects such as the distance between the wind farm and the utility grid to be connected, topography and geography of the selected area, security, and, of course, wind intensity over the year.

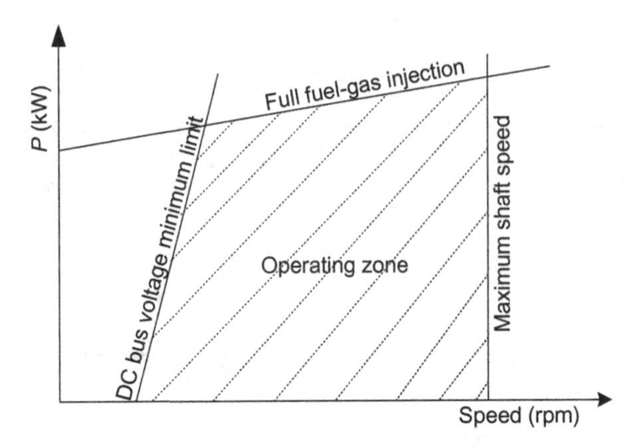

Figure 14.8 Power-speed characteristic of microturbines.

Figure 14.9 San Gorgino wind farm, Palm Springs, California, USA.

To decide the adequacy of wind farm area from the economical and technical point of view, the wind power intensity (W/m^2) must be evaluated. This power varies from season to season and from year to year. The wind mechanical power is proportional to the wind speed cubed as it can be found by applying Bernoulli's equation (typical speed–power curve is shown in Fig. 14.10):

$$P_{\text{w-mech}} = \frac{dk_{\text{e}}}{dt} = \frac{1}{2} v^2 \frac{dm}{dt} \text{ W/s,} \tag{14.1}$$

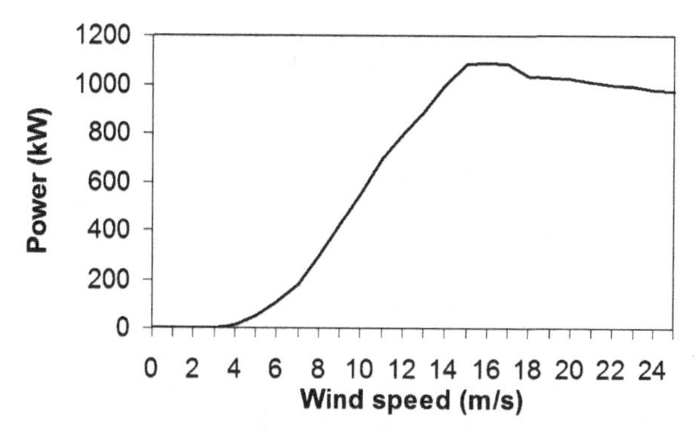

Figure 14.10 Typical speed–power curve of 1-MW wind turbine.

where

k_e = kinetic energy (usually in kg m/s = 9.81 W),
V = air velocity, and
$\frac{dm}{dt}$ = mass flow rate.

Also, $m = \rho V$ (ρ is the air density [kg/m^3] and V is the air volume [m^3]).

Thus, $dm/dt = \rho A\overline{v}$ (A is the surface area swept by rotor blades [m^2] and $\overline{v}$ is the average speed [m/s] of air reaching the turbine v_1 and leaving it v_2, i.e., $= (v_1 + v_2)/2$):

$$\frac{dm}{dt} = \frac{1}{2}\rho A(v_1 + v_2). \tag{14.2}$$

There is a difference in the KE of the wind and the net wind mechanical power of the turbine corresponding to the difference between the two speeds v_1 and v_2. Therefore, Equation 14.1 can be rewritten for this net power as

$$P_{\text{w-mech-net}} = \frac{1}{2}(v_1^2 - v_2^2)\frac{dm}{dt}. \tag{14.3}$$

From Equations 14.2 and 14.3, the net wind mechanical power is given as

$$\begin{aligned} P_{\text{w-mech-net}} &= \frac{1}{4}\rho A(v_1^2 - v_2^2)(v_1 + v_2) \\ &= \frac{1}{4}\rho A v_1^3\left[1 - \left(\frac{v_2}{v_1}\right)^2\right]\left[1 + \left(\frac{v_2}{v_1}\right)\right]. \end{aligned} \tag{14.4}$$

The total input mechanical power of the wind turbine, P_{input}, is a function of air velocity at its inlet, v_1:

$$P_{\text{input}} = \frac{1}{2}\rho A v_1^3. \tag{14.5}$$

Dividing Equation 14.4 by Equation 14.5 gives the rotor efficiency, η_{rotor}, as below:

$$\begin{aligned} \eta_{\text{rotor}} &= \frac{P_{\text{w-mech-net}}}{P_{\text{input}}} \\ &= \frac{1}{2}\left[1-\left(\frac{v_2}{v_1}\right)^2\right]\left[1+\left(\frac{v_2}{v_1}\right)\right]. \end{aligned} \tag{14.6}$$

It is seen that the rotor efficiency is a function of the speed ratio, $v_{\text{ratio}} = v_1/v_2$. Differentiating the function with respect to v_{ratio} and equating to zero gives the value of speed ratio at which the rotor efficiency is maximum:

$$\frac{d\eta_{\text{rotor}}}{dt} = 0 = (3v_{\text{ratio}} - 1)(v_{\text{ratio}} + 1).$$

Thus, $v_{\text{ratio}} = 1/3$ and η_{rotor} (max.) = 59.26%.

The rotor efficiency cannot reach 100% because this means that $v_2 = 0$; that is, the airflow is stopping immediately after leaving the rotor blades, which, of course, is impractical. The normal practical values of rotor efficiency range between 35% and 45%. The wind power available to be converted into electric power by the generator can be increased by increasing rotor efficiency. So, it is important to define the characteristic of η_{rotor}. This characteristic is determined by the variation of η_{rotor} against tip speed ratio λ, which is the ratio of rotor blade tip speed, ωR, to the free speed of wind, v. ω is the angular speed of turbine shaft and R is the length of turbine blade. Typical curves for η_{rotor} characteristics of wind turbines with different number of blades are shown in Figure 14.11.

On the other hand, it is also important to collect accurate data about wind speed distribution hour by hour, day by day, and year by year to be able to determine the calm periods. Calm periods are the periods during which the wind speed is less than 3 m/s and the wind system is completely stopped. Thus, power storage systems are needed. The extent of storage system need depends on the lengths of the calm periods [204].

The main components of wind turbines as depicted in Figure 14.12 are

- rotor that rotates by blades spinning,
- gearbox to raise shaft speed, and
- tower on which the turbine is mounted.

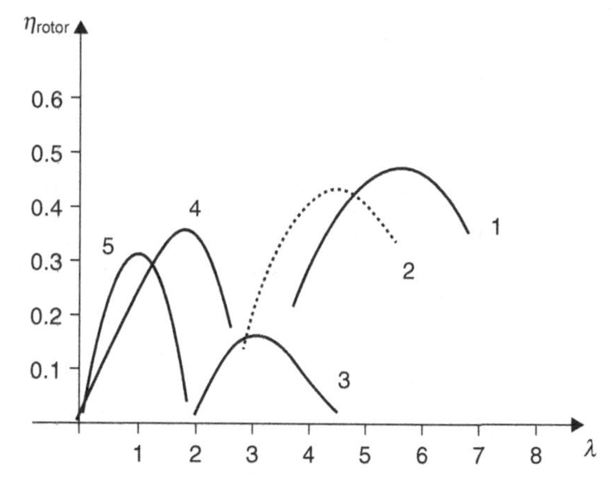

Figure 14.11 Rotor efficiency versus λ. 1 = Rotor with two blades; 2 = rotor with three blades; 3 = rotor with four blades; 4 = rotor with six blades; and 5 = rotor with multiple blades.

Figure 14.12 Wind turbine installations.

14.2.3 Hydroelectric Pumped Storage Systems

This type of storage system is widely used when large energy storage is required. The principle of operation is based on utilizing the potential energy (PE) of water at a specific head. The head is established by pumping process during periods of surplus electric power.

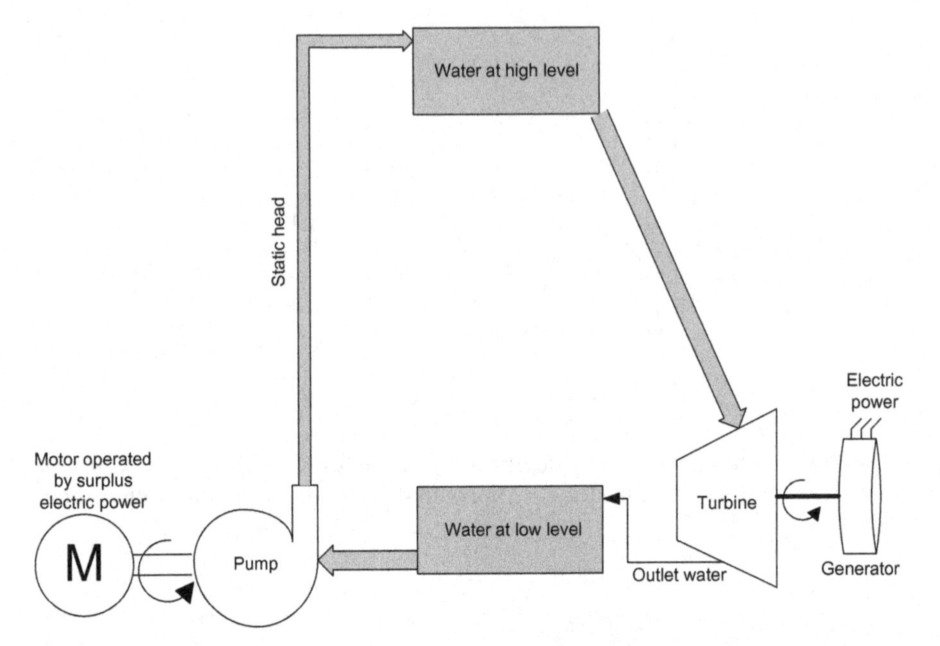

Figure 14.13 Pumping hydroelectric storage systems.

At periods of peak power demand, the water with PE is released to fall down through the turbine. The PE is converted into KE and enthalpy (ETH). If m_1 is a part of total mass, m, which is moving with velocity v to produce the KE, then, the remaining mass, $(m - m_1)$, multiplied by pressure, p, over water density, ρ, gives the ETH. Therefore, the energy balance is verified by the relation:

$$\begin{aligned} \text{PE} &= \text{KE} + \text{ETH} \\ &= \tfrac{1}{2} m_1 v^2 + (m - m_1) p / \rho. \end{aligned}$$

The cycle of operation is depicted in Figure 14.13. It is seen that the outlet water from the turbine can be supplied to the lower-level water tank.

14.2.4 Photovoltaic (PV) Devices

PV devices are also known as solar cells. Operation of PV devices is based on PV effect to convert solar radiation into electricity. PV devices incorporate a pn junction in a semiconductor across which the photovoltage is developed. Light absorption occurs in a semiconductor material, which has to be able to absorb a large part of the solar spectrum. As a result of this absorption, electron hole pairs are generated, and if their recombination is prevented by an electric field they can reach the junction separately. The pn junction separates the emitter and base layer. It is very close to the surface in order to have a high collection probability for the photo-generated charge carriers [205].

Each PV device delivers a voltage less than 1 V. So, these devices are connected in series and packaged into modules to deliver higher voltage and at the same time protect the devices from abnormal environmental conditions. The conversion efficiency of PV devices ranges from 13% to 16%.

The development of PV technology has received great interest from researchers to expand the applications of solar cells that have the following advantages:

- There is no noise, no pollution, and no mechanical moving parts.
- Long lifetime.
- Solar radiation is directly converted into electricity.
- There is no cost and expense for energy source (the sun), which is also inexhaustible.
- PV power ranges from microwatts to megawatts.
- PV power sources are not limited to certain geographic locations compared with wind energy.

So far, solar cells have several applications such as

- providing power for space vehicles;
- providing power source for remote installations; and
- supplying power from several milliwatts in consumer products to grid-connected systems in the kilowatt range up to central power plants of several megawatts. A module of solar cells used for small power applications, for example, water filtration, is shown in Figure 14.14.

On the other hand, in most locations, the intermittent nature of solar energy offers a major problem. The sun radiation differs from hour to hour during the day, from day to day during the season, and from season to season during the year. The remedy to this basic characteristic is to use energy storage systems, the cost of which depends on the type of application and duration of operation.

As any other DG system, PV systems applications are either stand-alone or utility grid-connected systems.

Stand-Alone Applications: Several applications use PV systems as source of energy starting from very small power products, for example, calculators, watches, and small lamps in gardens, to moderate power consumption such as houses to large buildings. The applications may be DC power consumed

Figure 14.14 Solar cells module for small power applications.

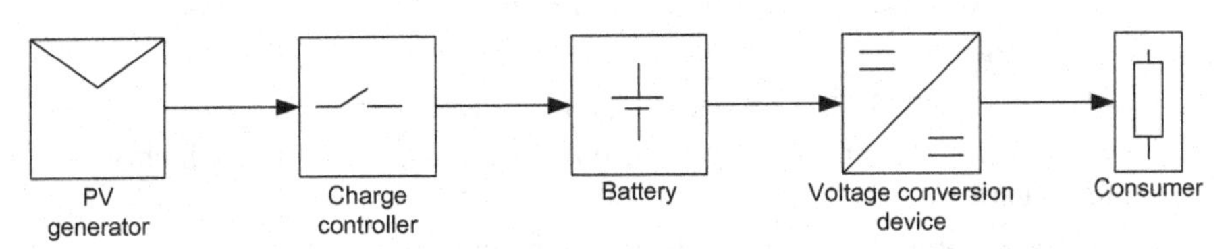

Figure 14.15 PV system feeding a DC consumer's load.

without storage systems (in case of using PV during sunny periods only) or with storage systems (e.g., batteries) if power is needed in the evening or in cloudy weather. PV systems with inverters are used for alternating current (AC) power applications. If the continuity of supply is crucial where the application is operated all the time, PV systems are associated with auxiliary power source (any other fossil-fuel generation). Therefore, the choice of PV system integrated with energy storage and associated with auxiliary power source depends on radiation conditions, required security, and economics.

For small DC power applications, the generated power by solar cells is delivered to batteries (storage system) via a charge controller. The batteries supply the load during periods of very low radiation. The charge controller protects the battery against overcharging and deep discharge to guarantee efficient operating conditions of batteries and PV system as a whole. Then, the power flows to the load at desired voltage through voltage conversion device (Fig. 14.15). If the load is an AC load, the same system is used by replacing the voltage conversion device by an inverter (Fig. 14.16).

For the consumer's load that requires energy source of high reliability and security, as mentioned before, the PV system is associated with an auxiliary supply, for example, natural-gas microturbine generator set. The auxiliary source can feed the load and charge the battery system simultaneously (Fig. 14.17).

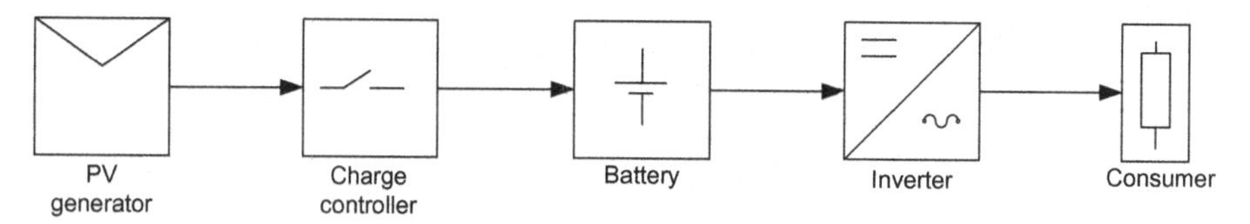

Figure 14.16 PV system feeding an AC consumer's load.

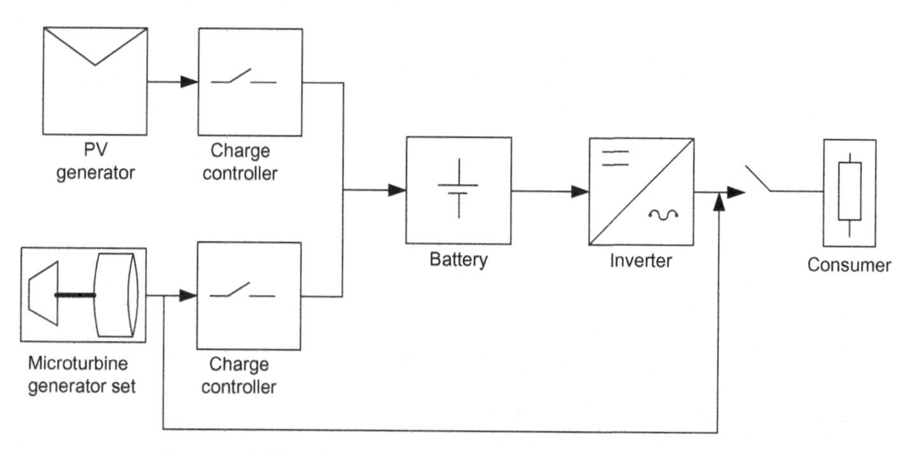

Figure 14.17 PV system associated with auxiliary source.

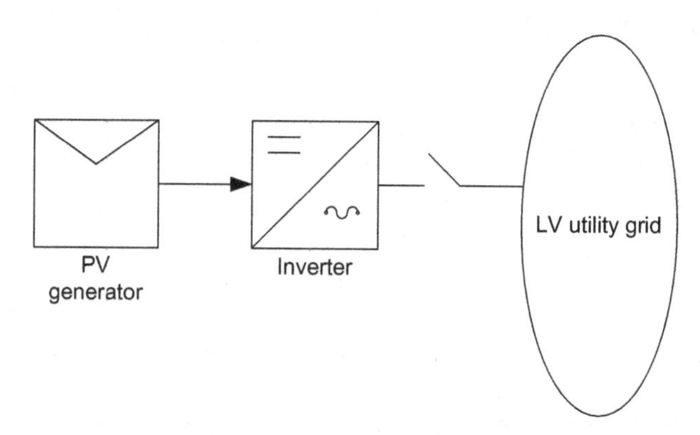

Figure 14.18 PV system connected to LV utility grid.

Grid-Connected Systems: Large power PV systems can be connected to LV utility grid through a suitable inverter [206] (Fig. 14.18) and the utility grid in turn supplies the consumer's loads.

The PV system can directly feed the local loads, which are very close to it along with the connected utility grid. Power consumption must be measured in the two directions by 2 m or one multifunction meter (Fig. 14.19).

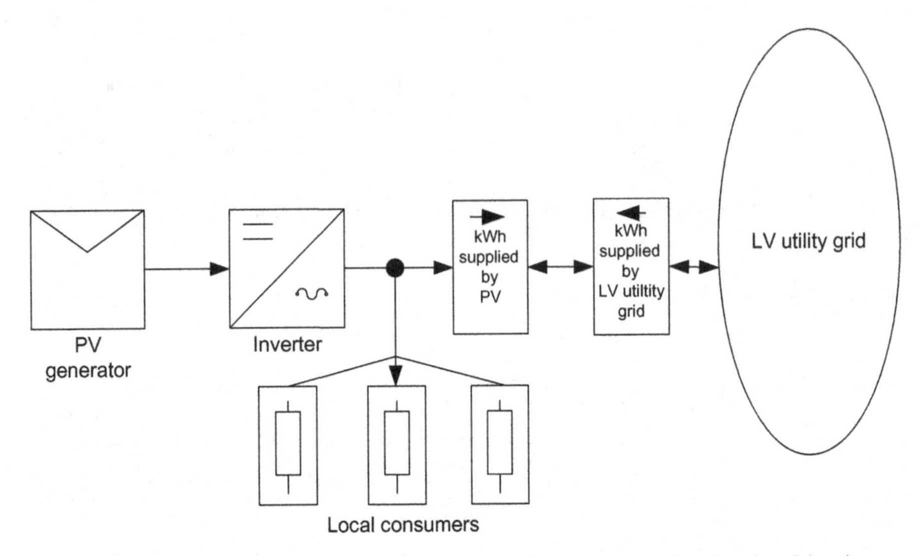

Figure 14.19 PV system connected to LV utility grid and feeding local loads.

14.2.5 Asynchronous Generators

Asynchronous generators are mainly used for small renewable energy power plants, in particular, small hydropower and wind systems. The asynchronous generator has, for this application of DG, the following advantages:

- The speed changes, resulting from the load changes, are easily absorbed by the rotor.
- Any current surge can be damped by the magnetization path of its iron core without fear of demagnetization.
- Lowest cost among generators.
- Easy for parallel operation with large power systems because the utility grid controls voltage and frequency.
- The construction features are robust, providing natural protection against short circuits and overcurrents. For instance, under these conditions, the self-excited asynchronous generator working in stand-alone mode may have a loss of residual magnetism and become unable to supply electric power.
- Widely available.
- Easy to start.

The principle of asynchronous (induction) generator operation, in brief, is based on the interaction between the rotor magnetic field and the stator magnetic field. A voltage is induced in the stator winding that delvers the output electric power. The induced voltage is proportional to the relative speed

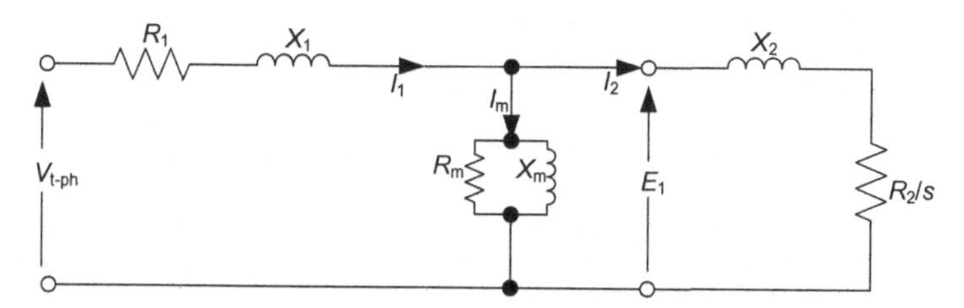

Figure 14.20 Per phase equivalent circuit of an asynchronous induction generator.

difference between the electric synchronous rotation speed n_s (rpm) and the mechanical rotational speed n_r (rpm). The relative speed difference, known as slip, s, is given as [207]

$$s = \frac{n_s - n_r}{n_s}. \tag{14.7}$$

By an approach similar to that of a transformer equivalent circuit, asynchronous generator equivalent circuit is as shown in Figure 14.20 [208], where R_1 and X_1 are the stator resistance and leakage reactance, respectively. R_2/s and X_2 are the rotor resistance and reactance, respectively, referred to the stator side. R_m is the equivalent core loss resistance and X_m is the magnetizing reactance.

Referring to Figure 14.20, the efficiency of a three-phase induction generator is calculated as below:

$$\begin{aligned} P_{\text{mech-input}} &= \text{mechanical input power into the rotor for a negative slip} \\ &= P_{\text{output}} + P_{\text{losses}}, \end{aligned} \tag{14.8}$$

where

$$\begin{aligned} P_{\text{losses}} &= \text{copper losses} + \text{iron losses} + \text{mechanical losses} \\ &= P_{\text{Cu}} + P_{\text{iron}} + P_{\text{mech}}. \end{aligned} \tag{14.9}$$

Copper losses arise from current flow through the resistance of both stator and rotor windings. Thus,

$$\begin{aligned} P_{\text{Cu-stator}} &= 3I_1^2 R_1 \\ P_{\text{Cu-rotor}} &= 3I_2^2 R_2. \end{aligned} \tag{14.10}$$

Power delivered through the air gap (P_{gap}) represents the active power in the second part of the circuit in Figure 14.20 and therefore,

$$P_{gap} = 3I_2^2 R_2/s$$
$$= P_{Cu\text{-}rotor}/s, \tag{14.11}$$

then,

$$P_{Cu\text{-}rotor} = sP_{gap}. \tag{14.12}$$

From Equations 14.10 and 14.12, the total copper loss is

$$P_{Cu} = 3I_1^2 R_1 + sP_{gap}, \tag{14.13}$$

$$P_{iron} = 3E_1^2/R_m, \tag{14.14}$$

The mechanical power converted into electric power P_{input}, which is the power developed by the rotor for a negative slip, is the difference between the power delivered through the air gap (P_{gap}) and the power dissipated in the rotor ($P_{Cu\text{-}rotor}$). Therefore,

$$P_{mech\text{-}input} = P_{gap} - P_{Cu\text{-}rotor} = (1-s)P_{gap} = \frac{n_r}{n_s} P_{gap}. \tag{14.15}$$

Thus, the efficiency η is given as

$$\eta = 1 - \frac{P_{losses}}{P_{mech\text{-}input}}. \tag{14.16}$$

The asynchronous generator needs a reasonable amount of reactive power to provide the magnetic field required to convert the input mechanical power of the rotor into electric power at generator terminals. In applications of interconnection to utility grid, the grid supplies such reactive power but in stand-alone applications, the reactive power must be supplied by the load itself or by electronic inverters or by applying capacitor banks at the machine terminals. These sources of reactive power are connected permanently at generator terminals to control the output voltage as well.

Therefore, asynchronous generators consume vars from the system. It is important to consider the addition of capacitors to improve the power factor and reduce reactive power draw.

When connecting asynchronous generators with a utility grid at speeds significantly below synchronous speed, they have a disadvantage of resulting in damaging inrush currents and associated torques.

14.2.6 Synchronous Generators

Most of the electric power is generated using AC synchronous machines. A synchronous machine normally operates at a constant rotational speed and in synchronism with the frequency of interconnected utility grid. The field excitation of such generators is supplied by a separate DC source; consequently, they can operate in both stand-alone mode and interconnected with utility grid. In the case of interconnection, the generator output is exactly in step with the voltage and frequency of the utility grid.

Synchronous generators must be equipped with more complex control than grid-connected asynchronous generators to be synchronized with the utility grid, and the field excitation needs to be controlled. Also, special protective equipment must be provided to isolate the generators from utility grid when faults occur. The main advantage of this type of generators is their capability of delivering power during the outage of utility grid. In addition, it allows the owners of DG to control the power factor by adjusting the DC field current at their facilities.

Most distributed synchronous generators use a voltage-following control mode, which differs from large generators. Distributed generators normally follow the utility voltage and inject a constant amount of real and reactive power, while large generators have controls to regulate voltage within the limits of their volt-ampere reactive capability [209].

Modeling and simulation of synchronous generators are necessary for short-circuit calculations and other studies such as load flow, stability, and harmonics studies. The synchronous generator can normally be represented by a voltage source behind a reactance, which is time varying (X''_d, X''_d, X_d), and accordingly, the sequence reactances are determined, particularly for short-circuit current calculations as explained in Chapter 4.

The values of synchronous generator parameters at different small ratings (normally used as DG units) are given in Table 14.1 from tests [210].

This table includes additional parameters needed for detailed simulations, for example, quadrature-axis parameters (denoted by subscript q), direct-axis parameters (denoted by subscript d), inertia constant H, open-circuit subtransient direct-axis time constant T''_{do}, open-circuit transient direct-axis time constant T'_{do}, open-circuit subtransient quadrature-axis time constant T''_q, and open-circuit direct and quadrature time constants T''_d and T''_q, respectively.

For fault calculations and in the absence of information, the following parameters may be assumed with machine rating as base: $X''_d = 25\%$, $X_2 = 25\%$, $X_o = 5\%$, and the voltage behind the subtransient reactance is normally in the range of 1.0–1.3 per unit (a typical value of 1.1) [209].

14.3 CASE STUDY

A study is described in this section to clarify what has been explained in this chapter and the more closely related topic of distribution system planning as well.

The U.S. Department of Energy has conducted a study of the potential benefits of cogeneration and small power production (DG) in the United States [211]. The benefits to be studied include those received directly or

TABLE 14.1 Synchronous Generator Parameters in per Unit on the Machine Base [210]

Machine rating (kVA)	69	156	781	1044
kW	55	125	625	835
V	240/480	240/480	240/480	240/480
rpm	1800	1800	1800	1800
H	0.26	0.20	0.40	0.43
Pf	0.80	0.80	0.80	0.80
X_d	2.02	6.16	2.43	2.38
X_d'	0.171	0.347	0.254	0.264
X_d''	0.087	0.291	0.207	0.201
X_q	1.06	2.49	1.12	1.10
X_q''	0.163	0.503	0.351	0.376
T_{do}' (s)	0.950	1.87	1.90	2.47
T_d' (s)	0.080	0.105	0.198	0.273
T_{do}'' (s)	0.078	0.013	0.024	0.018
T_d''(s)	0.004	0.011	0.020	0.014
T_{qo}'' (s)	0.045	0.020	0.016	0.009
T_q''(s)	0.007	0.004	0.005	0.003
r_a	0.011	0.034	0.017	0.013
X_o	0.038	0.054	0.051	0.074
X_2	0.125	0.375	0.279	0.260

indirectly by utilities, electric system planners, and operators. The study is concerned with specific areas of potential benefits, which include

- increase of system reliability,
- reduction of peak power requirements,
- provision of ancillary services including reactive power,
- improvements in power quality,
- reduction in land-use effects and rights-of-way acquisition costs, and
- reduction in vulnerability to terrorism and improvements in infrastructure resilience.

The study also analyzed the rate-related issue that may impede or discourage the expansion of DG facilities.

Situation in the United States at the time of this study included more than 12 million DG units installed with a total capacity of over 200 GW. In 2003, these units generated approximately 250,000 GWh. Over 99% of these units were small reciprocating engine generators or PV systems, installed with inverters that did not feed electricity directly into the distribution grid. However, as shown in Figure 14.21, this large number of small machines represents a relatively small fraction of the total installed capacity.

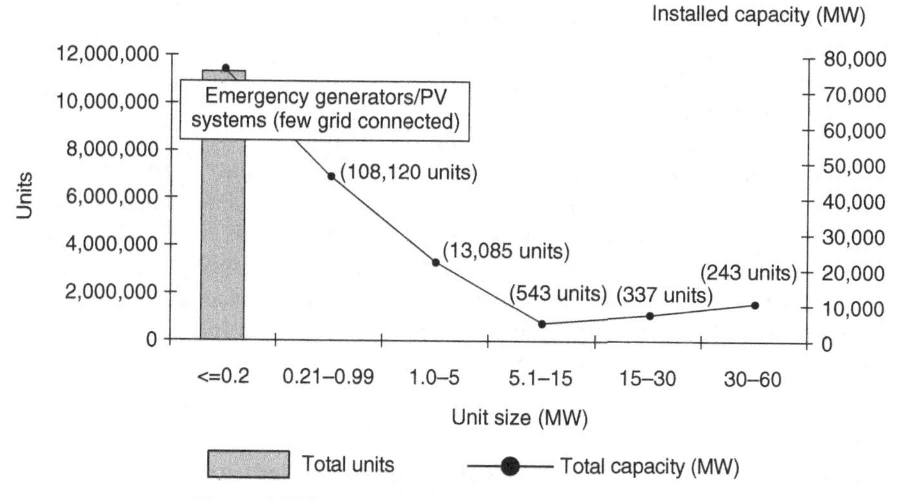

Figure 14.21 U.S. DG installed base (2003) [211].

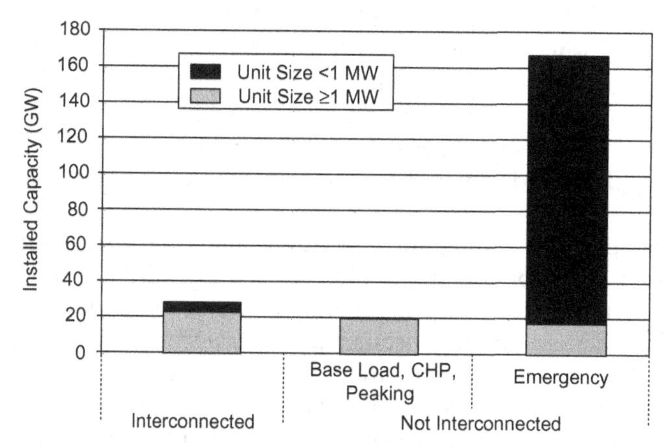

Figure 14.22 U.S. distributed generation capacity by application and interconnection status [211].

14.3.1 Distribution Generation Drivers

The introduction of smaller, more modular technologies capable of operating on a wide variety of fuels—or no fuel—offers direct material benefits to both the energy customer and utility service provider.

Risk-related benefits have driven growth in the DG market. As shown in Figure 14.22, the vast majority of DG units in the United States today are actually backup or emergency generators, installed to operate when grid-supplied electricity is not available. But September 11, 2001, Northeast blackout of August 2003, and hurricane Katrina have all impressed upon the planners the growing need to maintain secure civil operations during a catastrophic event. By changing the switchgear associated with an on-site CHP system, a hospital or other facility can use an integrated DG unit to reduce its electricity bill on a daily basis and to provide emergency power, heating, and cooling during a weather-related or human-induced disruption.

As 12 million DG units already installed attest, DG currently has a great attention in the nation's energy system [211]. However, a vast majority of these units have been installed by consumers to meet the needs for backup power during outages. While some power companies offer incentives to consumers to run their backup power units during peak load periods and other times of system need, DG today is primarily a consumer energy solution, and not one that is well integrated to meet the day-to-day planning and operational needs of the electric power system.

14.3.2 Potential Benefits of DG on Increased Electric System Reliability

Electric system reliability is a measure of the system's ability to meet the electricity needs of customers. It is a term used by electric system planners and operators to measure aggregate system conditions, and as aggregate measure, it generally applies to entire service territories of that system's component parts, including power plants, transmission lines, substations, and distribution feeder lines. To help ensure a reliable system, planners and operators prefer having as much redundancy in these components as can be justified economically [212].

The availability of redundant generating and transmission capacity has made those portions of the system more robust than the distribution system. However, the recent restructuring of electric power markets and regulations, and resulting increase in long-distance power transfers, have put pressure on traditional strategies and procedures for maintaining system reliability. For example, the number of times that the transmission grid was unable to transmit power for contracted transactions jumped from 50 in 1994 to 1494 in 2002.

In addition to redundant capacity, the electric system operators also use operating procedures to provide reliable service in the event of sudden disturbances. These procedures are needed because power flows reroute at close to the speed of light whenever power system conditions change. For example, operators count on sufficient "spinning" reserves to supply immediate replacement for any generation failure.

Problems in system operational reliability can usually be classified as faults and failures. Faults are caused by external events, such as tree contact, animal contact, lightning, automobile accidents, or vandalism. Failures are caused by an equipment malfunction or human error not linked to any external influence. Both faults and failures can cause outages. These outages can be short, lasting less than 15 s, and quickly resolved by automatic switching equipment. When a fault or failure results in a large outage, it typically involves damage to equipment such as a transformer that must be repaired before service can be restored. The time required for such remedies can range from hours to days to weeks. Faults and failures, rather than capacity deficiencies, are the causes of most outages. Outages created by faults and failures in generation are rare, while transmission faults are somewhat more common. As reported in Reference 212, "94% of all power outages are caused by faults and failures in the distribution system (Arthur D. Little, Inc. 2000)."

DG has the potential to be used by electric system planners and operators to improve system reliability, and there are a few examples of this being done currently. As discussed, DG is primarily used today as a customer-side energy resource for services such as emergency power, uninterruptible power, CHP, and district energy. Utilities could do more to use the DG already in place, and they could increase investment in DG resources themselves. However, successful business models for more widespread utility use of DG are limited to certain locations and certain conditions.

There are currently two primary mechanisms being used by the utilities to access customer-side DG for reliability purposes:

- Several utilities offer financial incentives to owners of emergency power units to make them available to grid operators during times of system need.
- Several regions offer financial incentives or price signals to customers to reduce demand during times of system need, and some participants use DG to maintain near-normal on-site operations while they reduce their demand for grid-connected power.

DG offers the potential to increase system reliability, but it can also cause reliability problems, depending on how it is used. In general, DG can increase the system adequacy by increasing the variety of generating technologies, increasing the number of generators, reducing the size of generators, reducing the distance between generators and loads, and reducing the loading on T&D lines. DG can also have a negative impact on reliability depending on a number of factors that include the local electric system composition as well as the DG itself. These factors include DG system size, location, control characteristics, reliability of supply, and reliability of the DG unit itself.

14.3.2.1 Reliability Indices System planners and operators use the reliability indices as a tool to improve the level of service to customers. Accordingly, the requirements for generation, transmission, and distribution capacity additions can be determined by the planners. Operators use these indices to emphasize the system robustness and withstanding possible failures without catastrophic consequences [209].

Generation reliability is measured by the index "loss-of-load probability" (LOLP). It is defined by the probability that generation will be insufficient to meet demand at some point over a specific period of time (hourly or daily, typically 1 day in 10 years). LOLP is actually an expected value; its calculation is based on a probabilistic analysis of generation resources and the peak loads.

Transmission reliability is measured by its availability (AVAIL), which refers to the number of hours the transmission is available to provide service divided by the total hours in the year:

$$\text{AVAIL} = \frac{\text{Available hours of equipment to be in service}}{\text{Total hours in a year}}.$$

Distribution reliability is measured by indices based on customer outage data. These data describe how often electric service was interrupted, how many customers were involved with each outage, how long the outages lasted, and how much load went unserved. The used indices are classified as below [213]:

Sustained Interruption Indices

System Average Interruption Frequency Index (SAIFI): It is the average frequency of sustained interruptions per customer over a predefined area:

$$\text{SAIFI} = \frac{\text{Total number of customers interrupted}}{\text{Total number of customers served}}.$$

System Average Interruption Duration Index (SAIDI): It measures the total duration of interruptions. It is cited in units of hours or minutes per year:

$$\text{SAIDI} = \frac{\text{Sum of all customer interruptions duration}}{\text{Total number of customers served}}.$$

Customer Average Interruption Duration Index (CAIDI): It is the average time needed to restore service to the average customer per sustained interruption:

$$\text{CAIDI} = \frac{\text{SAIDI}}{\text{SAIFI}} = \frac{\text{Sum of all customer interruptions duration}}{\text{Total number of customers interrupted}}.$$

Customer Total Average Interruption Duration Index (CTAIDI): It represents the total average time in the reporting period that customers who actually experienced an interruption were without power. This index is a hybrid of CAIDI and is similarly calculated except that those customers with multiple interruptions are counted only once:

$$\text{CTAIDI} = \frac{\text{Sum of customer interruptions duration}}{\text{Total number of customers interrupted}}.$$

Customer Average Interruption Frequency Index (CAIFI): It gives the average frequency of sustained interruptions for those customers experiencing sustained interruptions. The customer is counted once regardless of the number of times interrupted for this calculation:

$$\text{CAIFI} = \frac{\sum N_\text{i}}{N_\text{c}},$$

where

N_i = number of interrupted customers for each sustained interruption event during the reporting period and

N_c = total number of customers who have experienced a sustained interruption during the reporting period.

Average system availability index (ASAI): It represents the fraction of time (often in percentage) that a customer has received power during the defined reporting period:

$$\text{ASAI} = \frac{\text{Customer hours service availability}}{\text{Customer hours service demand}}.$$

Customers Experiencing Multiple Interruptions (CEMI_n): It indicates the ratio of individual customers experiencing more than n sustained interruptions to the total number of customers served:

$$\text{CEMI}_\text{n} = \frac{N_n}{\text{Total number of customers served}},$$

where N_n is the total number of customers that experience more than n sustained interruptions.

Momentary Indices

Momentary Average Interruption Frequency Index (MAIFI): It indicates the average frequency of momentary interruptions:

$$\text{MAIFI} = \frac{\sum \text{total number of customer momentary interruptions}}{\text{Total number of customers served}}.$$

Momentary Average Interruption Event Frequency Index (MAIFI_E): It has the same definition as MAIFI, but it does not include the events immediately preceding a lockout.

$$\text{MAIFI}_E = \frac{\sum \text{total number of customer momentary interruption events}}{\text{Total number of customers served}}.$$

Load-Based Indices

Average System Interruption Frequency Index (ASIFI): It includes the magnitude of the load unserved during an outage:

$$\text{ASIFI} = \frac{\sum kVA_{\text{sustained}}}{kVA_{\text{served}}}.$$

Average System Interruption Duration Frequency Index (ASIDI): It includes the magnitude of the load unserved during an outage:

$$\text{ASIDI} = \frac{\sum kVA_{\text{sustained}} D_{\text{sustained}}}{N_{\text{served}}}.$$

It is difficult to compare these indices from one location to another or from one utility to another because of differences in how they are calculated. Some utilities exclude outages due to major events, or normalize their results for adverse weather. For SAIDI calculation, some utilities consider an outage over when the substation is returned to service and others consider it over when the customer is returned to service. Some utilities use automatic data collection and analysis, while others rely on manual data entry and spreadsheet analysis.

Another common reliability index is referred to as "nines." This index is based on the expected minutes of power availability during the year. For example, if the expected outage is 50 min per year, the power is available 99.99% of the time or four nines. However, if this index is calculated using the LOLP, it would not reflect outages in the T&D systems. If the nines are calculated based on SAIDI, the nines index will give some indication of the average system availability, but not the availability for any particular customer.

14.3.2.2 DG and Electric System Reliability DG can be used by electric system planners and operators to improve reliability in both direct and indirect ways. For example, DG could be used directly to support local voltage levels and avoid an outage that would have occurred otherwise due to excessive voltage sag. DG can improve reliability by increasing the diversity of the power supply options. DG can improve reliability in indirect ways by reducing stress on grid components to the extent that the individual component reliability is enhanced. For example, DG could reduce the number of hours that a substation transformer operates at elevated temperature levels, which would

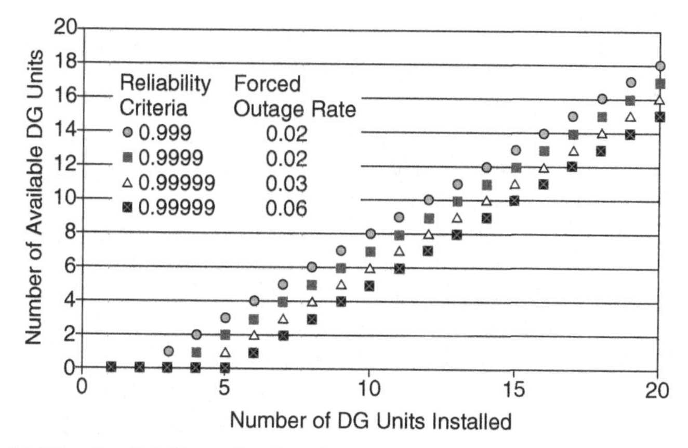

Figure 14.23 Availability of DG units as a function of number of units [211].

in turn extend the life of that transformer, thus improving the reliability of the component.

Multiple analyses have shown that a distributed network of smaller sources provides a greater level of adequacy than a centralized system with fewer large sources, reducing both the magnitude and duration of failures. This is because the failure of one large source (generator) will reduce the available capacity by a much larger percentage than if a small unit failed.

The capacity contribution that can be made by multiple DG units is shown in Figure 14.23 for a simplified case where all the DG units are the same size and have the same forced outage rate. This figure indicates that as the reliability criterion is relaxed from 99.99% to 99.9%, for an unchanged DG unit forced outage rate of 2%, the number of DG units that can be counted as "available" increases. It also shows that as the DG unit forced outage rate increases from 3% to 6% for fixed reliability criterion (99.99%), the number of DG units that can be counted "available" decreases.

As shown, the diversified system reliability is a function of the reliability of individual units, among other factors. A study of actual operating experience determines how DG units perform in the field. Study results include forced outage rates, scheduled outage factors, service factors, mean time between forced outages, and downtimes for a variety of DG technologies and duty cycles. The availability factors collected during this study are summarized in Figure 14.24. Although the sample size for the DG equipment was smaller than that for the central station equipment, the availability of the DG is generally comparable to that of the central station equipment.

An examination of systems with mixed centralized and DG shows that the potential reliability benefits depend on a mix of factors, particularly the reliability characteristics of the centralized generating technologies being replaced versus those being kept, the reliability characteristics of the distributed technology, and the degree of DG penetration.

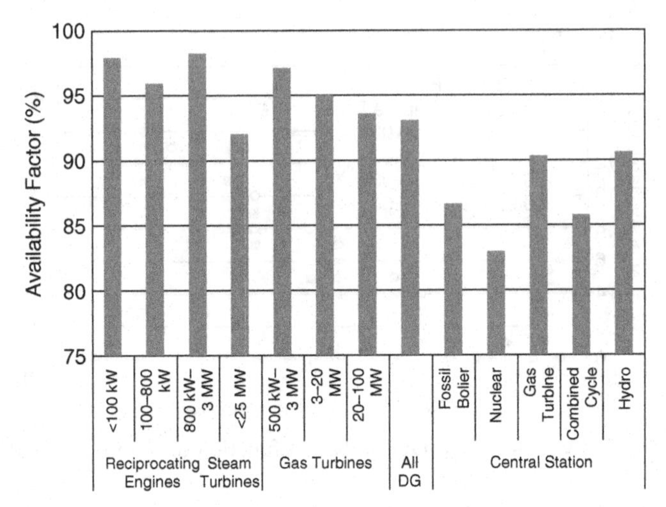

Figure 14.24 A comparison of availability factors for DG equipment and central station equipment. *Source*: NERC GAR (1997–2001) [211].

The economic benefit of using DG to improve electric system reliability can be estimated by determining the avoided costs of traditional forms of investment in electric reliability. Under this approach, the net benefit of installed DG to the utility is the benefit from deferred generation, transmission, and distribution investments, less the costs associated with installing, operating, maintaining, administering, coordinating, scheduling, and dispatching DG units. Not many utilities assess DG in this way when considering expansions and/or upgrades in T&D equipment. If many did, it is likely there would be more instances where the benefits of DG would outweigh the costs, although it is important to remember that the financial attractiveness of DG is highly dependent on local conditions, costs, and resources [214].

14.3.3 Potential Benefits of DG in Reducing Peak Power Requirements

Load reductions in peak demand on specific feeder lines will flow "upstream" and produce demand reductions on substations, transmission, distribution lines and equipment, and power plants, thus freeing up assets to serve other needs. The economic benefits from a reduction in peak power requirements are derived primarily from deferred investments in generation, transmission, and distribution capacity. Utilities make investment decisions for generation, transmission, and distribution capacity based on peak requirements. Thus, in the long run, any reduction in peak power requirements provides direct benefits to the utility in the form of deferred capacity addition/upgrade costs.

A common method for electric system planners and operators to produce demand reduction is by using demand response (DR) programs. DR has been defined as [215] "changes in electric usage by end-use customers from their normal patterns in response to changes in the price of electricity over time, or to incentive payments designed to induce lower electricity use at time of high wholesale market prices, or when system reliability is jeopardized."

DR programs are generally categorized as one of two types: (1) price-based programs such as real-time pricing, critical peak pricing, and time-of-use tariffs, or (2) incentive-based programs such as direct load control and interruptible rates. According to the North American Electric Reliability Council (NERC), about 2.5% of summer peak demand (20,000 MW) is affected by incentive-based programs [216]. DG can be effective in affecting customer responses to electricity demand. A study of DR programs operated by the New York Independent System Operator (NYISO) in 2002 showed that DG was an important factor in the ability of certain participating customers in successfully reducing their demand. DG enabled these customers to continue near-normal operations while they reduced their consumption of grid-connected power, thus reducing demand at NYISO [217].

14.3.3.1 Load Diversity and Congestion Not all electricity-using appliances and equipment demand power from the grid at the same time. For example, residential lighting loads are greatest in the morning and evening, while commercial lighting loads are greatest during business hours. Manufacturing loads vary according to the number of shifts used in any given factory and according to the electric equipment use schedule. Considering such demand diversity, the peak load is never the sum of all connected loads on a feeder or transmission line. One guideline shows that the peak load on a feeder is approximately one-half of the connected load, the peak load on a substation is approximately 45% of the connected load, and the peak load on a generating station is about 41% of the connected load as shown in Figure 14.25 (Department of the Army and the Air Force 1995). This trend shows that load diversity on any particular system component increases as the number of customers served by that component increases.

Just as there is diversity within the system, there is also "supply diversity." Central power plants are selected to provide power to the grid according to a dispatch order determined by their variable costs (fuel, variable operating costs, and emissions permits) subject to certain constraints. These constraints include start-up and shutdown costs, reliability implications, and maintenance requirements. For example, hydropower is almost always the lowest cost power, but its availability is limited by the amount of water stored behind the dam. Other plants operate outside of this dispatch order because they are outside the control of dispatchers, such as CHP plants, rooftop PV arrays, and other customer-owned DG. Plants that are called on for essentially continuous operation (either because of their low variable cost, and/or high start-up and shutdown costs, or because of their importance to reliability) are called base load plants. These typically include all nuclear and a major portion of coal plants. Plants are dispatched to meet the total load at any given time according

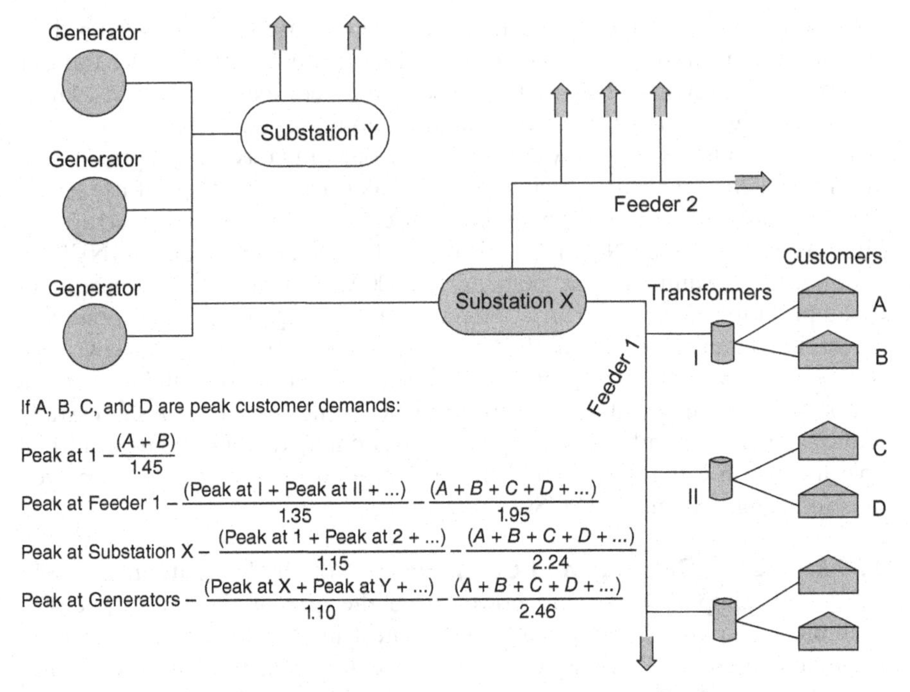

Figure 14.25 Electric demand flow diagram [217].

to this dispatch order so that most plants operate for only a portion of the year, and are the first removed from the system if the load is displaced by operations of DR programs.

Although multiple power plants and transmission lines are available to provide power to any given feeder, not all of them are running or fully loaded at any one point in time. The available capacity of the supply system is limited below the actual capacity of the lines, transmission equipment, and plants in service by the need to provide a contingency allowance and maintain operating reserves. A "contingency allowance" is a prudent operating strategy that holds transmission capacity in reserve in order to continue providing service in the event that any single transmission element in use were to fail. This is often called an "N-1" operating strategy.

With demand growth, peak demand eventually exceeds the capacity of the supply system, or the capacity and configuration of the supply system are insufficient to allow for the most economic system dispatch to meet demand. "Congestion" occurs when the demand for electricity within some geographic boundary is greater than the combined capacity of transmission lines serving that area and any generating stations located within that area, or when the capacity of any transmission system component prevents a dispatch that would otherwise be more economical than the constrained dispatch. Congestion is commonly manifested in the loss of economic efficiency rather than blackouts, but its effects are nonetheless significant.

14.3.3.2 Potential for DG to Reduce Peak Load Several utilities have evaluated use of DG to reduce peak load requirements, although it is not a very common practice. A variety of methodologies have been used for these evaluations, some of them using specific data for actual feeder lines and substations, and others using more generic information. An example of such an evaluation is provided below. In some of these evaluations, it is the case that DG is the most financially attractive option; in others, DG is not. Even in those instances where it has been determined that DG is the most financially attractive option, it is not always the case that investments are made in DG. This is due to a variety of issues, including a lack of familiarity with DG technologies, tools, and techniques, and the perceived likelihood that cost recovery will be less controversial with investments in traditional T&D equipment.

As reported in Reference 210, "a study focused on two real Southern California Edison (SCE) circuits, showed that adding DG would reduce peak demand on the two circuits enough to defer the need to upgrade circuit capacity (Kingston and Stovall 2006)." Results for a circuit that served a mix of commercial, small industrial, and residential customers are shown in Figure 14.26. If the DG installations are targeted optimally, the deferral could economically benefit SCE and its customers, with cost savings that outweigh the lost revenues due to lower sales of electricity.

14.3.4 Potential Benefits of DG from Ancillary Services

Ancillary services are those functions performed by the power system (generation, transmission, and distribution) to support energy transmission from primary resources to the consumers while maintaining reliable operation of the integrated transmission system. In the United States, Federal Energy Regulatory Commission (FERC) has defined ancillary services as "those ser-

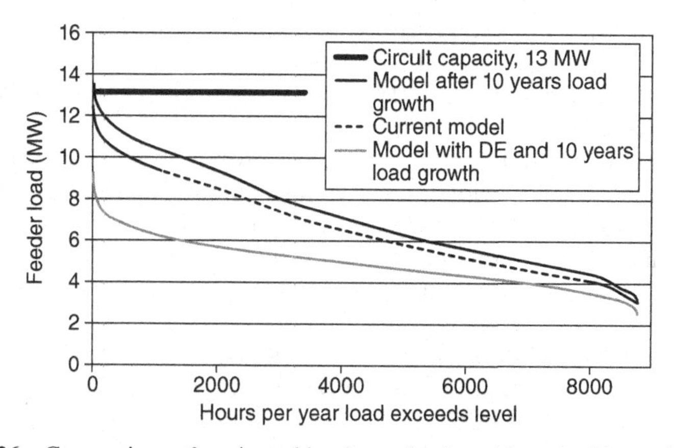

Figure 14.26 Comparison of projected load on a feeder with and without the addition of DG [211]. DE = distributed energy.

vices necessary to support the transmission of electric power from seller to purchaser given the obligations of control areas and transmitting utilities within those control area to maintain reliable operations of the interconnected transmission system." There are several categories of ancillary service, including voltage support, regulation, operating reserve, and backup supply.

Voltage support relates to ancillary service of ensuring that the line voltage is maintained within an acceptable range of its nominal value. Line voltage is strongly influenced by the power factor of the particular line (i.e., the amount of real and reactive power present in a power line). In turn, the power factor can be modified by the installation, removal, or adjustment of reactive power sources. Reactive power can be obtained from several sources, including electric generators, power electronics devices, shunt capacitors, static volt-ampere reactive compensators, synchronous condensers, or even from lightly loaded transmission lines.

Regulation deals with the minute-to-minute imbalances between system load and supply. Generation that provides regulation service must be equipped with automatic control systems capable of adjusting output many times per hour and must be online, providing power to the grid.

Operating reserve comes in two categories: spinning and nonspinning. *Spinning reserve* comes from generating equipment that is online and synchronized to the grid, that can begin to increase output immediately, and that can be fully available within 10 min. *Nonspinning reserve* does not have to be online when initially called but is typically required to fully respond within 10 min of the call to perform.

Backup supply services and supplemental reserves are very similar in function, differing in response time requirements. The response time requirements for backup supply vary across transmission control areas but are generally in the 30- to 60-min time frame. Because supplemental reserve and backup supply do not require a generation source to be already online when called, DG may be more likely to participate in these two ancillary service markets.

Black-start service is the procedure by which a generating unit self-starts without an external source of electricity thereby restoring power to the Independent System Operator (ISO) Controlled Grid following system or local area blackouts.

While not often used for the purpose of providing ancillary services, DG has the capability of providing local voltage support and backup or supplemental reserves, if the units are located on those portions of the grid where these ancillary services are needed, and if they are under the control of grid operators so that they can be called upon during times of system need.

14.3.4.1 Potential Benefits of the Provision of Reactive Power or Volt-Ampere Reactive (Voltage Support) Efficiency of the T&D network improves significantly when reactive power production from central station facilities is replaced by demand-side dynamic reactive power sources. Because sending reactive power to loads from central station facilities "takes up space" on transmission lines, providing reactive power locally frees up useful T&D system capacity for additional real power transfers from generation sources to loads. In addition, providing reactive power locally reduces real and reactive power losses, improving the efficiency of the T&D system.

Reactive power supply sources are broadly categorized as either dynamic or static. Dynamic reactive power resources include generators and dynamic volt-ampere reactive systems. Static reactive power resources include synchronous condensers, static volt-ampere reactive compensators, and capacitor banks. Dynamic sources such as generators are preferable to static sources mainly because their output responds dynamically to changing reactive power conditions. In contrast, static sources are incapable of rapidly responding to changing reactive power demand conditions. Thus, while static sources can provide reactive power service under normal operating conditions, under contingency conditions such as a transmission facility outage and/or a generating unit outage, static sources are more likely to fail when needed most. Therefore, under such contingency conditions, dynamic reactive power resources can rapidly respond to changing reactive power needs to maintain reliability. Thus, central station generators are a prime source of dynamic reactive power and are economically valuable in supporting T&D system and thereby maintaining system reliability.

However, using DG to provide for reactive power can save distribution line losses as well as transmission line losses. For example, distribution losses are the largest percentage of total system losses, comprising about 27% of total losses. When reactive power is supplied from a distributed energy resource (DER) such as a microturbine, losses on the distribution feeder can be reduced or even eliminated. Local power quality can also be significantly improved.

The complex behavior of transmission lines with respect to reactive power is shown in Figure 14.27. When the amount of power being transferred across a transmission line is low, the transmission line actually generates reactive power. On the other hand, at loading levels near the rated capacity of the transmission line, the transmission line consumes a significant amount of reactive power (several times the amount of real power losses in the transmission line). At these times of heavy transmission loading, a significant amount of reactive power is required from generation or other transmission sources simply to supply the transmission lines with the reactive power they require to maintain system voltage. Attempts to send additional reactive power to loads at these times are ineffective, since the additional reactive power transmitted increases the total load on the line, which in turn increases the amount of reactive losses in the line. Given this complex behavior of the transmission system, providing reactive power locally through the use of DG (or other means), when possible, allows system operators to avoid sending reactive power over heavily loaded transmission lines and incurring these avoidable reactive losses.

The location of dynamic reactive power resources is also very important, and this is another reason why DG units that are designed and operated to

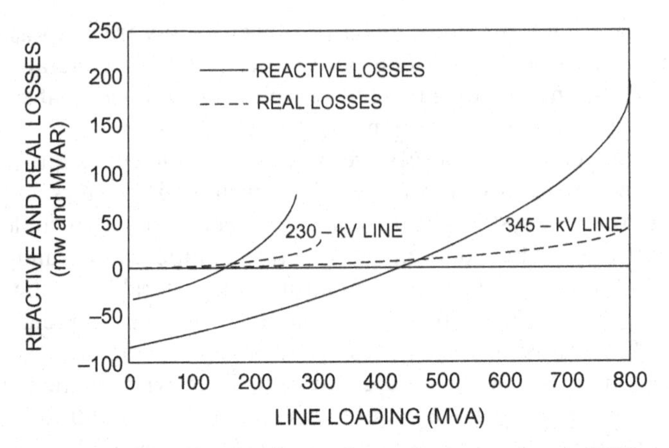

Figure 14.27 Line loading and reactive power losses [210].

produce or absorb reactive power can be even more economically valuable to the electric system. Unlike real power that can be economically transmitted from remote central station generating resources over long distances to demand locations, there are often significant transmission losses in transmitting reactive power from central station generating resources to demand locations. Therefore, under both normal and contingency conditions, it is good utility practice to have these dynamic reactive power resources distributed throughout a grid operator's footprint and closely located to load to ensure that local reactive power resources are available close to potential demand locations—hence the significance of the economic value of reactive power from DG.

14.3.4.2 Simulated DG Reactive Power Effects Reactive power analysis has been completed using a variety of grid simulation tools, and there are conflicting assessments of the ability of DG to reduce the system reactive power requirements. Two studies that include detailed grid analysis for strategic locations illustrate significant reactive power savings associated with DG. The first of these studies estimates that a 500-kW DG installation would save losses in the following amounts: 225 kvar on the distribution system, 113 kvar on the transformer, and 225 kvar on the transmission line. The second study examines specific feeders in Silicon Valley. Results show that sitting DG reactive sources close to the load in these geographic areas could reduce overall reactive power consumption by about 30%.

As it is stated in Reference 211, the voltage support available along a feeder line was calculated as a function of the DG location. The detailed circuit analysis demonstrated that the voltage support at any particular feeder location is the product of the DG plant current and the conductor impedance between the transformer and the point at which the lateral is attached to the line between the transformer and the DG. This shows that voltage support is independent of the total feeder current and is linearly related to DG plant output. For the purpose of formulating network design criteria, the interaction of multiple voltage support DG units has been modeled. The results from that model show that the impact of voltage support DG increases with the increase of size and/or number of voltage support DG units. Based on these results, the analysis was able to propose a design scheme for a voltage support DG controller based on voltage sensitivity that would correct the network voltage effectively.

These studies clearly show that in some locations, DG can improve the efficiency of the system such that significantly less reactive power is needed. However, another study that evaluated the impact of DG on reactive power requirements for California stated that "reactive power requirements for voltage support might be reduced with lower system peak loads. This effect would be extremely difficult to estimate and is likely to be small" [211].

14.3.4.3 Spinning Reserve, Supplemental Reserve, and Black Start

DG has not traditionally been considered as an attractive candidate for ancillary services. To explore contributions in this area, an in-depth examination of the ability of DG to provide other ancillary services was completed. Spinning reserve is a relatively high-priced service and may be an excellent candidate for DG. This is an especially good prospect for types of generation that can be operated in an idle mode or even shut down and then brought up to full load quickly. Some of the new microturbines can be started and ramped up very quickly, in a matter of seconds. If these microturbines were aggregated into meaningful generation blocks of 1 MW or more, they could be ideal sources for spinning reserve. One benefit of using small quick start generating units is that there is no environmental impact from the units idling online [211].

Smaller distributed generators may be designed to provide rapid, large power changes in response to frequency changes to help preserve system stability. While provision of spinning reserve would be a new concept for DG, it is likely to be put into effect in the future if DG constitutes a significant percentage of the total generation, that is, when large DG aggregations are capable of providing a few hundred megawatts of power. Distributed generators can provide this service relatively easily because the control signal (system frequency) is already available at each distributed generator. In the long term, DG may be used with power electronics to dampen and correct frequency oscillations, to regulate voltage, and so on.

The only distributed generators that are likely to be used for black start are larger units with capacities in the tens of megawatts that are already designed for blackout service. There are a large number of such units, at hospitals, airports, and other large installations; and they may be good candidates for black-start service.

Generation assets that provide regulation must be online, providing power to the grid. Customer-owned DG is unlikely to provide this ancillary service because (1) in most locations, the distributed generator is prohibited from providing power to the grid, and (2) the distributed generator operation would have to be controlled to meet the grid power needs rather than the customer's thermal or electric loads. However, regulation services could easily be provided by a utility-owned and operated DG resource.

14.3.4.4 Basis for Ancillary Services Valuations Valuation methodologies for ancillary services are not new. In the 1990s, when restructuring of electric power markets and regulations was being addressed across the United States, a number of studies were made to determine the appropriate market basis for services that had previously been within the traditional model for vertically integrated utilities.

Regulation and spinning reserves require generating units that are already online and synchronized to the grid, but that are operating at less than their maximum capacity. They therefore incur the following costs:

Opportunity and Redispatch Cost: If the generator's marginal cost is lower than the market price, the generator would earn profits operating at full capacity. Therefore, reduction in the energy output necessary to provide regulation is associated with the opportunity cost of foregone profits, roughly proportional to the difference between price and marginal cost of generation. If generator's marginal cost is higher than the energy market price, the redispatch cost of regulation is proportional to the difference between marginal cost and price.

Efficiency Penalty: In order to be able to ramp up quickly, a generator providing regulation or spinning reserve may have to operate at reduced efficiency. This "efficiency penalty" is especially pronounced for steam units.

Energy Cost: Regulation may require a generator to perform fast ramp-ups and ramp-downs. Thus, units offering regulation may incur energy costs associated with turbine acceleration and deceleration.

Wear-and-Tear Costs: For regulation, frequent output adjustments may incur additional wear-and-tear costs.

14.3.5 Value of Power Quality Improvements

The economic impact of poor power quality can be particularly large from an end-user perspective. For modern electronic-based businesses, it is not only outages that hurt but unstable power quality as well. Many high-tech businesses, from Web servers to biotech laboratories, need a very high level of power quality. Today, in the 24-hours-a-day, 7-days-a-week information age, many businesses operate computer-driven equipment with availabilities of

99.999% or even 99.9999%. Very brief sags in voltage or harmonic distortions that used to go entirely unnoticed by most customers can be devastating to customers using sensitive electronics. It requires as little as 8 ms to crash a computer system, often destroying data at the same time.

This means that power interruptions and the other power quality problems can cause severe financial losses for businesses through process disruptions, losses in finished products, equipment damage, lost productivity, and failure to meet customer needs. Redundant systems can often be very cost-effective means of ensuring the required power quality and reliability levels. An alternative way to achieve continuous and high-level power quality is to use redundant power sources (e.g., set of diesel engines and set of fuel cells) supplying the critical sensitive loads, while the other noncritical loads could be connected to separate electric feeders, installed from different substations. Therefore, DG units can provide the high reliability and power quality that some businesses need, particularly when combined with energy storage and power quality technologies.

14.3.6 Technical Specifications of DG and Utility Grid Interconnection

The trend of interconnecting distributed resources with utilities' grids is growing and its development is continuing further to fully realize the benefits and to avoid negative impact on system reliability and security.

Therefore, the methodology of distributed resource interconnection implementation must follow specific standards that define the technical requirements in a manner that can be universally adopted [218–221]. Some of the general requirements can be summarized as below:

Voltage Regulation: The service voltage at the points-of-common coupling (PCC) must be regulated to be within standard limits.

Synchronization: The parallel operation of distributed resource units with utility grid does not cause voltage fluctuations more than ±5%.

Integration with Utility Grid Earthing: Earthing system of distributed resource interconnection shall not cause overvoltages that exceed equipment rating and shall keep the effectiveness of earth-fault protection coordination.

Monitoring Provisions: Each distributed resource unit of 250 kVA or more or distributed resource aggregate of 250 kVA or more at a single PCC shall have provisions for monitoring its connection status (real and reactive power outputs and voltage at the point of distributed resource connection).

Inadvertent Energization of Utility Grid: The distributed resource shall not energize utility grid when the utility grid is de-energized.

Isolation Device: A readily accessible, lockable, visible-break isolation device shall be located between utility grid and distributed resource unit.

Interconnect Integrity

- Protection from electromagnetic interference
- Surge withstand performance
- Paralleling device that shall withstand 220% of the rated voltage

Response to Utility Grid Abnormal Conditions

Power Quality: It must be kept at the desired level.

Utility Grid Faults: The distributed resource unit shall cease to energize the utility grid for faults on grid to which it is connected.

Islanding: When an unintentional islanding is required, it should be within 2 s. For intentional islanding, the standard is to be released in the future.

Interconnection Test Specifications and Requirements

14.3.7 Planning Process

As discussed in this chapter, the main outlines of steps to be taken into account by planners are

- study of T&D capacity;
- determination of load demand performance (amount, duration, and location of peak loads);
- determination of distributed resources (capacity, location, and availability);
- augmentation required for interconnecting distributed resources with utility grid in compliance with standards;
- evaluation of interconnection potential benefits;
- adequacy of regulations and policy;
- factors affecting electricity pricing;
- cost/effect analysis to compromise between expenses and benefits;
- decision making to decide the usefulness of interconnection or feeding emergency loads by distributed resource individually; and
- implementation.

REFERENCES

1. Houston Advanced Research Center. *Guide to Electric Power in Texas*, 3rd ed. Institute for Energy, Law & Enterprise, 2003.
2. AGL. *Distribution System Planning Report.* Australian Gas and Electricity Co., Network Development Agility, 2003.
3. T. A. Short. *Electric Power Distribution Equipment and Systems*. CRC Press, 2005.
4. A. J. Pansini. *Guide to Electric Power Distribution Systems*, 6th ed. CRC Press, 2005.
5. A. R. Abo-Elwafa, F. A. Ghallab, A. A. Sallam, H. A. Zanaty, K. M. Rehab, and H. H. Saeed. Optimal distribution systems for demand continuity in different locations at Sinai Zone. Part #1, No. 98, Research Project, Academy of Scientific Research and Technology, Egypt, March 1999.
6. Y. Han and Y. H. Song. Condition monitoring techniques for electrical equipment——a literature survey. *IEEE Transactions on Power Delivery*, **18**(1):4–13, 2003.
7. M. Arshad, S. M. Islam, and A. Khaliq. Power transformer asset management. Proceedings of the IEEE International Conference on Power System Technology PowerCon '04, Singapore, November 21–24, 2004, Vol. 2, pp. 1395–1398.
8. A. E. B. Abu-Elanien and M. M. A. Salama. Survey on the transformer condition monitoring. Large Engineering Systems Conference on Power Engineering LESCOPE '07, Montreal, Canada, October 10–12, 2007, pp. 187–191.
9. M. Holzenthal, A. Osterholt, and U. Prause. Reliability based planning: reducing the re-investment needs of an urban utility. CIRED, 17th International Conference on Electricity Distribution, Barcelona, May 12–15, 2003, Session 5, Paper No. 34, pp. 1–5.

Electric Distribution Systems, First Edition. Abdelhay A. Sallam, Om P. Malik.

Published 2011 by John Wiley & Sons, Inc.

10. S. Honkapuro, J. Lassila, S. Viljainen, K. Tahvanainen, and J. Partanen. Regulatory effects on the investment strategies of electricity distribution companies. CIRED, 18th International Conference on Electricity Distribution, Turin, June 6–9, 2005, Session No. 5.
11. W. S. C. Moreira, F. L. R. Mussoi, and R. C. G. Teive. Investment prioritizing in distribution systems based on multi objective genetic algorithm. Available at http://www.labplan.ufsc.br/congressos/ISAP%202009/ISAP2009/PDFs/Paper_186.pdf.
12. W. M. Lin, Y. S. Su, and M. T. Tsay. Genetic algorithm for optimal distribution planning. International Conference on Power System Technology, Proceedings, August 1998, pp. 241–245.
13. P. Ngatchou, A. Zarel, and M. A. El-Sharkawi. Pareto multi objective optimization. International Conference on Intelligent Systems Application to Power Systems, Proceedings, November 2005, pp. 84–91.
14. IndEco Strategic Consulting Inc. Demand Side Management and Demand Response in Municipalities: Workshop Background Paper. January 27, 2004. Available at http://www.cleanairpartnership.org/pdf/1_backgrounder.pdf.
15. Ontario Energy Board. Demand side management and demand response in the Ontario electricity sector. Report to the Minister of Energy, March 1, 2004. Available at http://www.oeb.gov.on.ca/documents/cases/RP-2003-0144/pressrelease_report_finalwithappendices_030304.pdf.
16. C. Puret. MV public distribution networks throughout the world. Cahier Technique, no. 155, March 1992. Available at http://www.engineering.schneider-electric.dk/Attachments/ed/ct/MV_public_distribution_networks.pdf.
17. A. A. El-Desouky, R. Aggarwal, M. M. Elkateb, and F. Li. Advanced hybrid genetic algorithm for short-term generation scheduling. *IEE Proceedings, Generation, Transmission and Distribution*, **148**(6):511–517, 2001.
18. A. J. Wood and B. F. Wollenberg. *Power Generation Operation and Control*. John Wiley & Sons, 1996.
19. X. Wang and J. R. McDonald. *Modern Power System Planning*. McGraw-Hill, 1994.
20. S. Fans, K. Methaprayoon, and W. J. Lee. Multi-area load forecasting for system with large geographical area. Industrial and Commercial Power Systems Technical Conference, 2008, ICPS 2008, IEEE/IAS, May 4–8, 2008, pp. 1–8.
21. S. Fans, K. Methaprayoon, and W. J. Lee. Short-term load forecasting using comprehensive combination based on multi-meteorological information. Industrial and Commercial Power Systems Technical Conference, 2008, ICPS 2008, IEEE/IAS, May 4–8, 2008, pp. 1–7.
22. C. W. Gellings. *Demand Forecasting for Electric Utilities*. Fairmont Press, 1996.
23. D. W. Bunn and E. D. Farmer. *Comparative Models for Electrical Load Forecasting*. John Wiley & Sons, 1985.
24. G. E. P. Box, G. M. Jenkins, and G. C. Reinsel. *Time Series Analysis: Forecasting and Control*, 3rd ed. Prentice-Hall, 1994.
25. G. A. N. Mbamalu and M. E. El-Hawary. Load forecasting via suboptimal autoregressive models and iteratively reweighted least squares estimation. *IEEE Transactions on Power Delivery*, **8**(1):343–348, 1993.

26. S. Ruzic, A. Vuckovic, and N. Nikolic. Weather sensitive method for short-term load forecasting in electric power utility of Serbia. *IEEE Transactions on Power Systems*, **18**:1581–1586, 2003.
27. B. Abraham and J. Ledolter. *Statistical Methods for Forecasting*. John Wiley & Sons, 1983.
28. S. Rahman and O. Hazim. Load forecasting for multiple sites: development of an expert system-based technique. *Electric Power Systems Research*, **39**:161–169, 1996.
29. V. Miranda and C. Monteiro. Fuzzy inference in spatial load forecasting. *Proceedings of IEEE Power Engineering Winter Meeting*, **2**:1063–1068, 2000.
30. A. A. El Desouky and M. ElKateb. Hybrid adaptive techniques for electric-load forecast using ANN and ARIMA. *IEE Proceedings, Generation, Transmission and Distribution*, **147**(4):213–217, 2000.
31. H. Chen, C. A. Canizares, and A. Singh. ANN-based short-term load forecasting in electricity markets. *Proceedings of the IEEE Power Engineering Society Transmission and Distribution Conference*, **2**:411–415, 2001.
32. J.-S. Jang. ANFIS: Adaptive network-based fuzzy inference system. *IEEE Transactions on Systems, Man, and Cybernetics*, **23**(3):665–685, 1993.
33. A. A. El Desouky. Accurate fast weather dependent load forecasting for optimal generation scheduling in real time applications. Ph.D. Thesis, University of Bath, 2002.
34. H. L. Willis. *Spatial Electric Load Forecasting*. Marcel Dekker, 1996.
35. IEEE Std. 1100-1999. *Powering and Grounding Sensitive Electronic Equipment*. IEEE, 1999.
36. IEEE Std. 142-1991. *Recommended Practice for Grounding of Industrial and Commercial Power Systems*. Green Book. IEEE, 1991.
37. B. Scaddan. *17th Edition IEE Wiring Regulations: Design and Verification of Electrical Installations*, 6th ed. Elsevier, 2008.
38. D. Fulchiron. Overvoltages and insulation coordination in MV and HV. Cahier Technique, no. 151, February 1995. Available at http://www.scribd.com/doc/37921822/ect151.
39. IEEE Std. 80-2000. *Guide for Safety in AC Substation Grounding*. IEEE, 2000.
40. C. F. Dalziel. Effect of waveform on let-go currents. *AIEE Transactions*, **62**:739–744, 1943.
41. C. F. Dalziel and F. P. Massoglia. Let-go currents and volt-voltages. *AIEE Transactions*, **25**(part II):49–56, 1956.
42. C. F. Dalziel. Temporary paralysis following freezing to a wire. *AIEE Transactions*, **79**:174–175, 1960.
43. C. F. Dalziel and R. W. Lee. Lethal electric currents. *IEEE Spectrum*, **6**(2):44–50, 1969.
44. IEC. 479-1. *Effects of Currents Flowing through the Human Body*. 1984.
45. IEEE. 1048-1990. *IEEE Guide for Protective Grounding of Power Lines*. IEEE, 1990.
46. IEEE Std. 141-1993. *Recommended Practice for Electric Power Distribution for Industrial Plants*. Red Book, IEEE, 1993.

47. ANSI. C2-1990. *Grounding Methods for Electricity Supply and Communication Facilities*. National Electrical Safety Code, 1990 edition, Section 9, Article 96-A, 1990.
48. B. Lacroix and R. Calvas. Earthing systems worldwide and evolutions. Cahier Technique, no. 173, September 1995. Available at http://www.schneider-electric.com/documents/technical-publications/en/shared/electrical-engineering/dependability-availability-safety/low-voltage-minus-1kv/ect173.pdf.
49. IEC. 364-3. *Assessment of General Characteristics*. 1993.
50. B. Lacroix and R. Calvas. System earthings in LV. Cahier Technique, no. 172, December 2004. Available at http://www.schneider-electric.com/documents/technical-publications/en/shared/electrical-engineering/dependability-availability-safety/low-voltage-minus-1kv/ect172.pdf.
51. F. Jullien and I. Heritier. The IT earthing system (unearthed neutral) in LV. Cahier Technique, no. 178, June 1999. Available at http://www.schneider-electric.com/documents/technical-publications/en/shared/electrical-engineering/electrical-networks/low-voltage-minus-1kv/ect178.pdf.
52. IEEE Std. 242-1986. *IEEE Recommended Practice for Protection and Coordination of Industrial and Commercial Power Systems*. Buff Book, IEEE, 1986.
53. IEEE Std. 399-1997. *IEEE Recommended Practice for Industrial and Commercial Power Systems*. Brown Book, IEEE, 1997.
54. C. Prévé. *Industrial Network Protection Guide*. Schneider Electric, 1996.
55. ANSI/IEEE. C37.04-1979. *IEEE Standard Rating Structure for Alternating Current High-Voltage Circuit Breakers Rated on a Symmetrical Current Basis*. 1979.
56. J. Machowski, J. W. Bialek, and J. R. Bumby. *Power System Dynamics and Stability*. John Wiley & Sons, 1998.
57. IEC. 60909-0. *Short Circuit Currents in Three-Phase a.c. Systems—Calculation of Currents*, 1st ed. 2001.
58. J. Schlabbach. *Short Circuit Currents*. IET, 2005.
59. J. L. Blackburn. *Symmetrical Components for Power System Engineering*. Marcel Dekker, 1993.
60. IEC 909-4. *Short-Circuit Currents in Three-Phase a.c. Systems—Examples for the Calculation of Short-Circuit Currents*, 1st ed. 2000.
61. IEC. 909-2. *Electrical Equipment—Data for Short Circuit Current Calculations in Accordance with IEC 909 (1988)*, 1st ed. 1992.
62. A. E. Guile and W. Paterson. *Electrical Power Systems*. Vol. 1, 2nd ed. Pergamon Press, 1981.
63. A. F. Sleva. *Protective Relay Principles*. CRC Press, 2009.
64. A. T. Johns and S. K. Salman. *Digital Protection for Power Systems*. Peter Peregrinus, 1995.
65. J. M. Gers and E. J. Holmes. *Protection of Electricity Distribution Networks*, 2nd ed. IEE, 2004.
66. C. Seraudie. LV surges and surge arresters/LV insulation coordination. Cahier Technique, March 1999, Available at http://www.schneider-electric.com/documents/technical-publications/en/shared/electrical-engineering/electrical-environmental-constraints/low-voltage-minus-1kv/ect179.pdf.

67. A. J. Eriksson. The incidence of lightning strikes to power lines. *IEEE Transactions on Power Delivery*, **PWRD-2**(2):859–870, 1987.

68. IEEE Std. 1299/C62.22.1. *IEEE Guide for the Connection of Surge Arresters to Protect Insulated, Shielded Electric Power Cable Systems*. IEEE, 1996.

69. S. Rusck and R. H. Goldi. *Lightning Protection of Distribution Lines*. Academic Press, 1977.

70. IEEE Std. 1410. *IEEE Guide for Improving the Lightning Performance of Electric Power Overhead Distribution Lines*. IEEE, 1997.

71. IEC. 71-1. *Insulation Coordination: Definition, Principles and Rules*, 7th ed. 1993.

72. IEC. 71-2. *Insulation Coordination: Application Guide*, 3rd ed. 1996.

73. IEC. 60. *High Voltage Test Techniques—Part 2: Measuring Systems*, 2nd ed. 1994.

74. IEC. 99. *Surge Arresters—Part 4: Metal Oxide Surge Arresters Without Gaps for a.c. Systems*. 2001.

75. IEEE Std. C62.11. *IEEE Standard for Metal-Oxide Arresters for AC Power Circuits (>1 kV)*. IEEE, 2005.

76. IEEE Standards Collection. C62. *Surge Protection*, 1995 edition, 1995.

77. J. C. Tobias and D. J. Hull. The building blocks of a distribution tele-control system. Fourth International Conference on Trends in Distribution Switchgear, Pub. No. 400, London, November 1994, pp. 29–36.

78. J. C. Tobias, R. P. Leeuwerke, A. L. Brayford, and A. J. Robinson. The use of sectionalizing circuit breakers in urban MV distribution networks. IEE, Conference Publication No. 459, 1998.

79. IEEE Std. 37.48.1. *IEEE Guide for the Operation, Classification, Application, and Coordination of Current Limiting Fuses with Rated Voltages 1–38 kV*. IEEE, 2002.

80. IEEE. 242-2001. *IEEE Recommended Practice for Protection and Coordination of Industrial and Commercial Power Systems*. Buff Book, IEEE, 2001.

81. Hennig Gremmel for ABB AG, Mannheim. *ABB Switchgear Manual*, 11th Ed. Cornelsen Verlag Scriptor GmbH & Co., 2010.

82. IEEE. C37.91. *Approved Draft for Standard Requirements for Liquid-Immersed Distribution Substation Transformers*. Revision of C37. 90-1989, Superseded by C37.90-2005, 2005.

83. S. Stewart. *Distribution Switchgear*. IEE, 2004.

84. IEEE Std. 1015™-2006. *IEEE Recommended Practice for Applying Low Voltage Circuit Breakers Used in Industrial and Commercial Power Systems*. IEEE, 2006.

85. IEEE Std. 1458™-2005. *IEEE Recommended Practice for the Selection, Field Testing, and Life Expectancy of Molded Case Circuit Breakers for Industrial Applications*. IEEE Industry Applications Society, 2005.

86. R. Calvas and J. Delaballe. Cohabitation of high and low currents. Cahier Technique No. 187, April 1997. Available at http://www.google.com/search?q=cahier+technique+earth+electrodes&hl=en&client=firefox-a&hs=3ka&rls=org.mozilla:en-US:official&channel=s&prmd=ivns&ei=dYYGTYiZCKmK4gbPjfW_Bw&start=10&sa=N.

87. IEC. 364-5/1980. *Electrical Installations of Buildings: Selection and Erection of Electrical Equipment*. Amendment No. 1, July 1982.

88. IEC. 621-2A/1987. *Electrical Installations for Outdoor Sites Under Heavy Conditions (Including Open-Cast Mines and Quarries). Part 2: General Protection Requirements*. 1987.

89. R. Lee. Pressure developed by arcs. *IEEE Transactions on Industry Applications*, **IA-23**(4):760–764, 1987.

90. C. Davis, C. St. Pierre, D. Castor, R. Lue, and S. Shrestha. *Practical Solution Guide to Arc Flash Hazards*. ESA, 2003.

91. J. Weigel. Electric arc flash safety and risk management in healthcare facilities. Presented to WSSHE Puget Sound Chapter, March 2010. Available at www.wsshe.org/webinar-weigel-electricalsafety-presentation.pdf.

92. OSHA Construction cTool-Department of Labor. Arc Flash Hazards Photos: Photo Examples of Burns and Other Injuries. Occupational Safety & Health Administration (OSHA), U.S. Available at http://www.osha.gov/SLTC/etools/construction/electrical_incidents/burn_examples.html#electrical_burns.

93. NFPA. 70E. *Standard for Electric Safety in the Workplace*. NFPA, 2004.

94. IEEE Std. 1584-2002. *IEEE Guide for Performing Arc-Flash Hazard Calculations*. IEEE, 2002.

95. C. Inshaw and R. A. Wilson. Arc flash hazard analysis and mitigation. Western Protective Relay Conference, Spokane, WA, October 20, 2004.

96. R. Lee. The other electrical hazard: electrical arc blast burns. *IEEE Transactions on Industry Applications*, **IA-18**(3):246, 1982.

97. Data Bulletin. Arc-flash application guide: arc-flash energy calculations for circuit breakers and fuses class 100. Square D by Schneider Electric, 0100DB0402R3/06, March 2006.

98. Data Bulletin. Square D preferred methods for arc-flash incident energy reduction. Square D by Schneider Electric, 3000DB0810R6/08, June 2008.

99. T. Gönen. *Electric Power Distribution System Engineering*, 2nd ed. CRC Press, 2006.

100. B. W. Kennedy. *Power Quality Primer (Electrical Engineering Primers)*, 1st ed. McGraw-Hill Professional, 2000.

101. S. Swaminathan and R. K. Sen. Review of power quality applications of energy storage systems. Sandia National Laboratories, No. SAND98-1513 Unlimited, California, July 1998.

102. S. Santoso, H. W. Beaty, R. C. Dugan, and M. F. McGranaghan. *Electrical Power Systems Quality*, 2nd ed. McGraw-Hill, 2002.

103. EPRI. Power quality work book. TR-105500, April 1996.

104. S. T. Mak and S. E. Spencer. Power quality issues at utilities serving rural areas and smaller towns. CIRED, 15th International Conference and Exhibition on Electricity Distribution, June 1–4, 1999, Belgium, pp. 2_11-1–2_11-6.

105. J. Iglesias, G. Batrak, J. Gutierrez Iglesias, G. Bartak, N. Baumier, B. Defait, M. Dussart, F. Farrell, C. Graser, J. Sinclair, D. Start, and M. Mazzoni. Power quality in European electricity supply network. Eurelectric, Ref. No. 2002-2700-0005, February 2002.

106. EPRI. CU-7529. *Power Quality Assessment Procedures*. December 1991.

107. K. Chan, A. Kara, et al. Innovative system solutions for power quality enhancement. CIRED, 15th International Conference and Exhibition on Electricity Distribution, June 1–4, 1999, Belgium, pp. 2_25-1–2_25-8.

108. IEEE Std. C62.48-1995. *IEEE Guide on Interactions Between Power System Disturbances and Surge-Protective Devices*. IEEE, 1995.

109. ITI (CBEMA) curve application note. Published by Technical Committee3 (TC3) of the Information Technology Industry Council (ITI), Washington, DC, 2002, Updated 2010. Available at http://www.itic.org/technical/iticurv.pdf.

110. IEEE Std. 1159-1995. *IEEE Recommended Practice for Monitoring Electric Power Quality*. IEEE, 1995.

111. CEA. Canadian national power quality survey. Canadian Electricity Authority, Project 220D 711A, August 1995.

112. T. E. Grebe, D. D. Sabin, and M. F. McGranaghan. An assessment of distribution system power quality, volume 1: executive summary. EPRI Report TR-106249-V1, Palo Alto, CA, May 1996.

113. D. D. Sabin. An assessment of distribution system power quality, volume 2: statistical summary report. EPRI Report TR-106249-V2, Palo Alto, CA, May 1996.

114. D. L. Brooks and D. D. Sabin. An assessment of distribution system power quality, volume 3: the library of distribution system power quality monitoring case studies. EPRI Report TR-106249-V3, Palo Alto, CA, May 1996.

115. S. R. Whitesell. Proposal for ac power interruptions (clause 4.3.1.3). VTech Communications Report TR-41.7.4, Howell, NJ, January 2003.

116. C. Shimin. The economic analysis on power supply quality. CIRED, 15th International Conference and Exhibition on Electricity Distribution, June 1–4, 1999, Belgium, pp. 2_14-1–2_14-4.

117. M. Sullivan. Power interruption costs to industrial and commercial consumers of electricity. Commercial and Industrial Technology Conference, 1996.

118. J. N. Fiorina. Inverters and harmonics: case studies of non-linear loads. Cahier Technique, no. 159, September 1993. Available at http://www.schneider-electric.com/documents/technical-publications/en/shared/electrical-engineering/electrical-networks/low-voltage-minus-1kv/ect159.pdf.

119. J. N. Fiorina. Harmonics upstream of rectifiers in UPS. Cahier Technique, no. 160, December 1993. Available at http://www.schneider-electric.com/documents/technical-publications/en/shared/electrical-engineering/electrical-networks/low-voltage-minus-1kv/ect160.pdf.

120. A. Yagasaki. Characteristics of a special-isolation transformer capable of protecting from high-voltage surges and its performance. *IEEE Transactions on Electromagnetic Compatibility*, **43**(3):340–347, 2001.

121. J. Bluma, T. Stephanblome, M. Dauly, and P. Anderson. *Voltage Quality in Electric Power Systems*, 1st ed. IEEE Power and Energy Series 36, IEEE Press, 2001.

122. J. J. Burke. *Power Distribution Engineering: Fundamentals and Applications*. Marcel Dekker, 1994.

123. ANSI Std. C84.1-2006. *For Electric Power Systems and Equipment-Voltage Ratings (60 Hz)*. 2006.

124. IEC Std. 61000-4-11. *Voltage Dips, Short Interruptions and Voltage Variations Immunity Tests*, 2nd ed., 2004-03.

125. M. H. J. Bollen. *Understanding Power Quality Problems: Voltage Sags and Interruptions*. IEEE Press, 2000.
126. A. Baggini. *Handbook of Power Quality*. John Wiley & Sons, 2008.
127. P. Ferracci. Power Quality. Cahier Technique no. 199, October 2001. Available at http://www.schneider-electric.com/documents/technical-publications/en/shared/electrical-engineering/electrical-environmental-constraints/general-knowledge/ect199.pdf.
128. L. J. Kojovic and S. Hassler. Application of current limiting fuses in distribution systems for improved power quality and protection. *IEEE Transactions on Power Delivery*, **12**(2):791–800, 1997.
129. L. Zhang and M. H. J. Bollen. Characteristics of voltage dips (sags) in power systems. *IEEE Transactions on Power Delivery*, **15**(2):827–832, 2000.
130. M. E. Olivera and A. Padiha-Feltrin. A top-down approach for distribution loss evaluation. *IEEE Transactions on Power Delivery*, **24**(4):2117–2124, 2009.
131. C. A. Dortolina and R. Nadira. The loss that is unknown is no loss at all. *IEEE Transactions on Power Delivery*, **20**(2):1119–1125, 2005.
132. W. H. Kersting. *Distribution System Modeling and Analysis*, 2nd ed., CRC Press, 2006.
133. P. S. N. Ras and R. Deekshit. Energy loss estimation in distribution feeders. *IEEE Transactions on Power Delivery*, **21**(3):1092–1100, 2006.
134. R. Céspedes, H. Durán, H. Hernández, and A. Rodriguez. Assessment of electrical energy losses in the Colombian power system. *IEEE Transactions on Power Apparatus and Systems*, **PAS-102**(11):3509–3515, 1983.
135. D. L. Flaten. Distribution system losses calculated by percent loading. *IEEE Transactions on Power Systems*, **3**(3):1263–1269, 1988.
136. N. Vampati, R. R. Shoults, M. S. Chen, and L. Schwobel. Simplified feeder modeling for load-flow calculations. *IEEE Transactions on Power Systems*, **PWRS-2**(1):168–174, 1987.
137. R. Balchick and F. F. Wu. Approximation formulas for the distribution system: the loss function and voltage dependence. *IEEE Transactions on Power Delivery*, **6**(1):252–259, 1991.
138. H. D. Chiang, J. C. Wang, and K. N. Miu. Explicit loss formula, voltage formula and current flow formula for large scale unbalanced distribution systems. *IEEE Transactions on Power Systems*, **12**(3):1061–1067, 1997.
139. C. S. Chen, M. Y. Cho, and Y. W. Chen. Development of simplified loss models for distribution system analysis. *IEEE Transactions on Power Delivery*, **9**(3):1545–1551, 1994.
140. Y. Y. Hong and Z. T. Chao. Development of energy loss formula for distribution systems using FCN algorithm and cluster-wise fuzzy regression. *IEEE Transactions on Power Delivery*, **17**(3):794–799, 2002.
141. Schneider-Electric Group. Harmonic detection and filtering. No. D8TP152GUI/E, Schneider-Electric, France, 1999.
142. C. Collombet, J. M. Lupin, and J. Schonek. Harmonic disturbances in networks, and their treatment. Cahier Technique, no. 152, December 1999. Available at http://www.designers.schneider-electric.ru/Attachments/ed/ct/harmonic_disturbances.pdf.

143. J. Arrillaga and N. R. Watson. *Power System Harmonics*. John Wiley & Sons, 2003.

144. IEC. 61000-2-1. *Electromagnetic Compatibility (EMC)—Part 2: Environment—Section 1: Description of the Environment—Electromagnetic Environment for Low Frequency Conducted Disturbances and Signaling in Public Power Supply Systems*. Geneva: IEC, 1990.

145. ABB report. Power Quality Filter: Active Filtering Guide. © Asea Brown Boveri Jumet S.A. (BE) 1999. Available at http://www05.abb.com/global/scot/scot209.nsf/veritydisplay/0f61159109c44f58c1256f2e00445c57/$file/2gcs401012a0070.pdf.

146. ABB manual. Power Quality Filter PQFK. Document No. 2GCS213016A0070, October 2010, Rev. 6. Available at http://www05.abb.com/global/scot/scot209.nsf/veritydisplay/8a52e1016e556bdac12577d000481a9b/$File/2GCS213016A0070%20_Manual%20Power.

147. J. Delaballe. EMC: electromagnetic compatibility. Cahier Technique, no. 149, December 2001. Available at http://www.designers.schneider-electric.ru/Attachments/ed/ct/harmonic_disturbances.pdf.

148. IEEE Std. 519-1992. *IEEE Recommended Practices and Requirements for Harmonic Control in Electric Power Systems*. IEEE, 1993.

149. IEC Std. 61000-3-6. *Assessment of Emission Limits for the Connection of Distorting Installations to MV, HV and EHV Power Systems*. Edition 2.0, 2008-02.

150. IEC Std. 61000-3-7. *Assessment of Emission Limits for the Connection of Fluctuating Installations to MV, HV and EHV Power Systems*. Edition 2.0, 2008-02.

151. IEC Std. 61000-4-7. *Testing and Measurements Techniques—General Guide on Harmonics and Intraharmonics Measurements and Instrumentation for Power Supply Systems and Equipment Connected Thereto*, Edition 2.0, 2002-08.

152. IEC Std. 364-5-55. *Electrical Installations of Buildings: Selection and Erection of Electrical Equipment-Other Equipment*, Edition 1.1, 2002-05.

153. IEC Std. 61000-2-2. *Electromagnetic Compatibility (EMC)—Part 2-2: Environment-Compatibility Levels for Low frequency conducted disturbances and signaling in Public Low-Voltage Power Supply Systems*. 2002-03.

154. IEC Std. 61000-2-4. *Electromagnetic Compatibility (EMC)—Part 2-4: Environment Compatibility Levels for Low Frequency Conducted Disturbances*. 2002-01.

155. IEC Std. 61000-3-2. *Limits for Harmonic Current Emissions (Equipment Input Current ≤ 16A per phase)*. 1995.

156. European Std. EN 50160. *Voltage Characteristics of Electricity Supplied by Public Distribution Systems*. 1995.

157. J. Arrillaga, B. C. Smith, N. R. Watson, and A. R. Wood. *Power System Harmonic Analysis*. John Wiley & Sons, 2000.

158. E. Acha and M. Madrigal. *Power Systems Harmonics*. John Wiley & Sons, 2002.

159. D. Kim, M. Kang, and S. Rhee. Determination of optimal welding conditions with a controlled random search procedure. *Welding Research Council (WRC), Welding Journal*, August:125-s–130-s, 2005.

160. I. MacPherson, N. Larlee, R. Marios, and G. Thomas. Load forecast 2005–2015. Board of Commissioners of Public Utilities of New Brunswick, Energie NB Power, May 2005.

161. Hagler Bailly Consulting, Inc. Energy Conservation Services Program (ECSP), Energy Policy Development and Conservation Project. *Indonesia Demand-Side*

Management Action Plan. United States Agency for International Development, 2000.

162. R. Lee and J. Denlay. Demand-side management: energy efficiency potential in South Australia, Document Ref. 0001.doc, Issue No. 3, Energy SA, October 2002.
163. A. Eberhard, M. Lazarus, S. Bernow, C. Rajan, et al. *Electricity Supply and Demand-Side Management Options*. Report, Cape Town, South Africa: World Commission on Dams, 2000.
164. The E7 Experience. Demand-side management: an electric utility overview of policies and practices in demand-side management. The E7 Network of Expertise for the Global Environment, October 2000.
165. B. Finamore, H. Zhaoguang, L. Weizheng, L. Tijun, D. Yande, Z. Fuqiu, and Y. Zhirong. Demand-side management in China—benefits, barriers and policy recommendations. U.S. Natural Resources Defense Council, October 2003.
166. C. Crawford-Smith and M. Ellis. Development options involving demand-side management and local generation in the advance energy area. *TransGrid and Advance Energy*, March 2001.
167. D. Flahaut, P. Condols, and D. Aulin. Assessment of a demand-side management: programme in the provence-alpes-cote d'Azur region. Agence Re'gionale de l'Energie (ARENE), France, 1999.
168. D. S. Kirschen. Demand-side management view of electricity markets. *IEEE Transactions on Power Delivery*, **18**(2):520–527, 2003.
169. Z. Hu, D. Moskovitz, and J. Zhao. Demand side management in China's restructured power industry. Energy Sector Management Assistance Program (ESMAP), The World Bank, December 2005.
170. Energy Conservation and Environmental Protection Project. Demand side management. Development Research and Technological Planning Center, Tabbin Institute for Metallurgical Studies, Federation of Egyptian Industries, July 1994.
171. Electric load forecast 2005/06 to 2025/26. Market Forecasting, Power Planning and Portfolio Management, Distribution Line of Business, BC Hydro, December 2005.
172. Collins Electric Board. Benefits of demand-side management. U.S. Natural Resources Defense Council, 2004.
173. R. Belmans. Energy demand management. Report, Energy Institute K. U. Available at http://www.esat.kuleuven.be/electa/publications/fulltexts/pub_1274.pdf.
174. The Energy Research Institute, University of Cape Town, South Africa. How to save energy and money in electrical systems. Guide Book 7, 3E Strategy Series.
175. K. Chen. *Industrial Power Distribution and Illuminating Systems*. CRC Press, 1990.
176. A. de Almeida, P. Pertoldi, and W. Leonhard, eds. *Characterization of the Electricity Use in European Union and the Savings Potential in 2010*. Energy Efficiency Improvements in Electric Motors and Drives. Springer, 1997, pp. 19–36.
177. E. L. F. de Almeida. Energy efficiency and the limits of market forces: the example of the electric motor market in France. *Energy Policy*, **26**(8):643–653, 1998.
178. Union of the Electricity Industry "Eurelectric." Electricity for more efficiency: electric technologies and their energy savings potential. Report, Ref: 2004-440-0002, July 2004.
179. G. Davis. Guide to preparing feasibility studies for energy efficiency projects. California Energy Commission, February 2000.

180. J. Schonek. Energy efficiency: benefits of variable speed control in pumps, fans and compressors. Cahier Technique, no. CT214, May 2008. Available at http://www.engineering.schneider-electric.se/Attachments/ed/ct/CT214-variable-speed-pump-fan-compressor.pdf.

181. B. Massey. *Mechanics of Fluids*, 8th ed. Taylor & Francis, 2006.

182. A. M. Michael and S. D. Khepar. *Water Well and Pump Engineering*. 4th Reprint. Tata McGraw-Hill, 1997.

183. H. P. Bloch and A. R. Budris. *Pump Users Handbook—Life Extension*. The Fairmont Press and Marcel Dekker, 2004.

184. G. Clarke, D. Reynders, and E. Wright. *Practical Modern SCADA Protocols: DNP3, 60870.5 and Related Systems*. Elsevier, 2004.

185. D. Bailey and E. Wright. *Practical SCADA for Industry*. Elsevier, 2003.

186. D. J. Gaushell and W. R. Block. SCADA communication techniques and standards. *IEEE Computer Applications in Power*, **6**(3):45–50, 1993.

187. T. E. Dy-Liacco. Modern control centers and computer networking. *IEEE Computer Applications in Power*, October:17–22, 1994.

188. B. Qiu, H. B. Gooi, Y. Liu, and E. K. Chan. Internet-based SCADA display system. *IEEE Computer Applications in Power*, **15**(1):14–19, 2002.

189. R. Barillere et al. Results of OPC evaluation done within the JCOP for the control of the LHC experiments. Proceedings of the International Conference on Accelerator and Large Experimental Physics Control Systems, Trieste, 1999, p. 511.

190. K. Ghoshal and L. D. Douglas. GUI display guidelines drive winning SCADA projects. *IEEE Computer Applications in Power*, April:39–42, 1994.

191. D. Leslie, A. Hlushko, S. Abughazaleh, and F. Garza. Tailoring SCADA systems for standby power applications. *IEEE Computer Applications in Power*, **7**(2):20–23, 1994.

192. K. Ghoshal. Distribution automation: SCADA integration is key. *IEEE Computer Applications in Power*, January:31–35, 1997.

193. S. C. Sciacca and W. R. Block. Advanced SCADA concepts. *IEEE Computer Applications in Power*, January:23–28, 1995.

194. S. E. Collier. Ten steps to a smarter grid. IEEE Conference Papers, Paper No. 09 B2, 2009. Available at http://milsoft.com/download.php.

195. A. Mahmood, M. Aamir, and M. I. Anis. Design and implementation of AMR smart grid. IEEE Electrical Power & Energy Conference, 2008.

196. B. Meehan. Enterprise GIS and the smart electric grid. ESRI, 2008. Available at www.esri.com/electric.

197. European SmartGrids Technology Platform. Vision and strategy for Europe's electricity networks of the future. European Commission-Community Research, EUR 22040, 2006. Available at http://ec.europa.eu/research/energy/eu/publications/index_en.cfm.

198. L. Philipson and H. L. Willis. *Understanding Utilities and De-Regulation*, 2nd ed. Taylor & Francis, 2006.

199. H. L. Willis. *Power Distribution Planning Reference Book*, 2nd ed. Marcel Dekker, 2004.

200. L. Schwartz. *Distributed Generation in Oregon: Overview, Regulatory Barriers and Recommendations*. Oregon Public Utility Commission, 2005.

201. C. A. Lynch et al. *Technical Solutions to Enable Embedded Generation Growth.* IPSA Power Ltd., 2003.
202. A. Dexters, T. Loix, J. Driesen, and R. Belmans. A comparison of grounding techniques for distributed generators implemented in four-wire distribution grids, UPS systems and microgrids. CIRED, 19th International Conference on Electricity Distribution, Vienna, May 21–24, 2007, Paper No. 0638.
203. R. C. Dugan and T. E. McDermott. Distributed generation. *IEEE Industry Applications Magazine*, April:19–25, 2002.
204. F. A. Farret and M. G. Simões. *Integration of Alternative Sources of Energy*. IEEE Press-Wiley, 2006.
205. A. Goetzberger and V. U. Hoffmann. *Photovoltaic Solar Energy Generation.* Springer, 2005.
206. A. Ghosh and G. Ledwich. *Power Quality Enhancement Using Custom Power Devices (Power Electronics and Power Systems)*, 1st ed. Springer, 2002.
207. S. J. Chapman. *Electric Machinery Fundamentals*, 3rd ed. McGraw-Hill, 1999.
208. M. G. Simões and F. A. Farret. *Renewable Energy Systems: Design and Analysis with Induction Generators*. CRC Press, 2004.
209. T. A. Short. *Electric Power Distribution Handbook*. CRC Press, 2004.
210. W. B. Hish. Small induction generator and synchronous generator constants for DSG isolation studies. *IEEE Transactions on Power Delivery*, **PWRD-1**(2):231–239, 1986.
211. U.S. Department of Energy. The potential benefits of distributed generation and rate-related issues that may impede their expansions. A Study Pursuant to Section 1817 of the Energy Policy Act of 2005, February 2007.
212. W. H. Kersting. *Distribution System Modeling and Analysis*. CRC Press, 2001.
213. IEEE Std. 1366™-2003. *IEEE Guide for Electric Power Distribution Reliability Indices*. IEEE, 2003.
214. ORNL based on S. W. Handleg et al. Quantitative assessment of distributed energy resource benefits. ORNL/TM-2003/20, Oak Ridge National Laboratory, May 2003.
215. U.S. Department of Energy. Benefits of demand response in electricity markets and recommendations for achieving them. A report to the U.S. Congress Pursuant to Section 1252 of the Energy Policy Act of 2005, February 2006. Available at http://www.oe.energy.gov/DocumentsandMedia/congress_1252d.pdf.
216. North American Electric Reliability Council. 2006 long-term reliability assessment—the reliability of bulk power systems in North America. October 2006.
217. Lawrence Berkeley National Laboratory et al. How and why customers respond to electricity price variability: A study of NYISO and NYSERDA. 2002 PRL Program Performance, January 2003.
218. IEEE Std. 1547. *IEEE Standard for Interconnecting Distributed Resources with Electric Power Systems*. July 2003.
219. IEEE. P1547.1™. *Draft Standard for Conformance Test Procedures for Equipment Interconnecting Distributed Resources with Electric Power Systems*. July 2005.
220. IEEE. P1547.2™. *Draft Application Guide for IEEE Standard 1547-2003 for Interconnecting Distributed Resources with Electric Power Systems.*
221. IEEE. P1547.3™. *Draft Guide for Monitoring, Information Exchange, and Control of Distributed Resources Interconnected with Electric Power Systems*. 2007.

INDEX

Note: Page numbers in *italics* refer to Figures; those in **bold** to Tables.

Electric Distribution Systems, First Edition. Abdelhay A. Sallam, Om P. Malik.

Published 2011 by John Wiley & Sons, Inc.

Printed in the United States
By Bookmasters